Handbook of Experimental Pharmacology

Volume 173

Editor-in-Chief
K. Starke, Freiburg i. Br.

Editorial Board
G.V.R. Born, London
M. Eichelbaum, Stuttgart
D. Ganten, Berlin
F. Hofmann, München
W. Rosenthal, Berlin
G. Rubanyi, Richmond, CA

RNA Towards Medicine

Contributors

A. Adler, H. Akashi, J. Barciszewski, B. Berkhout, R. Czaja,
A.K. Deisingh, V.R. Dondeti, C. Ehresmann, A.D. Ellington,
G.C. Fanning, P. Fechter, T. Geissmann, A.M. Gewirtz,
J. Glökler, H.U. Göringer, T. Greiner-Stöffele, J. Haasnoot,
U. Hahn, M. Homann, E. Huntzinger, M. Janitz, S. Kainz,
A. Kalota, S. Kauppinen, L.A. Kirsebom, J. Kjems, J. Kurreck,
H. Lehrach, M. Lützelberger, S. Matsumoto, M. Menger,
N.E. Mikkelsen, M. Milovnikova, N. Miyano-Kurosaki,
K.V. Morris, R. Narayanaswamy, E. Nielsen, M. Possedko,
A. Rich, M. Rimmele, P. Romby, J.J. Rossi, S. Schubert,
M. Sioud, M. Sprinzl, G. Symonds, M. Szymański, K. Taira,
H. Takaku, H. Ulrich, D. Vanhecke, B. Vester, A. Virtanen,
C.S. Voertler, J. Wengel, M. Zacharias

Editors
Volker Erdmann, Jürgen Brosius and Jan Barciszewski

Springer

Professor Dr. Volker Erdmann
Free University Berlin
Institute of Chemistry/Biochemistry
Thielallee 63
14195 Berlin, Germany
erdmann@chemie.fu-berlin.de

Professor Dr. Jürgen Brosius
Institute of Experimental Pathology
Molecular Neurobiology (ZMBE)
University of Münster
Von-Esmarch-Str. 56
48149 Münster, Germany
brosius@pop.uni-muenster.de

Prof. Dr. hab. Jan Barciszewski
Institute of Bioorganic Chemistry
Polish Academy of Scienes
Noskowskiego 12/14
61-704 Poznan, Poland
jan.barciszewski@ibch.poznan.pl

With 76 Figures and 17 Tables

ISSN 0171-2004

ISBN-10 3-540-27261-5 Springer Berlin Heidelberg New York

ISBN-13 978-3-540-27261-8 Springer Berlin Heidelberg New York

This work is subject to copyright. All rights reserved, whether the whole or part of the material is concerned, specifically the rights of translation, reprinting, reuse of illustrations, recitation, broadcasting, reproduction on microfilm or in any other way, and storage in data banks. Duplication of this publication or parts thereof is permitted only under the provisions of the German Copyright Law of September 9, 1965, in its current version, and permission for use must always be obtained from Springer. Violations are liable for prosecution under the German Copyright Law.

Springer is a part of Springer Science + Business Media
springer.com

© Springer-Verlag Berlin Heidelberg 2006
Printed in Germany

The use of general descriptive names, registered names, trademarks, etc. in this publication does not imply, even in the absence of a specific statement, that such names are exempt from the relevant protective laws and regulations and therefore free for general use.

Product liability: The publishers cannot guarantee the accuracy of any information about dosage and application contained in this book. In every individual case the user must check such information by consulting the relevant literature.

Editor: S. Rallison
Editorial Assistant: S. Dathe
Cover design: *design&production* GmbH, Heidelberg, Germany
Typesetting and production: LE-TEX Jelonek, Schmidt & Vöckler GbR, Leipzig, Germany
Printed on acid-free paper   27/3100-YL - 5 4 3 2 1 0

Preface

Developments over the past few years have revealed the remarkable versatility of RNA in any compartment of the cell, tasks that had been thought to be exclusively in the realm of proteins and even beyond (see the introductory chapter by A. Rich). For example, important aspects of organismal development are controlled by very short, so-called micro (mi)RNAs (Ambros 2004). It became apparent that not only proteins or small molecules could be potential targets or effectors of diagnostic and therapeutic intervention but also various classes of RNA molecules. The chapters in this book provide insight into various promising avenues where RNA and nucleic acid derivatives—including antisense RNAs, miRNAs, and amplification/selection (SELEX)-generated aptamers, as well as ribozymes—are at the threshold of impacting medicine.

Although a minority of molecular biologists sensed the fundamental significance of RNA in the evolution of life (Woese 1967) and a few pioneers focused on RNAs other than messenger (m)RNAs, transfer (t)RNAs, or ribosomal (r)RNAs (Hodnett and Busch 1968; Nakamura et al. 1968; Prestayko and Busch 1968; Zieve and Penman 1976; see the introductory chapter by A. Rich), only the finding that RNAs can have catalytic activity raised a certain level of general interest in them (Guerrier-Takada et al. 1983; Kruger et al. 1982). Nevertheless, at the dawn of genomic sequencing a decade ago (Fleischmann et al. 1995; Fraser et al. 1995), the scientific community had to be alerted not to ignore the treasures of non-messenger RNAs that are hidden in every genome (Brosius 1996). This was followed by efforts both in computational and experimental RNomics (Barciszewski and Erdmann 2003; Filipowicz 2000; Eddy 2002). A while back, two unusual RNAs of approximately 22 nucleotides in length were considered mere oddities despite the fact that their functions in *Caenorhabditis elegans* development were striking (Lee et al. 1993; Reinhart et al. 2000). Meanwhile, cloning and biocomputational predictions have revealed close to 200 experimentally proven miRNAs—and the number approaches 1,000 if biomathematically predicted candidates are added (Lagos-Quintana 2001; Berezikov et al. 2005). Almost weekly we learn about their stunning functions in regulation of many cellular programs

(Ambros 2004). Studies on processing and interactions with proteins as well as the action of another class of small antisense RNAs (siRNAs) that were processed from larger double-stranded RNAs revealed important modes of action that revived earlier efforts to use antisense RNA in medical applications. Today, the excitement about functional RNAs has heated up the field of RNA biology to such a degree that a function is suspected for any transcript, any chunk of RNA, and a certain level of moderation is being called for (Brosius 2005). In any event, the macromolecular class of untranslated (ut)RNAs (Brosius and Tiedge 2004) cannot be ignored any longer, neither in understanding basic biological principles, nor as effectors or targets in molecular medicine.

While in many lineages of multicellular organisms the RNA/RNP world is still a major part of cellular componentry, bacteria are apparently more advanced in that their reliance on regulatory RNA is no longer as pronounced. Nevertheless, bacteria do express a fair number of untranslated regulatory RNAs that appear to play major regulatory roles in adaptation to changing environments as well as mediators of virulence in bacterial pathogens (dealt with in the T. Geissmann et al. chapter). These findings add this class of utRNAs to the list of potential targets for future antibiotic strategies. As an example, targeting functional RNAs with aminoglycosides is discussed in the chapter by L.A. Kirsebom et al. In addition to targeting cells, viruses are "attractive" targets, e.g., for RNA interference (RNAi) applications (see the chapters by K.V. Morris and J.J. Rossi; J. Haasnoot and B. Berkhout; N. Miyano-Kurosaki and H. Takaku; S. Schubert and J. Kurreck). Unicellular eukaryotic parasites may also be targeted, for example by RNA aptamers (see H.U. Göringer and colleagues' contribution). Multicellular organisms including humans exhibit a rich world of functional, in particular, regulatory RNAs in their cells (contribution by M. Szymański and J. Barciszewski). This has initiated serious efforts to study RNA function and, in particular, multidisciplinary efforts to understand the mechanisms of RNAi, antisense RNA including siRNA, and miRNA. These RNAs, in turn, can be used to study the function of almost any protein in the cell by targeting their transcripts and translation (chapters by M. Janitz et al. and S. Matsumoto et al.). Even small molecules can be targeted by RNA aptamers and thus may complement or even surpass antibody technologies (M. Menger et al.). Knowledge of mechanisms of RNA action and their use in regulating protein expression invites exploration of RNAs as targets and effectors for therapeutic exploitation. The chapters by A. Kalota et al. and M. Sioud discuss progress in the development of nucleic acid therapeutics and stumbling blocks to be overcome. This volume also examines various related aspects, from identification of potential therapeutic target sites in RNA (M. Lützelberger and J. Kjems) to delivery or toxicity (G.C. Fanning and G. Symonds) and engineering of RNA-based circuits

(R. Narayanaswamy and A.D. Ellington). Several chapters deal with stability and bioavailability issues of RNA (H. Ulrich; P.E. Nielsen; S. Kauppinen et al.; S. Kainz et al.). There is also well-founded hope to use RNA for analytic and diagnostic purposes (M. Sprinzl et al.) including as biosensors (A.K. Deisingh).

Berlin, V.A. Erdmann
Münster, J. Brosius
Poznan, J. Barciszewski
June 2005

References

Ambros V (2004) The functions of animal microRNAs. Nature 431:350–355
Barciszewski J, Erdmann VA (eds) (2003) Non-coding RNAs: molecular biology and molecular medicine. Landes Bioscience, Georgetown, pp 1–292
Berezikov E, Guryev V, van de Belt J, Wienholds E, Plasterk RH, Cuppen E (2005) Phylogenetic shadowing and computational identification of human microRNA genes. Cell 120:21–24
Brosius J (1996) More Haemophilus and Mycoplasma genes. Science 271:1302
Brosius J (2005) Waste not, want not-transcript excess in multicellular Eukaryotes. Trends Genet 21:287–288
Brosius J, Tiedge H (2004) RNomenclature. RNA Biol 1:81–83
Eddy SR (2002) Computational genomics of noncoding RNA genes. Cell 109:137–140
Filipowicz W (2000) Imprinted expression of small nucleolar RNAs in brain: time for RNomics. Proc Natl Acad Sci U S A 97:14035–14037
Fleischmann RD, Adams MD, White O, Clayton RA, Kirkness EF, Kerlavage AR, Bult CJ, Tomb JF, Dougherty BA, Merrick JM, el al (1995) Whole-genome random sequencing and assembly of Haemophilus influenzae Rd. Science 269:496–512
Fraser CM, Gocayne JD, White O, Adams MD, Clayton RA, Fleischmann RD, Bult CJ, Kerlavage AR, Sutton G, Kelley JM, Fritchman RD, Weidman JF, Small KV, Sandusky M, Fuhrmann J, Nguyen D, Utterback TR, Saudek DM, Phillips CA, Merrick JM, Tomb JF, Dougherty BA, Bott KF, Hu PC, Lucier TS, Peterson SN, Smith HO, Hutchison CA, Venter JC (1995) The minimal gene complement of Mycoplasma genitalium. Science 270:397–403
Guerrier-Takada C, Gardiner K, Marsh T, Pace N, Altman S (1983) The RNA moiety of ribonuclease P is the catalytic subunit of the enzyme. Cell 35:849–857
Hodnett JL, Busch H (1968) Isolation and characterization of uridylic acid-rich 7 S ribonucleic acid of rat liver nuclei. J Biol Chem 243:6334–6342
Kruger K, Grabowski PJ, Zaug AJ, Sands J, Gottschling DE, Cech TR (1982) Self-splicing RNA: autoexcision and autocyclization of the ribosomal RNA intervening sequence of Tetrahymena. Cell 31:147–157
Lagos-Quintana M, Rauhut R, Lendeckel W, Tuschl T (2001) Identification of novel genes coding for small expressed RNAs. Science 294:853–858
Lee RC, Feinbaum RL, Ambros V (1993) The C. elegans heterochronic gene lin-4 encodes small RNAs with antisense complementarity to lin-14. Cell 75:843–854

Nakamura T, Prestayko AW, Busch H (1968) Studies on nucleolar 4 to 6 S ribonucleic acid of Novikoff hepatoma cells. J Biol Chem 243:1368–1375

Prestayko AW, Busch H (1968) Low molecular weight RNA of the chromatin fraction from Novikoff hepatoma and rat liver nuclei. Biochim Biophys Acta 169:327–337

Reinhart BJ, Slack FJ, Basson M, Pasquinelli AE, Bettinger JC, Rougvie AE, Horvitz HR, Ruvkun G (2000) The 21-nucleotide let-7 RNA regulates developmental timing in Caenorhabditis elegans. Nature 403:901–906

Woese CR (1967) The genetic code: the molecular basis for genetic expression. Harper and Row, New York, pp 179–195

Zieve G, Penman S (1976) Small RNA species of the HeLa cell: metabolism and subcellular localization. Cell 8:19–31

List of Contents

Why RNA and DNA Have Different Structures 1
  A. Rich

Regulatory RNAs as Mediators of Virulence Gene Expression in Bacteria 9
  T. Geissmann, M. Possedko, E. Huntzinger, P. Fechter, C. Ehresmann,
  P. Romby

Regulatory RNAs in Mammals 45
  M. Szymański, J. Barciszewski

Aminoglycoside Interactions with RNAs and Nucleases 73
  L. A. Kirsebom, A. Virtanen, N. E. Mikkelsen

High-Throughput RNA Interference in Functional Genomics 97
  M. Janitz, D. Vanhecke, H. Lehrach

Antiviral Applications of RNAi 105
  K. V. Morris, J. J. Rossi

RNA Interference: Its Use as Antiviral Therapy 117
  J. Haasnoot, B. Berkhout

Gene Silencing of Virus Replication by RNA Interference 151
  N. Miyano-Kurosaki, H. Takaku

Progress in the Development of Nucleic Acid Therapeutics 173
  A. Kalota, V. R. Dondeti, A. M. Gewirtz

Screening and Determination of Gene Function Using
Randomized Ribozyme and siRNA Libraries 197
  S. Matsumoto, H. Akashi, K. Taira

Ribozymes and siRNAs: From Structure to Preclinical Applications . . 223
  M. Sioud

Strategies to Identify Potential Therapeutic Target Sites in RNA 243
  M. Lützelberger, J. Kjems

Oligonucleotide-Based Antiviral Strategies 261
  S. Schubert, J. Kurreck

Gene-Expressed RNA as a Therapeutic: Issues to Consider,
Using Ribozymes and Small Hairpin RNA as Specific Examples 289
  G. C. Fanning, G. Symonds

RNA Aptamers: From Basic Science Towards Therapy 305
  H. Ulrich

RNA Aptamers Directed Against Oligosaccharides 327
  M. Sprinzl, M. Milovnikova, C. S. Voertler

Aptamer-Based Biosensors: Biomedical Applications 341
  A. K. Deisingh

Application of Aptamers in Therapeutics
and for Small-Molecule Detection 359
  M. Menger, J. Glökler, M. Rimmele

RNA Aptamers as Potential Pharmaceuticals Against Infections
with African Trypanosomes 375
  H. U. Göringer, M. Homann, M. Zacharias, A. Adler

RNA Targeting Using Peptide Nucleic Acid 395
  E. Nielsen

Locked Nucleic Acid: High-Affinity Targeting of Complementary RNA
for RNomics 405
  S. Kauppinen, B. Vester, J. Wengel

Engineering RNA-Based Circuits 423
  R. Narayanaswamy, A. D. Ellington

Selection of RNase-Resistant RNAs 447
  S. Kainz, R. Czaja, T. Greiner-Stöffele, U. Hahn

Subject Index 457

List of Contributors

Addresses given at the beginning of respective chapters

Adler, A. 375
Akashi, H. 197

Barciszewski, J. 45
Berkhout, B. 117

Czaja, R. 447

Deisingh, A.K. 341
Dondeti, V.R. 173

Ehresmann, C. 9
Ellington, A.D. 423

Fanning, G.C. 289
Fechter, P. 9

Geissmann, T. 9
Gewirtz, A.M. 173
Glökler, J. 359
Göringer, H.U. 375
Greiner-Stöffele, T. 447

Haasnoot, J. 117
Hahn, U. 447
Homann, M. 375
Huntzinger, E. 9

Janitz, M. 97

Kainz, S. 447
Kalota, A. 173
Kauppinen, S. 405
Kirsebom, L.A. 73
Kjems, J. 243
Kurreck, J. 261

Lehrach, H. 97

Lützelberger, M. 243

Matsumoto, S. 197
Menger, M. 359
Mikkelsen, N.E. 73
Milovnikova, M. 327
Miyano-Kurosaki, N. 151
Morris, K.V. 105

Narayanaswamy, R. 423
Nielsen, E. 395

Possedko, M. 9

Rich, A. 1
Rimmele, M. 359
Romby, P. 9
Rossi, J.J. 105

Schubert, S. 261
Sioud, M. 223
Sprinzl, M. 327
Symonds, G. 289
Szymański, M. 45

Taira, K. 197
Takaku, H. 151

Ulrich, H. 305

Vanhecke, D. 97
Vester, B. 405
Virtanen, A. 73
Voertler, C.S. 327

Wengel, J. 405

Zacharias, M. 375

Why RNA and DNA Have Different Structures

A. Rich

Department of Biology,
Massachusetts Institute of Technology, Cambridge MA, 02139, USA
cbeckman@mit.edu

1	Introduction	1
2	Can RNA Form a Double Helix?	3
3	Can DNA and RNA Form a Hybrid Helix?	3
4	Double-Stranded RNA at Atomic Resolution	4
5	Different Roles in Evolution	5
References		7

Abstract In the early years of molecular biology—over 50 years ago—we were faced with many unknowns. A significant one at the time was the relationship between DNA and RNA, both in terms of structure and function. Function is often a reflection of structure. Here I outline some of the early research in this area, especially for RNA structure, which was completely unknown when we started.

Keywords RNA · DNA · Double helix · X-ray structure

1
Introduction

We learned about the double helix over 50 years ago with publication of the Watson–Crick formulation (Watson and Crick 1953) and the fiber X-ray diffraction patterns of groups led by Maurice Wilkins (Wilkins and Randall 1953) and Rosalind Franklin (Franklin and Gosling 1953). Analysis of the diffraction pattern, especially the fibers of the hydrated B form, could be immediately interpreted as consistent with a double helix. The weakness of the first-layer line relative to the second and the virtual absence of the fourth-layer line clearly suggested two chains wrapping around each other with the phosphate groups on the outside. More complex and not answered at the time was the question of why there were two forms. What was the nature of the less-hydrated fibers that produced the better oriented and crystalline A form that could convert to the B form? In those days a half-century ago, fiber diffraction was the only way such large, elongated molecules could be studied. Generally, the patterns had rotational disorder around the fiber axis, which could be at the molecular level in the case of the B form and often involved crystalline segments in the A form. The diffraction patterns were limited in resolution, but it could be said that they were consistent with the formulation. Over the next several years, work by Maurice Wilkins and his colleagues gradually refined the nature of the double-helical model that could give rise to the increasingly detailed diffraction patterns. However, the diffraction pattern could not "prove" the structure of the molecule, as there were too little data.

The ribose sugar ring contains five atoms, but they cannot all lie in one plane, and at least one atom must be out-of-plane (Fig. 1). With the continued analysis of the fiber patterns, it became clear that the B form contained a ribose ring pucker in which the C2′ atom was out-of-plane on the same side as the base (C2′ endo). Because of that pucker, the phosphate groups were nearly 7 Å apart, yielding an extended polynucleotide chain. Study of the more complex A form led to the conclusion that the C3′ atom was out-of-plane (C3′ endo). In that conformation, the phosphate groups were about 5.8–6 Å apart. Thus, the sugar phosphate backbone was shortened, leading to a double helix in which the base pairs were slightly displaced from the center of the helix to produce a flatter helix and a somewhat thicker molecule. A relative scarcity of water molecules stabilized that conformation. It became clear that the normal conformation in the hydrated in vivo environment involved the C2′ endo sugar pucker of B form DNA.

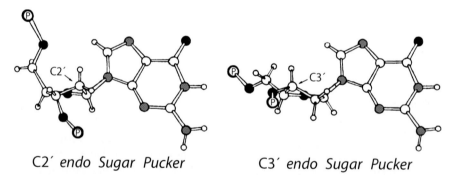

C2′ endo Sugar Pucker   C3′ endo Sugar Pucker

Fig. 1 The major sugar puckers are shown for the nucleic acids. The C2′ endo pucker (*left*) is found in B-DNA, while the C3′ endo pucker (*right*) is found in A-DNA or in RNA. In the ribose of RNA, an oxygen atom is found on C2′, pointing down in the C3′ endo pucker. The distance between successive phosphate groups is close to 7.0 Å in C2′ endo and shortens to 5.8–6 Å in the C3′ endo pucker. Thus, there is an elastic element in the backbone. Nucleic acids can convert from one pucker to the other, but it takes greater energy for conversion of ribonucleotides, due to the fact that the oxygen atom on C2′ is in too close van der Waals contact with the oxygen on C3′ when it is in the C2′ endo conformation

2
Can RNA Form a Double Helix?

Watson and Crick (Watson and Crick 1953) pointed out that it was unlikely that RNA could form their proposed structure because the added O2′ would make too close a contact. Starting in 1954, attempts were made to study RNA fibers to see if they would also form a double helix. The results were ambiguous. Slightly oriented patterns could be obtained, but they all looked the same, independent of the base composition of the material (Rich and Watson 1954). A breakthrough came in 1956 when, together with David Davies, we discovered that mixing polyuridylic acid (polyU) and polyadenylic acid (polyA) would form a double helix, as indicated by a well-oriented fiber diffraction pattern (Rich and Davies 1956). Unlike the DNA diffraction patterns, there was no change in the pattern with changing hydration, and the first-layer line was stronger than the second-layer line. In addition, the diameter of the molecule was greater than that of B-DNA. Thus, it was a different molecule. Eventually, it was discovered that the double-stranded (ds)RNA molecule adopted a conformation similar to the A form of DNA, exclusively using a C3′ *endo* sugar pucker (Fig. 1). The reason for this adherence to the C3′ *endo* sugar pucker in RNA becomes apparent on looking at the position of the additional oxygen that would be present at the C2′ position of ribose (See Fig. 1). In the C3′ *endo* conformation of dsRNA, there is adequate separation between the oxygen on C2′ and the oxygen on C3′ in contrast to a van der Waals crowding that occurs if the dsRNA sugar pucker were C2′ *endo*. Because of the unfavorable energetic situation of ribose in the C2′ *endo* conformation, RNA molecules are usually found in the C3′ *endo* conformation. There is an energy barrier between the two puckers for ribose; in contrast, the deoxyribose ring has very little barrier.

Finding that polyA reacted with polyU to form a double helix was surprising at the time, and it represented a paradigm shift. This was the discovery of the first hybridization reaction in which long polynucleotide chains formed a double helix based on the specificity of hydrogen bonds. Within a year, together with Gary Felsenfeld, we discovered that a second strand of polyU could be taken into the helix to make a triple helix of polyA plus two polyU (Felsenfeld et al. 1957). Since there was no increase in the diameter of the helix, we suggested that the additional uracil residue was bound by two hydrogen bonds to the amino group and N7 of adenine. This interpretation was considerably strengthened 2 years later by Hoogsteen's single-crystal analysis of 1-methyl thymine complexed to 9-methyl adenine (Hoogsteen 1959).

3
Can DNA and RNA Form a Hybrid Helix?

The availability of short polynucleotides in the deoxy series chemically synthesized by Khorana and colleagues (Tener et al. 1958) made it possible to ask whether a hybrid helix could be made with one RNA strand and one DNA strand. It was known at that time that the conformation of B-DNA was quite different from that seen in the RNA–RNA duplexes; thus, it was not obvious that they could combine. In order to combine, one strand would have to change conformation. In 1960, I could show that a two-stranded molecule would form with an RNA chain (polyA) and a DNA chain (polydeoxy thymidylic acid or poly dT) (Rich 1960). This reaction was important in two respects. It was the first DNA–RNA hybridization [which is still used today in isolating messenger (m)RNA through its polyA tails], and it also represented a model for how RNA polymerase might make an RNA strand by forming a hybrid duplex with a single DNA strand as a template. One year later a purified preparation of DNA-dependent RNA polymerase revealed that this was precisely how the enzyme worked (Furth et al. 1961). It was only later that it was possible to show in single-crystal X-ray analysis that the hybrid helix DNA strand conformed to the RNA pucker so that both strands had the C3′ *endo* sugar pucker (Wang et al. 1982; Egli et al. 1992; Fig. 1). The important consideration was that it was energetically costly to change the conformation of the RNA strand but relatively easy to change the conformation of the DNA strand. This largely explained why the melting temperature of an RNA duplex is higher than a DNA duplex with the same sequence, and the melting temperature of the RNA—DNA hybrid is intermediate.

4
Double-Stranded RNA at Atomic Resolution

A number of single-crystal diffraction studies of purine-pyrimidine co-crystals were carried out in the 1960s. A disturbing trend was found in that all co-crystals of adenine derivatives with uracil or thymine derivatives had Hoogsteen pairing, but none had Watson–Crick base pairs (Voet and Rich 1970). This led to a Hoogsteen model of the DNA double helix (Arnott et al. 1965) that did not fit the diffraction data as well as the Watson–Crick model, but one could rely on the fiber diffraction data only to a limited extent. The question remained, what was the structure of a double helix?

The first single-crystal structures of a double helix were solved in my laboratory in 1973. This was before it was possible to obtain oligonucleotides in quantities suitable for crystallographic experiments. However, we succeeded in crystallizing and solving two dinucleoside phosphates, the RNA oligomers GpC (Day et al. 1973) and ApU (Rosenberg et al. 1973). Furthermore, the resolution of the diffraction pattern was 0.8 Å. At atomic resolution, we could visualize not only the sugar phosphate backbone in the form of a double helix but also the position of ions and water molecules. By extending the structure using the symmetry of the two base pairs, it was possible to generate RNA double helices that were quite similar to the structures that had been deduced from studies of double-helical fibers of RNA.

The GpC structure had the anticipated base pairs connected by three hydrogen bonds. However, the ApU structure showed for the first time that Watson–Crick base pairs were formed when the molecule was constrained in a double helix, as opposed to the Hoogsteen base pairs that were favored in single-crystal complexes of bases. The significance of the double helix at atomic resolution was recognized by the editors of *Nature*, who in their "News and Views" commentary called this the "missing link" of nucleic acid structure and recognized "the many pearls offered" to help resolve one of the major uncertainties in nucleic acid conformation (Anonymous 1973). These structures capped the effort started some 20 years earlier asking whether RNA could form a double helix. It had been pursued in earnest with the recognition that polyA and polyU would form a double helix. High-resolution crystallographic analyses of larger fragments of the double helix (DNA or RNA) did not emerge for another 6–7 years after the dinucleotide phosphate structures, with the availability of chemically synthesized oligonucleotides.

The complex manner in which the RNA double helix can fold into novel conformations was first visualized in the structure of transfer (t)RNA. The rate-limiting step was obtaining a crystal of sufficient resolution. In 1971, we found that the addition of spermine to yeast tRNA^{phe} led to a crystal that diffracted to 2.3 Å resolution (Kim et al. 1971). By 1973 at 4 Å resolution, the folding of the polynucleotide chain could be traced through visualization of the electron-dense phosphate groups. This revealed an unusual L-shaped structure in which double helical segments were organized to form an L-shaped structure with the anticodon at one end and the amino acid acceptor at the other (Kim et al. 1973). The detailed nature of the complexity was visualized a year later at 3 Å resolution in which a number of interactions were found beyond the Watson–Crick base pairs in the double helical segments (Kim et al. 1974; Robertus et al. 1974). Although the nucleotides were predominantly in the C3′ *endo* ribose conformation, there were places where the chain had to span a longer distance (Rich et al. 1980). In one segment, the chain was lengthened with two adjacent nucleotides adopting the C2′ *endo* conformation. This was a good illustration of the manner in which an otherwise inflexible RNA strand would adopt the less favorable conformation locally to yield an overall stable structure.

5
Different Roles in Evolution

In the early 1960s, I was pondering problems of evolution and the origin of life. The most popular ideas about the origin of life in the 1940s was that espoused by the Russian scientist Oparin, who expressed the belief that life began in coacervates of polypeptide chains that formed specialized environments, leading to the production of enzymatic activity and eventually to living systems. I felt this was likely to be incorrect since it did not explain the fundamental role of the nucleic acids in providing the information needed for specifying biological systems.

In 1962, I put down a number of thoughts about the origin of living systems in an article entitled "On the Problems of Evolution and Biochemical Information Transfer" (Rich 1962), which presented a brief overview of the way information was transmitted from DNA to RNA and eventually to directing the synthesis of proteins through the interaction of tRNA molecules with mRNA in ribosomes. I stressed the fact that life could not have originated with protein molecules, since it could not explain how nucleic acids came to control protein synthesis. It was more likely that polynucleotides were the origin of living systems. A primitive environment was postulated in which polynucleotide chains were able to act as a template or as a somewhat inefficient catalyst for promoting the polymerization of the complementary nucleotide residues to build up an initial two-stranded molecule. Such an inefficient system could be followed by denaturation of the nucleic acid duplex, and continuation of the process would ultimately lead to an increasingly larger number of nucleic acid polymers. Various ways were outlined in which the polymerization of nucleic acids might be coupled with an inefficient polymerization of amino acids. Of key importance in this view was the development of primitive activating enzymes that would begin to relate a specific nucleic acid sequence to the assembly of specific amino acids. Thus, "life" was viewed as starting with a coupling of nucleic acid polymerization and amino acid polymerization, although it was stressed that the prototype of this reaction might have been in a form quite different from that which we observe today.

In another section I asked, Why are there two nucleic acids in contemporary biological systems? It seemed reasonable to believe that both RNA and DNA stemmed from a common precursor. It was apparent in contemporary biological systems that DNA seems to act solely as a major carrier of genetic information, while the RNA molecule is used to convert the genetic information into actual protein molecules. However, I noted that RNA molecules are also able to carry genetic information, as in RNA-containing viruses. Thus, it seemed reasonable to speculate that the first polynucleotide molecule was initially an RNA polymer that was able to convey genetic information as well as organize amino acids into specific sequences to make proteins.

This article, published in 1962, was probably the first statement to suggest that RNA was the fundamental nucleic acid involved in the origin of living systems. In this same essay, I discussed the method by which the newly discovered mRNA was made. I suggested the possibility that mRNA might be made in vivo as complementary copies of one or both strands of DNA. If both strands are active, then the DNA would produce two RNA strands, and only one of these might be active as messenger RNA in protein synthesis. The other strand, I speculated, might be a component of a control or regulatory system. This was probably the first statement of an anti-sense function for RNA molecules. It also suggested the possibility that RNA could have other regulatory functions.

These thoughts were published over 40 years ago. Today, we have a wealth of information that strengthens the role of RNA in the early evolution of life. The discovery of ribozymes and the more recent discovery of micro-RNAs with a variety of functions in controlling the development of biological systems suggests that these might be trace evidence of what has been called the "RNA world," meaning an era in early evolution in which RNA played a dominant role in both replication and in carrying out a number of chemical modifications leading to the organization of present-day biological systems.

It is likely that we will not be able to define a precise event that led to the origin of life. Rather, it is likely that there was a growing level of molecular complexity that eventually yielded a system we would call "living," but for which it would be very hard to define a unique point at which we can say that life began. RNA, rather than DNA probably played a major role at that time. Perhaps the increased rigidity of RNA structures due to the strongly preferred sugar pucker played an important role in the catalytic functions that had to develop. Of course, a key element is the extent to which all of these processes in early evolutionary history were error prone. These errors provided the substrate for Darwinian selection since, among the errors in the system, some would create efficiencies that form the basis for selection that eventually provided the direction for molecular evolution.

References

Anonymous (1973) News and views. Nature 243:114

Arnott S, Wilkins MHF, Hamilton LD, Langridge R (1965) Fourier synthesis studies of lithium DNA. III. Hoogsteen models. J Mol Biol 11:391–402

Day RO, Seeman NC, Rosenberg JM, Rich A (1973) A crystalline fragment of the double helix: the structure of the dinucleoside phosphate guanylyl-3′, 5″-cytidine. Proc Natl Acad Sci USA 70:849–853

Egli M, Usman N, Zhang S, Rich A (1992) Crystal structure of an Okazaki fragment at 2-Å resolution. Proc Natl Acad Sci USA 89:534–538

Felsenfeld G, Davies DR, Rich A (1957) Formation of a three-stranded polynucleotide molecule. J Am Chem Soc 79:2023–2024

Franklin RE, Gosling RG (1953) Molecular configuration in sodium thymonucleate. Nature 171:740–741

Furth JJ, Hurwitz J, Goldmann M (1961) The directing role of DNA in RNA synthesis. Biochem Biophys Res Commun 4:362–367

Hoogsteen K (1959) The crystal and molecular structure of a hydrogen-bonded complex between 1-methylthymine and 9-methyladenine. Acta Crystallogr 12:822–823

Kim S-H, Quigley G, Suddath FL, Rich A (1971) High resolution X-ray diffraction patterns of tRNA crystals showing helical regions of the molecule. Proc Natl Acad Sci USA 68:841–845

Kim S-H, Quigley GJ, Suddath FL, McPherson A, Sneden D, Kim JJ, Weinzierl J, Rich A (1973) Three-dimensional structure of yeast phenylalanine transfer RNA: folding of the polynucleotide chain. Science 179:285–288

Kim SH, Suddath FL, Quigley GJ, McPherson A, Sussman JL, Wang AH, Seeman NC, Rich A (1974) Three-dimensional tertiary structure of yeast phenylalanine transfer RNA. Science 185:435–439

Rich A (1960) A hybrid helix containing both deoxyribose and ribose polynucleotides and its relation to the transfer of information between the nucleic acids. Proc Natl Acad Sci USA 46:1044–1053

Rich A (1962) On the problems of evolution and biochemical information transfer. In: Kasha M, Pullman B (eds) Horizons in biochemistry. Academic Press, New York, pp 103–126

Rich A, Davies DR (1956) A new two-stranded helical structure: polyadenylic acid and polyuridylic acid. J Am Chem Soc 78:3548

Rich A, Watson JD (1954) Some relations between DNA and RNA. Proc Natl Acad Sci USA 40:759–764

Rich A, Wang AHJ, Quigley GJ (1980) Diversity of sugar pucker in the nucleic acids. In: Ananchenko SN (ed) Frontiers of bioorganic chemistry and molecular biology (International Union of Pure and Applied Chemistry). Pergamon Press, Oxford, pp 327–343

Robertus JD, Ladner JE, Finch JT, Rhodes D, Brown RS, Clark BF, Klug A (1974) A structure of yeast phenylalanine tRNA at 3 Å resolution. Nature 250:546–551

Rosenberg JM, Seeman NC, Kim JJP, Suddath FL, Nicholas HB, Rich A (1973) Double helix at atomic resolution. Nature 243:150–154

Tener GM, Khorana HG, Markham R, Pol EH (1958) Studies on polynucleotides. II. The synthesis and characterization of linear and cyclic thymidine oligonucleotides. J Am Chem Soc 80:6223–6230

Voet D, Rich A (1970) The crystal structures of purine, pyrimidines and their intermolecular complexes. Prog Nucleic Acid Res Mol Biol 10:183–265

Wang AH, Fujii S, van Boom JH, van der Marel GA, van Boeckel SA, Rich A (1982) Molecular structure of r(GCG)d(TATACGC): a DNA-RNA hybrid helix joined to double helical DNA. Nature 299:601–604

Watson JD, Crick FHC (1953) A structure for deoxyribose nucleic acid. Nature 171:737–738

Wilkins MHF, Randall JT (1953) Molecular structure of deoxypentose nucleic acids. Nature 171:738–740

Regulatory RNAs as Mediators of Virulence Gene Expression in Bacteria

T. Geissmann · M. Possedko · E. Huntzinger · P. Fechter · C. Ehresmann (✉) · P. Romby (✉)

UPR 9002 CNRS, Université Louis Pasteur, Institut de Biologie Moléculaire et Cellulaire, 15 rue R. Descartes, 67084 Strasbourg cedex, France
Chantal.Ehresmann@ibmc.u-strasbg.fr
p.romby@ibmc.u-strasbg.fr

1	Introduction	10
2	*cis*-mRNA Elements as Mediators of Virulence	11
3	*trans*-Acting RNAs as Mediators of Virulence	14
3.1	Regulatory Antisense RNAs	15
3.1.1	sRNAs and Iron Metabolism	15
3.1.2	sRNAs and Virulence in *V. cholerae*	21
3.2	Modulation of Protein Activity by Regulatory RNAs	22
3.2.1	The *E. coli* CsrA/CsrB System	22
3.2.2	Homologous Systems in Bacterial Pathogens	23
3.3	Multifunctional RNAs	23
3.3.1	The Case of *S. aureus* RNAIII	23
3.3.2	Other Regulatory RNAs	26
3.4	Other Types of RNAs Influencing Pathogenicity	26
3.4.1	The Case of tRNA Modifications	26
3.4.2	The Case of tmRNA	27
3.4.3	Location of Pathogenicity Islands	27
4	A Link Between Regulatory RNAs and Signal Transduction	28
4.1	RNAs as Effectors of Quorum-Sensing Systems	28
4.2	Other Systems	30
5	Concluding Remarks	31
5.1	RNAs Regulate Directly or Indirectly Multiple Genes	31
5.2	A Direct Coupling Between the Structure and the Regulatory Activity?	32
5.3	Stable or Unstable RNAs in Cells?	33
5.4	Some Regulatory RNAs Require *trans*-Acting Proteins	34
5.5	Functional Redundancies and How to Find Them?	35
	References	35

Abstract Bacteria exploit functional diversity of RNAs in a wide range of regulatory mechanisms to control gene expression. In last few years, small RNA molecules have been discovered at a staggering rate in bacteria, mainly in *Escherichia coli*. While functions of many of these RNA molecules are still not known, several of them behave as key effectors of adaptive responses, such as environmental cue recognition, stress response, and virulence control. Most fascinating, perhaps, is the discovery that mRNAs behave as direct sensors of small molecules or of environmental cues. The astonishing diversity of RNA-dependent regulatory mechanisms is linked to the dynamic properties and versatility of the RNA structure. In this review, we relate several recent studies in different bacterial pathogens that illustrate the diverse roles of RNA to control virulence gene expression.

Keywords Prokaryotes · Virulence · Regulation · Regulatory RNAs · Riboswitch

1
Introduction

Bacteria inhabit a variety of ecological niches and meet continuous environmental challenges. As highly adaptable organisms, they have evolved a plethora of sensory systems in order to express the appropriate genes in response to a given environment and to preserve energy. This is in particular the case with many bacterial pathogens, which turn on virulence genes in response to the host signals and to evade the host defense system. These accessory factors are usually subject to tight and coordinated regulation. A diverse array of extracellular signals that modulate virulence gene expression has been identified in vitro. Many of these signals are ubiquitous and involve physical or chemical cues. In other cases, bacteria have evolved to recognize specific metabolic products generated by their respective hosts. Furthermore, diffusible signals can also be produced by the bacteria through the ability to measure cell density.

Recent advances in molecular microbiology have also enabled comprehensive genetic screening for gene products required for full virulence (Mecsas 2002). As an example, *Staphylococcus aureus* virulence is determined by cell wall-associated proteins and secreted toxins (Novick 2003). More recent studies have identified a broad range of additional genes that are essential to maintain bacterial survival within the host (Benton et al. 2004). These factors include signal transduction systems, as well as enzymes involved in the biosynthesis of metabolites, and *trans*-membrane transporters. Thus, virulence expression is directly coupled to an adaptation of the bacteria to their different niches and environmental stimuli. In many cases, this is directed by signaling using regulatory mechanisms similar to those of the control genes that are not specific to pathogenesis. One of the most encountered families of virulence regulatory proteins is the two-component system of signal transducers. These complexes respond to the signals through usually regulated phosphotransfer reactions. Although there are many variations on bacterial two-component regulatory systems, they constantly sense an external environment and transmit this information to a bacterial interior.

Recently, it has been shown that RNAs are key players in stress-response and pathogenesis processes, which brings an additional level of complexity to the cell physiology. In these last few years, more than 50 small non-coding RNAs have been experimentally discovered in *Escherichia coli*, representing 1%–2% of the number of protein-coding genes (Wagner and Vogel 2003; Gottesman 2004; Storz et al. 2004). Even though the assigned functions of many of these RNAs remain to be identify, some of them are involved in regulatory networks, which are required for appropriate response to environmental and stress condition changes (Wagner and Vogel 2003). The most prevalent class involves small RNAs (or sRNAs), which target specific mRNA through base pairings. Regulatory activity of these sRNAs requires the RNA chaperone protein Hfq (Brescia and Sledjeski 2003). Recent observations point to the role of RNAs in the establishment of virulence in several bacterial pathogens, such as *Erwinia carotovora*, *Vibrio cholerae*, *Clostridium perfringens*, *Listeria monocytogenes*, *S. aureus*, and *Streptococcus pyogenes* (Johansson and Cossart 2003). Most of these regulatory RNAs are *trans*-acting RNAs. Others are messenger (mRNAs) that carry *cis*-acting regulatory elements ("riboswitches"), which respond to environmental cues or to metabolite concentration. These RNAs activate or inhibit virulence gene expression at the transcriptional or post-transcriptional levels using a diversity of mechanisms (i.e., antisense RNA, sequestration of regulatory proteins, riboswitches, Fig. 1). Precise knowledge of these regulatory mechanisms and how they control virulence factor expression would open up new perspectives for antimicrobial chemotherapy.

This review focuses on new features that emerged recently on the pathogenesis-related RNAs. A particular emphasis is given to the links existing between the structure and the functional activity of these regulatory RNAs, and the regulatory networks in which they are involved. A number of recent reviews have also covered various aspects of the regulatory RNAs, mainly in non-pathogenic bacteria (Wagner and Vogel 2003; Gottesman 2004; Storz et al. 2004).

2
cis-mRNA Elements as Mediators of Virulence

The increase of temperature is one of the problems pathogens have to face when infecting warm-blooded hosts. Temperature sensing is often acquired by proteins or DNA (Hurme and Rhen 1998). However, there are now accumulating data suggesting that an mRNA secondary structure can also act as a thermosensor (Table 1; Narberhaus 2002). The first evidence for temperature-mediated gene regulation via alternative mRNA structures was first provided by Altuvia et al. (1989) in the lytic pathway of phage λ. Since then, the cellular level of several chaperones and heat-shock or cold-shock proteins, which is tightly controlled at both transcriptional and post-transcriptional levels, was shown to be also regulated by self-induced conformational changes in their mRNAs (Table 1; Narberhaus 2002). A good example of translational thermoregulation

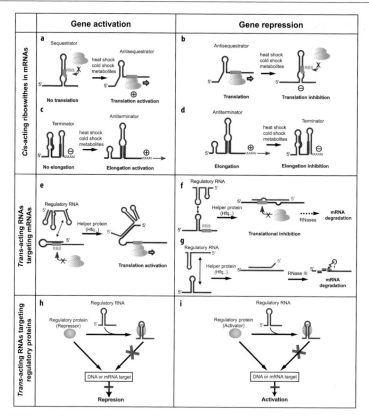

Fig. 1a–i Different RNA-mediated mechanisms used in virulence gene regulation. *Trans*-acting regulatory RNAs are shown by *red lines* and mRNAs by *blue lines*, with their polarity indicated. Hypothetical secondary structures are schematized by *stem-loops*. Complementary sequences forming alternative pairing in mRNAs are shown in *magenta*. The ribosome-binding site (*RBS*) is in *green*, the ribosome in *pale gray*, and regulatory proteins in *pale green*. The regulation mechanisms can be ranged into three classes, depending of the nature of the regulatory RNAs. (1) *cis*-Acting elements in mRNAs building riboswitches. The riboswitch can adopt alternative structures, depending on changes in temperature or metabolite concentration. These conformational switches can modulate the accessibility of the RBS (**a, b**) or formation of a transcriptional terminator or antiterminator (**c, d**), thus regulating translational initiation or transcription termination, respectively. (2) *trans*-Acting RNAs directly targeting mRNAs. These RNAs have variable lengths and display more or less extended regions of complementarity with their mRNA targets. Annealing with the complementary mRNA sequence can unmask the RBS, thus activating translational initiation (**e**). Annealing can also occlude the RBS, thus inhibiting translation (**f**), resulting in the degradation of the untranslated mRNA. The interaction with the antisense RNA can also unmask or induce a site for RNase III, generally leading to the degradation of both mRNA and regulatory RNA (**g**). The interaction between antisense RNAs and their mRNA targets are most often facilitated by helper proteins, such as Hfq. (3) *trans*-Acting RNAs targeting regulatory proteins. These RNAs trap a regulatory protein, the binding of which to DNA or mRNA represses or activates transcription or translation. The regulatory RNA indirectly acts as an activator or repressor of gene expression, by sequestrating the regulatory protein (**e, f**)

Table 1 Examples of *cis*-mRNA elements acting as thermosensors. Temperature shift induces conformational changes of mRNA leader regions, which affect initiation of translation

Signal	Bacterial species	Regulated gene(s)	Function	Reference(s)
Heat	Escherichia coli	λcIII	Regulation of lysogeny	Altuvia et al. 1989
	Escherichia coli	rpoH	σ-Factor	Morita et al. 1999a,b
	Bradyrhizobium japonicum	α-Heat shock genes	Chaperone	Nocker et al. 2001
	Caulobacter crescentus	dnaK	Chaperone	Avedissian et al. 1995
	Haemophilus ducreyi	dnaJ	Chaperone	Parsons et al. 1999
	Listeria monocytogenes	prfA	Virulence gene activator	Johansson et al. 2002
	Streptomyces albus	hsp18	Chaperone	Servant and Mazodier 2001
	Yersinia pestis	lcrF	Virulence gene activator	Hoe and Goguen 1993
Cold	Escherichia coli	cspA	RNA chaperone	Yamanaka et al. 1999

is illustrated by the *E. coli* heat shock σ-factor RpoH. At low temperature, the *rpoH* mRNA is translationally inactive and not abundant. Induction of the heat shock response also includes its translational activation. Thermoregulation relies on two *cis*-acting mRNA segments that can form alternative structures depending on temperature (Kamath-Loeb and Gross 1991; Nagai et al. 1991). There is a clear correlation between the stability of the mRNA structure and *rpoH* expression, indicating that the binding of 30S subunits can only occur at high temperature, when the structure is destabilized (Morita et al. 1999a,b).

Remarkably, pathogens can exploit mRNA thermosensors to detect that they entered the body of warm-blooded host, thus allowing transmission of a signal to turn on virulence functions (Hurme and Rhen 1998). Such a mechanism was first suggested in *Yersinia pestis*, in which *lcrF*, encoding the transcriptional activator involved in virulence gene regulation, is transcribed at equal rates at 26 °C and 37 °C, while the synthesis of LcrF is thermally controlled (Hoe et al. 1992). A thermally sensitive post-transcriptional mechanism is directly exerted on mRNA (Hoe and Goguen 1993). A model was proposed in which an mRNA secondary structure element sequestering the Shine-Dalgarno sequence (SD) is melted at high temperature, allowing translation to proceed (Fig. 1a).

That model was recently strongly supported by the thermoregulated expression of the pleiotropic transcriptional regulator PrfA in *L. monocytogenes*, which activates the synthesis of several virulence proteins. At temperature below 30°C, the low expression of virulence genes coincides with the absence of

PrfA protein, although the PrfA gene is still transcribed (Renzoni et al. 1997). It was shown that the stability of PrfA was unaffected by low and high temperature, while its expression was thermoregulated at the translational level in *L. monocytogenes* (Johansson et al. 2002). The proposed mechanism involves a self-induced structural switch at the 5′ leader region of the *prfA* mRNA, promoted by temperature changes. At low temperature, the ribosome binding site is sequestered into a secondary structure that hinders ribosome binding. An absence of translation might in turn favor degradation of *prfA* mRNA. Upon increasing temperature (i.e., in a host organism), the structure is destabilized, rendering accessible the ribosome loading site, thus triggering translation initiation (Fig. 1a). This mechanism is sustained by chemical base probing data and mutational analyses (Johansson et al. 2002). Indeed, base substitutions expected to weaken the stability of the putative RNA structure lead to constitutive expression of PrfA, even at a low temperature. Conversely, mutations designed to reinforce base-pairing of the SD sequence lower PrfA synthesis. Moreover, the *prfA* regulatory region can control the expression of the green fluorescence protein when fused upstream of the *gfp* mRNA in *E. coli* (Johansson et al. 2002). Taken together, those experiments indicate that the *prfA*-5′ untranslated region (UTR) acts as a thermosensor to regulate translation efficiency. Although regulation appears to rely on the intrinsic property of the mRNA to adopt dynamic alternative conformations, a possible involvement of any general *trans*-acting factors (such as Hfq protein) remains an open question.

Remarkably, there is increasing evidence that mRNAs can also sense physiological signals other than temperature. The unique feature of these systems, also called "riboswitches," is that they can sense directly small metabolites, such as vitamins, amino-acids, or purine precursors (Narberhaus 2002; Barrick et al. 2004; Vitreschak et al. 2004). Riboswitches are generally found in the leader part of bacterial operons to control either transcription or translation by forming alternative structures, which can induce or prevent the formation of intrinsic terminators or ribosome binding sites (Fig. 1a–d). They exploit an astonishing dynamic property of RNA structure and provide economical as well as fast-reacting solutions, allowing adaptation to environmental cues. Their utilization most likely extends much beyond the currently known examples. This is a reasonable assumption, since these metabolic functions were recently shown to be required for full virulence of *S. aureus*. Also, they are important in permitting the bacteria to survive and replicate in distinct in vivo environments (Benton et al. 2004).

3
trans-Acting RNAs as Mediators of Virulence

A subset of *trans*-acting RNAs has been characterized as mediator of virulence gene expression (Johansson and Cossart 2003). Due to the nature of RNA, a prevalent group includes the regulatory RNAs, which act by base-pairings with target mRNAs. Several of them target multiple mRNAs, or regulate the expression of mRNA encoding global transcriptional regulatory proteins that in turn control many genes. (This paper does not tackle antisense RNAs that control essential functions in plasmids, phages, and transposons, see reviews Wagner and Simons 1994; Wagner et al. 2002.)

Another class of regulatory RNAs is capable to mimic structure elements of mRNA. This is the case of *E. carotovora* RsmB (repressor of secondary metabolite B), which carries multiple purine-rich repeats similar to the RsmA protein-binding sequence found in several mRNAs. As a consequence, the RNA traps the global post-transcriptional regulatory protein.

These RNAs are usually at the central point of regulatory networks and may indirectly induce pleiotropic effects by targeting regulatory proteins or mRNA encoding regulatory proteins (Fig. 1). The *trans*-acting pathogenesis-related RNAs are listed in Table 2, and some of the most striking examples are described below. It is worth noting that different RNA names have been used such as: ncRNA (non-coding RNA), sRNA (small RNA), snmRNA (small non-messenger RNA), eRNA (effector RNA), or regulatory RNAs. Since the pathogenesis-related RNAs are rather heterogenous (Tables 1–2), we will mainly refer to them as regulatory RNAs, and sRNAs for the small non-coding RNAs.

3.1
Regulatory Antisense RNAs

3.1.1
sRNAs and Iron Metabolism

The iron acquisition process is one of the major determinants as to whether a microorganism can colonize an organism and is able to maintain itself therein, as well as to escape direct attack from the host defense mechanisms. The pathogen can acquire rather easily many of the nutrients from the host tissue except for iron, which is poorly soluble in its oxidized form. Thus, bacteria use specific mechanisms for the iron sequestration from their host, e.g., the siderophores (Ratledge and Dover 2000; Schaible and Kaufmann 2004). While iron is required for bacteria survival, the iron-catalyzed production of reactive oxygen species (ROS) could lead to severe cellular damages (Halliwell and Gutteridge 1984; Keyer and Imlay 1996). Consequently, bacteria have evolved tightly regulated systems for both uptake and sequestration of this essential element. In *E. coli* and in many other bacteria, the global regulatory protein Fur (Ferric uptake regulator) maintains this equilibrium. Fur senses iron concentration and represses the transcription of all genes involved in iron acquisition (Bagg and Neilands 1987). If iron is abundant, Fur binds to a specific sequence overlapping the RNA polymerase binding sites in a Fe^{2+}-dependent manner, inhibiting the initiation of transcription (Escolar et al. 1999).

Table 2 Occurrence, mechanisms, and properties of prokaryotic *trans*-acting RNAs as mediators of virulence. The target mRNAs encode the following proteins: aconitase A (*acnA*), iron superoxide dismutase (*sodB*), succinate dehydrogenase (*sdhD*), fumarase A (*fumA*), bacterioferritin (*bfr*), ferritin (*ftn*), bacterial iron transport gene (*fatB*), hemolysin α (*hla*), protein A (*spa*), and two transcriptional regulators (*hapR, luxR*). Non-coding RNAs with a similar function to CsrB/RsmB were also predicted in genomes of other pathogenic bacteria such as *Y. pestis, Shigella flexneri,* and *V. cholerae* (Griffiths-Jones et al. 2003)

Regulatory RNA (nts)	Bacteria species	Known targets	Function	Regulatory mechanism	References
RyhB (90)	*Escherichia coli*	*acnA* mRNA *sodB* mRNA *sdhD* mRNA *fumA* mRNA *bfr* mRNA *ftn* mRNA	Iron metabolism	Inhibition of translation and rapid mRNA degradation	Massé and Gottesman 2002; Massé et al. 2003
PrrF1/PrrF2 (110)	*Pseudomonas aeruginosa*	*sodB* mRNA *sdhD* mRNA *bfr* mRNA	Iron metabolism	Inhibition of translation and rapid mRNA degradation	Wilderman et al. 2004
RNAα (650)	*Vibrio anguillarum*	*fatB* mRNA	Iron metabolism	mRNA degradation ?	Salinas et al. 1993; Waldbeser et al. 1995
Qrr (1–4) sRNA (96 to 108)	*Vibrio cholerae Vibrio harveyi*	*hapR* mRNA *luxR* mRNA	Virulence Bioluminescence	Inhibition of translation and rapid mRNA degradation	Lenz et al. 2004
RNAIII (514)	*Staphylococcus aureus*	*hla* mRNA *spa* mRNA Others?	Hemolysin synthesis Host-pathogen interaction Exoenzymes, proteases	Activation of translation Inhibition of translation and rapid mRNA degradation ?	Novick et al. 1993; Morfeldt et al. 1995; Huntzinger et al. 2005
FasX (250)	*Streptococcus pyogenes*	?	Virulence (secreted factors)	?	Kreikemeyer et al. 2001

Table 2 (continued)

Regulatory RNA (nts)	Bacteria species	Known targets	Function	Regulatory mechanism	References
Pel (459)	Streptococcus pyogenes	?	Virulence (protease, M protein, streptokinase....)	?	Li et al. 1999; Mangold et al. 2004
VR RNA (386)	Clostridium perfringens	?	Virulence (secreted toxins)	?	Shimizu et al. 2002
VirX (400)	Clostridium perfringens	?	Virulence (secreted toxins)	?	Ohtani et al. 2002
CsrB/CsrC (366/245)	Escherichia coli	CsrA protein	Glycogen biosynthesis, biofilm formation, host-bacteria interaction	Protein sequestration	Liu et al. 1997; Weilbacher et al. 2003
RsmB (479)	Erwinia carotovora	RsmA protein	Cell-wall degrading enzymes	Protein sequestration	Liu et al. 1998
CsrB (350)	Salmonella typhimurium	CsrA protein	Virulence	Protein sequestration	Altier et al. 2000
RsmZ (PrrB)/RsmY (130/118)	Pseudomonas fluorescens	RsmA homolog	Exoenzymes, secondary metabolites	Protein sequestration	Aarons et al. 2000; Valverde et al. 2003
RsmZ/RsmB (120/240)	Pseudomonas aeruginosa	RsmA homolog	Elastase, pyocyanin, secondary metabolites	Protein sequestration	Heurlier et al. 2004; Burrowes et al. 2005
tRNA$_5^{Leu}$ (75)	Escherichia coli	–	Virulence (hemolysins...)	Decoding properties	Ritter et al. 1995; Dobrindt and Hacker 2001
tmRNA (363)	Salmonella typhimurium	?	Virulence *ivi* gene expression	Protein degradation?	Julio et al. 2000

?, unknown target or regulatory mechanism; nts, nucleotides

In addition, Fur couples the production of superoxide dismutases (SODs) with iron metabolism (Niederhoffer et al. 1990; Tardat and Touati 1991). The production of SOD is modulated and adapted to the environmental threat of oxidative stress, keeping superoxide low enough to prevent damage (Compan and Touati 1993). SODs are key enzymes of the multiple defense system that protects organisms against the deleterious effects of naturally produced oxygen radicals (Touati 1997). Two cytoplasmic SODs from *E. coli*, require different metals for activity, Mn and Fe. Iron starvation induces production of MnSOD (*sodA*), but decreases production of FeSOD (*sodB*), suggesting this antagonistic regulation permits the cell to maintain sufficient SOD levels, regardless of the iron availability. While Fur-dependent transcriptional repression of MnSOD production is well characterized, the mechanism by which Fur induced *sodB* expression remained elusive until recently (Dubrac and Touati 2000, 2002). The clue came from the work of Massé and Gottesman (2002), who identified a small regulatory RNA, RyhB (Figs. 2, 3a). It displays a Fur-dependent regulation and acts on the *sodB* mRNA as an antisense RNA (Massé et al. 2003; Geissmann and Touati 2004). Since complementarity between RyhB and the ribosome binding site of *sodB* mRNA is restricted, the global regulatory protein Hfq is required to facilitate their interaction (Geissmann and Touati 2004). As a consequence, RyhB blocks the translation initiation site of *sodB* and triggers the rapid and concomitant degradation of RyhB and *sodB* mRNA via RNase E (Massé et al. 2003).

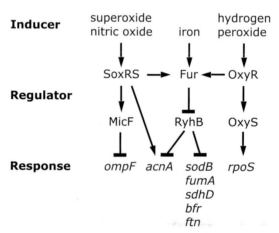

Fig. 2 Connections between global regulators involved in iron metabolism and in protection mechanisms against oxidative damages in *E. coli*. SoxRS, Fur and OxyR are global transcriptional regulatory proteins that directly sense signals, and that regulate directly the transcription of many target genes (not shown). They also regulate synthesis of the small non-coding RNAs MicF, RhyB, and OxyS, respectively, and as a consequence control indirectly many downstream genes at the post-transcriptional level. *RpoS* encoded the stationary phase σ-factor and is also regulated by different sRNAs (Gottesman 2004). Activation is denoted by *arrows* and repression by *lines*

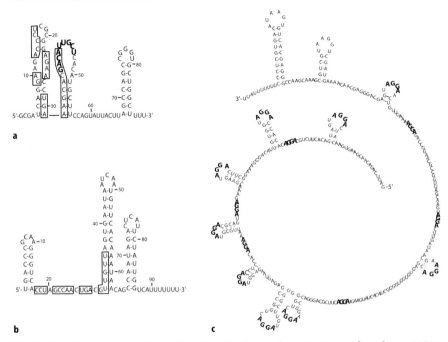

Fig. 3a–c Predicted and experimentally determined secondary structures of regulatory RNAs. **a** *E. coli* RyhB RNA regulates mRNAs encoding proteins involved in iron metabolism. Nucleotides in **bold characters** are complementary to the ribosome binding site of *sodB* mRNA, and nucleotides complementary to the 5′ leader sequence of *sdh* mRNA are *framed* (Massé and Gottesman 2002). The structure was defined using chemical and enzymatic probing (Geissmann and Touati 2004). **b** Predicted secondary structure of *V. cholerae* Qrr1 RNA (Lenz et al. 2004). Nucleotides complementary to the ribosome binding site of *hapR* mRNA are *framed*. **c** Secondary structure model, mainly based on sequence alignment of *E. coli* CsrB RNA and *E. carotovora* RsmB RNA (Griffiths-Jones et al. 2003). The consensus sequence AGGA of the repeats recognized by the global regulatory protein CsrA/RsmA is shown in *bold characters*

Under the conditions of iron starvation, RyhB is produced and represses the expression of six other genes encoding both iron-storage proteins (ferritin and bacterioferritin) and iron-requiring enzymes (aconitase A, fumarase A, succinate dehydrogenase, and iron superoxide dismutase). Conversely, when the concentration of iron increases, Fur inhibits *ryhB* transcription initiation, thus derepressing an expression of the iron-storage proteins and enzymes (Fig. 2). Thus, Fur acts via two different mechanisms: It represses directly transcription initiation of numerous genes and regulates indirectly the expression of several genes at the post-transcriptional level via the regulatory RNA RyhB (Fig. 2). This double level of regulation in *E. coli* is probably largely used to control iron metabolism in response to iron availability in the environment. Many other genes that encode proteins involved in iron-rich complexes have been identified to be regulated by Fur (McHugh et al. 2003; Fig. 2). Whether these genes are also under RyhB control remains to be tested.

The promoter region and the 90-nucleotide (nt) sequence of *ryhB* are well conserved in *E. coli*, *Salmonella typhimurium*, and *Klebsiella pneumoniae*, whereas a core sequence of 27 nts within RyhB is conserved in *Y. pestis* and *V. cholerae*. A RyhB homolog was further identified in *S. typhimurium* and *Y. pestis* (Massé and Gottesman 2002). Although no RyhB sequence homologs were found in *Pseudomonas aeruginosa*, two other sRNAs (PrrF1 and PrrF2) involved in iron metabolism were recently evidenced (Wilderman et al. 2004). As for RyhB, their synthesis is negatively controlled by Fur when iron is present in excess. The induction of the PrrF sRNAs leads to the rapid degradation of mRNAs *sodB*, *sdh* (succinate dehydrogenase), and a gene encoding a bacterioferritin. These data suggest that sRNAs are largely involved in controlling bacterial iron homeostasis and defense against oxidative stress (Fig. 2). One may expect that genome-wide screens for sRNAs in other bacteria will reveal more Fur-dependent regulated RNAs.

Another interesting example of iron metabolism regulation came from the highly virulent strain of the fish pathogen *V. anguillarum*, which possesses a very efficient plasmid-mediated high-affinity iron uptake system. In the operon *fatDCBA*, two of the genes, which encode the transport proteins *fatA* and *fatB*, are downregulated at the transcriptional level by Fur protein and at the post-transcriptional level by the plasmid-encoded antisense RNAα (Salinas et al. 1993). This 650-nt RNA is transcribed in the opposite direction of the *fatB* coding region, and is preferentially expressed under iron-rich conditions. The interaction of RNAα and *fatB* mRNA is thought to enhance degradation of *fatB* mRNA and to decrease the expression of *fatA* (Waldbeser et al. 1993; 1995), possibly by inducing a conformational switch of *fatB* mRNA, rendering the mRNA accessible to ribonucleases. The endogenous-encoded Fur protein is also required for RNAα synthesis at the transcription initiation level, but independently of the iron status of the cell (Chen and Crosa 1996). Unexpectedly, iron regulates RNAα synthesis post-transcriptionally by stabilizing RNAα rather than enhancing transcription initiation. One intriguing feature is the absence of a "Fur box" in the RNAα promoter region, indicating either the existence of a unique regulation mechanism or that Fur exerts an indirect effect on the transcription initiation of the regulatory RNA. Interestingly, *Acinetobacter baumannii*, which causes serious human bloodstream infections, displays both siderophore-mediated iron-acquisition, and *fat* operon systems, very similar to those found in *V. anguillarum* (Dorsey et al. 2004). However, the presence of a regulatory antisense RNA has not been evidenced yet.

3.1.2
sRNAs and Virulence in *V. cholerae*

V. cholerae uses quorum sensing to regulate virulence gene expression in response to changes in cell density. HapR protein appears to be the main regulator of the quorum-sensing regulon in *V. cholerae* that allows the expression of several virulence genes at low cell density (Zhu et al. 2002). It was recently shown that synthesis of HapR protein is regulated at the post-transcriptional level and depends concomitantly of the Hfq protein and of four small non-coding RNAs, named Qrr (Lenz et al. 2004). Phylogenetic analysis indicated that these four sRNAs present extensive similarities, suggesting that they derived from duplication of a single sRNA gene from the ancestral organism (Lenz et al. 2004). In addition, they all carry partial complementarities with the SD region of the target *hapR* mRNA (Fig. 3b). Moreover, the regulatory protein Hfq is required for virulence-factor expression and the action of the four Qrr RNAs (Lenz et al. 2004).

Computer analysis revealed that five homologous Qrr RNAs are also present in *V. parahaemolyticus*, *V. vulnificus*, and *V. harveyi* (Lenz et al. 2004). Conservation of genes across species is in general a good indication of similar functions (Michel et al. 2000), suggesting that these sRNAs may also be involved in analogous quorum-sensing circuits. Indeed in *V. harveyi*, Qrr RNAs, and Hfq were shown to control in a coordinated way the density-dependant light production by inducing the rapid degradation of *luxR* mRNA at low cell density (Lenz et al. 2004). Strikingly, while the four Qrr RNAs are required for quorum-sensing control in *V. cholerae*, the overexpression of a single sRNA is capable to ensure virulence factor expression (Lenz et al. 2004). Such a redundancy may be required for a fine-tuning transition between the low- and high-density cell levels. Alternatively, each Qrr sRNA may act at different regulatory circuits, i.e., metabolism or nutrient utilization, by targeting different mRNAs. Differences in sequences may also lead to pairing and degradation of mRNAs with different efficiencies.

While the mechanism of action of Qrr RNAs resembles that of the *E. coli* RyhB sRNA (Massé and Gottesman 2002; Geissmann and Touati 2004; Fig. 1f), several questions still remain open about the effect of Qrr RNAs on ribosome binding to *hapR* mRNA, on the RNases involved in the degradation pathway, and on the exact function of the Hfq protein in regulatory processes. Indeed, *hapR* mRNA is rapidly degraded when the synthesis of Qrr RNAs is induced. Conversely, the steady-state levels of Qrr RNAs increase in the absence of the target gene, which has been interpreted as a mutual degradation of sRNA and target (Lenz et al. 2004). Thus, sRNAs can be rapidly synthesized under appropriate conditions, but can also be turned off quickly once they have served.

The use of sRNAs instead of regulatory proteins provides an ultrasensitive and rapid response, which is particularly well appropriate for adaptive processes such as quorum sensing.

3.2
Modulation of Protein Activity by Regulatory RNAs

3.2.1
The *E. coli* CsrA/CsrB System

Another class of regulatory RNAs includes those that indirectly regulate gene expression by targeting regulatory proteins and modulating their functions (Fig. 1h,i; Gottesman 2004). One of the most characteristic examples is *E. coli* CsrA, the carbon storage regulatory protein (Romeo et al. 1993). When reaching the stationary phase, non-sporulating bacteria readjust the gene-expression pattern, improving stress resistance and enhancing the ability to scavenge substrate from the medium. The activation of glycogen biosynthesis is one of the essential routes used to redirect carbon utilization. The RNA-binding protein CsrA, which is constantly expressed during growth, acts both as a translational repressor of glycogen biosynthesis and as an activator of glycolysis in *E. coli*. The phenotypes of *E. coli csrA* mutants led to the prediction that the protein affects the interactions of enteric bacteria species with mammalian hosts, and motility (Romeo 1998). Thus, CrsA can be considered as a global regulator. Strikingly, the regulatory activity of CsrA is antagonized by two non-coding RNAs, CsrB and CsrC (Liu et al. 1997; Weilbacher et al. 2003). Whereas their sizes vary, they carry several purine-rich repeats (5'-CAGGA(U/A/C)G-3') which are located either in single-stranded regions or in hairpin loops (Fig. 3c, Table 2). These repeats are suggested to mimic the binding sites of CsrA present in the 5' UTR of target mRNAs. These sites are bipartite, and comprise the SD sequence and a short upstream hairpin (Baker et al. 2002), but their nucleotide determinants remain to be defined. *E. coli* CsrB contains 18 repeats (while CsrC contains 9), and binds 18 CsrA molecules in a cooperative way, thus facilitating the sequestration of the protein within a compact ribonucleoprotein complex (Liu et al. 1997). The expression of CsrB is growth-phase dependent, and the RNA accumulates as the culture enters the stationary phase of growth (Gudapaty et al. 2001). A model has been suggested in which the activity of CsrA would be controlled by an equilibrium between the free protein and the CsrA–CsrB ribonucleoprotein complex. Thus, these regulatory RNAs would compete with mRNA for binding to CsrA, and would induce the release of the regulatory protein from the target mRNAs, thereby allowing their translation (Romeo 1998).

3.2.2
Homologous Systems in Bacterial Pathogens

The role of CsrA homologs in bacterial virulence was first established in the plant pathogenic *Erwinia* species. In *E. carotovora*, the CsrA homolog RsmA represses the expression of several virulence factors such as pectate lyase, cellulases, and proteases (Chatterjee et al. 1995; Cui et al. 1995). These enzymes degrade the cell wall of plant, allowing the dissemination of the bacteria within the host tissue. As in the case of *E. coli*, a regulatory activity of RsmA is counteracted by a non-coding RNA, RsmB (Liu et al. 1998). More recently, homologs of CrsA were also found in human pathogens. In *P. aeruginosa*, RsmA inhibits the translation of mRNAs encoding virulence factors including proteases and elastase (Pessi et al. 2001), and two corresponding regulatory RNAs were recently identified (Heurlier et al. 2004; Burrowes et al. 2005). In *S. enterica* an homologous CsrA/CsrB system was shown to control SPI1 genes (Altier et al. 2000). Finally, CsrA exerts a broad role in regulating the physiology of *Helicobacter pylori* in response to environmental stimuli, and facilitates the adaptation of the pathogen to the different environments encountered during colonization of the gastric mucosa (Barnard et al. 2004). However, a homolog to CsrB has not yet been identified.

Whereas the protein components can be easily identified by homology searches in many gram-negative bacteria, the riboregulators are not strictly homologous. Their sizes, sequences, and predicted secondary structures diverge despite the fact that all these riboregulators perform essentially the same function, i.e., sequestration of the regulatory proteins. As an example, in different *Pseudomonas* species, the regulatory RNAs are smaller than those from *E. coli* and the consensus sequence of the repeats are shorter (Valverde et al. 2004; Table 2). It is quite probable that these RNAs have evolved independently.

3.3
Multifunctional RNAs

3.3.1
The Case of *S. aureus* RNAIII

In *S. aureus*, the expression of virulence factors is tightly regulated in response to cell density (quorum sensing), energy availability, and various environmental signals (Novick 2003). Signal receptors are the primary regulatory mediators for the expression of the virulon in *S. aureus*. Among these receptors, the *agr* system, composed of two divergent transcription units, functions as a sensor of the population density. The involvement of *agr* operon in pathogenesis has been demonstrated in several infection models (Bunce et al. 1992; Abdelnour et al. 1993; Cheung et al. 1994; Wesson et al. 1998; Vuong et al. 2000; Novick 2003). Transcriptome analysis has also revealed that the *agr* operon regulates not only virulence-related factors, but also several proteins involved in basic metabolism (lipid, amino acid, nucleotides) and transport processes (Dunman et al. 2001). Some of these proteins were shown to be essential to maintain bacterial survival within the host (Benton et al. 2004). Thus, virulence and adaptation to stress or environmental signals are linked processes.

The *agr* system encodes a regulatory RNA (RNAIII) that plays a key role in the quorum sensing-dependent regulatory circuit and coordinately regulates several virulence-associated genes (Novick 2003). The expression of RNAIII is maximal in the late-logarithmic and stationary phase cultures, and thus participates in the switch of gene expression occurring when *S. aureus* reaches the stationary phase. This rather long RNA (514 nts) has the property of acting as a messenger RNA-encoding *hld* (δ-hemolysin), and having a variety of regulatory functions: repression of the expression of surface proteins such as protein A during the exponential phase, and activation of the expression of extracellular toxins and enzymes during the post-exponential phase (Janzon and Arvidson 1990; Kornblum et al. 1990; Novick 2003). Thus, RNAIII participates in the switch between the expression of surface proteins and excreted toxins, which may reflect a dichotomy between colonization and disease (Novick 2003).

The secondary structure of RNAIII is characterized by several stem-loop structures (Fig. 4a), which represents potential binding sites for *trans*-acting factors such as mRNAs or proteins (Benito et al. 2000). RNAIII regulates virulence gene expression at both the transcriptional and post-transcriptional levels. The mechanism by which RNAIII controls transcription is still unknown. Regulation of transcription of virulence genes involves a complex interplay between positive and negative regulatory proteins (Novick 2003). Thus, one possible role of RNAIII would be to modulate the activity of transcriptional activators or inhibitors (Said-Salim et al. 2003). Alternatively, RNAIII could indirectly affect transcription through inhibition of mRNAs that encode transcriptional regulatory factors.

The mechanism of regulation at the post-transcriptional level has been elucidated for two targets, the *hla* mRNA (encoding α-hemolysin) and *spa* mRNA (encoding protein A) (Morfeldt et al. 1995; Benito et al. 2000; Novick 2003; Huntzinger et al. 2005). Two distinct domains of RNAIII act as antisense to activate translation of *hla* mRNA and to inhibit translation of *spa* mRNA (Fig. 4). Binding of the 5' end of RNAIII with a segment of *hla* mRNA prevents the formation of an intramolecular RNA secondary structure that sequesters the *hla* ribosomal binding site (Morfeldt et al. 1995; Novick 2003). The 3' domain of RNAIII, partially complementary to the ribosome binding site of *spa* mRNA, efficiently anneals to *spa* mRNA. Although RNAIII binding is sufficient to inhibit in vitro the formation of the translation initiation complex, the coordinated action of the double-stranded RNase III is essential in vivo to degrade *spa* mRNA and irreversibly arrest translation (Huntzinger et al. 2005). The RNase III cleaves the formed duplex and is also involved in the degradation pathway of *spa* mRNA. In *E. coli*, most of the small non-coding RNAs, which

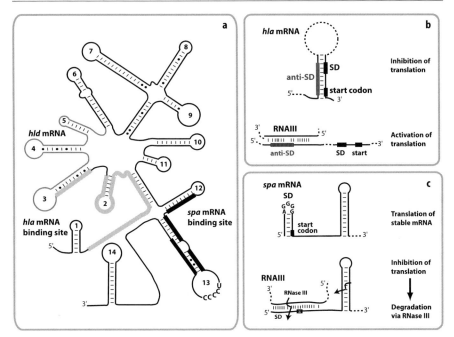

Fig. 4a–c *S. aureus* RNAIII and regulatory mechanisms on two target mRNAs. **a** Experimentally determined secondary structure of RNAIII (Benito et al. 2000). The 5′ region of RNAIII complementary to *hla* mRNA (encoding hemolysin α) is shown in *gray lines* and regions complementary to the ribosome binding site of *spa* mRNA (encoding protein A) in the 3′ domain are shown in *black lines*. The open reading frame encoding *hld* mRNA (hemolysin δ) comprises the hairpins 3 to 5. **b** The 5′ region of RNAIII binds to a 5′ segment of *hla* mRNA, and sequesters the anti-Shine and Dalgarno sequence, activating *hla* translation (Morfeldt et al. 1995). **c** The 3′ domain of RNAIII binds to the ribosome binding site of *spa* mRNA, inhibits translation, and induces rapid degradation via the double-stranded endoribonuclease III (Huntzinger et al. 2005). The initial contacts probably involved a loop–loop interaction (the sequences are shown in **a** and **c**)

target specific mRNAs, require Hfq protein (Storz et al. 2004). *S. aureus* RNAIII was also found to be a putative target of Hfq protein, but the function of this interaction remains to be clarified (Huntzinger et al. 2005).

RNAIII homologs were found in other pathogenic *Staphylococcus* species, including *S. epidermidis*, *S. simulans*, *S. warneri* (Tegmark et al. 1998), and *S. lugdunensis* (Vandenesch et al. 1993). Except for *S. lugdunensis* RNAIII, all these RNAs encode one or two toxins. These RNAs are able to restore, at least partially, the expression of virulence factors in *agr* (−) mutant of *S. aureus*. Sequence alignment has revealed a high conservation of the 3′ end domain (Benito et al. 2000). This domain in *S. aureus* RNAIII carries multiple regulatory functions, since it also induces the synthesis of several exoproteins (M. Possedko and P. Fechter, unpublished results; Novick 2003).

3.3.2
Other Regulatory RNAs

The case of *S. aureus* RNAIII is not unique, and other regulatory RNAs are also the effectors of a two-component system shown to be involved in virulence gene expression in other pathogens (Table 2). In *S. pyogenes*, FasX RNA (for fibronectin/fibrinogen binding/haemolytic activity/streptokinase regulator) and Pel RNA (for pleiotropic effector locus) control the expression of several virulence-associated genes. The expression of both RNAs is growth-phase dependent and reaches the maximum during the transition at the stationary phase. FasX RNA positively regulates the expression of several secreted virulence factors like streptokinase and streptolysin S, and probably inhibits the expression of fibronectin- and fibrinogen-binding proteins (Kreikemeyer et al. 2001). Like *S. aureus* RNAIII, Pel RNA contains an open reading frame-encoding streptolysin S that is not required for the regulatory function (Mangold et al. 2004). The RNA positively regulates the expression of streptokinase, secreted cysteine protease (SpeB), M protein, and streptococcal inhibitor of complement (Li et al. 1999; Mangold et al. 2004). Regulation occurs both at the transcriptional and post-transcriptional levels (Mangold et al. 2004). Thus, FasX and Pel RNAs present some functional redundancies. It is worth noting that the deletion of FasX results in increased Pel RNA expression (Kreikemeyer et al. 2001). An analogous case was found in *C perfringens*, where two non-coding RNAs (VR RNA, and VirX) activate the expression of several chromosome- and plasmid-encoded secreted toxins (Ohtani et al. 2002, 2003; Shimizu et al. 2002). In all these systems, the mode of action of these regulatory RNAs remains to be defined.

3.4
Other Types of RNAs Influencing Pathogenicity

3.4.1
The Case of tRNA Modifications

Housekeeping RNAs such as transfer (t)RNAs have also been described to be involved in adaptive responses. In *S. flexneri*, a random Tn5 mutation analysis of the genome allowed Durand et al. (1994) to identify modifications in tRNAs as essential elements for virulence expression. Indeed, an insertion of Tn5 in the gene coding for the tRNA guanosine transglycosylase, catalyzing the Q34 modification of some tRNAs, reduced the virulence of *S. flexneri* to 50% of that of the wildtype, while the growth of the bacteria was not affected. The presence of other modified nucleosides in the anticodon loop of several tRNAs (mainly at position 37) has also been shown to be required for the expression of virulence factors in *Agrobacterium tumefaciens* (Gray et al. 1992), *P. aeruginosa* (Sage et al. 1997), and *P. syringae* (Brégeon et al. 2001; Kinscherf and Willis 2002). The modified nucleosides in the anticodon stem-loop of tRNAs are of importance for modulation of codon recognition as well as for maintenance of the correct reading frame during translation on the ribosome. Indeed, modified nucleosides at position 34 restrict or extend a wobble base pairing and thereby the decoding pattern of the tRNA. Modifications at the ubiquitous purine at position 37 plays a major role in stabilization of the codon–anticodon interaction, and because most of the modified nucleosides cannot form base-pairing, their presence directly 3' to the anticodon restricts the pairing to the in-frame codon–anticodon pair, thus avoiding frame shifting (Motorin and Grosjean 1999). While the exact role of these modifications in virulence gene expression remains to be addressed, the degree and/or nature of modification of tRNAs can also be considered as a regulatory device to modulate the translation of specific mRNAs as a response to specific environmental changes.

3.4.2
The Case of tmRNA

The primary function of transfer-messenger (tm)RNA is to specifically target abnormal proteins for further degradation. This RNA, which acts as both a tRNA and an mRNA, adds to truncated mRNA a specific RNA tag that encodes for a protease motif (Keiler et al. 1996). Interestingly, this RNA is overexpressed under stress conditions in different pathogenic bacteria [e.g., in *S. pyogenes* (Steiner and Malke 2001) and *S. enterica* serovar Typhimurium (Julio et al. 2000)]. Most likely, tmRNA is one of the effectors that counteract the perturbation of the transcriptional and translation machinery due to stress conditions. For example, amino acid starvation induces incomplete translation of mRNAs, and reactive oxidants are one of the major sources of nucleic acids and protein damage that leads to the production of potentially toxic proteins. Recently, it was shown that tmRNA is required for full virulence in *S. enterica* serovar Typhimurium and that it affects the expression of specific genes induced during infection (Julio et al. 2000). This may be related to more specific functions of tmRNA (Withey and Friedman 2002), including regulation of an alternative protease activity (Kirby et al. 1994) or the availability of lactose (Abo et al. 2000). In this latter case, it was shown that binding of LacI repressor to *lacI* mRNA cleaved the mRNA that is in turn targeted by the tmRNA to induce rapid degradation of the truncated protein (Abo et al. 2000).

3.4.3
Location of Pathogenicity Islands

An interesting observation is that many tRNA or tRNA-like encoded genes flank many of the genes associated with bacterial pathogenicity (Hou 1999). This is the case of the *ssrA* gene encoding tmRNA that is frequently associated with pathogenicity islands of several human pathogens such as *V. cholerae* (Karaolis et al. 1998), and *S. typhimurium* (Julio et al. 2000). In other cases, minor tRNA genes have been found located at such strategic positions [e.g. *leuX* in *E. coli* (Ritter et al. 1995), *asnT* in *Y. enterocolitica* (Carniel et al. 1996), *valV* in *S. typhimurium* (Shea et al. 1996), and *serV* in *Dichelobacter nodosus* (Cheetham et al. 1995)]. Encoded tRNAs may be used to read minor codons for the synthesis of specific virulence-associated proteins. A well-documented example is the uropathogenic *E. coli* strain 536 that carries loci *leuX*, encoding the minor $tRNA_5^{Leu}$. This tRNA allows the expression of different proteins necessary for *E. coli* to fully express its virulence (Ritter et al. 1995). The transcription of this minor tRNA is also specifically induced by the heat shock-specific σ-factor RpoH after an increase of temperature (Dobrindt and Hacker 2001). This tRNA was located at the insertion site of the pathogenic island II. Given the number of tRNA genes in the bacterial genome, this would also favor an amplification of the insertion of pathogenicity islands. While the mechanism of inheritance remains unclear, bacterial pathogenicity islands exhibit features of mobile genetic elements. Because these elements are expected to participate in the horizontal transfer of bacterial genes, their association with tRNA elements may suggest a general role for tRNA or tRNA-like genes in the development of new bacterial variants (Hou 1999).

4
A Link Between Regulatory RNAs and Signal Transduction

4.1
RNAs as Effectors of Quorum-Sensing Systems

The transcription of many of the pathogenesis-related RNAs is usually growth phase-dependent. Their promoters are tightly regulated, frequently as part of well-understood regulons that respond to specific signals. In several pathogenic bacteria, the secretion of the virulence factors is regulated in response to changes in cell-population density. This process, called quorum sensing, enables bacteria to communicate using secreted signaling molecules. Quorum sensing systems have been divided into three major classes mainly based on the type of auto-inducer signals (Cotter and Miller 1998; Hoch 2000; Henke and Bassler 2004; Podbielski and Kreikemeyer 2004). The first class involves the LuxI-type enzyme, which synthesizes an acylated homoserine lactone auto-inducer (AHL), and the LuxR-type protein, which binds to the auto-inducer and in turn controls the transcription of many genes. The LuxI/R system is used by gram-negative bacteria, such as *P. aeruginosa* and *E. carotovora*, to control virulence gene expression.

For the two other classes, regulatory RNAs were shown to be the main effectors of the quorum sensing systems mainly in *S. aureus* and in *V. cholerae*. The second class found in many gram-positive bacteria uses signals synthesized as precursor peptides, which are subsequently processed and secreted

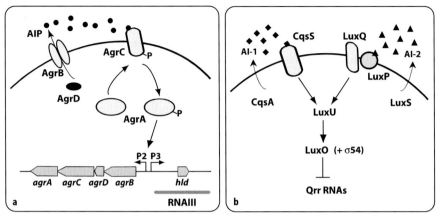

Fig. 5a, b A model of *S. aureus* and *V. cholerae* quorum sensing systems. **a** The *S. aureus agr* system, adapted from Novick (2003). The pro-auto-inducer peptide (AIP) is processed and secreted by AgrB. The AIP binds to an extracellular loop of the receptor-HPK (histidine phospho-kinase) AgrC, and activates its autophosphorylation. This in turn induces phosphorylation of AgrA, which, in conjunction with SarA, activates the two *agr* promoters P2 and P3, leading to the synthesis of the regulatory RNAIII. **b** In *V. cholerae*, two quorum-sensing systems function in parallel to regulate virulence gene expression (Henke and Bassler 2004). *Diamonds* and *triangles* represent the auto-inducers 1 and 2 (AI-1, AI-2), respectively. At low cell density, phospho-LuxO binds to σ54 factor and activates the transcription of four regulatory RNAs (Qrr sRNA). These RNAs bind to *hapR* mRNA and induce rapid mRNA degradation (Lenz et al. 2004). Conversely, at high cell density, Lux-O protein is not phosphorylated and the synthesis of Qrr RNAs is repressed

(Novick 2003). In *S. aureus*, a sensor of the population density is encoded by the *agr* system, which is composed of two divergent transcription units driven by promoters P2 and P3 (Fig. 5a). The P2 operon combines a density-sensing cassette (*agrD* and *B*) and a two-component signal transduction system (*agrA* and *C*). The sensor histidine kinase AgrC detects the extracellular auto-inducer and transmits information via phosphorylation of the response regulator AgrA protein. This modification is supposed to induce changes in the DNA binding properties of AgrA, allowing the autocatalytic activation of the P2 promoter and transcriptional activation of the P3 operon encoding for RNAIII, the intracellular effector of the *agr* regulon (Novick 2003). Recently, an *agr*-like locus was also identified in *L. monocytogenes* (Autret et al. 2003). The response regulator AgrA shared 41% of amino acid identity with *S. aureus* AgrA. Inactivation of *agrA* causes a significant attenuation of the virulence in a mouse model. However, in that case no RNAIII homolog was identified as the main effector of virulence expression.

Finally, the third class of quorum sensing system has been initially described to regulate bioluminescence in *V. harveyi* and virulence gene expression in *V. cholerae* (Fig. 5b). These bacteria produce at least two different signals

(AHL, and a furanosyl borate diester), which are detected by specific two-component signal transduction proteins (Henke and Bassler 2004). The sensory information converges to a response regulatory protein LuxO, which, in its phosphorylated form, interacts with the alternative σ-factor σ^{54} (Lilley and Bassler 2000). This complex further induces the activation of the Qrr RNAs synthesis (Lenz et al. 2004). Interestingly, the gene encoding one of the Qrr RNAs was shown to be located immediately upstream of the *luxOU* operon in all vibrios, suggesting that this locus represents an ancient evolutionary unit (Lenz et al. 2004).

4.2
Other Systems

Many other pathogenesis-related RNAs were shown to be activated by the two-component signal transduction systems (TCSTS). In *S. pyogenes*, the *fas* operon encodes two sensor kinases and only one response regulator (Kreikemeyer et al. 2001). The presence of two co-transcribed sensor kinases suggested that two different signal molecules might activate the same regulatory pathway. This two-component system is required to activate a synthesis of FasX RNA, which was shown to be the main effector of the *fas* operon. Although this RNA is regulated in a growth phase-dependant manner, its regulation is not driven by a quorum-sensing mechanism (Kreikemeyer et al. 2001). A second pathogenesis-related RNA encoded by the *pel* locus in *S. pyogenes* is probably part of a signal-transduction cascade that remains to be identified (Mangold et al. 2004). Indeed, transcription of Pel is growth phase-dependent, and the addition of conditioned media to early logarithmic cells triggered *pel* transcription, a phenomenon characteristic of the existence of a soluble auto-inducer. However, instead of *agr* and *fas* operons where the regulatory RNA is encoded by the same locus, the *pel* system does not encode for such a two-component transduction system. In *C. perfringens*, transcription of VR-RNA is activated by the two-component system VirS/VirR (Shimizu et al. 2002). In that case, the VirR protein either directly regulates the transcription of the virulence factors by binding to the promoter region (i.e., τ-toxin), or acts indirectly via transcription activation of VR-RNA (Ba-Thein et al. 1996).

The CsrA/RsmA-binding RNAs were also shown to be positively regulated by two-component signal transduction systems (Aarons et al. 2000; Altier et al. 2000; Cui et al. 2001; Suzuki et al. 2002; Valverde et al. 2003). Direct binding of the corresponding response regulator to CsrB promoter was demonstrated in *S. typhimurium* (Teplitski et al. 2003). It is of interest that orthologous RNAs are all regulated by homologous TCSTS (known as BarA/UvrY in *E. coli*, GacA/GacS in *P. aeruginosa*, and *E. carotovora*, BarA/SirA in *S. typhimurium*). The signal to which the sensor kinase responds is not known, but it must be growth-phase dependent, since the concentration of these regulatory RNAs reaches the maximum as cells enter the stationary phase.

Even if the general organization of two-component or quorum sensing systems that are involved in the activation of transcription of regulatory RNAs are different, many response regulators (*S. pyogenes* FasA, *S. aureus* AgrA, *C. perfringens* VirR) responsible for their activation share a homologous DNA-binding domain (Nikolskaya and Galperin 2002). This observation suggests that all these systems have evolved from a common ancestor two-component system.

5
Concluding Remarks

RNA is now considered a key effector of gene regulation, playing a universal function in all living organisms. This is illustrated by the recent discovery of the miRNAs (developmentally regulated) in eukaryotes (Ambros 2004), of small non-coding RNAs in archaebacteria (Tang et al. 2002) and in bacteria (Wagner and Vogel 2003; Gottesman 2004; Storz et al. 2004). In bacteria, most of the regulatory RNAs mediate a rapid and reversible switch in response to different signals and large environmental changes. They are appropriate for adaptive processes such as stress responses, environmental cueing, quorum sensing, and virulence. Today regulatory RNAs have evolved to fulfill biological functions, and their properties reflect particular requirements of the regulatory system. Even though only a few pathogenesis-related RNAs have been identified, several general rules can nevertheless be deduced.

5.1
RNAs Regulate Directly or Indirectly Multiple Genes

Many bacterial pathogens produce a variety of extra-cellular toxins, which are tightly regulated. A high diversity is also seen in the organization of the virulence control networks. These regulatory circuits appear to be quite dynamic in an evolutionary respect, and have the ability to adapt existing regulatory networks to control horizontally acquired genes. Many of the virulence genes, for which a tight regulation of expression might be critical, are in general regulated both at the transcriptional level and at the post-transcriptional level (Novick 2003). Why pathogens have come to control their virulence through regulatory RNA molecules? One advantage is to provide a rapid response and to preserve energy. Furthermore, RNAs can simultaneously downregulate many genes. In principle, RNA-dependent regulation is driven by different mechanisms (Fig. 1). However, due to the nature of the RNA, most of the mechanisms involve sequence-specific binding to a target mRNA. In many examples, the regions of complementarity appear to be rather short and often non-contiguous (Fig. 3), thus allowing functional interactions with more than one mRNA target. Moreover, these regulatory RNAs can achieve multiple gene regulation indirectly. Several of these regulatory RNAs target either mRNAs encoding transcriptional regulators, or sequestering post-transcriptional regulatory proteins (Table 2). Riboswitches can also occur in mRNAs encoding transcriptional activators of virulence factors (Table 1). As a consequence, cascades of downstream genes can be affected. One can also speculate that several regulatory RNAs can be made, under different experimental conditions, to regulate a single target. This would allow the bacteria to integrate many environmental signals. Several *E. coli* sRNAs produced under different stress conditions regulate *rpoS* translation (Gottesman 2004; Storz et al. 2004). The studies of several of these pathogenesis-related RNAs revealed unexpected links between metabolism, defense mechanisms, and virulence (see Fig. 2). Other connections between virulence and housekeeping networks will certainly appear as we learn more about these regulatory RNAs.

5.2
A Direct Coupling Between the Structure and the Regulatory Activity?

The astonishing variety of RNA performance is imputable to its peculiar structure capability, and its potentiality to form stable or transient interactions. Thus, RNA function is linked to its dynamic properties and versatile structure. Structure–function relationships have not yet been addressed for many regulatory RNAs. It is expected that their stability and regulatory activities are dictated by their structure. The size of the pathogenesis-related RNAs varies considerably (from 50 to 700 nt). Many of them carry several stem-loop structures connected by unpaired regions. The target sequences (in proteins or mRNAs) are generally exposed in an appropriate structural context (Figs. 3, 4). As an example, many antisense RNAs recognize their target mRNA via a loop–loop or a loop–single-stranded regions which are appropriate motifs for fast recognition (Brunel et al. 2002; Wagner et al. 2002). In several cases, chaperone proteins, such as Hfq, are needed for facilitating pairings between the two RNAs.

Another important aspect, often neglected, is the fact that sequential formation of RNA interactions during transcription can modify a folding pathway and ultimately determine the functional state of the RNA transcript (Heilman-Miller and Woodson 2003). Indeed, the stability of some of the intermediate structures formed during transcription determines the time required to rearrange to the final structure. The rate of RNA elongation and the extent of pausing may also alter the distribution of the folding intermediates by limiting the conversion of one structure to another one. Thus, kinetically trapped intermediates can create a time window for gene regulation. This is particularly true for riboswitches in mRNAs which control transcription (Figs. 1c, d). The refolding of a metastable mRNA structure regulates translation of a plasmid-encoded killer mRNA (Nagel et al. 1999). *trans*-Acting RNAs may also bind to mRNA during transcription and trap either a translationally active or inactive form of the mRNA (Figs. 1, 4b).

Another particularity of RNA molecules is for them to adopt a vast range of alternative conformations that extend from local (i.e., position of bulged nucleotides within helical groove) to drastic rearrangements of the secondary and tertiary folding. The mRNA biosensor belongs to those "switchable" RNA structures associated with protein synthesis on/off switches. The dynamic equilibrium between two RNA structures can be influenced by environmental cues (temperature), small metabolites, or by *trans*-acting factors. Interestingly some of these structural elements are highly conserved across phylogeny and can be detected by bioinformatics analysis (Abreu-Goodger et al. 2004; Barrick et al. 2004; Vitreschak et al. 2004). The recent determination of the structure of a purine-responsive riboswitch represented one major advance towards the understanding of how RNA switches function at the molecular level (Batey et al. 2004; Serganov et al. 2004).

5.3
Stable or Unstable RNAs in Cells?

In contrast to the unstable plasmid-encoded antisense RNAs, in which decay pathways have been elucidated (Wagner et al. 2002), little is known about the stability of the regulatory RNAs. Many of them were found to be stable when induced, having half-lives of 20 to 60 min, such as *S. aureus* RNAIII (Novick 2003; Huntzinger et al. 2005).

Most of the regulatory RNAs, which act as *trans*-acting antisense RNAs, induce degradation of stable mRNA targets (Massé et al. 2003; Novick 2003; Lenz et al. 2004; Huntzinger et al. 2005). These mRNAs carry either short 5′ UTR or a stem-loop structure at their 5′ end. It was recently proposed that in gram-negative bacteria, some of the sRNAs are self-limiting their action. They act stoichiometrically on their target RNAs and are degraded together with their target, probably via RNase E for which the mechanism of action is not clearly elucidated (Massé et al. 2003). In the gram-positive bacteria *S. aureus*, RNAIII mediates the inhibition of translation and degradation of the stable *spa* mRNA through the double-strand-specific endoribonuclease III (Huntzinger et al. 2005). It was suggested that RNase III might also be recruited for targeting the paired RNAs, since the enzyme cleaves both the free RNA species and the formed duplex.

Thus, this mutual degradation together with the fact that the synthesis of the regulatory RNAs is induced under specific signals, provide an ultrasensitive and transient response (Gottesman 2004). Indeed, if the synthesis of the regulatory RNA exceeds the rate of synthesis of its target message, the mRNA levels can be reduced considerably. Conversely, the target mRNA accumulates if its rate of synthesis exceeds that of the regulatory RNA. It remains, however, to be seen whether this model can be universally applied to all these antisense-like RNAs. The situation may be more complex for regulatory RNAs that targeted both protein and mRNA.

Regulatory signals for degradation of the non-coding RNAs, which trap regulatory proteins, are just beginning to be defined (Weilbacher et al. 2003). The half-life of *E. coli* CrsB was shown to be relatively short (around 2 min), allowing CsrA activity to respond rapidly to conditions that alter CsrB levels (Weilbacher et al. 2003). In *E. carotovora*, an increased stability of the regulatory RNA RsmB was observed in the presence of RsmA protein (Chatterjee et al. 2003).

5.4
Some Regulatory RNAs Require *trans*-Acting Proteins

The *E. coli* Hfq protein, a highly abundant protein (50,000 copies per cell in logarithmic growth) is linked to the action of several small regulatory RNAs that use base-pairings to regulate the expression of their target mRNAs (e.g., Zhang et al. 2002). This protein has been proposed to facilitate base pairings (Møller et al. 2002; Zhang et al. 2002; Geissmann and Touati 2004) and to protect sRNAs against RNase E degradation in *E. coli* (Massé et al. 2003). Recent work shows that RNase E forms a ribonucleoprotein complex with Hfq-sRNA to initiate rapid degradation of targeted mRNAs (Morita et al. 2005). The crystallographic structure of *S. aureus* Hfq reveals that the protein forms a hexameric ring (Schumacher et al. 2002). The Hfq sequence is well conserved among most gram-negative and several gram-positive bacteria, and some analogies are detected with eukaryotic Sm proteins (Brescia and Sledjeski 2003). A model of how Hfq enhances RNA–RNA interactions has been recently proposed in which two Hfq hexamers bind to form a dodecamer in order to bring into close proximity the two RNA molecules (Brescia and Sledjeski 2003). It is noteworthy that in *Brucella abortus* (Robertson and Roop 1999), *L. monocytogenes* (Christiansen et al. 2004), *P. aeruginosa* (Sonnleitner et al. 2003), and *V. cholerae* (Ding et al. 2004) mutations in Hfq produced virulence defects. As shown for *V. cholerae* (Lenz et al. 2004), regulatory RNAs may also be recruited for virulence expression in these pathogens. In *Legionella pneumophila*, Hfq was shown to play a major function in exponential-phase regulatory cascades, but induced slight defects in virulence in macrophage infection models (McNealy et al. 2005). *S. aureus* RNAIII is recognized by Hfq protein in vivo and in vitro (Huntzinger et al. 2005), suggesting that the protein might also be required for full virulence.

One cannot exclude that other abundant proteins, such as the RNA chaperone StpA (Waldsich et al. 2002), the histone-like protein HU (Balandina et al. 2002), and the transcriptional regulator H-NS (Cusick and Belfort 1998), may also affect the regulatory function of these RNAs. Other RNA-binding proteins might also be critical for the function of regulatory RNAs, mainly ribonucleases or modification enzymes. It would not be so surprising that these RNAs carry post-transcriptional modifications, which may contribute to the stabilization of the RNA structure, or to specific recognition of ligands.

5.5
Functional Redundancies and How to Find Them?

The few examples described here show that the pathogenesis-related RNAs are much more prevalent than previously anticipated. One interesting feature that emerges is the functional redundancy of some of these regulatory RNAs in different pathogenic bacteria (Table 2). It is conceivable that these RNAs are differentially expressed upon different environmental/niches conditions or that they have evolved independent regulatory functions (i.e., different sequences can generate new mRNA targets).

To identify new pathogenesis-related RNAs, different experimental and global strategies (i.e., biocomputational approaches, functional screens, cloning, co-purification with proteins) can be used (Vogel and Wagner 2005). Broad criteria will be certainly required since this class of regulatory RNA is rather heterogeneous and no specific criteria (size, genomic organization, structure, presence or not of open reading frames, riboswitches) can be used to classify them into specific functional categories. So far, many of the direct targets of the identified regulatory RNAs have also not yet been assigned. For regulatory RNAs acting as antisenses, bioinformatics has proved to be useful in searching for putative target sequences. Proteomics can also be used to monitor the changes of protein synthesis from strains deleted of the regulatory RNAs, suggesting their involvement in regulatory networks or in metabolic pathways (Wagner and Vogel 2005). The ultimate goal in understanding bacterial virulence gene regulation is to assess an essential role of these RNAs in the adaptation of bacteria to their host and in developing the virulence programs in animal models, and to determine the dynamics of gene expression throughout the infectious cycle. Beyond the obvious fundamental interest of these RNA-dependent regulatory mechanisms, a number of useful applications can be envisioned, including therapeutic purpose (Johansson et al. 2002). Identification of the regulatory RNA networks required for the establishment of the pathogenicity may reveal new potential targets for bacterial treatment.

Acknowledgements We thank B. Ehresmann for constant support and interest. The authors gratefully acknowledge that funding was from the Centre National de la Recherche Scientifique, the Ministère de la Recherche (ACI microbiologie), the european community (contract 005120, FOSRAK), the Ligue Régionale contre le Cancer, and the Fondation pour la Recherche Médicale. M.P. was supported by a fellowship from the Fondation pour la Recherche Médicale, and Alsace Region.

References

Aarons S, Abbas A, Adams C, Fenton A, O'Gara F (2000) A regulatory RNA (PrrB RNA) modulates expression of secondary metabolite genes in Pseudomonas fluorescens F113. J Bacteriol 182:3913–3919

Abdelnour A, Arvidson S, Bremell T, Ryden C, Tarkowski A (1993) The accessory gene regulator (agr) controls Staphylococcus aureus virulence in a murine arthritis model. Infect Immun 61:3879–3885

Abo T, Inada T, Ogawa K, Aiba H (2000) SsrA-mediated tagging and proteolysis of LacI and its role in the regulation of lac operon. EMBO J 19:3762–3769

Abreu-Goodger C, Ontiveros-Palacios N, Ciria R, Merino E (2004) Conserved regulatory motifs in bacteria: riboswitches and beyond. Trends Genet 20:475–479

Altier C, Suyemoto M, Ruiz AI, Burnham KD, Maurer R (2000) Characterization of two novel regulatory genes affecting Salmonella invasion gene expression. Mol Microbiol 35:635–646

Altuvia S, Kornitzer D, Teff D, Oppenheim AB (1989) Alternative mRNA structures of the cIII gene of bacteriophage l determine the rate of its translation initiation. J Mol Biol 210:265–280

Ambros V (2004) The functions of animal microRNAs. Nature 431:350–355

Autret N, Raynaud C, Dubail I, Berche P, Charbit A (2003) Identification of the agr locus of Listeria monocytogenes: role in bacterial virulence. Infect Immun 71:4463–4471

Avedissian M, Lessing D, Gober JW, Shapiro L, Gomes SL (1995) Regulation of the Caulobacter crescentus dnaKJ operon. J Bacteriol 177:3479–3484

Ba-Thein W, Lyristis M, Ohtani K, Nisbet IT, Hayashi H, Rood JI, Shimizu T (1996) The virR/virS locus regulates the transcription of genes encoding extracellular toxin production in Clostridium perfringens. J Bacteriol 178:2514–2520

Bagg A, Neilands JB (1987) Ferric uptake regulation protein acts as a repressor, employing iron (II) as a cofactor to bind the operator of an iron transport operon in Escherichia coli. Biochemistry 26:5471–5477

Baker CS, Morozov I, Suzuki K, Romeo T, Babitzke P (2002) CsrA regulates glycogen biosynthesis by preventing translation of glgC in Escherichia coli. Mol Microbiol 44:1599–1610

Balandina A, Kamashev D, Rouviere-Yaniv J (2002) The bacterial histone-like protein HU specifically recognizes similar structures in all nucleic acids. DNA, RNA, and their hybrids. J Biol Chem 277:27622–27628

Barnard FM, Loughlin MF, Fainberg HP, Messenger MP, Ussery DW, Williams P, Jenks PJ (2004) Global regulation of virulence and the stress response by CsrA in the highly adapted human gastric pathogen Helicobacter pylori. Mol Microbiol 51:15–32

Barrick JE, Corbino KA, Winkler WC, Nahvi A, Mandal M, Collins J, Lee M, Roth A, Sudarsan N, Jona I, Wickiser JK, Breaker RR (2004) New RNA motifs suggest an expanded scope for riboswitches in bacterial genetic control. Proc Natl Acad Sci U S A 101:6421–6426

Batey RT, Gilbert SD, Montange RK (2004) Structure of a natural guanine-responsive riboswitch complexed with the metabolite hypoxanthine. Nature 432:411–415

Benito Y, Kolb FA, Romby P, Lina G, Etienne J, Vandenesch F (2000) Probing the structure of RNAIII, the Staphylococcus aureus agr regulatory RNA, and identification of the RNA domain involved in repression of protein A expression. RNA 6:668–679

Benton BM, Zhang JP, Bond S, Pope C, Christian T, Lee L, Winterberg KM, Schmid MB, Buysse JM (2004) Large-scale identification of genes required for full virulence of Staphylococcus aureus. J Bacteriol 186:8478–8489

Brégeon D, Colot V, Radman M, Taddei F (2001) Translational misreading: a tRNA modification counteracts a +2 ribosomal frameshift. Genes Dev 15:2295–2306

Brescia CC, Sledjeski DD (2003) We are legion: Noncoding regulatory RNAs and Hfq. In: Barciszewski J, Erdmann VA (eds) Noncoding RNAs: molecular biology and molecular medicine. Kluwer Academic/Plenum Publishers, New York, pp 259–269

Brunel C, Marquet R, Romby P, Ehresmann C (2002) RNA loop–loop interactions as dynamic functional motifs. Biochimie 84:925–944

Bunce C, Wheeler L, Reed G, Musser J, Barg N (1992) Murine model of cutaneous infection with gram-positive cocci. Infect Immun 60:2636–2640

Burrowes E, Abbas A, O'neill A, Adams C, O'gara F (2005) Characterisation of the regulatory RNA RsmB from Pseudomonas aeruginosa PAO1. Res Microbiol 156:7–16

Carniel E, Guilvout I, Prentice M (1996) Characterization of a large chromosomal "high-pathogenicity island" in biotype 1B Yersinia enterocolitica. J Bacteriol 178:6743–6751

Chatterjee A, Cui Y, Liu Y, Dumenyo CK, Chatterjee AK (1995) Inactivation of rsmA leads to overproduction of extracellular pectinases, cellulases, and proteases in Erwinia carotovora subsp. carotovora in the absence of the starvation/cell density-sensing signal, N-(3-oxohexanoyl)-L-homoserine lactone. Appl Environ Microbiol 61:1959–1967

Chatterjee A, Cui Y, Yang H, Collmer A, Alfano JR, Chatterjee AK (2003) GacA, the response regulator of a two-component system, acts as a master regulator in Pseudomonas syringae pv. tomato DC3000 by controlling regulatory RNA, transcriptional activators, and alternate sigma factors. Mol Plant Microbe Interact 16:1106–1117

Cheetham BF, Tattersall DB, Bloomfield GA, Rood JI, Katz ME (1995) Identification of a gene encoding a bacteriophage-related integrase in a vap region of the Dichelobacter nodosus genome. Gene 162:53–58

Chen Q, Crosa JH (1996) Antisense RNA, Fur, iron, and the regulation of iron transport genes in Vibrio anguillarum. J Biol Chem 271:18885–18891

Cheung AL, Eberhardt KJ, Chung E, Yeaman MR, Sullam PM, Ramos M, Bayer AS (1994) Diminished virulence of a sar-/agr-mutant of Staphylococcus aureus in the rabbit model of endocarditis. J Clin Invest 94:1815–1822

Christiansen JK, Larsen MH, Ingmer H, Søgaard-Andersen L, Kallipolitis BH (2004) The RNA-binding protein Hfq of Listeria monocytogenes: role in stress tolerance and virulence. J Bacteriol 186:3355–3362

Compan I, Touati D (1993) Interaction of six global transcription regulators in expression of manganese superoxide dismutase in Escherichia coli K-12. J Bacteriol 175:1687–1696

Cotter PA, Miller JF (1998) In vivo and ex vivo regulation of bacterial virulence gene expression. Curr Opin Microbiol 1:17–26

Cui Y, Chatterjee A, Liu Y, Dumenyo CK, Chatterjee AK (1995) Identification of a global repressor gene, rsmA, of Erwinia carotovora subsp. carotovora that controls extracellular enzymes, N-(3-oxohexanoyl)-L-homoserine lactone, and pathogenicity in soft-rotting Erwinia spp. J Bacteriol 177:5108–5115

Cui Y, Chatterjee A, Chatterjee AK (2001) Effects of the two-component system comprising GacA and GacS of Erwinia carotovora subsp. carotovora on the production of global regulatory rsmB RNA, extracellular enzymes, and harpinEcc. Mol Plant Microbe Interact 14:516–526

Cusick ME, Belfort M (1998) Domain structure and RNA annealing activity of the Escherichia coli regulatory protein StpA. Mol Microbiol 28:847–857

Ding Y, Davis BM, Waldor MK (2004) Hfq is essential for Vibrio cholerae virulence and downregulates σE expression. Mol Microbiol 53:345–354

Dobrindt U, Hacker J (2001) Regulation of tRNA5Leu-encoding gene leuX that is associated with a pathogenicity island in the uropathogenic Escherichia coli strain 536. Mol Genet Genomics 265:895–904

Dorsey CW, Tomaras AP, Connerly PL, Tolmasky ME, Crosa JH, Actis LA (2004) The siderophore-mediated iron acquisition systems of Acinetobacter baumannii ATCC 19606 and Vibrio anguillarum 775 are structurally and functionally related. Microbiology 150:3657–3667

Dubrac S, Touati D (2000) Fur positive regulation of iron superoxide dismutase in Escherichia coli: functional analysis of the sodB promoter. J Bacteriol 182:3802–3808

Dubrac S, Touati D (2002) Fur-mediated transcriptional and post-transcriptional regulation of FeSOD expression in Escherichia coli. Microbiology 148:147–156

Dunman PM, Murphy E, Haney S, Palacios D, Tucker-Kellogg G, Wu S, Brown EL, Zagursky RJ, Shlaes D, Projan SJ (2001) Transcription profiling-based identification of Staphylococcus aureus genes regulated by the agr and/or sarA loci. J Bacteriol 183:7341–7353

Durand JM, Okada N, Tobe T, Watarai M, Fukuda I, Suzuki T, Nakata N, Komatsu K, Yoshikawa M, Sasakawa C (1994) vacC, a virulence-associated chromosomal locus of Shigella flexneri, is homologous to tgt, a gene encoding tRNA-guanine transglycosylase (Tgt) of Escherichia coli K-12. J Bacteriol 176:4627–4634

Escolar L, Perez-Martin J, de Lorenzo V (1999) Opening the iron box: transcriptional metalloregulation by the Fur protein. J Bacteriol 181:6223–6229

Geissmann TA, Touati D (2004) Hfq, a new chaperoning role: binding to messenger RNA determines access for small RNA regulator. EMBO J 23:396–405

Gottesman S (2004) The small RNA regulators of Escherichia coli: roles and mechanisms. Annu Rev Microbiol 58:303–328

Gray J, Wang J, Gelvin SB (1992) Mutation of the miaA gene of Agrobacterium tumefaciens results in reduced vir gene expression. J Bacteriol 174:1086–1098

Griffiths-Jones S, Bateman A, Marshall M, Khanna A, Eddy SR (2003) Rfam: an RNA family database. Nucleic Acids Res 31:439–441

Gudapaty S, Suzuki K, Wang X, Babitzke P, Romeo T (2001) Regulatory interactions of Csr components: the RNA binding protein CsrA activates csrB transcription in Escherichia coli. J Bacteriol 183:6017–6027

Halliwell B, Gutteridge JM (1984) Role of iron in oxygen radical reactions. Methods Enzymol 105:47–56

Heilman-Miller SL, Woodson SA (2003) Effect of transcription on folding of the Tetrahymena ribozyme. RNA 9:722–733

Henke JM, Bassler BL (2004) Bacterial social engagements. Trends Cell Biol 14:648–656

Heurlier K, Williams F, Heeb S, Dormond C, Pessi G, Singer D, Camara M, Williams P, Haas D (2004) Positive control of swarming, rhamnolipid synthesis, and lipase production by the posttranscriptional RsmA/RsmZ system in Pseudomonas aeruginosa PAO1. J Bacteriol 186:2936–2945

Hoch JA (2000) Two-component and phosphorelay signal transduction. Curr Opin Microbiol 3:165–170

Hoe NP, Goguen JD (1993) Temperature sensing in Yersinia pestis: translation of the LcrF activator protein is thermally regulated. J Bacteriol 175:7901–7909

Hoe NP, Minion FC, Goguen JD (1992) Temperature sensing in Yersinia pestis: regulation of yopE transcription by lcrF. J Bacteriol 174:4275–4286

Hou YM (1999) Transfer RNAs and pathogenicity islands. Trends Biochem Sci 24:295–298

Huntzinger E, Boisset S, Saveanu C, Benito Y, Geissmann T, Namane A, Lina G, Etienne J, Ehresmann B, Ehresmann C, Jacquier A, Vandenesch F, Romby P (2005) Staphylococcus aureus RNAIII and the endoribonuclease III coordinately regulate spa gene expression. EMBO J 24:825–835

Hurme R, Rhen M (1998) Temperature sensing in bacterial gene regulation-what it all boils down to. Mol Microbiol 30:1–6

Janzon L, Arvidson S (1990) The role of the δ-lysin gene (hld) in the regulation of virulence genes by the accessory gene regulator (agr) in Staphylococcus aureus. EMBO J 9:1391–1399

Johansson J, Cossart P (2003) RNA-mediated control of virulence gene expression in bacterial pathogens. Trends Microbiol 11:280–285

Johansson J, Mandin P, Renzoni A, Chiaruttini C, Springer M, Cossart P (2002) An RNA thermosensor controls expression of virulence genes in Listeria monocytogenes. Cell 110:551–561

Julio SM, Heithoff DM, Mahan MJ (2000) ssrA (tmRNA) plays a role in Salmonella enterica serovar Typhimurium pathogenesis. J Bacteriol 182:1558–1563

Kamath-Loeb AS, Gross CA (1991) Translational regulation of σ^{32} synthesis: requirement for an internal control element. J Bacteriol 173:3904–3906

Karaolis DK, Johnson JA, Bailey CC, Boedeker EC, Kaper JB, Reeves PR (1998) A Vibrio cholerae pathogenicity island associated with epidemic and pandemic strains. Proc Natl Acad Sci U S A 95:3134–3139

Keiler KC, Waller PR, Sauer RT (1996) Role of a peptide tagging system in degradation of proteins synthesized from damaged messenger RNA. Science 271:990–993

Keyer K, Imlay JA (1996) Superoxide accelerates DNA damage by elevating free-iron levels. Proc Natl Acad Sci U S A 93:13635–13640

Kinscherf TG, Willis DK (2002) Global regulation by gidA in Pseudomonas syringae. J Bacteriol 184:2281–2286

Kirby JE, Trempy JE, Gottesman S (1994) Excision of a P4-like cryptic prophage leads to Alp protease expression in Escherichia coli. J Bacteriol 176:2068–2081

Kornblum J, Kreiswirth B, Projan SJ, Ross H, Novick R (1990) Agr: a polycistronic locus regulating exoprotein synthesis in Staphylococcus aureus. In: Novick R (ed) Molecular biology of the staphylococci. Wiley-VCH, New York, pp 373–401

Kreikemeyer B, Boyle MD, Buttaro BA, Heinemann M, Podbielski A (2001) Group A streptococcal growth phase-associated virulence factor regulation by a novel operon (Fas) with homologies to two-component-type regulators requires a small RNA molecule. Mol Microbiol 39:392–406

Lenz DH, Mok KC, Lilley BN, Kulkarni RV, Wingreen NS, Bassler BL (2004) The small RNA chaperone Hfq and multiple small RNAs control quorum sensing in Vibrio harveyi and Vibrio cholerae. Cell 118:69–82

Li Z, Sledjeski DD, Kreikemeyer B, Podbielski A, Boyle MD (1999) Identification of pel, a Streptococcus pyogenes locus that affects both surface and secreted proteins. J Bacteriol 181:6019–6027

Lilley BN, Bassler BL (2000) Regulation of quorum sensing in Vibrio harveyi by LuxO and sigma-54. Mol Microbiol 36:940–954

Liu MY, Gui G, Wei B, Preston JFr, Oakford L, Yuksel U, Giedroc DP, Romeo T (1997) The RNA molecule CsrB binds to the global regulatory protein CsrA and antagonizes its activity in Escherichia coli. J Biol Chem 272:17502–17510

Liu Y, Cui Y, Mukherjee A, Chatterjee AK (1998) Characterization of a novel RNA regulator of Erwinia carotovora ssp. carotovora that controls production of extracellular enzymes and secondary metabolites. Mol Microbiol 29:219–234

Mangold M, Siller M, Roppenser B, Vlaminckx BJ, Penfound TA, Klein R, Novak R, Novick RP, Charpentier E (2004) Synthesis of group A streptococcal virulence factors is controlled by a regulatory RNA molecule. Mol Microbiol 53:1515–1527

Massé E, Gottesman S (2002) A small RNA regulates the expression of genes involved in iron metabolism in Escherichia coli. Proc Natl Acad Sci U S A 99:4620–4625

Massé E, Escorcia FE, Gottesman S (2003) Coupled degradation of a small regulatory RNA and its mRNA targets in Escherichia coli. Genes Dev 17:2374–2383

McHugh JP, Rodriguez-Quinones F, Abdul-Tehrani H, Svistunenko DA, Poole RK, Cooper CE, Andrews SC (2003) Global iron-dependent gene regulation in Escherichia coli. A new mechanism for iron homeostasis. J Biol Chem 278:29478–29486

McNealy TL, Forsbach-Birk V, Shi C, Marre R (2005) The Hfq homolog in Legionella pneumophila demonstrates regulation by LetA and RpoS and interacts with the global regulator CsrA. J Bacteriol 187:1527–1532

Mecsas J (2002) Use of signature-tagged mutagenesis in pathogenesis studies. Curr Opin Microbiol 5:33–37

Michel F, Costa M, Massire C, Westhof E (2000) Modeling RNA tertiary structure from patterns of sequence variation. Methods Enzymol 317:491–510

Møller T, Franch T, Hojrup P, Keene DR, Bachinger HP, Brennan RG, Valentin-Hansen P (2002) Hfq: a bacterial Sm-like protein that mediates RNA-RNA interaction. Mol Cell 9:23–30

Morfeldt E, Taylor D, von Gabain A, Arvidson S (1995) Activation of a-toxin translation in Staphylococcus aureus by the trans-encoded antisense RNA, RNAIII. EMBO J 14:4569–4577

Morita M, Kanemori M, Yanagi H, Yura T (1999a) Heat-induced synthesis of σ^{32} in Escherichia coli: structural and functional dissection of rpoH mRNA secondary structure. J Bacteriol 181:401–410

Morita MT, Tanaka Y, Kodama TS, Kyogoku Y, Yanagi H, Yura T (1999b) Translational induction of heat shock transcription factor σ^{32}: evidence for a built-in RNA thermosensor. Genes Dev 13:655–665

Morita MT, Maki K, Aiba H (2005) RNake E-based ribonucleoprotein complexes: mechanical basis of mRNA destabilization mediated by bacterial non-coding RNAs. Genes Dev 19:2176–2186

Motorin Y, Grosjean H (1999) Transfer RNA modification. In: Nature encyclopedia of life sciences. John Wiley and Sons, Chichester. (http://www.els.net.gate1.inist.fr/, cited 22 June 2005). [doi:10.1038/npg.els.0000528]

Nagai H, Yuzawa H, Yura T (1991) Interplay of two cis-acting mRNA regions in translational control of s32 synthesis during the heat shock response of Escherichia coli. Proc Natl Acad Sci U S A 88:10515–10519

Nagel JH, Gultyaev AP, Gerdes K, Pleij CW (1999) Metastable structures and refolding kinetics in hok mRNA of plasmid R1. RNA 5:1408–1418

Narberhaus F (2002) mRNA-mediated detection of environmental conditions. Arch Microbiol 178:404–410

Niederhoffer EC, Naranjo CM, Bradley KL, Fee JA (1990) Control of Escherichia coli superoxide dismutase (sodA and sodB) genes by the ferric uptake regulation (fur) locus. J Bacteriol 172:1930–1938

Nikolskaya AN, Galperin MY (2002) A novel type of conserved DNA-binding domain in the transcriptional regulators of the AlgR/AgrA/LytR family. Nucleic Acids Res 30:2453–2459

Nocker A, Hausherr T, Balsiger S, Krstulovic NP, Hennecke H, Narberhaus F (2001) A mRNA-based thermosensor controls expression of rhizobial heat shock genes. Nucleic Acids Res 29:4800–4807

Novick RP (2003) Autoinduction and signal transduction in the regulation of staphylococcal virulence. Mol Microbiol 48:1429–1449

Novick RP, Ross HF, Projan SJ, Kornblum J, Kreiswirth B, Moghazeh S (1993) Synthesis of staphylococcal virulence factors is controlled by a regulatory RNA molecule. EMBO J 12:3967–3975

Ohtani K, Bhowmik SK, Hayashi H, Shimizu T (2002) Identification of a novel locus that regulates expression of toxin genes in Clostridium perfringens. FEMS Microbiol Lett 209:113–118

Ohtani K, Kawsar HI, Okumura K, Hayashi H, Shimizu T (2003) The VirR/VirS regulatory cascade affects transcription of plasmid-encoded putative virulence genes in Clostridium perfringens strain 13. FEMS Microbiol Lett 222:137–141

Parsons LM, Waring AL, Limberger RJ, Shayegani M (1999) The dnaK/dnaJ operon of Haemophilus ducreyi contains a unique combination of regulatory elements. Gene 233:109–119

Pessi G, Williams F, Hindle Z, Heurlier K, Holden MT, Camara M, Haas D, Williams P (2001) The global posttranscriptional regulator RsmA modulates production of virulence determinants and N-acylhomoserine lactones in Pseudomonas aeruginosa. J Bacteriol 183:6676–6683

Podbielski A, Kreikemeyer B (2004) Cell density-dependent regulation: basic principles and effects on the virulence of Gram-positive cocci. Int J Infect Dis 8:81–95

Ratledge C, Dover LG (2000) Iron metabolism in pathogenic bacteria. Annu Rev Microbiol 54:881–941

Renzoni A, Klarsfeld A, Dramsi S, Cossart P (1997) Evidence that PrfA, the pleiotropic activator of virulence genes in Listeria monocytogenes, can be present but inactive. Infect Immun 65:1515–1518

Ritter A, Blum G, Emody L, Kerenyi M, Bock A, Neuhierl B, Rabsch W, Scheutz F, Hacker J (1995) tRNA genes and pathogenicity islands: influence on virulence and metabolic properties of uropathogenic Escherichia coli. Mol Microbiol 17:109–121

Robertson GT, Roop RMJ (1999) The Brucella abortus host factor I (HF-I) protein contributes to stress resistance during stationary phase and is a major determinant of virulence in mice. Mol Microbiol 34:690–700

Romeo T (1998) Global regulation by the small RNA-binding protein CsrA and the non-coding RNA molecule CsrB. Mol Microbiol 29:1321–1330

Romeo T, Gong M, Liu MY, Brun-Zinkernagel AM (1993) Identification and molecular characterization of csrA, a pleiotropic gene from Escherichia coli that affects glycogen biosynthesis, gluconeogenesis, cell size, and surface properties. J Bacteriol 175:4744–4755

Sage AE, Vasil AI, Vasil ML (1997) Molecular characterization of mutants affected in the osmoprotectant-dependent induction of phospholipase C in Pseudomonas aeruginosa PAO1. Mol Microbiol 23:43–56

Said-Salim B, Dunman PM, McAleese FM, Macapagal D, Murphy E, McNamara PJ, Arvidson S, Foster TJ, Projan SJ, Kreiswirth BN (2003) Global regulation of Staphylococcus aureus genes by Rot. J Bacteriol 185:610–619

Salinas PC, Waldbeser LS, Crosa JH (1993) Regulation of the expression of bacterial iron transport genes: possible role of an antisense RNA as a repressor. Gene 123:33–38

Schaible UE, Kaufmann SH (2004) Iron and microbial infection. Nat Rev Microbiol 2:946–953

Schumacher MA, Pearson RF, Møller T, Valentin-Hansen P, Brennan RG (2002) Structures of the pleiotropic translational regulator Hfq and an Hfq-RNA complex: a bacterial Sm-like protein. EMBO J 21:3546–3556

Serganov A, Yuan YR, Pikovskaya O, Polonskaia A, Malinina L, Phan AT, Hobartner C, Micura R, Breaker RR, Patel DJ (2004) Structural basis for discriminative regulation of gene expression by adenine- and guanine-sensing mRNAs. Chem Biol 11:1729–1741

Servant P, Mazodier P (2001) Negative regulation of the heat shock response in Streptomyces. Arch Microbiol 176:237–242

Shea JE, Hensel M, Gleeson C, Holden DW (1996) Identification of a virulence locus encoding a second type III secretion system in Salmonella typhimurium. Proc Natl Acad Sci U S A 93:2593–2597

Shimizu T, Yaguchi H, Ohtani K, Banu S, Hayashi H (2002) Clostridial VirR/VirS regulon involves a regulatory RNA molecule for expression of toxins. Mol Microbiol 43:257–265

Sonnleitner E, Hagens S, Rosenau F, Wilhelm S, Habel A, Jäger KE, Bläsi U (2003) Reduced virulence of a hfq mutant of Pseudomonas aeruginosa O1. Microb Pathog 35:217–228

Steiner K, Malke H (2001) relA-Independent amino acid starvation response network of Streptococcus pyogenes. J Bacteriol 183:7354–7364

Storz G, Opdyke JA, Zhang A (2004) Controlling mRNA stability and translation with small, noncoding RNAs. Curr Opin Microbiol 7:140–144

Suzuki K, Wang X, Weilbacher T, Pernestig AK, Melefors O, Georgellis D, Babitzke P, Romeo T (2002) Regulatory circuitry of the CsrA/CsrB and BarA/UvrY systems of Escherichia coli. J Bacteriol 184:5130–5140

Tang TH, Bachellerie JP, Rozhdestvensky T, Bortolin ML, Huber H, Drungowski M, Elge T, Brosius J, Huttenhofer A (2002) Identification of 86 candidates for small non-messenger RNAs from the archaeon Archaeoglobus fulgidus. Proc Natl Acad Sci U S A 99:7536–7541

Tardat B, Touati D (1991) Two global regulators repress the anaerobic expression of MnSOD in Escherichia coli:Fur (ferric uptake regulation) and Arc (aerobic respiration control). Mol Microbiol 5:455–465

Tegmark K, Morfeldt E, Arvidson S (1998) Regulation of agr-dependent virulence genes in Staphylococcus aureus by RNAIII from coagulase-negative staphylococci. J Bacteriol 180:3181–3186

Teplitski M, Goodier RI, Ahmer BM (2003) Pathways leading from BarA/SirA to motility and virulence gene expression in Salmonella. J Bacteriol 185:7257–7265

Touati D (1997) Superoxide dismutases in bacteria and pathogen protists. In: Scandalios J (ed) Oxidative stress and the molecular biology of antioxidant defenses. Cold Spring Harbor Laboratory, Cold Spring Harbor, pp 447–493

Valverde C, Heeb S, Keel C, Haas D (2003) RsmY, a small regulatory RNA, is required in concert with RsmZ for GacA-dependent expression of biocontrol traits in Pseudomonas fluorescens CHA0. Mol Microbiol 50:1361–1379

Valverde C, Lindell M, Wagner EGH, Haas D (2004) A repeated GGA motif is critical for the activity and stability of the riboregulator RsmY of Pseudomonas fluorescens. J Biol Chem 279:25066–25074

Vandenesch F, Projan SJ, Kreiswirth B, Etienne J, Novick RP (1993) agr-related sequences in Staphylococcus lugdunensis. FEMS Microbiol Lett 111:115–122

Vitreschak AG, Rodionov DA, Mironov AA, Gelfand MS (2004) Riboswitches: the oldest mechanism for the regulation of gene expression. Trends Genet 20:44–50

Vogel J, Wagner EGH (2005) Approaches to identify novel non-messenger RNAs in bacteria and to investigate their biological functions: RNA mining. In: Hartmann RK, Bindereif A, Schön A, Westhof E (eds) Handbook of RNA biochemistry. Wiley-VCH, Weinheim, pp 595–613

Vuong C, Saenz HL, Gotz F, Otto M (2000) Impact of the agr quorum-sensing system on adherence to polystyrene in Staphylococcus aureus. J Infect Dis 182:1688–1693

Wagner EGH, Simons RW (1994) Antisense RNA control in bacteria, phages, and plasmids. Annu Rev Microbiol 48:713–742

Wagner EGH, Vogel J (2003) Noncoding RNAs encoded by bacterial chromosomes. In: Barciszewski J, Erdmann VA (eds) Noncoding RNAs: molecular biology and molecular medicine. Kluwer Academic / Plenum Publishers, New York, pp 242–258

Wagner EGH, Vogel J (2005) Approaches to identify novel non-messenger RNAs in bacteria and to investigate their biological functions: functional analysis of identified non-mRNAs. In: Hartmann RK, Bindereif A, Schön A, Westhof E (eds) Handbook of RNA biochemistry. Wiley-VCH, Weinheim, pp 614–654

Wagner EGH, Altuvia S, Romby P (2002) Antisense RNAs in bacteria and their genetic elements. Adv Genet 46:361–398

Waldbeser LS, Tolmasky ME, Actis LA, Crosa JH (1993) Mechanisms for negative regulation by iron of the fatA outer membrane protein gene expression in Vibrio anguillarum 775. J Biol Chem 268:10433–10439

Waldbeser LS, Chen Q, Crosa JH (1995) Antisense RNA regulation of the fatB iron transport protein gene in Vibrio anguillarum. Mol Microbiol 17:747–756

Waldsich C, Grossberger R, Schroeder R (2002) RNA chaperone StpA loosens interactions of the tertiary structure in the td group I intron in vivo. Genes Dev 16:2300–2312

Weilbacher T, Suzuki K, Dubey AK, Wang X, Gudapaty S, Morozov I, Baker CS, Georgellis D, Babitzke P, Romeo T (2003) A novel sRNA component of the carbon storage regulatory system of Escherichia coli. Mol Microbiol 48:657–670

Wesson CA, Liou LE, Todd KM, Bohach GA, Trumble WR, Bayles KW (1998) Staphylococcus aureus Agr and Sar global regulators influence internalization and induction of apoptosis. Infect Immun 66:5238–5243

Wilderman PJ, Sowa NA, FitzGerald DJ, FitzGerald PC, Gottesman S, Ochsner UA, Vasil ML (2004) Identification of tandem duplicate regulatory small RNAs in Pseudomonas aeruginosa involved in iron homeostasis. Proc Natl Acad Sci U S A 101:9792–9797

Withey JH, Friedman DI (2002) The biological roles of trans-translation. Curr Opin Microbiol 5:154–159

Yamanaka K, Mitta M, Inouye M (1999) Mutation analysis of the 5′ untranslated region of the cold shock cspA mRNA of Escherichia coli. J Bacteriol 181:6284–6291

Zhang A, Wassarman KM, Ortega J, Steven AC, Storz G (2002) The Sm-like Hfq protein increases OxyS RNA interaction with target mRNAs. Mol Cell 9:11–22

Zhu J, Miller MB, Vance RE, Dziejman M, Bassler BL, Mekalanos JJ (2002) Quorum-sensing regulators control virulence gene expression in Vibrio cholerae. Proc Natl Acad Sci U S A 99:3129–3134

Regulatory RNAs in Mammals

M. Szymański (✉) · J. Barciszewski

Institute of Bioorganic Chemistry of the Polish Academy of Sciences, Noskowskiego 12, 61-704 Poznan, Poland
mszyman@ibch.poznan.pl

1	Mammalian Genomes and Their Contents	46
2	Non-protein-Coding RNAs	47
2.1	NpcRNAs as Epigenetic Regulators	49
2.1.1	RNAs in X-Chromosome Inactivation	49
2.1.2	RNA Regulators of Imprinted Genes	51
2.1.3	Regulation of DNA Methylation	53
2.2	NpcRNAs in Transcriptional Regulation	53
2.2.1	Steroid Receptor Activator RNA	54
2.2.2	7SK RNA	55
2.2.3	B2 RNA	56
2.3	Antisense Transcripts	56
2.4	Post-Transcriptional Regulation by npcRNAs	57
2.4.1	Expressed Processed Pseudogenes	57
2.4.2	Regulation of Translation	58
3	The Medical Perspective	60
3.1	Noncoding RNAs in Developmental Disorders	61
3.1.1	Angelman and Prader-Willi Syndromes	61
3.1.2	DiGeorge Syndrome	61
3.2	NpcRNAs in Cancer	62
4	Concluding Remarks	64
	References	65

Abstract Recent years have brought a dramatic change in our understanding of the role of ribonucleic acids (RNAs) within the cell. In addition to the already well-known classes of RNAs that take part in the transmission of genetic information from DNA to proteins, a new highly heterogeneous group of RNA molecules has emerged. The regulatory non-protein-coding RNAs (npcRNAs) have been shown to be involved in modulation of gene expression on both the transcriptional and post-transcriptional level. They participate in mechanisms of chromatin modification, regulation of transcription factor activity, and influencing mRNA stability, processing, and translation. npcRNAs are key factors in genetic imprinting, dosage compensation of X-chromosome-linked genes, and many processes of differentiation and development.

Keywords Noncoding RNA · Imprinting · Riboregulation

1
Mammalian Genomes and Their Contents

The draft human genome assembly completed in 2001 revealed that the human genome contains 25,000–35,000 protein-coding genes (Lander et al. 2001; Venter et al. 2001). Since then, attempts have been made to reevaluate these numbers, and several proposals have been put forward with the estimates reaching up to 65,000–75,000 (Wright et al. 2001). Currently available data obtained using both computational and experimental methods indicate, however, that the actual number of protein-coding genes in the human genome is between 25,000 and 30,000 (Southan 2004). These figures are again well below the estimates, ranging from 50,000 to 140,000 genes, that were proposed before the completion of genome sequencing (Antequera and Bird 1993; Fields et al. 1994; Roest Crollius et al. 2000). Upon completion of the mouse genome, it turned out that apart from the difference in size (2.9×10^9 bp in human vs 2.5×10^9 bp in mouse) the two genomes share similarities in both their content and linear organization of genes along the chromosomes (Waterston et al. 2002). A comparative analysis showed that only about 1% of mouse genes have no identifiable homologues in humans, and 96% of homologous genes are localized within conserved syntenic regions of the human genome. The protein gene count, however, does not reflect the whole repertoire of possible protein variants that can be produced in the cell. When compared with the number of genes, the proteome diversity is much greater than the gene diversity due to alternative splicing, which, according to recent estimates, affects over half of human genes (Lareau et al. 2004).

A common feature of all mammalian genomes is that a very small fraction (1.5%–2%) of genomic DNA constitutes actual open reading frames (ORFs) of the protein-coding genes (Fig. 1). This contrasts with the protein-coding potential observed in eubacteria and archaea in which the protein-coding portions account for over 90% of genomic DNA. In lower eukaryotes, the contribution of translated sequences varies from 10%–30% in invertebrates to 60%–90% in unicellular organisms like yeast or *Encephalitozoon cuniculi*.

One can easily notice that the more complex organisms tend to reduce the amount of genomic DNA that actually encodes proteins. This is accompanied by a significant accumulation of noncoding (i.e., not translated) regions. The vast majority of a non-protein-coding DNA in mammalian genomes is made up of repetitive sequences (Fig. 1). In human and mouse, these elements constitute 46% and 38% of chromosomal DNA, respectively (Lander et al. 2001; Waterston et al. 2002). The non-translated regions of protein-coding genes, 5'- and 3'-untranslated regions (UTRs), and introns account for approximately a quarter of noncoding DNA in mammalian genomes (Venter et al. 2001). The functional significance of the majority of the remaining portions of genome is largely unknown. It is, however, clear that many elements play a key role in spatial and temporal coordination of gene expression. Within these regions,

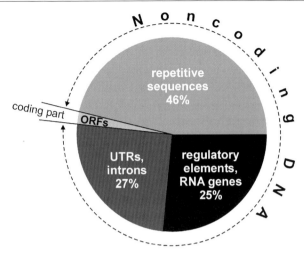

Fig. 1 Coding and noncoding DNA in the human genome. Based on protein-coding capacity, genomic DNA can be divided into two parts. The coding part (~2%) consists of the open reading frames (*ORFs*). The noncoding part (~98%) comprises untranslated parts of protein coding genes—introns, untranslated regions (*UTRs*)—(27%) and repetitive sequences (46%). The remaining 25% of the genome is a repository of regulatory elements including noncoding RNA genes

there are promoter and enhancer sequences which, together with their respective DNA-binding proteins, are directly involved in transcription regulation. Approximately 0.3%–1.0% of total mammalian DNA constitutes conserved non-genic sequences (CNGs) that show remarkable evolutionary conservation between all mammalian species examined so far. Some of the CNGs seem to be regulatory elements of nearby genes, but the majority shows random distribution within intergenic regions and does not depend on the distance from neighboring genes, and their role remains obscure (Dermitzakis et al. 2004).

One should keep in mind that the numbers shown above are based on the assumption that any given fragment of genomic DNA is transcribed only in one direction, and do not take into account bi-directional transcription of certain regions of genomic DNA, a phenomenon that is quite common in eukaryotic genomes (Yelin et al. 2003; Kiyosawa et al. 2003; Shendure and Church 2002).

2
Non-protein-Coding RNAs

Based on protein-coding potential, the whole transcriptional output from the genome can be divided into two major groups: protein-coding messenger (m)RNAs and non-protein-coding RNAs (npcRNAs). Among the latter, there are constitutively expressed housekeeping RNAs that play essential roles for correct functioning of the cell. They include RNAs performing crucial roles in protein biosynthesis [ribosomal (r)RNA, transfer (t)RNA], processing and modifications of precursor RNAs [small nuclear (sn)RNA, small nucleolar (sno)RNA, RNase P RNA, and guide (g)RNA], synthesis of telomeres (telomerase RNA), quality control of translation (tmRNA) or components of other ribonucleoprotein complexes (4.5S RNA, vault (v)RNA). Housekeeping RNAs are phylogenetically conserved, and most of them are present in all organisms.

The second group of noncoding RNAs includes the transcripts that perform regulatory functions. Such RNAs have been found both in prokaryotes as well as in eukaryotes. Most of those RNAs were initially identified as transcripts of genes activated in response to the changes in the environment, developmental signals, or as cell-/tissue-specific transcripts. In eukaryotes, a majority of these RNAs resembles mRNAs. They are transcribed by RNA polymerase II, capped at the 5'-end and polyadenylated. Like primary transcripts from protein-coding genes, they are often spliced. The only distinctive feature is the lack of ORF and thus protein-coding ability. The size of mammalian regulatory RNAs varies considerably from approximately 20 nt in the case of micro (mi)RNAs to over 100-kb transcripts.

At first, regulatory npcRNAs were regarded as a kind of curiosity, but in the last decade, a significant number of findings clearly demonstrated that RNA-dependent regulatory mechanisms are widespread in all organisms and they are involved in the precise modulation of expression of many genes (Szymanski and Barciszewski 2003). The regulatory pathways employing regulatory RNAs affect practically all levels of expression of genetic information (Fig. 2). Noncoding RNAs have been shown to participate in setting up epigenetic features and regulation of transcription, as well as certain aspects of post-transcriptional control of gene expression.

The employment of RNAs as regulatory factors has several advantages. Unlike some other signaling molecules, npcRNAs are encoded in the genomic DNA and are themselves subject to a strict control of expression. The synthesis of RNA is energetically more favorable than the biosynthesis of proteins, and the turnover rate of RNA in the cell is much faster than that of proteins. This allows for a quick response and limits the time of the signaling action. RNAs can adopt a variety of higher order structures that can bind proteins or small molecules. Activity of a large number of npcRNAs depends on interactions with other RNA molecules via complementary base pairing. They ensure very high specificity, which in the case of proteins would require a complex RNA-binding domain. Moreover, RNA-coding genes are less sensitive to point mutations than protein-coding ones.

In addition to clearly distinct npcRNA-coding genes, introns excised from precursor mRNAs can also be perceived as a potential source of regulatory RNAs (Mattick and Gagen 2001). In some cases, introns are further processed and produce functional RNAs (Runte et al. 2001). It has been postulated that introns may constitute an important element of regulatory networks, providing

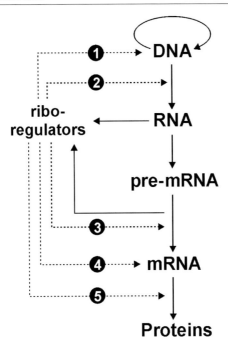

Fig. 2 Regulation of gene expression with noncoding RNAs. Non-protein-coding regulatory RNAs can influence expression of genes on many levels. RNAs have been shown to participate in modification of chromatin, establishing and maintenance of imprinted status of genes (*1*), direct modulation of transcription by interactions with transcription factors (*2*), regulation of alternative splicing (*3*), mRNA stability (*4*), and translation (*5*)

information about gene expression status (Mattick 1994, 2001, 2003; Mattick and Gagen 2001).

2.1
NpcRNAs as Epigenetic Regulators

2.1.1
RNAs in X-Chromosome Inactivation

Non-protein-coding RNAs play a key role in dosage compensation whereby the expression of X-chromosome genes is equalized between male and female cells. In mammals, it is accomplished by the process of X-chromosome inactivation (XCI), which results in transcriptional silencing of one of the X chromosomes in XX females. In fact, the silencing operates according to the 'n-1 rule,' which inactivates all but one X chromosomes in the cell. A key element in that process is spliced and polyadenylated Xist/XIST RNA expressed from the *XIC* locus (X-inactivation center). Mature Xist transcripts range in size from 12 kb in *Microtus* to 19 kb in human (Chureau et al. 2002). Upon X-inactivation initiation, Xist RNA coats the X chromosome from which it is expressed, thereby marking it for silencing (Plath et al. 2002). The role of Xist RNA in XCI was confirmed by the observations that X chromosomes with Xist gene deletion cannot undergo inactivation (Newall et al. 2001). Moreover, it is possible to initiate inactivation of the autosome that expresses *Xist* from a transgene during embryonic stem cells differentiation (Lee and Jaenisch 1997).

Two distinct functions of Xist RNA have been analyzed in detail. The induction of silencing and X-chromosome localization were attributed to different domains of the transcript. The 5'-region involved in transcriptional silencing, contains the so-called A-repeats that can form two short hairpin structures that can function as binding sites for heteronuclear ribonucleoprotein (hnRNP) C1/C2. Mutant Xist RNA with 5'-deletions associates with X chromosome but does not undergo inactivation. The X-chromosome localization activity depends on dispersed, poorly conserved, and functionally redundant sequence elements spread along the entire length of Xist RNA (Wutz et al. 2002). Interestingly, an association of Xist RNA with the chromatin requires expression of BRCA1 protein, known as a breast and ovarian cancer tumor suppressor (Ganesan et al. 2002). Xist RNA is responsible for modification of chromatin structure primarily by the changes in histone composition. These include association with macroH2A (Chadwick and Willard 2004) and hypoacetylation of histones H3 and H4, (Keohane et al. 1996), as well as methylation of lysines K9 and K27 in H3 (Heard et al. 2001; Rougeulle et al. 2004). Inactive X chromosome also accumulates ubiquitinated histone H2A (Smith et al. 2004).

In mouse, expression of Xist RNA from the active X chromosome is negatively regulated by an antisense Tsix RNA that is initiated approximately 40 kb downstream from the *Xist* gene. Various models have been proposed to account for the downregulation of Xist expression by Tsix, including an RNA interference (RNAi) mechanism, destabilization by preventing formation of complexes with proteins, and interference with *Xist* transcription. All of the alternatively spliced isoforms of Tsix overlap the A-repeats within the 5'-terminal domain of Xist RNA, which may prevent proper folding and association with proteins (Shibata and Lee 2003).

The human TSIX gene is also expressed exclusively in embryos, and its transcript is partially antisense to *XIST* (Migeon et al. 2001). However, *Tsix* is not conserved in mammals (Chureau et al. 2002). Moreover, the expression patterns of human and murine *TSIX/Tsix* and their organization show significant differences. Unlike *Tsix*, *TSIX* is expressed from both X chromosomes, does not overlap the entire length of *XIST*, and does not repress its expression. There is also no differentially methylated CpG island in *TSIX*, which plays a role in regulation of expression of *Tsix*. Thus the two undoubtedly homologous genes, cannot be regarded as functional equivalents (Migeon et al. 2002).

2.1.2
RNA Regulators of Imprinted Genes

Noncoding RNAs also play a role in establishing and maintaining imprinted expression of genes. A 3.7-kb imprinting control element (ICE) known as Region 2, located within the second intron of insulin-like growth factor type-2 receptor (*Igf2r*) gene, is responsible for monoallelic, maternal expression of three genes (*Igf2r*, *Slc22a2*, and *Slc22a3*) on mouse chromosome 17 (Fig. 3a). Region 2 contains a maternally methylated CpG island (Wutz et al. 1997), which constitutes a promoter for the *Air* (antisense *Igf2r* RNA) gene transcribed from the antisense strand of *Igf2r*. The *Air* product is a 108 kb long and rich in repetitive elements and unspliced RNA expressed exclusively from the paternal allele (Sleutels et al. 2002). Air RNA overlaps roughly 30 kb of *Igf2r*, including the promoter region, and extends over 70 kb further upstream. Interestingly, the expression of Air RNA is also required for the silencing of the other two genes of the cluster, *Slc22a2* and *Slc22a3*, located 110 and 155 kb downstream from *Igf2r*, respectively. Deletion of Region 2 led to biallelic expression of the

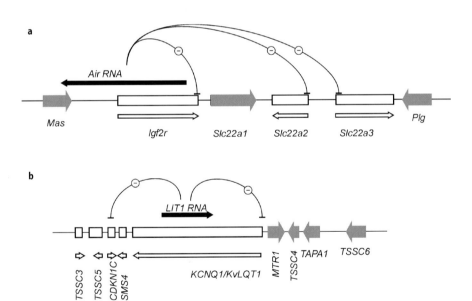

Fig. 3a, b Regulation of imprinted genes by RNA. Paternally and maternally expressed genes are represented by *black* and *white arrows*, respectively. Genes with biallelic expression are shown as *gray arrows*. **a** An imprinted region of mouse chromosome 17. Paternal alleles of maternally expressed Igf2r, Slc22a2, and Slc22a3 genes are repressed by an antisense transcript (Air RNA) originating from the differentially methylated CpG island within intron 2 of Igf2r. **b** Paternally expressed antisense LIT1 RNA originating from the differentially methylated CpG island represses expression from paternal alleles of both KvLQT1 and an upstream gene for cyclin-dependent kinase inhibitor 1C (CDKN1C/p57Kip2)

cluster, which suggested that it represents a bi-directional silencer for the approximately 400-kb region of the paternal chromosome (Zwart et al. 2001). The silencing activity of Region 2, in fact, requires expression of Air RNA. Mutant *Air*, with a premature polyadenylation site terminating downstream from the second exon of *Igf2r* when inherited paternally, results in a phenotype identical to deletion of paternally inherited Region 2. The truncated Air RNA was transcribed, but the *Igf2r*, *Slc22a2*, and *Slc22a3* showed biallelic expression (Sleutels et al. 2002). Two different mechanisms were proposed that could account for the bi-directional silencing of the genes within the imprinted cluster. According to the two-step mechanism, the antisense transcription through the *Igf2r* promoter region induces a silent chromatin state. Bi-directional spreading of silencing modifications could account for transcriptional inactivation of the genes located upstream, *Air*, *Slc22a2*, and *Slc22a3* (Sleutels et al. 2002). The two-step mechanism requires the presence of *Igf2r* promoter that was shown to be dispensable for correct imprinting patterns of *Slc22a2* and *Slc22a3*, which seem to be affected by Air RNA directly (Sleutels et al. 2003). These results suggested that Air RNA has intrinsic promoter-independent *cis* silencing properties and may play a similar role as Xist RNA in X-chromosome inactivation. Its association with the chromatin in *cis* would provide a signal recruiting silencing factors. The difference is in the extent of silencing. In the case of Xist RNA, it is the entire X chromosome with a few genes that escape inactivation. Air RNA-dependent silencing operates locally, repressing susceptible gene promoters within a limited chromosomal domain.

Similar bi-directional silencing activity associated with noncoding RNA was observed in the case of a cluster of imprinted genes associated with Beckwith-Wiedemann syndrome (BWS) on human chromosome 11p15 (Fig. 3b). This 1-Mbp region contains 13 maternally and 4 paternally expressed genes organized within two independently regulated imprinted domains. The genetic and epigenetic defects affecting this region have been implicated, apart from BWS, in several human cancers. The most frequent alteration found in BWS patients is the absence of methylation at the maternal allele of *KvDMR1*, an intronic CpG island within the *KCNQ1* gene, deletion of which affects imprinting of multiple genes in 11p15.5 (DeBaun et al. 2002; Diaz-Meyer et al. 2004). Within the *KvDMR1* there is a promoter for a paternally expressed antisense RNA, LIT1 (KCNQ1OT1, KvLQT1-AS) (Horike et al. 2000; Fitzpatrick et al. 2002; Mancini et al. 2003). In mouse it has been demonstrated that a deletion of a paternal *KvDMR1* and lack of LIT1 RNA transcription causes activation of *CDKN1C* and five other silent genes on the paternal chromosome, which in turn leads to developmental defects (Fitzpatrick et al. 2002; Thakur et al. 2004). Therefore, it functions as a bi-directional imprinting control region in a manner similar to Air RNA. Both LIT1 and Air RNAs are antisense transcripts, and their expression depends on the methylation status of the differentially methylated regions (DMRs) localized within the introns of reciprocally imprinted protein coding genes.

There is a feature that is common for RNA-dependent silencing of imprinted genes and X-chromosome inactivation that suggests that these processes are closely related. In both cases it has been demonstrated that a maintenance of inactive state of the repressed genes requires Polycomb group Eed protein, which is a component of the embryonic Polycomb repressive complex responsible for histone H3 methylation (Wang et al. 2001; Mager et al. 2003). It is, however, unclear what the precise role of RNAs in these mechanisms might be.

2.1.3
Regulation of DNA Methylation

Noncoding RNA that affects gene expression via an epigenetic mechanism was also found in rat *Sphk1* locus encoding sphingosine kinase-1. This enzyme is responsible for production of sphingosine 1-phosphate, which acts as an extracellular mediator affecting, in a receptor-dependent manner, various cellular responses including mitogenesis, differentiation, and apoptosis. Moreover, sphingosine 1-phosphate is an intracellular second messenger involved in calcium mobilization and activation of non-receptor tyrosine kinases (Pyne and Pyne 2000). Rat *Sphk1* transcription and splicing produces six subtypes (*Sphk1a–f*) that differ in their 5′-UTRs, but encode identical proteins. These differences are due to usage of five alternative transcription start sites and splicing of the alternative first exons encoded within a 3.7-kb CpG island that harbors a tissue-dependent, differentially methylated region (T-DMR). The changes in the methylation status of the T-DMR were proposed to be responsible for tissue- and developmental stage-specific expression of various mRNA variants (Imamura et al. 2001). It has been shown that an antisense transcript, Khps1, overlapping the T-DMR and initiated within the *Sphk1* CpG island induces demethylation of CG sites and methylation of non-CG sites (Imamura et al. 2004). It has been proposed that specific variants of the antisense transcript may regulate differential methylation patterns of the T-DMR that may be responsible for tissue-specific expression of *Sphk1* subtypes. The analysis of CpG islands on a genomic scale revealed a large number of T-DMRs (Imamura et al. 2001). Methylation and, in consequence, tissue-specific expression of different transcripts driven by such elements may be regulated by an analogous, antisense RNA-dependent mechanism.

2.2
NpcRNAs in Transcriptional Regulation

One of the mechanisms of gene expression regulation by RNA at the level of transcription is modulation of an activity of protein transcription factors. In bacteria, transcriptional regulation by RNA has been described for 6S RNA. This RNA is expressed in the stationary phase and tightly binds RNA polymerase associated with the σ^{70} subunit responsible for the transcription of genes expressed in the logarithmic phase. It does not bind σ^{38}-containing polymerase, thus giving a preference to the expression of stationary phase-specific genes (Wassarman and Storz 2000). In mammals, three RNAs have been shown to play a role of specific regulators of proteins constituting the transcriptional machinery.

2.2.1
Steroid Receptor Activator RNA

The first npcRNA affecting the activity of transcription factors described in mammals is steroid receptor activator RNA (SRA RNA). It was identified as a polyadenylated, spliced transcript that acts as a coactivator of nuclear receptors of steroid hormones. SRA RNA associates with the SRC-1 (steroid receptor coactivator 1) and is a strong coactivator of receptors for progestins, estrogens, androgens, and glucocorticoids (Lanz et al. 1999). SRA RNA primary transcripts are subject to alternative splicing that produces cell type-specific isoforms with a common core sequence flanked by variable regions. The core region contains five secondary structure elements crucial for coactivator function. The mutations that affect these structures reduce transcriptional coactivation by SRA RNA. On the other hand, disruption of the putative ORF had no adverse effect on the RNA's activity (Lanz et al. 2002). It has been found that SRA RNA binds a subfamily of DEAD-box RNA-binding proteins, p72/p68, that probably mediate the association of SRA RNA with SRC-1 (Watanabe et al. 2001). Interestingly, this RNA can also bind to the RNA-recognition motifs within a hormone-induced transcriptional repressor, SHARP (SMRT/HDAC1 associated repressor protein), a protein responsible for recruitment of histone deacetylases. The interactions of SRA RNA with both transcriptional activator and repressor suggest that a competition between SHARP and steroid receptors for SRA RNA may be responsible for fine-tuning expression of hormone-responsive genes (Shi et al. 2001).

Although the role of SRA RNA in the cell is still not fully understood, it has been found that its transcription is increased in breast, uterus, and ovarian cancers (Lanz et al. 1999). SRA RNA overexpression in mouse induced proliferation, inflammation, and apoptosis in steroid hormone responsive tissues, but the transgenic animals lines had a lifespan comparable to that of wildtype and did not develop cancer. Thus, the overexpression of SRA RNA was ruled out as a causative agent in tumorigenesis. The suppression of *ras*-induced tumor formation suggested that an overexpression of SRA RNA might constitute an element of the defense system that counters excessive proliferation associated with tumor growth (Lanz et al. 2003).

Interestingly, although the first identified splicing isoforms of SRA do not code for proteins, new translated variants of the SRA transcript were isolated and shown to encode nuclear protein in vivo. The coding variant differs from the SRA RNA by the presence of and extension at the 5'-end, which provides an initiation codon. These new isoforms yield protein products when translated in an in vitro system or expressed in transfected human breast cancer cells. SRA is therefore the first example of a gene that can produce the transcripts functioning both as an mRNA and npcRNA (Emberley et al. 2003).

2.2.2
7SK RNA

Another npcRNA involved in the regulation of protein function in mammals is 7SK RNA. Although it was discovered over two decades ago as an abundant small nuclear RNA, it escaped functional characterization. Unlike most of the mammalian regulatory npcRNAs, 7SK RNA is a product of RNA polymerase III (PolIII). It has been demonstrated that this RNA plays a role as an inhibitor of positive transcription elongation factor b (P-TEFb) (Nguyen et al. 2001; Yang et al. 2001). P-TEFb is a PolII cofactor responsible for a transition from abortive to productive elongation. It also plays a role as a host cofactor for human immunodeficiency virus (HIV)-1 Tat protein and is required for HIV-1 transcription (Jeang et al. 1999; Price 2000). One of the P-TEFb subunits is a cyclin-dependent kinase 9 (Cdk9) that phosphorylates the C-terminal domain of the largest subunit of PolII. 7SK RNA forms a complex with P-TEFb and HEXIM1/MAQ1 protein, inhibiting the kinase activity of Cdk9. Under stress conditions, P-TEFb is released from the complex and becomes available for transcription of stress-induced genes (Yik et al. 2003). Binding of HEXIM1 to 7SK RNA is necessary for the suppression of the P-TEFb kinase activity (Yik et al. 2004). Since the HEXIM1 nuclear localization signal (NLS) required for 7SK RNA binding resembles the arginine-rich motif found in HIV-1 Tat protein, it has been suggested that the formation of both complexes, TAR:Tat:P-TEFb and 7SK RNA:HEXIM1:P-TEFb, may depend on similar recognition between RNA and protein components.

7SK RNA also regulates expression of the *c-myc* protooncogene, which is involved in regulation of cells' proliferation, growth arrest, and differentiation. Deregulation of *c-myc* expression is a common feature in many tumors. Mammalian *c-myc* genes have four promoter regions, P0 through P3. In normal cells, the most active promoter, P2, produces 75%–90% of c-myc's RNA. P1 is responsible for 10%–25%, and the combined activity of P0 and P3 gives rise to about 5% of transcripts. Upon the transformation induced by SV40, a transient three- to fivefold increase in the activity of promoters P1 and P3 is observed that correlates with a significant up-regulation of 7SK RNA transcription. An increase in P1 and P3 promoter-driven transcription was suppressed by antisense oligonucleotides targeted at single-stranded regions of 7SK RNA (Luo et al. 1997).

The regulation of gene expression by npcRNAs interacting with transcription factors may be more common than is currently evident from a limited number of well-studied cases. The expression of human protease-activated receptor (PAR)-1 is positively regulated by a regulatory element located approximately 4 kb upstream from its gene. From this region, a novel (~450 nt long) npcRNA, called ncR-uPAR (noncoding RNA upstream of the PAR-1 gene), is expressed. It has been proposed that this RNA acts *in trans* regulating expression of PAR-1. A possible mechanism involves yet unidentified interactions with transcription factors, since it has been shown that ncR-uPAR can bind nuclear proteins (Madamanchi et al. 2002).

2.2.3
B2 RNA

Noncoding B2 RNA transcribed by PolIII from repetitive short interspersed elements (SINEs) plays a role in the cellular response to heat shock. One of the responses to heat shock in mouse cells is an increased transcription of B2 RNA by PolIII. This is accompanied by repression of expression of PolII-dependent genes (Allen et al. 2004). The mechanism of B2 RNA-dependent PolII repression was analyzed in an in vitro system. B2 RNA was shown to bind tightly to an RNA docking site on core PolII before preinitiation complexes assembly. The transcription reaction is inhibited either at an initiation step or by preventing formation of open complex. It has been also proposed that there exists a factor responsible for removal of B2 RNA from preinitiation complexes to restore PolII transcription when cells recover from heat shock (Espinoza et al. 2004). Although there are no human RNAs similar to mouse B2, a good candidate seems to be Alu RNA. It is an abundant PolIII transcript originating from SINEs, which is also induced by heat shock and other stress conditions (Allen et al. 2004).

Interestingly, a spliceosomal U1 snRNA was also found to regulate PolII transcription. This RNA forms a stable stoichiometric complex with PolII transcription factor TFIIH, specifically stimulates TFIIH-dependent abortive transcription initiation, and increases the rate of reinitiation of productive transcription. The latter effect depends on the presence of the proximal 5′-splice site and may implicate splicing of the first intron, but the precise mechanism underlying this process is unknown (Kwek et al. 2002).

2.3
Antisense Transcripts

The thyroid hormone receptor (TR)α locus encodes two functionally distinct isoforms (TRα1 and TRα2) with different C-terminal regions. Each isoform shows tissue- and developmental stage-specific distribution reflecting the cell's responsiveness to thyroid hormone. TRα2 does not bind hormone, does not possess activation function, and is thought to represent an antagonist of hormone responsive genes. It has been found that the alternative splicing of pre-mRNA encoding TRα isoforms is regulated by the overlapping antisense transcript encoding a nuclear receptor RevErb. The 3′-end of the RevErb transcript overlaps sequences coding for TRα2 and suppresses alternative splicing, increasing production of TRα1 mRNA (Hastings et al. 2000). The changes in a TRα1/TRα2 ratio significantly alter the cell's responsiveness to thyroid hormone. Thus, natural antisense transcripts (protein-coding and noncoding) can be involved in the regulation of expression of alternative protein products. A coordinated change in the relative amounts of TRα1, TRα2, and RevErb will also influence the overall profile of the expressed genes regulated by these receptors.

Natural antisense RNAs have been also shown to influence the expression of developmentally regulated genes. Mammalian cardiac myosin heavy chain (MHC) is encoded by two adjacent specifying two isoforms α and β. In rodents, during the first 3 weeks after birth, the expression of βMHC, which accounts for over 90% of MHC at birth, is almost totally suppressed, and αMHC becomes a predominant isoform throughout adult life. The increase in βMHC is associated with aging, but can also accompany certain pathological conditions like hypothyroidism and diabetes. It has been found that the expression of βMHC is regulated by an antisense transcript that shows coordinated expression with the αMHC gene and is positively regulated by thyroid hormone by the common promoter element (Haddad et al. 2003). The precise mechanism of βMHC gene suppression is unknown. It could operate at the processing step or by the interference with transcription of the sense transcript as proposed for the antisense RNA-regulation of *HOXA11* expression, where a reciprocal progesterone-dependent correlation between sense and antisense transcripts has been observed (Chau et al. 2002).

2.4
Post-Transcriptional Regulation by npcRNAs

2.4.1
Expressed Processed Pseudogenes

Potential a source of npcRNAs in the genome, processed pseudogenes originate from reintegration of complementary (c)DNAs produced by reverse transcription mRNAs into the genome. Unlike nonprocessed pseudogenes derived from duplicated regions of the genome, processed pseudogenes are intronless and possess poly-A tails. They have been found only in metazoan animals and flowering plants (Zhang et al. 2003). In the recently published database of processed pseudogenes, there are 5,206 human and 3,428 mouse sequences (Adel et al. 2005), although it is possible that actual numbers are significantly higher (Zhang et al. 2003). It has been estimated that approximately 2%–3% of human and 0.5%–1% of mouse processed pseudogenes is expressed (Yano et al. 2004). Currently, there is one example of a functional expressed pseudogene that is involved in the regulation of expression of its homologous protein-coding gene. In mouse, *Makorin1-p1* is an expressed pseudogene of *Makorin1*. Decreased expression or deletion of *Makorin1-p1* results in an increased turnover rate of the *Makorin1* mRNA and is lethal. The degradation of *Makorin1* mRNA depends on the *cis*-acting RNA decay element within its 5′-UTR. This sequence is also conserved in the transcript originating from the pseudogene. Thus, the stability of *Makorin1* mRNA is regulated by the expression of its processed pseudogene that competes with its corresponding gene's decay element for the protein factors (Hirotsune et al. 2003). A functionality of *Makorin1-p1* is also evident from the observation that its transcribed region shows significantly lower substitution rates than the untranscribed portions (Podlaha and Zhang 2004). Although there are no other documented examples of functional pseudogenes, it is possible that the mechanism analogous to that observed for *Makorin1-p1* is more general and may apply to other gene-processed pseudogene pairs (Yano et al. 2004).

2.4.2
Regulation of Translation

2.4.2.1
MicroRNAs

One of the most spectacular findings in the noncoding RNA field was a discovery of small (20–25 nt) RNAs called micro RNAs (miRNA) that are involved in the regulation of mRNA stability and modulation of translation. There are two distinct pathways of miRNA action, both of which depend on their binding to specific sites on target mRNAs. The interaction between miRNA and mRNA can induce either cleavage of the message or inhibition of translation without affecting mRNA integrity (Bartel 2004; Bartel and Chen 2004). In contrast to plants, a majority of animal miRNAs works through repression of translation. This is accomplished by binding of miRNA to specific sites within 3′-UTRs of target mRNAs, but the precise mechanism of translation arrest is not known. Although it was previously believed that all animal miRNAs act as translational repressors, it has been recently shown that mouse mir-196 and Epstein-Barr virus miR-BART2 direct cleavage of mRNAs encoding Hox-B8 and viral DNA polymerase, respectively (Yekta et al. 2004; Pfeffer et al. 2004).

miRNA-induced cleavage of mRNA requires a fully complementary target site binding to which triggers cleavage by Dicer endonuclease, an RNase III-like enzyme involved in RNAi pathway and maturation of miRNA precursors. Thus, to shut off gene expression, a single highly complementary binding site for one miRNA is sufficient.

Translational arrest by miRNAs is more complex and requires more than one miRNA target site within the mRNA 3′-UTR. These binding sites are not fully complementary with corresponding miRNAs, and the resulting RNA duplexes show numerous bulges, interior loops, and mismatches. In fact, these unpaired nucleotides within the miRNA–mRNA duplex are crucial for miRNA regulatory activity (Vella et al. 2004). The effect of miRNAs binding to the 3'-UTRs is probably not limited to inhibition of translation resulting in decrease in protein production. Microarray analysis demonstrated that miRNAs can also reduce the levels of their target transcripts (Lim et al. 2005).

There is a growing number of reports on miRNA-regulated genes, yet, for most miRNAs, their functions and role in molecular mechanisms are not known. Known examples of verified miRNA–mRNA pairs has allowed for the derivation of certain rules, including free energy and the extent of complementarity of RNA duplexes that may help in identification of putative miRNA binding sites (Doench and Sharp 2004). An analysis of 3'-UTRs of human mRNAs demonstrated that over 2,000 genes show conservation of miRNA target sites with other mammalian species. Furthermore, about 10% of human genes have two or more potential miRNA binding sequences. A majority of miRNAs targets a limited number of genes, but there are miRNA species that may be involved in the regulation of hundreds of mRNAs. It is also evident that some of the mRNAs may be regulated by more than one miRNA (John et al. 2004). These features resemble transcriptional regulation by transcription factors (Hobert 2004). In both cases, gene expression is affected by interaction of *trans*-acting factors (transcription factors or miRNAs) with *cis*-regulatory elements (promoters, or miRNA target sequence in mRNA). Expression of transcription factors is usually cell type-specific, which in turn defines the repertoire of genes that can be turned on or off by their action. The same applies to miRNAs, which often display tissue- or cell type-specific patterns of expression with a defined range of target genes. In fact, transcription of a miRNA gene in a tissue-specific manner requires an appropriate transcription factor (Sun et al. 2004). On the other hand, some of the miRNAs have been shown to target mRNAs encoding transcription factors, thus changing gene expression profiles associated with differentiation (Yekta et al. 2004). It is also possible that, in the same way that a specific set of transcription factors determines pattern of transcribed genes in a given cell, a complementary set of miRNAs is responsible for their translational regulation. Moreover, different miRNAs may work cooperatively to inhibit the expression of a given mRNA. This means that certain mRNAs may be susceptible to miRNA-dependent regulation only in cells in which a specific set of miRNAs is expressed. Hobert (2004) proposed the existence of *miRNA-codes* analogous to *transcription factor codes* specific for each cell type. Together, the combinations of various transcription factors and miRNAs affecting the repertoire of expressed genes can be regarded as crucial determinants of cellular complexity.

2.4.2.2
Brain-Specific Transcripts BC1 and BC200

Another group of npcRNAs involved in the regulation of translation in rodents and primates comprises brain-specific PolIII transcripts BC1 and BC200 (Martignetti and Brosius 1995). Both BC200 and BC1 RNAs associate with proteins forming ribonucleoprotein particles 11.4S and 8.7S, respectively (Cheng et al. 1996). In the nervous system, BC1 and BC200 RNA expression is restricted to neurons where these RNAs are found in cell bodies and in dendrites (Tiedge et al. 1991, 1993). BC1 RNA is a translational repressor (Wang et al. 2002) that affects translation in the initiation phase by preventing formation of a stable 48S preinitiation complex. It was demonstrated that BC1 inhibits eIF4-dependent translation, and the protein that is targeted by BC1 is a helicase eIF4A. BC1 forms stable complexes with eIF4A and the poly(A)-binding protein (PABP) (Wang et al. 2002). It is not known how the BC1-mediated repression of translation could be reversed when the expression of a given messenger is needed. BC1 constitutes only one of the elements that govern complex networks of gene expression patterns responsible for neuronal functions, and its levels are tightly regulated (Nimmrich et al. 2005). At synapses, BC1 RNA forms complexes with FMRP (fragile X mental retardation associated protein). This interaction is crucial for the function of FMRP as a translational repressor of specific mRNAs. BC1 probably plays a role as a mediator facilitating FMRP binding to the mRNA. It has been shown that in the absence of other factors, BC1 can bind mRNAs that are repressed by FMRP (Zalfa et al. 2003).

The expression of BC1 and BC200 RNAs is crucial for normal brain functions. It has been reported that the BC200 level in Alzheimer disease-affected brains was approximately 70% lower than in normal and non-Alzheimer-demented brains (Lukiw et al. 1992). Mutant BC1-deficient mice, despite normal phenotypes and brain morphology, show behavioral changes like reduced exploration and increased anxiety (Lewejohann et al. 2004).

Interestingly, BC1 and BC200 RNAs are also expressed in certain human and mouse cancers (Chen et al. 1997a,b). The expression of BC200 RNA is associated with invasive forms of breast cancer, but no expression was detected in benign tumors (Iacoangeli et al. 2004).

3
The Medical Perspective

A number of human npcRNAs have been implicated in human diseases including neurobehavioral and developmental disorders as well as certain forms of cancer. The changes of expression levels or genetic and epigenetic alterations affecting the npcRNAs accompanying malignant processes strongly support the functional role of RNA, especially in cases when the chromosomal aberrations disrupt the npcRNA genes. Noncoding transcripts, or at least a majority of them, cannot then be considered as only non-functional products of spurious transcription.

3.1
Noncoding RNAs in Developmental Disorders

Many mammalian noncoding RNAs are transcribed from imprinted genes. Their abnormal patterns of expression can result in severe congenital disorders including Prader-Willi, Beckwith-Wiedemann, and Angelman syndromes. The genetic aberrations responsible for these disorders affect complex imprinted loci and cause a number of developmental defects often associated with mental retardation and other neurobehavioral phenotypes (Nicholls 2000). Although the role of most imprinted npcRNAs is still enigmatic, their integrity and correct expression seem to be crucial for the maintenance of transcriptional activity of protein-coding genes located within the imprinted clusters.

3.1.1
Angelman and Prader-Willi Syndromes

Within the chromosomal region 15q11–q13, where defects linked to Angelman syndrome (AS) and Prader-Willi syndrome (PWS) reside, there are 11 paternally and 1 maternally expressed genes. One of the genes, expression of which is inhibited in the PWS, is *IPW* (imprinted in Prader-Willi syndrome), located about 180 kb from the imprinting control element. This gene encodes a 2.3-kb, polyadenylated npcRNA expressed exclusively from the paternal allele in all human tissues (Wevrick et al. 1994). Interestingly, the mouse counterpart *Ipw* shows tissue-specific patterns of expression with high levels present in the brain (Wevrick and Francke 1997). The AS results from a disruption of maternal expression of a single gene, *UBE3A*. The imprinted expression of *UBE3A* is probably regulated by the antisense paternal transcript UBE3A-AS (Chamberlain and Brannan 2001).

3.1.2
DiGeorge Syndrome

Another developmental defect associated with non-protein-coding genes is DiGeorge syndrome (DGS), which correlates with deletions within human chromosome 22q11. DGS critical region (DGCR) was identified as a shortest region of deletion overlap (SRO) containing the ADU breakpoint that is disrupted by a balanced translocation associated with DGS. The analysis of transcripts expressed from the breakpoint region revealed several alternatively spliced cDNAs expressed during human and murine embryogenesis. None of those transcripts seems to encode proteins. These RNAs are encoded by a gene, *DGCR5*, that is disrupted by the ADU breakpoint. Within the first intron of the *DGCR5* there is a protein-coding gene, *DGCR3*, transcribed in the same direction as *DGCR5* (Sutherland et al. 1996).

Another gene deleted in DiGeorge syndrome is *HIRA*. Antisense, alternatively spliced transcripts originating from the first intron of *HIRA*, were detected in the cytoplasm of neurons of both central and peripheral nervous systems. The gene 22k48 consists of three exons and contains several tandemly arranged repeated elements (Alu, LINEs, CA_n) that surround a unique sequence. Although the function of this RNA is not known, it has been shown to bind several cytoplasmic proteins, which led to the speculation that 22k48 haploinsufficiency may contribute to the pathogenesis of DiGeorge syndrome (Pizzuti et al. 1999).

3.2
NpcRNAs in Cancer

Expression of specific npcRNAs has been associated with certain forms of cancer. The functions of the majority of npcRNAs expressed during cancer development are unknown, and it is difficult to judge if their overexpression or lack of expression is the cause or consequence of changes in the cells' gene expression program that ultimately brings about cancer.

One of the best-studied npcRNAs is a product of imprinted H19 gene expressed exclusively from the maternal allele. During development, *H19* is transcribed at very high levels in placenta, embryo, and in most of fetal tissues. After birth, its expression is significantly reduced. Reciprocal expression patterns of H19 and the adjacent insulin-like growth factor (IGF)2 gene during development suggest different roles played by the products of these genes. IGF2 is a growth factor stimulating cell proliferation, while H19 RNA may be responsible for differentiation (Looijenga et al. 1997). The loss of imprinted expression is often associated with cancer, and biallelic expression of H19 and/or IGF2 may lead to malignant growth (Ulaner et al. 2003). The role of H19 expression in cancers is not clear. It was described as a tumor suppressor gene capable of reducing tumorigenicity and growth of certain malignant cell lines (Hao et al. 1993), but its elevated expression in lung, breast, and bladder cancers suggested that it may have oncogenic properties (Lottin et al. 2002, 2005). However, various effects of H19 expression may depend on the type or the developmental stage of cells, its alternatively spliced isoforms (Matouk et al. 2004), or the expression of specific RNA-binding proteins (Ioannidis et al. 2004).

Prostate tumors show specific expression of two noncoding RNAs. PCGEM1 is a highly specific transcript that is expressed exclusively in prostate glandular epithelial cells. The analysis of PCGEM1 cDNA revealed the longest ORF encoding a 35-aa peptide. The context of the initiation codon does not match Kozak's consensus, and the in vitro transcription/translation of PCGEM1 cDNA did not yield a detectable protein. PCGEM1 RNA is expressed at low levels in normal prostate tissue, but its transcription is significantly elevated in tumor cells. Depending on the method used, the association of PCGEM1 overexpression with cancer between matched tumor/normal specimens was between 56% and 84% (Srikantan et al. 2000). An overexpression of PCGEM1 RNA in cultured cells (LNCaP and in NIH3T3) was shown to promote cell proliferation and colony formation, which suggested that this RNA is involved in the regulation of cell growth and, at least in some cases, it may contribute to the pathogenesis of prostate cancer (Petrovics et al. 2004)

Another prostate-specific transcript that shows significant overexpression in prostatic cancers when compared with normal tissue is PCA3 (prostate cancer antigen 3, DD3). The PCA3 gene consists of four exons and the primary transcripts are subject to alternative splicing and alternative polyadenylation, giving rise to differently sized mature transcripts ranging from 0.6 to 4 kb in size. There are no significant ORFs and the gene shows no similarity with other genes. In the human genome, PCA3 was mapped to chromosome 9q21-22, and it has been shown to be conserved in several mammalian species except for rodents (Bussemakers et al. 1999). PCA3 is one of the most prostate cancer-specific genes, which makes it a promising marker for early diagnosis. Over 95% of analyzed prostate cancer samples showed up to 66-fold increase in PCA3 transcription. Such a high level of expression makes the PCA3 transcripts excellent markers for early detection of prostatic tumors (Hessels et al. 2003; de Kok et al. 2002). However, studies by Gandini et al. (2003) demonstrated that the prostate specific exon is only exon 4 of PCA3 (DD3). Using RT-PCR analysis with primers corresponding to exons 1 and 3, it was possible to detect PCA3 expression in all analyzed samples of normal and malignant human tissues.

Colon carcinoma cells show significant overexpression of OCC-1 RNA (overexpressed in colon carcinoma 1). The two 1.2- and 1.3-kb transcripts with different 5′ and 3′ ends show tissue-specific expression in kidney, pancreas, and placenta. The *OCC-1* was detected in a subset of colon carcinoma cells, but it is absent or expressed at very low levels in normal mucosa (Pibouin et al. 2002). The expression of 8-kb MALAT-1 RNA (metastasis associated in lung adenocarcinoma transcript 1) was shown to be associated with metastasis in non-small cell lung cancer (Ji et al. 2003). Over 50% of B-cell chronic lymphocytic leukemia (B-CLL) and over 60% of mantle cell lymphoma cases are associated with deletions of the chromosomal region 13q14.3 harboring a 560-kb BCMS gene (B-cell neoplasia-associated gene with multiple splicing). Alternative splicing of at least 50 exons produces a large number of mature variants that show tissue-specific distribution, none of which have a protein coding potential. The functions of these transcripts and their possible involvement in tumor suppression are not known (Wolf et al. 2001).

npcRNA expression patterns can differ for various subtypes of cancer. In rhabdomyosarcoma (RMS), the two histological subtypes can be distinguished based on the expression of the 1.25-kb ncRNA called *NCRMS* (noncoding RNA in RMS). Significantly higher levels of NCRMS in the alveolar RMS, when compared with the embryonal subtype as well as its presence in neuroblastoma and synovial sarcoma, may indicate a deregulation of gene expression within the large chromosomal region, including several genes associated with muscle development (Chan et al. 2002).

Certain forms of cancer may be also associated with changes in expression of miRNAs. A genome-wide analysis of chromosomal localization of human miRNAs revealed that approximately half of them occur in fragile sites and regions associated with cancer (Calin et al. 2004). miRNAs have been shown to play a pivotal role in the regulation of certain processes related to development in all eukaryotes. Their potential for a specific posttranscriptional regulation of gene expression combined with their small size and evolutionary conservation make them ideal candidates for agents controlling complex gene expression networks governing cell growth and differentiation. Altered patterns of expression of miRNAs may be therefore responsible for changes in the cells' genetic program, which in turn results in malignant growth. They may be responsible for the maintenance of cell homeostasis by inhibiting expression of genes involved in carcinogenesis. Various cancer cell lines display qualitative and quantitative differences in miRNAs expression. For example, in colorectal cancers, the levels of miR-24-2 showed a 50-fold difference between samples (Schmittgen et al. 2004). A reduced expression or deletion of miR-15 and miR-16 coding genes is often observed in B cell chronic lymphocytic leukemias. It is unclear whether any of these miRNAs play a role in B cell differentiation. A putative target for miR-16 is arginyl-tRNA synthetase, expression of which correlates with levels of miR-16, suggesting a post-transcriptional regulatory mechanism. Deletion of the region containing the two miRNAs also occurs in other types of cancer, including prostate tumors and multiple myeloma (Calin et al. 2002). An increased or reduced expression of miRNAs was also associated with Burkitt lymphoma (miR-155) and colorectal cancer (miR-143 and miR-145) (Metzler et al. 2004; Michael et al. 2003). A large-scale computational prediction of potential targets for miRNAs demonstrated that significant numbers of candidate mRNAs are encoded by known cancer-related genes (John et al. 2004).

4
Concluding Remarks

The growing number of reports of novel npcRNAs in mammals prompted questions about the number of genes that encode them. In contrast to proteins—where their related genes can be predicted with high confidence using the genomic sequences—it is much more difficult to predict RNA-coding regions. The commonly used gene-finding algorithms were designed to search for the features characteristic for protein-coding genes (ORFs or codon usage) and cannot be generally applied to npcRNA searching. Another problem associated with computational identification of npcRNAs is the fact that some of them may be actually processed from introns, as is the case of some snoRNAs and miRNAs (Cai et al. 2004; Rodriguez et al. 2004; Runte et al. 2001).

Recently, significant progress has been made in the attempt to define human and mouse transcriptomes in two large projects aimed at annotation and functional characterization of full-length cDNA sequences. The human H-InvDB (Human Full-length cDNA Annotation Invitational Database) (Imanishi et al. 2004) and mouse FANTOM (Functional Annotation of Mouse) databases (Okazaki et al. 2002) provide comprehensive catalogs of genes and their transcripts, including alternative splicing information. In the set of representative human transcripts from 21,037 loci, 1,377 (6.5%) have been classified as non-protein-coding (Imanishi et al. 2004). Based on the genomic localization and the presence or absence of neighboring protein-coding transcripts as well as poly-A signals or tails, 269 human transcripts were annotated as putative non-coding RNAs. The transcriptional analyses of individual human chromosomes also indicate that non-protein-coding transcripts constitute a significant fraction of all RNA produced in the cell. Within the male-specific region of the human Y chromosome, 78 out of 156 identified transcription units are likely to produce npcRNAs (Skaletsky et al. 2003). Also, a distribution of transcription factor binding sites and expression analysis using microarrays suggested that a significant fraction of transcriptional output from human chromosomes 21 and 22 are ncRNAs (Cawley et al. 2004). It is therefore becoming increasingly evident that the number of non-protein-coding genes may actually be equal or even exceed the number of protein-coding ones.

Our present-day knowledge of the roles of npcRNAs and RNA-directed regulation of cellular processes is still very superficial. Only a small fraction of mammalian npcRNAs has been partially characterized in terms of function or expression patterns. It seems, however, that unraveling the intricacies of RNA-based regulation will be crucial for our understanding of the functioning and evolution of complex biological systems.

References

Adel K, Laurent D, Dominique M (2005) HOPPSIGEN: a database of human and mouse processed pseudogenes. Nucleic Acids Res 33 Database Issue:D59–D66

Allen TA, Von Kaenel S, Goodrich JA, Kugel JF (2004) The SINE-encoded mouse B2 RNA represses mRNA transcription in response to heat shock. Nat Struct Mol Biol 11:816–821

Antequera F, Bird A (1993) Number of CpG islands and genes in human and mouse. Proc Natl Acad Sci USA 90:11995–11999

Bartel DP (2004) MicroRNAs: Genomics, biogenesis, mechanism, and function. Cell 116:281–297

Bartel DP, Chen CZ (2004) Micromanagers of gene expression: the potentially widespread influence of metazoan microRNAs. Nat Rev Genet 5:396–400

Bussemakers MJ, van Bokhoven A, Verhaegh GW, Smit FP, Karthaus HF, Schalken JA, Debruyne FM, Ru N, Isaacs WBl (1999) DD3:A new prostate-specific gene, highly overexpressed in prostate cancer. Cancer Res 59:5975–5979

Cai X, Hagedorn CH, Cullen BR (2004) Human microRNAs are processed from capped, polyadenylated transcripts that can also function as mRNAs. RNA 10:1957–1966

Calin GA, Dumitru CD, Shimizu M, Bichi R, Zupo S, Noch E, Aldler H, Rattan S, Keating M, Rai K, Rassenti L, Kipps T, Negrini M, Bullrich F, Croce CM (2002) Frequent deletions and down-regulation of micro-RNA genes miR15 and miR16 at 13q14 in chronic lymphocytic leukemia. Proc Natl Acad Sci U S A 99:15524–15529

Calin GA, Sevignani C, Dumitru CD, Hyslop T, Noch E, Yendamuri S, Shimizu M, Rattan S, Bullrich F, Negrini M, Croce CM (2004) Human microRNA genes are frequently located at fragile sites and genomic regions involved in cancers. Proc Natl Acad Sci U S A 101:2999–3004

Cawley S, Bekiranov S, Ng HH, Kapranov P, Sekinger EA, Kampa D, Piccolboni A, Sementchenko V, Cheng J, Williams AJ, Wheeler R, Wong B, Drenkow J, Yamanaka M, Patel S, Brubaker S, Tammana H, Helt G, Struhl K, Gingeras TR (2004) Unbiased mapping of transcription factor binding sites along human chromosomes 21 and 22 points to widespread regulation of noncoding RNAs. Cell 116:499–509

Chadwick BP, Willard HF (2004) Multiple spatially distinct types of facultative heterochromatin on the human inactive X chromosome. Proc Natl Acad Sci USA 101:17450–17455

Chamberlain SJ, Brannan CI (2001) The Prader-Willi syndrome imprinting center activates the paternally expressed murine Ube3a antisense transcript but represses paternal Ube3a. Genomics 73:316–322

Chan AS, Thorner PS, Squire JA, Zielenska M (2002) Identification of a novel gene NCRMS on chromosome 12q21 with differential expression between Rhabdomyosarcoma types. Oncogene 21:3029–3037

Chau YM, Pando S, Taylor HS (2002) HOXA11 silencing and endogenous HOXA11 antisense ribonucleic acid in the uterine endometrium. J Clin Endocrinol Metab 87:2674–2680

Chen W, Heierhorst J, Brosius J, Tiedge H (1997a) Expression of neural BC1 RNA: induction in murine tumours. Eur J Cancer 33:288–292

Chen W, Bocker W, Brosius J, Tiedge H (1997b) Expression of neural BC200 RNA: in human tumours. J Pathol 183:345–351

Cheng JG, Tiedge H, Brosius J (1996) Identification and characterization of BC1 RNP particles. DNA Cell Biol 15:549–559

Chureau C, Prissette M, Bourdet A, Barbe V, Cattolico L, Jones L, Eggen A, Avner P, Duret L (2002) Comparative sequence analysis of the X-inactivation center region in mouse, human, and bovine. Genome Res 12:894–908

de Kok JB, Verhaegh GW, Roelofs RW, Hessels D, Kiemeney LA, Aalders TW, Swinkels DW, Schalken JA (2002) DD3(PCA3), a very sensitive and specific marker to detect prostate tumors. Cancer Res 62:2695–2698

DeBaun MR, Niemitz EL, McNeil DE, Brandenburg SA, Lee MP, Feinberg AP (2002) Epigenetic alterations of H19 and LIT1 distinguish patients with Beckwith-Wiedemann syndrome with cancer and birth defects. Am J Hum Genet 70:604–611

Dermitzakis ET, Kirkness E, Schwarz S, Birney E, Reymond A, Antonarakis SE (2004) Comparison of human chromosome 21 conserved nongenic sequences (CNGs) with the mouse and dog genomes shows that their selective constraint is independent of their genic environment. Genome Res 14:852–859

Diaz-Meyer N, Day CD, Khatod K, Maher ER, Cooper W, Reik W, Junien C, Graham G, Algar E, Der Kaloustian VM, Higgins MJ (2003) Silencing of CDKN1C (p57KIP2) is associated with hypomethylation at KvDMR1 in Beckwith-Wiedemann syndrome. J Med Genet 40:797–801

Doench JG, Sharp PA (2004) Specificity of microRNA target selection in translational repression. Genes Dev 18:504–511

Emberley E, Huang GJ, Hamedani MK, Czosnek A, Ali D, Grolla A, Lu B, Watson PH, Murphy LC, Leygue E (2003) Identification of new human coding steroid receptor RNA activator isoforms. Biochem Biophys Res Commun 301:509–515

Espinoza CA, Allen TA, Hieb AR, Kugel JF, Goodrich JA (2004) B2 RNA binds directly to RNA polymerase to repress transcript synthesis. Nat Struct Mol Biol 11:822–829

Fields C, Adams MD, White O, Venter JC (1994) How many genes in the human genome? Nat Genet 7:345–346

Fitzpatrick GV, Soloway PD, Higgins MJ (2002) Regional loss of imprinting and growth deficiency in mice with a targeted deletion of KvDMR1. Nat Genet 32:426–431

Gandini O, Luci L, Stigliano A, Lucera R, Di Silverio F, Toscano V, Cardillo MR (2003) Is DD3 a new prostate-specific gene? Anticancer Res 23:305–308

Ganesan S, Silver DP, Greenberg RA, Avni D, Drapkin R, Miron A, Mok SC, Randrianarison V, Brodie S, Salstrom J, Rasmussen TP, Klimke A, Marrese C, Marahrens Y, Deng CX, Feunteun J, Livingston DM (2002) BRCA1 supports XIST RNA concentration on the inactive X chromosome. Cell 111:393–405

Haddad F, Bodell PW, Qin AX, Giger JM, Baldwin KM (2003) Role of antisense RNA in coordinating cardiac myosin heavy chain gene switching. J Biol Chem 278:37132–37138

Hao Y, Crenshaw T, Moulton T, Newcomb E, Tycko B (1993) Tumour-suppressor activity of H19 RNA. Nature 365:764–767

Hastings ML, Ingle HA, Lazar MA, Munroe SH (2000) Post-transcriptional regulation of thyroid hormone receptor expression by cis-acting sequences and a naturally occurring antisense RNA. J Biol Chem 275:11507–11513

Heard E, Rougeulle C, Arnaud D, Avner P, Allis CD, Spector DL (2001) Methylation of histone H3 at Lys-9 is an early mark on the X chromosome during X inactivation. Cell 107:727–738

Hessels D, Klein Gunnewiek JM, van Oort I, Karthaus HF, van Leenders GJ, van Balken B, Kiemeney LA, Witjes JA, Schalken JA (2003) DD3(PCA3)-based molecular urine analysis for the diagnosis of prostate cancer. Eur Urol 44:8–15

Hirotsune S, Yoshida N, Chen A, Garrett L, Sugiyama F, Takahashi S, Yagami K, Wynshaw-Boris A, Yoshiki A (2003) An expressed pseudogene regulates the messenger-RNA stability of its homologous coding gene. Nature 423:91–96

Hobert O (2004) Common logic of transcription factor and microRNA action. Trends Biochem Sci 9:462–468

Horike S, Mitsuya K, Meguro M, Kotobuki N, Kashiwagi A, Notsu T, Schulz TC, Shirayoshi Y, Oshimura M (2000) Targeted disruption of the human LIT1 locus defines a putative imprinting control element playing an essential role in Beckwith-Wiedemann syndrome. Hum Mol Genet 9:2075–2083

Iacoangeli A, Lin Y, Morley EJ, Muslimov IA, Bianchi R, Reilly J, Weedon J, Diallo R, Bocker W, Tiedge H (2004) BC200 RNA in invasive and preinvasive breast cancer. Carcinogenesis 25:2125–2133

Imamura T, Ohgane J, Ito S, Ogawa T, Hattori N, Tanaka S, Shiota K (2001) CpG island of rat sphingosine kinase-1 gene: tissue-dependent DNA methylation status and multiple alternative first exons. Genomics 76:113–125

Imamura T, Yamamoto S, Ohgane J, Hattori N, Tanaka S, Shiota K (2004) Non-coding RNA directed DNA demethylation of Sphk1 CpG island. Biochem Biophys Res Commun 322:593–600

Imanishi T, Itoh T, Suzuki Y, et al (2004) Integrative annotation of 21,037 human genes validated by full-length cDNA clones. PLoS Biol 2:e162

Ioannidis P, Kottaridi C, Dimitriadis E, Courtis N, Mahaira L, Talieri M, Giannopoulos A, Iliadis K, Papaioannou D, Nasioulas G, Trangas T (2004) Expression of the RNA-binding protein CRD-BP in brain and non-small cell lung tumors. Cancer Lett 209:245–250

Jeang KT, Xiao H, Rich EA (1999) Multifaceted activities of the HIV-1 transactivator of transcription, Tat. J Biol Chem 274:28837–28840

Ji P, Diederichs S, Wang W, Boing S, Metzger R, Schneider PM, Tidow N, Brandt B, Buerger H, Bulk E, Thomas M, Berdel WE, Serve H, Muller-Tidow C (2003) MALAT-1, a novel noncoding RNA, and thymosin beta4 predict metastasis and survival in early-stage non-small cell lung cancer. Oncogene 22:8031–8041

John B, Enright AJ, Aravin A, Tuschl T, Sander C, Marks DS (2004) Human microRNA targets. PLoS Biol 2:e363

Keohane AM, O'neill LP, Belyaev ND, Lavender JS, Turner BM (1996) X-Inactivation and histone H4 acetylation in embryonic stem cells. Dev Biol 180:618–630

Kiyosawa H, Yamanaka I, Osato N, Kondo S, Hayashizaki Y, RIKEN GER Group, GSL Members (2003) Antisense transcripts with FANTOM2 clone set and their implications for gene regulation. Genome Res 13:1324–1334

Kwek KY, Murphy S, Furger A, Thomas B, O'Gorman W, Kimura H, Proudfoot NJ, Akoulitchev A (2002) U1 snRNA associates with TFIIH and regulates transcription initiation. Nat Struct Biol 9:800–805

Lander ES, Linton LM, Birren B et al. (2001) Initial sequencing and analysis of the human genome. Nature 409:860–921

Lanz R, Razani B, Goldberg AD, O'Malley BW (2002) Distinct RNA motifs are important for coactivation of steroid hormone receptors by steroid receptor RNA activator (SRA). Proc Natl Acad Sci USA 99:16081–16086

Lanz RB, McKenna NJ, Onate SA, Albrecht U, Wong J, Tsai SY, Tsai MJ, O'Malley BW (1999) A steroid receptor coactivator, SRA, functions as an RNA and is present in an SRC-1 complex. Cell 97:7–27

Lanz RB, Chua SS, Barron N, Soder BM, DeMayo F, O'Malley BW (2003) Steroid receptor RNA activator stimulates proliferation as well as apoptosis in vivo. Mol Cell Biol 23:7163–7176

Lareau LF, Green RE, Bhatnagar RS, Brenner SE (2004) The evolving roles of alternative splicing. Curr Opin Struct Biol 14:273–282

Lee JT, Jaenisch R (1997) Long-range cis effects of ectopic X-inactivation centres on a mouse autosome. Nature 386:275–279

Lewejohann L, Skryabin BV, Sachser N, Prehn C, Heiduschka P, Thanos S, Jordan U, Dell'Omo G, Vyssotski AL, Pleskacheva MG, Lipp HP, Tiedge H, Brosius J, Prior H (2004) Role of a neuronal small non-messenger RNA: behavioural alterations in BC1 RNA-deleted mice. Behav Brain Res 154:273–289

Lim LP, Lau NC, Garrett-Engele P, Grimson A, Schelter JM, Castle J, Bartel DP, Linsley PS, Johnson JM (2005) Microarray analysis shows that some microRNAs downregulate large numbers of target mRNAs. Nature 433:769–773

Looijenga LH, Verkerk AJ, De Groot N, Hochberg AA, Oosterhuis JW (1997) H19 in normal development and neoplasia. Mol Reprod Dev 46:419–439

Lottin S, Adriaenssens E, Dupressoir T, Berteaux N, Montpellier C, Coll J, Dugimont T, Curgy JJ (2002) Overexpression of an ectopic H19 gene enhances the tumorigenic properties of breast cancer cells. Carcinogenesis 23:1885–1895

Lottin S, Adriaenssens E, Berteaux N, Lepretre A, Vilain MO, Denhez E, Coll J, Dugimont T, Curgy JJ (2005) The human H19 gene is frequently overexpressed in myometrium and stroma during pathological endometrial proliferative events. Eur J Cancer 41:168–177

Lukiw WJ, Handley P, Wong L, Crapper McLachlan DR (1992) BC200 RNA in normal human neocortex, non-Alzheimer dementia (NAD), and senile dementia of the Alzheimer type (AD). Neurochem Res 17:591–597

Luo Y, Kurz J, MacAfee N, Krause MO (1997) C-myc deregulation during transformation induction: involvement of 7SK RNA. J Cell Biochem 64:313–327

Madamanchi NR, Hu ZY, Li F, Horaist C, Moon SK, Patterson C, Runge MS, Ruef J, Fritz PH, Aaron J (2002) A noncoding RNA regulates human protease-activated receptor-1 gene during embryogenesis. Biochim Biophys Acta 1576:237–245

Mager J, Montgomery ND, de Villena F, Magnuson T (2003) Genome imprinting regulated by the mouse polycomb group protein Eed. Nat Genet 33:502–507

Mancini-DiNardo D, Steele SJ, Ingram RS, Tilghman SM (2003) A differentially methylated region within the gene Kcnq1 functions as an imprinted promoter and silencer. Hum Mol Genet 12:283–294

Martignetti JA, Brosius J (1995) BC1 RNA: transcriptional analysis of a neural cell-specific RNA polymerase III transcript. Mol Cell Biol 15:1642–1650

Matouk I, Ayesh B, Schneider T, Ayesh S, Ohana P, de-Groot N, Hochberg A, Galun E (2004) Oncofetal splice-pattern of the human H19 gene. Biochem Biophys Res Commun 318:916–919

Mattick JS (1994) Introns: evolution and function. Curr Opin Genet Dev 4:823–831

Mattick JS (2001) Non-coding RNAs: the architects of eukaryotic complexity. EMBO Rep 2:986–991

Mattick JS (2003) Introns and noncoding RNAs the hidden layer of eukaryotic complexity. In: Barciszewski J, Erdmann VA (eds) Noncoding RNAs: molecular biology and molecular medicine. Landes Bioscience, Georgtown, pp 12–33

Mattick JS, Gagen MJ (2001) The evolution of controlled multitasked gene networks: the role of introns and other noncoding RNAs in the development of complex organisms. Mol Biol Evol 18:1611–1630

Metzler M, Wilda M, Busch K, Viehmann S, Borkhardt A (2004) High expression of precursor microRNA-155/BIC RNA in children with Burkitt lymphoma. Genes Chromosomes Cancer 39:167–169

Michael MZ, O'Connor SM, van Holst Pellekaan NG, Young GP, James RJ (2003) Reduced accumulation of specific microRNAs in colorectal neoplasia. Mol Cancer Res 1:882–891

Migeon BR, Chowdhury AK, Dunston JA, McIntosh I (2001) Identification of TSIX, encoding an RNA antisense to human XIST, reveals differences from its murine counterpart: implications for X inactivation. Am J Hum Genet 69:951–960

Migeon BR, Lee CH, Chowdhury AK, Carpenter H (2002) Species differences in TSIX/Tsix reveal the roles of these genes in X-chromosome inactivation. Am J Hum Genet 71:286–293

Newall AE, Duthie S, Formstone E, Nesterova T, Alexiou M, Johnston C, Caparros ML, Brockdorff N (2001) Primary non-random X inactivation associated with disruption of Xist promoter regulation. Hum Mol Genet 10:581–589

Nguyen VT, Kiss T, Michels AA, Bensaude O (2001) 7SK small nuclear RNA binds to and inhibits the activity of CDK9/cyclin T complexes. Nature 414:322–325

Nicholls RD (2000) The impact of genomic imprinting for neurobehavioral and developmental disorders. J Clin Invest 105:413–418

Nimmrich V, Hargreaves EL, Muslimov IA, Bianchi R, Tiedge H (2005) Dendritic BC1 RNA: modulation by kindling-induced afterdischarges. Brain Res Mol Brain Res 133:110–118

Okazaki Y, Furuno M, Kasukawa T et al. (2002) Analysis of the mouse transcriptome based on functional annotation of 60,770 full-length cDNAs. Nature 420:563–573

Petrovics G, Zhang W, Makarem M, Street JP, Connelly R, Sun L, Sesterhenn IA, Srikantan V, Moul JW, Srivastava S (2004) Elevated expression of PCGEM1, a prostate-specific gene with cell growth-promoting function, is associated with high-risk prostate cancer patients. Oncogene 23:605–611

Pfeffer S, Zavolan M, Grasser FA, Chien M, Russo JJ, Ju J, John B, Enright AJ, Marks D, Sander C, Tuschl T (2004) Identification of virus-encoded microRNAs. Science 304:734–736

Pibouin L, Villaudy J, Ferbus D, Muleris M, Prosperi MT, Remvikos Y, Goubin G (2002) Cloning of the mRNA of overexpression in colon carcinoma-1: a sequence overexpressed in a subset of colon carcinomas. Cancer Genet Cytogenet 133:55–60

Pizzuti A, Novelli G, Ratti A, Amati F, Bordoni R, Mandich P, Bellone E, Conti E, Bengala M, Mari A, Silani V, Dallapiccola B (1999) Isolation and characterization of a novel transcript embedded within HIRA, a gene deleted in DiGeorge syndrome. Mol Genet Metab 67:227–235

Plath K, Mlynarczyk-Evans S, Nusinow DA, Panning B (2002) Xist RNA and the mechanism of X chromosome inactivation. Annu Rev Genet 36:233–278

Podlaha O, Zhang J (2004) Nonneutral evolution of the transcribed pseudogene Makorin1-p1 in mice. Mol Biol Evol 21:2202–2209

Price DH (2000) P-TEFb, a cyclin-dependent kinase controlling elongation by RNA polymerase II. Mol Cell Biol 20:2629–2634

Pyne S, Pyne NJ (2000) Sphingosine 1-phosphate signalling in mammalian cells. Biochem J 349:385–402

Rodriguez A, Griffiths-Jones S, Ashurst JL, Bradley A (2004) Identification of mammalian microRNA host genes and transcription units. Genome Res 14:1902–1910

Roest Crollius H, Jaillon O, Bernot A, Dasilva C, Bouneau L, Fischer C, Fizames C, Wincker P, Brottier P, Quetier F, Saurin W, Weissenbach J (2000) Estimate of human gene number provided by genome-wide analysis using Tetraodon nigroviridis DNA sequence. Nat Genet 25:235–238

Rougeulle C, Chaumeil J, Sarma K, Allis CD, Reinberg D, Avner P, Heard E (2004) Differential histone H3 Lys-9 and Lys-27 methylation profiles on the X chromosome. Mol Cell Biol 24:5475–5484

Runte M, Huttenhofer A, Gross S, Kiefmann M, Horsthemke B, Buiting K (2001) The IC-SNURF-SNRPN transcript serves as a host for multiple small nucleolar RNA species and as an antisense RNA for UBE3A. Hum Mol Genet 10:2687–2700

Schmittgen TD, Jiang J, Liu Q, Yang L (2004) A high-throughput method to monitor the expression of microRNA precursors. Nucleic Acids Res 32:e43

Shendure J, Church GM (2002) Computational discovery of sense-antisense transcription in the human and mouse genomes. Genome Biol 3:research0044

Shi Y, Downes M, Xie W, Kao HY, Ordentlich P, Tsai CC, Hon M, Evans RM (2001) SHARP, an inducible cofactor that integrates nuclear receptor repression and activation. Genes Dev 15:1140–1151

Shibata S, Lee JT (2003) Characterization and quantitation of differential Tsix transcripts: implications for Tsix function. Hum Mol Genet 12:125–136

Skaletsky H, Kuroda-Kawaguchi T, Minx PJ et al. (2003) The male-specific region of the human Y chromosome is a mosaic of discrete sequence classes. Nature 423:825–837

Sleutels F, Zwart R, Barlow DP (2002) The non-coding Air RNA is required for silencing autosomal imprinted genes. Nature 415:810–813

Sleutels F, Tjon G, Ludwig T, Barlow DP (2003) Imprinted silencing of Slc22a2 and Slc22a3 does not need transcriptional overlap between Igf2r and Air. EMBO J 22:3696–3704

Smith KP, Byron M, Clemson CM, Lawrence JB (2004) Ubiquitinated proteins including uH2A on the human and mouse inactive X chromosome: enrichment in gene rich bands. Chromosoma 113:324–335

Southan C (2004) Has the yo-yo stopped? An assessment of human protein-coding gene number. Proteomics 4:1712–1726

Srikantan V, Zou Z, Petrovics G, Xu L, Augustus M, Davis L, Livezey JR, Connell T, Sesterhenn IA, Yoshino K, Buzard GS, Mostofi FK, McLeod DG, Moul JW, Srivastava S (2000) PCGEM1, a prostate-specific gene, is overexpressed in prostate cancer. Proc Natl Acad Sci USA 97:12216–12221

Sun Y, Koo S, White N, Peralta E, Esau C, Dean NM, Perera RJ (2004) Development of a microarray to detect human and mouse microRNAs and characterization of expression in human organs. Nucleic Acids Res 32:e188

Sutherland HF, Wadey R, McKie JM, Taylor C, Atif U, Johnstone KA, Halford S, Kim UJ, Goodship J, Baldini A, Scambler PJ (1996) Identification of a novel transcript disrupted by a balanced translocation associated with DiGeorge syndrome. Am J Hum Genet 59:23–31

Szymanski M, Barciszewski J (2003) Regulation by RNA. Int Rev Cytol 231:197–258

Thakur N, Tiwari VK, Thomassin H, Pandey RR, Kanduri M, Gondor A, Grange T, Ohlsson R, Kanduri C (2004) An antisense RNA regulates the bidirectional silencing property of the Kcnq1 imprinting control region. Mol Cell Biol 24:7855–7862

Tiedge H, Fremeau RT Jr, Weinstock PH, Arancio O, Brosius J (1991) Dendritic location of neural BC1 RNA. Proc Natl Acad Sci USA 88:2093–2097

Tiedge H, Chen W, Brosius J (1993) Primary structure, neural-specific expression, and dendritic location of human BC200 RNA. J Neurosci 13:2382–2390

Ulaner GA, Vu TH, Li T, Hu JF, Yao XM, Yang Y, Gorlick R, Meyers P, Healey J, Ladanyi M, Hoffman AR (2003) Loss of imprinting of IGF2 and H19 in osteosarcoma is accompanied by reciprocal methylation changes of a CTCF-binding site. Hum Mol Genet 12:535–549

Vella MC, Reinert K, Slack FJ (2004) Architecture of a validated microRNA:target interaction. Chem Biol 11:1619–1623

Venter JC, Adams MD, Myers EW et al. (2001) The sequence of the human genome. Science 291:1304–1351

Wang H, Iacoangeli A, Popp S, Muslimov IA, Imataka H, Sonenberg N, Lomakin IB, Tiedge H (2002) Dendritic BC1 RNA: functional role in regulation of translation initiation. J Neurosci 22:10232–10241

Wang J, Mager J, Chen Y, Schneider E, Cross JC, Nagy A, Magnuson T (2001) Imprinted X inactivation maintained by Polycomb group gene. Nat Genet 28:371–375

Wassarman KM, Storz G (2000) 6S RNA regulates E. coli RNA polymerase activity. Cell 101:613–623

Watanabe M, Yanagisawa J, Kitagawa H, Takeyama K, Ogawa S, Arao Y, Suzawa M, Kobayashi Y, Yano T, Yoshikawa H, Masuhiro Y, Kato S (2001) A subfamily of RNA-binding DEAD-box proteins acts as an estrogen receptor a coactivator through the N-terminal activation domain (AF-1) with an RNA coactivator, SRA. EMBO J 20:1341–1352

Waterston RH, Lindblad-Toh K, Birney E et al. (2002) Initial sequencing and comparative analysis of the mouse genome. Nature 420:520–562

Wevrick R, Francke U (1997) An imprinted mouse transcript homologous to the human imprinted in Prader-Willi syndrome (IPW) gene. Hum Mol Genet 6:325–332

Wevrick R, Kerns JA, Francke U (1994) Identification of a novel paternally expressed gene in the Prader-Willi syndrome region. Hum Mol Genet 3:1877–1882

Wolf S, Mertens D, Schaffner C, Korz C, Dohner H, Stilgenbauer S, Lichter P (2001) B-cell neoplasia associated gene with multiple splicing (BCMS): the candidate B-CLL gene on 13q14 comprises more than 560 kb covering all critical regions. Hum Mol Genet 10:1275–1285

Wright FA, Lemon WJ, Zhao WD, Sears R, Zhuo D, Wang JP, Yang HY, Baer T, Stredney D, Spitzner J, Stutz A, Krahe R, Yuan B (2001) A draft annotation and overview of the human genome. Genome Biol 2:RESEARCH0025

Wutz A, Smrzka OW, Schweifer N, Schellander K, Wagner EF, Barlow DP (1997) Imprinted expression of the Igf2r gene depends on an intronic CpG island. Nature 389:745–749

Wutz A, Rasmussen TP, Jaenisch R (2002) Chromosomal silencing and localization are mediated by different domains of Xist RNA. Nat Genet 30:167–174

Yang Z, Zhu Q, Luo K, Zhuo Q (2001) The 7SK small nuclear RNA inhibits the CDK9/cyclin T1 kinase to control transcription. Nature 414:317–322

Yano Y, Saito R, Yoshida N, Yoshiki A, Wynshaw-Boris A, Tomita M, Hirotsune S (2004) A new role for expressed pseudogenes as ncRNA: regulation of mRNA stability of its homologous coding gene. J Mol Med 82:414–422

Yekta S, Shih IH, Bartel DP (2004) MicroRNA-directed cleavage of HOXB8 mRNA. Science 304:594–596

Yelin R, Dahary D, Sorek R, Levanon EY, Goldstein O, Shoshan A, Diber A, Biton S, Tamir Y, Khosravi R, Nemzer S, Pinner E, Walach S, Bernstein J, Savitsky K, Rotman G (2003) Widespread occurrence of antisense transcription in the human genome. Nat Biotechnol 21:379–386

Yik JH, Chen R, Nishimura R, Jennings JL, Link AJ, Zhou Q (2003) Inhibition of P-TEFb (CDK9/Cyclin T) kinase and RNA polymerase II transcription by the coordinated actions of HEXIM1 and 7SK snRNA. Mol Cell 12:971–982

Yik JH, Chen R, Pezda AC, Samford CS, Zhou Q (2004) A human immunodeficiency virus type 1 Tat-like arginine-rich RNA-binding domain is essential for HEXIM1 to inhibit RNA polymerase II transcription through 7SK snRNA-mediated inactivation of P-TEFb. Mol Cell Biol 24:5094–5105

Zalfa F, Giorgi M, Primerano B, Moro A, Di Penta A, Reis S, Oostra B, Bagni C (2003) The fragile X syndrome protein FMRP associates with BC1 RNA and regulates the translation of specific mRNAs at synapses. Cell 112:317–327

Zhang Z, Harrison PM, Liu L, Gerstein M (2003) Millions of years of evolution preserved: a comprehensive catalogue of the processed pseudogenes in the human genome. Genome Res 13:2541–2558

Zwart R, Sleutels F, Wutz A, Schinkel AH, Barlow DP (2001) Bidirectional action of the Igf2r imprint control element on upstream and downstream imprinted genes. Genes Dev 15:2361–2366

Aminoglycoside Interactions with RNAs and Nucleases

L. A. Kirsebom[1] (✉) · A. Virtanen[1] · N. E. Mikkelsen[2]

[1]Department of Cell and Molecular Biology, Biomedical Center, Uppsala University, Box 596, Uppsala, Sweden
Leif.Kirsebom@icm.uu.se

[2]Department of Molecular Biology, Biomedical Center, The Swedish Agricultural University, Box 590, Uppsala, Sweden

1	Introduction	74
1.1	Aminoglycosides	75
1.2	Resistance Toward Aminoglycosides	75
1.3	Properties and Functions of RNA	79
2	Small Ligands and RNA	80
2.1	RNA, Metal Ions and Small Molecules	80
2.2	Aminoglycoside Binding to RNA and Displacement of Divalent Metal Ions	81
2.3	RNA as a Drug Target	85
2.3.1	tRNA	86
2.3.2	Small Stable RNAs	86
2.3.3	Other RNA Molecules Interacting with Aminoglycosides	88
3	Aminoglycoside Inhibition of Metalloenzymes	89
4	Concluding Remarks and Future Aspects	90
	References	91

Abstract One of the major challenges in medicine today is the development of new antibiotics as well as effective antiviral agents. The well-known aminoglycosides interact and interfere with the function of several noncoding RNAs, among which ribosomal RNAs (rRNAs) are the best studied. Aminoglycosides are also known to interact with proteins such as ribonucleases. Here we review our current understanding of the interaction between aminoglycosides and RNA. Moreover, we discuss briefly mechanisms behind the inactivation of aminoglycosides, a major concern due to the increasing appearance of multiresistant bacterial strains. Taken together, the general knowledge about aminoglycoside and RNA interaction is of utmost importance in the process of identifying/developing the next generation or new classes of antibiotics. In this perspective, previously unrecognized as well as known noncoding RNAs, apart from rRNA, are promising targets to explore.

Keywords RNA · Aminoglycosides · Metal ions · Small ligands · Antibiotics

1
Introduction

Until relatively recently it was believed that the long struggle for control over infectious disease was almost over. Smallpox was eradicated 1980, vaccine programs were in place to protect the world's children against major killer diseases, and a variety of antibiotic drugs were effectively suppressing countless microbial infections. However, cautious optimism has been overtaken by a fatal complacency that is costing millions of lives, and threatening global socioeconomic development. According to the World Health Report 1996 by the WHO, infectious diseases are still the leading cause of death in the world, killing at least 17 million people every year (http://www.who.int/whr/1996/en/). Diseases such as tuberculosis and malaria once believed to be under control are re-emerging with renewed ferocity. Another major challenge today is the developing resistance to antibiotics, and some infections are virtually untreatable due to the occurrence of multiresistant bacteria (see, for example, Davies 1994; Davies and Wright 1997). Other important aspects are the appearance of "new" infectious agents such as the severe acute respiratory syndrome (SARS) virus, the geographical allocation of infectious agents due to, e.g., increased traveling, and the use of medications that suppress the immune system. Thus, far from being over, the struggle to control infectious diseases has become increasingly difficult, and this situation has resulted in increased costs for healthcare for the society worldwide. Consequently, one of the major challenges in medicine today is the development of new antibiotics as well as effective antiviral agents.

The plethora of properties and functions associated with RNA molecules, i.e., non-coding RNA (ncRNA), has led to the realization that RNA molecules frequently are associated with the development and progression of diseases: genetic disorders, tumor progression, autoimmune diseases (Sullenger and Gilboa 2002). Numerous RNA molecules are also essential for the growth of microbial pathogens (e.g., bacteria, virus, parasites, etc.) and thereby are essential for the progression of infectious diseases (Gottesman 2004). Thus, RNA is indeed a potential drug target and this is witnessed by the fact aminoglycosides interact with RNA and interfere with its function (e.g., von Ahsen et al. 1991 reviewed in Davies et al. 1993).

In this chapter we will review what is currently known about the interaction between aminoglycosides and RNA. However, first we will briefly give an overview about aminoglycosides, resistance against aminoglycosides, and the interaction between RNA and metal ions and other small ligands. Finally, we will discuss the interaction between aminoglycosides and protein enzymes that depend on metal ions for activity followed by a brief outline of future perspectives.

1.1
Aminoglycosides

Aminoglycosides are secondary metabolites that are produced and secreted by the producer to ensure a growth advantage in relation to its neighbors (Davies 1994; Davies and Wright, 1997; Zembower et al. 1998; for further information about classification and biosynthesis of aminoglycosides and other antibiotics, we refer the reader to a recent and excellent book: Walsh 2003). Streptomycin was identified and isolated as early as 1944, and since then many other aminoglycosides have been identified. Also, semisynthetic aminoglycosides such as amikacin and tobramycin have been generated. Aminoglycosides show predictable pharmacokinetics and have been used over the years in the clinic for the treatment of infections caused by both Gram negatives and positives including *Mycobacterium tuberculosis*. In many cases, aminoglycosides work in synergy with other antibiotics. Although many of them are very potent drugs, they are also associated with high toxicity as exemplified by neomycin B. It is well established that aminoglycoside treatment is associated with nephro- and ototoxicity, where the latter is irreversible (Mingeot-Leclercq and Tulkens 1999; Hutchin and Cortopassi 1994; Begg and Barclay 1995). Noteworthy is that chemical approaches have had little effect addressing these toxicity-associated problems. The aminoglycosides are divided into two main classes, the 2-deoxystreptamine-containing and streptomycin antibiotics. The former class includes neomycin B and kanamycin A and B, while the latter is exemplified by streptomycin (Fig. 1). The 2-deoxystreptamine class is further divided into 4,6-disubstituted deoxystreptamine (e.g., kanamycin A and B) and 4,5-disubstituted deoxystreptamine (e.g., neomycin B and paromomycin).

Antibiotics such as aminoglycosides are mainly produced by bacteria, and in the case of aminoglycosides, the main producers are found among the actinomycetes. These carbohydrate antibiotics are also referred to as aminocyclitol, and for their synthesis the activity of a large number of gene products are required. In the case of streptomycin, approximately 30 genes in *Streptomyces griseus* are turned on in response to changes in the environment such as changes in nutrient supply or stationary growth and the outcome is secretion of streptomycin. Thus, antibiotic-producing bacteria have put a large investment into their production.

1.2
Resistance Toward Aminoglycosides

Resistance toward naturally occurring aminoglycosides is an essential property for the bacteria that synthesizes a particular aminoglycoside. This simple fact is probably a key reason why antibiotic resistance has become such a growing medical problem. Here we will briefly review some aspects of the mechanisms behind aminoglycoside resistance in bacteria. For a more extensive discussion

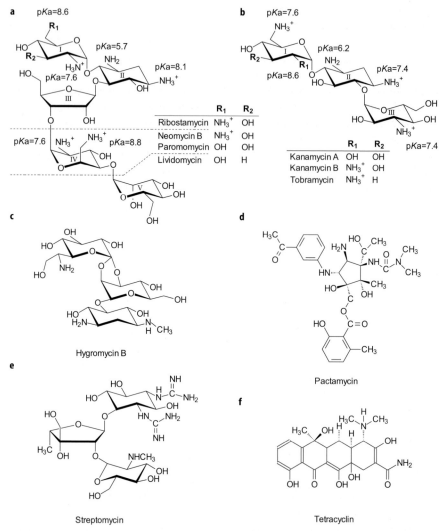

Fig. 1a–f Molecular structures of antibiotics. **a** The neomycin family, **b** kanamycin family, **c** hygromycin B, and **e** streptomycin are all representatives of the aminoglycoside family. Positions where they differ are indicated with R_1 and R_2 and pK_a values for the ammonium groups are indicated. The structures of two other RNA-binding antibiotics that are discussed here are also illustrated: **d** pactamycin and **f** tetracyclin

of the topic, we refer to several recent and excellent reviews (see for example Walsh 2003; Mingeot-Leclercq et al. 1999; Kotra et al. 2000; Dessen et al. 2001; Stewart and Costerton 2001).

Streptomycin is active as a free substance, and in order to ensure that the producer, i.e., *S. griseus*, does not kill itself during synthesis, streptomycin does not become an active substance until the secretion process, during which it is activated via two chemical reactions. Thus, natural systems exist that modify aminoglycosides, as well as other naturally occurring antibiotics, which lead to either activation or inactivation of the antibiotic. With respect to the resistance problem, three main classes of enzymes are involved in chemical and covalent modification of the amino and hydroxyl groups of aminoglycosides: *O*-phosphotransferases, O-nucleotidyltransferases and *N*-acetyltransferases. *O*-Phosphotransferases, also referred to as APH, use ATP as phosphate donor and modify specific hydroxyl groups on the aminoglycoside. *O*-Nucleotidyltransferases (ANT) also use ATP as a donor, resulting in adenylation of hydroxyl groups. The *N*-acetyltransferases (AAC) use acetyl-CoA as donor and modify the amino groups. The genes encoding these enzymes are often carried by transposable genetic elements or plasmids. Within each class, several aminoglycoside-modifying enzymes with different regiospecificities have been identified: Seven APHs, four ANTs, and four AACs are currently known. The crystal structures of some enzymes have been solved, and this has given mechanistic information and generated strategies of how to circumvent the resistance problem. The binding pocket encompassing kanamycin A and the active site of kanamycin nucleotidyltransferase is illustrated in Fig. 2.

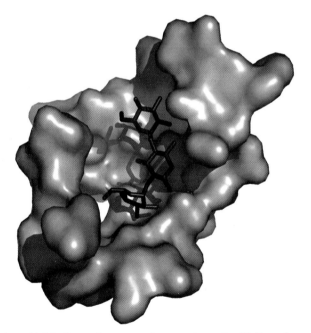

Fig. 2 Aminoglycoside binding to the enzyme kanamycin nucleotidyltransferase. The surface representation shows the interacting region within a radius of 10 Å surrounding the bound kanamycin A (as *stick model*). The figure was made using the molecular graphics program Pymol (DeLano 2002). The structure is according to Pedersen et al. (1995). PDB code 1KNY

Another strategy used by microorganisms that results in resistance (or increased tolerance) is by modification of the targets. In the context of RNA, this means modification of the base or the ribose as well as replacing the entire nucleotide with another nucleotide, i.e., introducing a mutation. For example, *Micromonospora purpurea*, which produces gentamicin, protects itself by methylation of its 16 S ribosomal (r)RNA (Thompson et al. 1985). Also, eukaryotic ribosomes show a decreased affinity ($\geq$tenfold; Table 1) toward aminoglycosides due to substitution of A1408, with G resulting in a G1408/A1493 base pair not present in the bacterial rRNA. In addition, the eukaryotic ribosome lacks the C1409/G1491 base pair (Recht et al. 1998; Vincens and Westhof 2001, 2002, 2003; see also Walter et al. 1999).

Table 1 Summary of apparent inhibition constants for a selected number of aminoglycosides inhibiting different RNA activities (concentrations are given in micromolars). Given K_i values are defined as the concentration resulting in 50% inhibition, with the exception of values that are appK_i[#] values and calculated K_i values[##]

RNA	Aminoglycosides				Reference
	Neomycin B	Paromomycin	Kanamycin A	Tobramycin	
sunY td group I intron	1.3[#]	100	≫1,000	nd	von Ahsen et al. 1992
Hammerhead	13.5				Stage et al. 1995
Hairpin	190	600	nd	nd	Earnshaw and Gait 1998
E. coli RNase P RNA −C5 protein	35	190	nd	nd	Mikkelsen et al. 1999
E. coli RNase P RNA +C5 protein	60	nd	nd	nd	Mikkelsen et al. 1999
Human RNase P	≥600	nd	nd	nd	Eubank et al. 2002
Charging of *E. coli* tRNA[Phe]	300	nd	nd	nd	Mikkelsen et al. 2001
Charging of yeast tRNA[Asp]				0.036[#]	Walter et al. 2002
E. coli tmRNA	70	225	1,400	1,600	Corvaisier et al. 2002
Genomic HDV RNA	28[#]	1,000	1,000	nd	Rogers et al. 1996
A-site (bacterial) 16S rRNA	nd	0.11	nd	2	Griffey et al. 1999
A-site (eukaryotic) 18S rRNA	nd	>20	nd	1.4	Griffey et al. 1999
Affinity to[1] RRE RNA[##]	0.9	nd	100	10	Luedtke et al. 2003

nd, not determined. [1]Measured as Rev peptide displacement from the HIV-1 Rev response element (RRE)

Antibiotic resistance can also be achieved by active transport or efflux of the antibiotic out of the bacteria, resulting in a concentration that is too low to cause harm to the bacteria. This strategy to achieve resistance is very common among bacteria, and for further discussion on this topic we refer to Walsh (2003). Finally, we would like to mention that antibiotic resistance is also manifested as a result of the formation of bacterial biofilms. Here the mechanisms of resistance are different in relation to the discussion above, and we refer to a recent review covering this topic (Stewart and Costerton 2001).

1.3
Properties and Functions of RNA

It has become evident during the past few years that RNA plays a much more vital role in all living organisms than initially anticipated, when it was believed that the only role of RNA was to physically convey genetic information stored in DNA to functionally acting proteins. Today it is clear that RNA, besides being the physical link between DNA and protein, plays several other key roles, i.e., structural, functional, regulatory, and informational. A large number of ncRNAs have recently been identified, and today we know that such ncRNA molecules have several fundamental functions essential for cell growth, survival, and development. The functions that RNA carries out or participates in include RNA processing and protein translation, acting as structural scaffolds, transporters, gene regulators, and biocatalysts. In fact, most likely RNA and not protein constitutes the active center where peptide bond formation takes place (Ban et al. 2000; Nissen et al. 2000). Moreover, the rapid increase in available genome sequences has permitted researchers to search for and analyze regulatory RNAs, which used to be impossible. In the last 2 years, several novel and biologically important small RNAs have been discovered in a variety of organisms from bacteria to mammalian cells. In bacteria, these small RNAs are sometimes referred to as sRNA, while the novel small RNAs in eukaryotes include, for example, micro (mi)RNA and small interfering (si)RNA. In eukaryotes the large collection of novel small and ncRNAs have been demonstrated to be involved in gene silencing via RNA and to play essential roles in controlling all steps of gene expression, including transcription, chromatin modification, epigenetic memory, and alternative splicing (Mattick 2003). The recognition that many of the recently discovered ncRNAs in both bacteria and eukaryotes possibly act as regulators of gene expression has led to the initiation of vigorously pursued research efforts worldwide. Moreover, we note that the role of small RNAs in immunity was recently discussed by McManus (2004).

2
Small Ligands and RNA

2.1
RNA, Metal Ions and Small Molecules

Our increased knowledge of RNA function/structure has lead to the realization that divalent metal ions such as Mg^{2+} play crucial roles for RNA function, being both structurally and/or catalytically important. On average, there is one Mg^{2+}-ion bound per 3–4 nucleotides of the negatively charged RNA. It has been demonstrated that binding of Mg^{2+} to RNA is important for RNA folding, RNA–RNA interactions, RNA–protein interactions, and various catalytic processes such as cleavage of RNA, transfer (t)RNA charging, and codon–anticodon interaction (Gesteland et al. 1999). Other biologically relevant divalent metal ions, such as Ca^{2+} and Mn^{2+}, also bind to RNA. The former binds with approximately the same affinity to RNA as Mg^{2+}, while the latter binds 3–4 times stronger (Brännvall et al. 2001). Addition of, for example, Ca^{2+} to RNase P RNA cleavage and $tRNA^{Ala}$ alanyl-tRNA synthetase charging reactions result in reduced activities (Brännvall and Kirsebom 2001 and references therein). This raises the interesting possibility that biocatalysts that depend on RNA for activity are up- or down-regulated depending on the intracellular concentrations of Mg^{2+} and Ca^{2+}. For example, the flux of Ca^{2+} is perturbed in tumor cells (Berridge et al. 1998) and in bacterially infected cells (Uhlen et al. 2000). Another possibility is that binding of different metal ions influences the interaction between RNA and other cellular factors such as proteins (Brännvall et al. 2004).

Addition of Mn^{2+} influences the accuracy of, for example, RNA-mediated cleavage of RNA and protein-mediated cleavage of DNA (Brännvall and Kirsebom 2001; Hsu and Berg 1978). In this perspective, it is interesting to note that certain bacteria, for example *Borrelia burgdorferi*, an intracellular parasite causing Lyme disease (Posey and Gherardini 2000) and certain *Lactobacillus* spp. (Archibald and Duong 1984) have elevated intracellular concentration of Mn^{2+}. Clearly this emphasizes the importance of understanding the interaction between RNA and metal ions from a biological perspective, as well as investigating in detail the biological consequences as a result of the interaction between RNA and different metal ions. Noteworthy in this context is the recent finding that other ligands, such as vitamin B12 and thiamine, interact with specific structural motifs of RNA and thereby influence the expression of specific genes. These structural motifs are referred to as riboswitches, i.e., structural domains in the non-coding regions of mRNAs, acting as metabolite-responsive genetic switches (see for example, Nahvi et al. 2002; Winkler et al. 2002; Mandal et al. 2003; Vitreschak et al. 2004). Moreover, tRNA has been demonstrated to be involved in regulation of many "amino-acid-related" genes (e.g., amino acyl tRNA synthetase genes) in gram-positive bacteria (for a recent review see Grundy and Henkin 2004).

An interesting aspect regarding the function of small RNAs is their potential role for virulence. In fact, recently, an RNA involved in regulation of the expression of an *E. coli* toxin gene was identified (Vogel et al. 2004), and we foresee many others yet to be identified. For the interaction between RNA and other small ligands, we refer to Hermann (2003). Thus, small ligands of various classes, from simple metal ions to more complex organic compounds, interfere with and regulate RNA function.

Taken together, an increased knowledge of the way metal ions and other small ligands such as aminoglycosides (see the following sections) interact with RNAs and carry out their functions is fundamental and necessary in order to understand the mechanism of action of the different RNA molecules that exist in a cell. This knowledge can be exploited to identify small ligands that bind a given RNA specifically and interfere with its function. Needless to say, these ligands can subsequently be used as leads to develop novel drugs/antibiotics.

2.2
Aminoglycoside Binding to RNA and Displacement of Divalent Metal Ions

The bacterial ribosome is a primary target for various antibiotics such as the aminoglycosides, which bind to 16 S rRNA in the A-site and interfere with the decoding process. Aminoglycosides also bind to other RNAs and interfere with their functions, for example: inhibition of several ribozymes including RNase P RNA; inhibition of tRNA charging and inhibition of splicing (Table 2; see Sect. 2.3). Currently, the understanding of how aminoglycosides interact with RNA and interfere with its function has become an increasingly important research field that is exploited worldwide to identify novel aminoglycoside-based antibiotics.

Structural studies and biochemical probing analysis have been used to obtain information regarding the interaction between RNA and various aminoglycosides (for reviews see, for example, Kotra et al. 2000; Walter et al. 1999; Vicens and Westhof 2003; Yonath and Bashan 2004). Here we will only highlight some aspects of the RNA-aminoglycoside interaction based primarily on the high-resolution RNA-aminoglycoside structure complexes that recently have been reported (e.g., Vicens and Westhof 2001, 2002, 2003b; Mikkelsen et al. 2001; Brodersen et al. 2000; Carter et al. 2000; Ogle et al. 2000; Schlünzen et al. 2000; Piolleti et al. 2001 for binding of several other antibiotics to the ribosome).

Crystal studies and nuclear magnetic resonance (NMR) spectroscopy of different complexes containing either 4,5- (e.g., tobramycin and geneticin) or 4,6-disubstituted (e.g., paromomycin and neomycin B) (see, for example, Fourmy et al. 1996, 1998a,b) have revealed that aminoglycosides often bind in the deep groove: in the A-site of the ribosomal 16S rRNA (Fig. 3) and below

Table 2 A compilation of novel RNA drug targets as indicated

RNA	Function	Bacteria	Essential	Human	Structural differences	Potential drug target
Group I intron	Gene regulation expression	In certain bacteria		No	Not relevant	Possibly
RNase P RNA	tRNA biosynthesis	Yes	Yes	Yes	Yes	Yes
tRNA charging	Translation	Yes	Yes	Yes	Yes	Yes
tmRNA	Scavenging	Yes	No[##]	No	Not Relevant	Yes
4.5S RNA	Protein export secretion	Yes	Yes	Yes	Yes[#]	Possibly
Spot 42 RNA	Gene regulation	Yes	No	No	Not relevant	Not known[###]
6S rRNA	Inhibitor of RNA polymerase	Yes	No	No	Not relevant	Not known

[#]Not including low GC gram-positive bacteria [##]Has been demonstrated to be essential in *Neisseria* spp. (see text) but not in *E. coli* under laboratory conditions [###]Might be a model system to use to exploit antisense RNA as a drug target

the D-loop in yeast tRNA^{Phe} (Fig. 4). Based on these and biochemical studies (see Sect. 2.3), it is clear that (1) electrostatic interactions and (2) hydrogen bond formation directly between RNA residues/backbone and amino/hydroxyl groups on the aminoglycoside and water-mediated interactions are crucial to achieve high affinity. In the case of binding to the 16S rRNA A-site, the dissociation constant has been determined to be as low as 1.5 μM for both 4,5- and 4,6-disubstituted aminoglycosides (Hyun Ryu and Rando 2001) while for binding to yeast tRNA^{Phe} the affinity is higher (Table 1; Mikkelsen et al. 2001). However, increasing pH above the physiological level results in reduced binding and no inhibition due to deprotonation of amino groups on the aminoglycoside (Fig. 5). The latter observation emphasizes experimentally the importance of electrostatic interactions between the positively charged aminoglycoside and the negatively charged RNA. Moreover, it is the functional groups of the neamine domain (Fig. 1), which is shared between these aminoglycosides, that play a decisive role in the interaction with RNA.

Studying the interaction between yeast tRNA^{Phe} and aminoglycoside demonstrated that aminoglycoside binding to the RNA resulted in displacement of a divalent metal ion, and hence it was suggested that aminoglycosides could be considered as "metal mimics" (Fig. 4). That aminoglycoside binding could

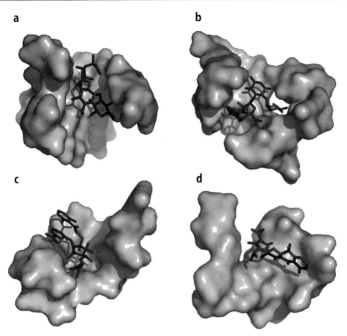

Fig. 3a–d Specific binding pockets for different antibiotics in *Thermus thermophilus* 30S ribosomal subunit. All surface representations show the interacting region within a radius of 10 Å surrounding the bound antibiotic. **a** Hygromycin B (Brodersen et al. 2000—PDB file 1HNZ). **b** Paromomycin (Ogle et al. 2001—PDB file 1IBK). **c** Tetracyclin (Brodersen et al. 2000—PDB file 1HNW). **d** Pactamycin (Brodersen et al. 2000—PDB file 1HNX). The figures where made using the molecular graphics program Pymol (DeLano 2002)

result in displacement of functionally important metal ions was originally suggested by Hermann and Westhof (1998), and the structural analysis as well as biochemical studies of aminoglycoside binding to various RNA is in keeping with this (see the following section). In fact, displacement of Mg^{2+} also occurs when paromomycin binds to the 30S ribosomal subunit (Fig. 4), providing further evidence that divalent metal ion and aminoglycoside binding sites in RNA overlap (Carter et al. 2000). This is in keeping with recent studies using small RNA model molecules representing the 16S rRNA A-site (Mikkelsen et al. 2001; Summers et al. 2002). Whether Mg^{2+} plays a direct functional role in the decoding process is not clear, but Porschke and coworkers have discussed the role of Mg^{2+} in the decoding process (e.g., Labuda et al. 1984). Nonetheless, from the structural studies of the 16S rRNA A-site, it is evident that addition of aminoglycoside stabilizes the bulging conformation of two functionally important adenines, A1492 and A1493. These adenines play an essential role in the decoding process during translation. Thus, this gives a structural reason as to why addition of aminoglycosides results in increased errors during translation, as has been shown be Ehrenberg and coworkers, among others (for a review see

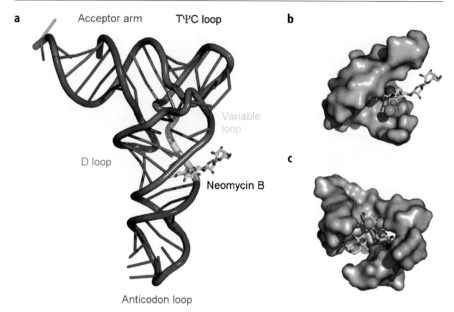

Fig. 4a–c Displacement of divalent metal-ions by aminoglycosides. **a** 3D ribbon-and-stick model of neomycin B overlapping the metal ion-binding pocket in the deep groove below the D loop in yeast tRNA^{Phe} (Mikkelsen et al. 2001—PDB file 1I9 V). Individual regions of the molecule colored according to Jovine et al. (2000). **b** Divalent metal ions from yeast tRNA^{Phe}-Pb^{2+} (and Mg^{2+}) structures superimposed on the yeast tRNA^{Phe}-neomycin B structure (Brown et al. 1983—PDB file 1TN1; Brown et al. 1985—PDB file 1TN2; Jovine et al. 2000—PDB file 1EVV; Shi and Moore 2000—PDB file 1EHZ; Mikkelsen et al. 2001). Yeast tRNA^{Phe} represented as a surface model surrounding the bound neomycin B at a 10 Å radius. Magnesium and lead ions are colored in *cyan* and *magenta*, respectively. **c** *Thermus thermophilus* 30S-Paromomycin structure superimposed with overlapping divalent metal ion binding sites (Ogle et al. 2000—PDB codes 1IBK and 1IBM). Surface representation of *Thermus thermophilus* 30S surrounding the bound paromomycin and magnesium ions 433 and 437 at a 10 Å radius. The figures where made using the molecular graphics program Pymol (DeLano 2002) and in generating the tRNA stick model, the program Nuccyl (Jovine 2003; www.mssm.edu/students/jovinl02/research/nuccyl.html) was used

Schroeder et al. 2000). Moreover, both crystal and NMR studies have revealed structural information about resistance against various aminoglycosides. This information is crucial for the work aiming at identifying the next generation of aminoglycosides.

In this context, we would like to mention that studies where aminoglycoside RNA aptamers have been studied have also provided valuable information and we refer to various articles covering this aspect of RNA-aminoglycoside interaction (apart from the references already mentioned, see Patel et al. 1997; Herrmann and Patel 2000).

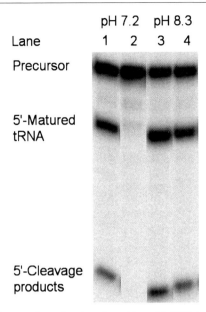

Fig. 5 The electrostatic interaction of aminoglycosides with RNA is pH dependent. Monitoring M1 RNA cleavage activity of a tRNA precursor, pSu3, at two different pH conditions at 37 °C. *Lanes 1* and *3* show M1 RNA cleavage activity in the absence of Neomycin B whereas *lanes 2* and *4* show cleavage in the presence of 1 mM neomycin B. The time of cleavage in all cases was 2 min. Concentration of M1 RNA and pSu3 were 82 nM and 5.2 nM, respectively (for details see Mikkelsen et al. (1999))

2.3
RNA as a Drug Target

Given that RNA is essential for bacterial growth, and the possible importance of RNA for virulence makes RNA a suitable and promising drug target. This potential is witnessed by the fact that many antibiotics targeting the ribosome have been and are still used in the clinic today. As discussed above, high resolution structural studies of the bacterial ribosome in complex with a number of different antibiotics have revealed that drugs bind with high specificity to bacterial rRNA (16S and 23S) and tRNA. For example, there are drugs that interact with or close to the decoding region on the 30S subunit as well as the peptidyl transfer center on the 50S subunit (Fig. 3). Although the ribosome is the primary RNA target known today, other RNA molecules, both naturally occurring RNA and in vitro-selected RNA aptamers, have been demonstrated to interact with antibiotics such as aminoglycosides as well as other classes of drugs (see above). In this section, we will discuss other RNAs, primarily bacterial RNAs that have been demonstrated to interact with aminoglycosides and as such are potential drug targets when searching for novel antibiotics. In addition, we will mention some other functional RNAs that potentially could be targeted and/or used in the process to identify novel lead compounds. These RNA molecules/targets are summarized in Table 2.

To function as a suitable drug target the overall criteria is that the targeted RNA has to be unique for the infectious agent (e.g., bacteria, virus, or fungi). Alternatively, the structural differences comparing the targeted RNA in the infectious agent with that of the host homolog are large enough to result in higher tolerance toward the drug for the host RNA.

2.3.1
tRNA

tRNA plays a central role in protein synthesis by bringing the amino acid to the ribosome. Prior to this, the various tRNAs have to be charged with the correct amino acid, and the tRNA structure is essential in this process. During the charging process, different aminoacyl-tRNA synthetases recognize and charge the respective cognate tRNA. Thus, tRNA aminoacylation is a cellular process that can be targeted when searching for novel bacterial drugs. Actually, one anti-infective drug based on pseudomonic acid is directed against and inhibits isoleucyl-tRNA synthetase (IleRS) derived from several gram-negative and gram-positive bacteria without affecting human IleRS (Sutherland et al. 1985; Ward and Campoli-Richards 1986). Besides pseudomonic acid, it has also been demonstrated that various aminoglycosides interact with tRNA^{Phe} and tRNA^{Asp} and inhibit aminoacylation. In the case of *E. coli*-tRNA^{Phe}, the inhibition is most likely due to an overlap between the aminoglycoside binding site and the positive identity elements that are essential for aminoacylation by phenylalanyl-tRNA synthetase (Mikkelsen et al. 2001) while inhibition of tRNA^{Asp} is a consequence of an aminoglycoside-induced destabilization of the L-shape tRNA structure (Walter et al. 2002).

2.3.2
Small Stable RNAs

In bacteria, several small stable RNAs, apart from tRNA, have been identified; for example, RNase P RNA, transfer messenger (tm)RNA, 6S RNA, 4.5S RNA, and Spot 42 RNA [see Table 2; for a recent review regarding small RNAs in *E. coli* see Gottesman (2004)]. Among these, only RNase P RNA and tmRNA are known to be inhibited by aminoglycosides. Beside these two stable RNAs, it is known that aminoglycosides both bind and inhibit the function of a large variety of different small RNA molecules or structures, including group I introns and viral RNAs. Thus, it is conceivable that also other stable RNAs interact with aminoglycosides or similar compounds. Therefore, it will be of interest to elucidate if aminoglycosides inhibit the function of, for example, small novel regulatory RNAs that recently have been identified in various bacteria.

2.3.2.1
RNase P

RNase P is ubiquitous and responsible for generating the 5' end of almost all tRNA in bacteria, archaea, and eukarya. In bacteria, RNase P consists of the catalytic RNA subunit and a basic protein, the C5 protein, in a 1:1 ratio (Guerrier-Takada et al. 1983; Vioque et al. 1988; see also Fang et al. 2001). This endoribonuclease plays an essential role in the processing of tRNA as well as other RNA transcripts, e.g., mRNA (Altman and Kirsebom 1999). Based on secondary structure comparison, bacterial and RNase P RNA of mammalian origin show distinct structural differences (Kirsebom and Virtanen 2001 and references therein). Moreover, human RNase P consists of at least nine protein subunits and is therefore more complex in its protein composition relative to bacterial RNase P, which only has one protein component (Kovrigina et al. 2003). These differences make bacterial RNase P with its catalytic RNA subunit a suitable target. So far bacterial RNase P is not known to be targeted by any of the antibiotics used in the clinic today. However, as discussed above, aminoglycosides do indeed interact with and inhibit RNase P RNA from various bacterial origins and, in the case of *E. coli* RNase P, even in the presence of the C5 protein (Table 1; Mikkelsen et al. 1999; Eubank et al. 2002). Based on Pb^{2+}-induced cleavage studies, it was suggested that addition of aminoglycosides interferes with the binding of functionally important metal ions. Gopalan and coworkers showed that aminoglycosides do not inhibit the action of human RNase P to the same extent as in the case of bacterial RNase P (Eubank et al. 2002). Taken together, this clearly indicates that RNase P is a suitable drug target.

2.3.2.2
Transfer Messenger RNA

tmRNA is of approximately the same size as RNase P RNA. Apirion and coworkers observed tmRNA as early as 1978 and they referred to it as 10S A RNA (*E. coli* RNase P RNA was referred to as 10S B RNA; Gegenheimer and Apirion 1981). Its function was not apparent until 1996 when it was shown that it rescued stalled ribosomes by acting as a transfer-messenger (tm)RNA. This results in the synthesis of a tag at the C-terminus on the growing polypeptide that subsequently is recognized by proteases resulting in degradation of the tagged polypeptide. Thus, tmRNA plays a role in degradation of unwanted and incorrect polypeptides (Keiler et al. 1996; Withey and Friedman 2003). tmRNA is apparently unique to bacteria since no homolog in mammalians has been identified. Its presence in *E. coli* is not essential, since strains lacking tmRNA are viable under various laboratory conditions, although it cannot be excluded that tmRNA is essential under more natural conditions (or its presence gives a growth advantage). It should be noted that in *Neisseria gonorrhoeae* tm-

RNA is essential for growth (Huang et al. 2000). This raises the possibility that tmRNA may also be essential for growth of other bacteria. Felden and coworkers demonstrated that aminoglycosides bind to tmRNA and interfere with aminoacylation of tmRNA. The inhibition of aminoacylation is not a result of binding of the aminoglycoside to the aminoacyl-acceptor stem where the major determinant for alanyl-tRNA-synthetase, rather the data, suggested that aminoglycoside binding perturbed the conformation of the aminoacyl-acceptor stem of tmRNA (Corvaisier et al. 2002). These data clearly suggest that tmRNA is a potential and suitable drug target, if not in the case of all bacteria, at least to some.

2.3.2.3
Spot 42 RNA, 6S RNA, and 4.5S RNA

Beside RNase P RNA and tmRNA, no other "classical" stable non-coding RNAs, i.e., 4.5S RNA, 6S RNA, and Spot 42 RNA, have so far, to our knowledge, been demonstrated to interact with aminoglycosides. However, given the generality for the RNA/aminoglycoside interaction, it is expected that these RNAs should also interact with aminoglycosides. Thus, it would definitely be of interest to investigate whether aminoglycosides indeed bind to these and inhibit their cellular function. Noteworthy is that the 4.5S RNA is involved in secretion and is essential for viability. Moreover, comparing the structures of bacterial (excluding low GC content gram-positive bacteria e.g., *Mycoplasma* spp.) 4.5S RNA and the corresponding human signal recognition particle (SRP) RNA reveals structural differences that clearly would be possible to explore in the quest for novel drug targets/candidates. Likewise, the 6S RNA is an interesting candidate as a drug target, since a homolog of 6S RNA has not been identified in mammals. The bacterial 6S RNA functions as an RNA regulator that inhibits RNA polymerase, and it is up-regulated during the stationary phase of bacterial growth (Wassarman and Storz 2000).

The Spot 42 RNA (Ikemura and Dahlberg 1973) functions in sugar metabolism where it is involved in translational regulation of *galK* (Møller et al. 2002). Spot 42 RNA represent a class of RNA involved in translational regulation by an antisense RNA mechanism, and as such might be of interest as a model system to identify inhibitors that interfere with anti-sense RNA. For a recent review of other small RNAs that have been identified in bacteria, we recommend Gottesman (2004).

2.3.3
Other RNA Molecules Interacting with Aminoglycosides

Other RNA-based activities inhibited by aminoglycosides include both a number of ribozymes and viral RNAs. Among the ribozymes, the group I intron is one of the best-studied examples. In their pioneering work, Schroeder and coworkers (von Ahsen et al. 1991) demonstrated that aminoglycosides bind and inhibit group I intron self-splicing (for a review see Schroeder et al. 2000). Given the existence of self-splicing RNA group I introns in bacteria, these are indeed potential drug targets to explore. This is also apparent based on the knowledge that group I introns have not been identified in mammalian cells. Other ribozymes known to be inhibited by aminoglycosides are the self-cleaving hammerhead ribozyme, the hairpin ribozyme (Stage et al. 1995; Earnshaw and Gait 1998), and the self-cleaving hepatitis delta virus HDV RNA (Rogers et al. 1996; Chia et al. 1997). In this context, we note that the aminoglycoside streptomycin is apparently a better inhibitor of nuclear pre-mRNA splicing compared to inhibition of the group I intron splicing (Hertweck et al. 2002). Since no other aminoglycosides have been tested for inhibition of nuclear pre-mRNA splicing, it would be of value to study this in more detail. This is not only important in the evaluation of the group I intron as a potential target, but this information is also of significance in order to be able to ensure that the next generation of aminoglycosides is more specific/efficient against chosen bacterial RNA targets.

Aminoglycosides are known to bind to human immunodeficiency virus (HIV) RNA and, in some cases, even to prevent viral replication. In a classic study by Zapp et al. (1993), it was shown that aminoglycosides, in particular neomycin B, bound to the Rev response element (RRE) of HIV-1 and selectively blocked Rev protein binding to the RRE, resulting in inhibition of viral growth. Later, it was shown that aminoglycosides also bind to the HIV Tat-responsive element (TAR) and prevent binding of Tat protein. In this case, both structural and molecular dynamic simulations suggest that aminoglycoside binding induces conformational changes in the RNA and thereby prevent Tat protein binding (Mei et al. 1995; Hermann and Westhof 1999; Faber et al. 2000). These findings emphasize that aminoglycoside also targets viral RNA and influences viral replication by perturbing essential RNA protein interactions. These studies, therefore, highlight the possibility that RNA protein interactions, which are essential for viral growth, serve as potential drug targets to explore for drug development. Of course, any essential ribonucleoprotein complexes present in any infectious agent (e.g., viral, bacterial, parasitic) would serve as a potential target.

3
Aminoglycoside Inhibition of Metalloenzymes

It has been observed that some aminoglycosides inhibit enzymes involved in cleavage and formation of phosphodiester bonds (Lazarus and Kitron, 1973; McDonald and Mamrack, 1995; Woegerbauer et al. 2000). However, the mechanism of inhibition has not been fully understood. Based on the studies performed on aminoglycoside inhibition of RNA function, it has become clear that the RNA-binding property of aminoglycosides is a key reason why they inhibit RNA function. Most likely, as discussed above, the binding results in a disturbed structure of the RNA and/or displacement of functionally important divalent metal ions. To investigate if a related mechanism of inhibition was also plausible for enzymes involved in cleavage and formation of phosphodiester bonds, we investigated if aminoglycosides could interact with and inhibit the polymerizing activity of *E. coli* DNA polymerase I and/or the exoribonuclease activity of human poly(A)-specific ribonuclease (PARN) (Ren et al. 2002). These enzymes encompass active sites that form negatively charged binding pockets that resemble aminoglycoside binding sites in RNA. Moreover, both active sites of these two enzymes depend on and coordinate divalent metal ions. Our studies (Ren et al. 2002) showed that aminoglycosides inhibited the activity of both enzymes and suggested that this was caused by the aminoglycoside distorting the active sites and/or displacing functionally important divalent metal ions. Thus, the property of aminoglycosides to bind to negatively charged binding pockets seems to be a generally applicable property of aminoglycosides. As a matter of fact, this property is even relevant for the binding of kanamycin B to the active site of the bacterial kanamycin-modifying enzyme kanamycin nucleotidyltransferase (Fig. 2), as this active site encompasses a negatively binding pocket and binds divalent metal ions (Pedersen et al. 1995; Sakon et al. 1993). Thus, proteins that depend on metal ions for activity are potential targets for aminoglycosides. Therefore, in the process to develop RNA-specific drugs based on aminoglycosides, it is essential to ensure that these do not interact and interfere with protein function, such as various nucleases/polymerases that depend on metal ions for activity. In addition, these findings identify nucleases and polymerases as potential targets in the process of identifying novel drugs/lead compounds.

4
Concluding Remarks and Future Aspects

Over the last decades, very few new classes of antibiotics have been introduced. The recent increase in the appearance of multiresistant bacteria emphasizes the urgent need for new substances that can be used in the process of identifying novel antibiotics. Here we have discussed the potential of aminoglycosides and related molecules or derivatives thereof as potential starting molecules for the search of novel antibiotics. We have primarily focused on the RNA-binding property of aminoglycosides and discussed some potential RNA molecules that could serve as drug targets.

Aminoglycosides were introduced in the early 1950s and have been an important class of antibiotic used in the clinic since then. As discussed here, functional and structural studies of RNA in the last 15 years have demonstrated that RNA is a primary target for aminoglycosides. Importantly, aminoglycoside binding to RNA results in conformational changes and/or displacement of divalent metal ions. Therefore, the identification of metal ion binding sites can be explored to identify novel aminoglycoside derivatives that bind RNA.

In the process of identifying novel drugs, several laboratories worldwide have used chemistry to modify existing aminoglycosides. For example, such laboratories have (1) used the synthesis of aminoglycoside conjugates with intercalators, such as acridine attached to them, to make dimeric aminoglycosides, (2) replaced existing sugar moieties with different sugar derivatives, or (3) introduced arginine residues at specific positions on the aminoglycoside (Agnelli et al. 2004; Cheng et al. 2001; Luedtke et al. 2003; Litovchick et al. 2001; Sucheck and Shue 2001; Yao et al. 2004). Many of these new aminoglycoside derivatives have been demonstrated to bind various RNAs with improved affinities and to be more resistant toward enzymatic modification. However, a difficult problem to address yet is toxicity. Nonetheless, this demonstrates that it is possible to use aminoglycosides in the search for new and more efficient drugs against a number of various pathogens such as bacteria and viruses.

Acknowledgements We thank our colleagues over the years for a pleasant and stimulating work atmosphere, and Dr. L.W. Riley for comments on the manuscript. This work was supported by the Wallenberg Consortium North, the Strategic Research Foundation, and the Swedish Research Council to L.A.K. and A.V.

References

Agnelli F, Sucheck SJ, Marby KA, Rabuka D, Yao S-L, Sears PS, Liang F-S, Wong C-H (2004) Dimeric aminoglycosides as antibiotics. Angew Chem Int Ed Engl 43:1562–1566

Altman S, Kirsebom LA (1999) Ribonuclease P. In: Gesteland R, Cech T, Atkins J (eds) RNA world II. Cold Spring Harbor Laboratory Press, Cold Spring Harbor, New York, pp 351–380

Ambrose V (2001) MicroRNAs: tiny regulators with great potential (minireview). Cell 107:823–826

Archibald S, Duong M (1984) Manganese acquisition by Lactobacillus plantarum. J Bacteriol 158:1–6

Ban N, Nissen P, Hansen J, Moore PB, Steitz TA (2000) The complete atomic structure of the large ribosomal subunit at 2.4 Å resolution. Science 289:905–920

Begg EJ, Barclay ML (1995) Aminoglycosides—50 years on. Br J Clin Pharmacol 39:597–603

Berridge MJ, Bootman MD, Lipp P (1998) Calcium—a life and death signal. Nature 395:645–648

Brännvall M, Kirsebom LA (2001) Metal ion cooperativity in ribozyme cleavage of RNA. Proc Natl Acad Sci USA 98:12943–12947

Brännvall M, Mikkelesen NE, Kirsebom LA (2001) Monitoring the structure of Escherichia coli RNase P RNA in the presence of various metal ion. Nucleic Acids Res 29:1426–1432

Brännvall M, Kikovska E, Kirsebom LA (2004) Cross talk in RNase P RNA mediated cleavage. Nucleic Acids Res 32:5418–5429

Brodersen DE, Clemons WM Jr, Carter AP, Morgan-Warren RJ, Wimberly BT, Ramakrishnan V (2000) The structural basis for the action of the antibiotics tetracycline, pactamycin, and hygromycin B on the 30S ribosomal subunit. Cell 103:1143–1154

Brown RS, Hingerty BE, Dewan JC, Klug A (1983) Pb(II)-catalysed cleavage of the sugar-phosphate backbone of yeast tRNAPhe—implications for lead toxicity and self-splicing RNA. Nature 303:543–546

Brown RS, Dewan JC, Klug A (1985) Crystallographic and biochemical investigation of the lead(II)-catalyzed hydrolysis of yeast phenylalanine tRNA. Biochemistry 24:4785–4801

Carter AP, Clemons WM, Brodersen DE, Morgan-Warren RJ, Wimberly BT, Ramakrishnan V (2000) Functional insights from the structure of the 30S ribosomal subunit and its interaction with antibiotics. Nature 407:340–348

Cheng AC, Calabro V, Frankel AD (2001) Design of RNA-binding proteins and ligands. Curr Opin Struct Biol 11:478–484

Chia JS, Wu HL, Wang HW, Chen DS, Chen PJ (1997) Inhibition of hepatitis delta virus genomic ribozyme self-cleavage by aminoglycosides. J Biomed Sci 4:208–216

Corvaisier S, Bordeau V, Felden B (2002) Inhibition of transfer messenger RNA aminoacylation and trans-translation by aminoglycoside antibiotics. J Biol Chem 278:14788–14797

Davies J (1994) New pathogens and old resistance genes. Microbiologica 10:9–12

Davies J, Wright GD (1997) Bacterial resistance to aminoglycoside antibiotics. Trends Microbiol 5:234–240

Davies J, von Ahsen U, Schroeder R (1993) Antibiotics and the RNA world: a role for low-molecular-weight effectors in biochemical evolution? In: Gesteland RF, Atkins JF (eds) The RNA world. Cold Spring Harbor Laboratory Press, Cold Spring Harbor, pp 185–204

DeLano WL (2002) The PyMOL molecular graphics system on the World Wide Web: http://www.pymol.org. Cited 10 June 2005

Dessen A, Di Guilmi AM, Vernet T, Dideberg O (2001) Molecular mechanisms of antibiotic resistance in gram-positive pathogens. Curr Drug Targets Infect Disord 1:63–77

Earnshaw DJ, Gait MJ (1998) Hairpin ribozyme cleavage catalysed by aminoglycoside antibiotics and the polyamine spermine in the absence of metal ions. Nucleic Acids Res 26:5551–5561

Eubank TD, Biswas R, Jovanovic M, Litovchick A, Lapidot A, Gopalan V (2002) Inhibition of bacterial RNase P by aminoglycoside-arginine conjugates. FEBS Lett 511:107–112

Faber C, Sticht H, Schweimer K, Rösch P (2000) Structural rearrangements of HIV-1 Tat-responsive RNA upon binding of neomycin B. J Biol Chem 275:20660–20666

Fang XW, Yang XJ, Littrell K, Niranjanakumari S, Thiyagarajan P, Fierke CA, Sosnick TR, Pan T (2001) The Bacillus subtilis RNase P holoenzyme contains two RNase P RNA and two RNase P protein subunits. RNA 7:233–241

Fourmy D, Recht MI, Blanchard SC, Puglisi JD (1996) Structure of the A site of E. coli 16 S rRNA complexed with an aminoglycoside antibiotic. Science 274:1367–1371

Fourmy D, Recht MI, Puglisi JD (1998a) Binding of neomycin-class aminoglycoside antibiotics to the A-site of 16 S rRNA. J Mol Biol 277:347–362

Fourmy D, Yoshizawa S, Puglisi JD (1998b) Paromomycin binding induces a local conformational change in the A-site of 16 S rRNA. J Mol Biol 277:333–345

Gegenheimer P, Apirion D (1981) Processing of prokaryotic ribonucleic acid. Microbiol Rev 45:502–541

Giegé R, Sissler M, Florentz C (1998) Universal rules and idiosyncratic features in tRNA identity. Nucleic Acids Res 26:5017–5035

Gottesman S (2004) The small RNA regulators of Escherichia coli: Roles and mechanisms. Annu Rev Microbiol 58:303–338

Griffey RH, Hofstadler SA, Sannes-Lowery KA, Ecker DJ, Crooke ST (1999) Determinants of aminoglycoside-binding specificity for rRNA by using mass spectrometry. Proc Natl Acad Sci USA 96:10129–10133

Grundy FJ, Henkin TM (2004) Regulation of gene expression by effectors that bind to RNA. Curr Opin Microbiol 7:126–131

Guerrier-Takada C, Gardiner K, Marsh T, Pace N, Altman S (1983) The RNA moiety of ribonuclease P is the catalytic subunit of the enzyme. Cell 35:849–857

Hermann T (2003) Chemical and functional diversity of small molecule ligands for RNA. Biopolymers 70:4–18

Hermann T, Patel DJ (2000) Adaptive recognition by nucleic acid aptamers. Science 287:820–825

Hermann T, Westhof E (1998) Aminoglycoside binding to the hammerhead ribozyme: a general model for the interaction of cationic antibiotics with RNA. J Mol Biol 276:903–912

Hermann T, Westhof E (1999) Docking of cationic antibiotics to negatively charged pockets in RNA folds. J Med Chem 42:1250–1251

Hershberg R, Altuvia S, Margalit H (2003) A survey of small RNA-encoding genes in Escherichia coli. Nucleic Acids Res 31:1813–1820

Hertweck M, Hiller R, Mueller MW (2002) Inhibition of nuclear pre-mRNA splicing by antibiotics in vitro. Eur J Biochem 269:175–183

Hsu M, Berg P (1978) Altering the specificity of restriction endonuclease: effect of replacing Mg^{2+} with Mn^{2+}. Biochemistry 17:131–138

Huang C, Wolfgang MC, Withey J, Kommey M, Friedman DI (2000) Charged tmRNA but not tmRNA-mediated proteolysis is essential for Neisseria gonorrhoeae viability. EMBO J 19:1098–1107

Hutchin T, Cortopassi G (1994) Proposed molecular and cellular mechanism for aminoglycoside ototoxicity. Antimicrob Agents Chemother 38:2517–2520

Hyun Ryu D, Rando RR (2001) Aminoglycoside binding to human and bacterial A-site rRNA decoding region constructs. Bioorg Med Chem 9:2601–2608

Ikemura T, Dahlberg JE (1973) Small ribonucleic acids of Escherichia coli. II. Noncoordinate accumulation during stringent control. J Biol Chem 258:5033–5041

Jovine L (2003) Nuccyl http://www.mssm.edu/students/jovinl02/research/nuccyl.html. Cited 10 June 2005

Jovine L, Djordjevic S, Rhodes D (2000) The crystal structure of yeast phenylalanine tRNA at 2.0 Å resolution: cleavage by Mg^{2+} in 15-year-old crystals. J Mol Biol 301:401–414

Keiler KC, Waller PR, Sauer RT (1996) Role of tagging system in degradation of proteins synthesized from damaged messenger RNA. Science 271:990–993

Kirsebom LA, Virtanen A (2001) Inhibition of RNase P processing. In: Schroeder R, Wallis MG (eds) RNA-binding antibiotics. Molecular Biology Intelligence Unit 13, Eurekah.com, Austin. Landes Biosciences, Georgetown, pp 56–72

Kotra LP, Haddad J, Mobashery S (2000) Aminoglycosides: perspectives on mechanisms of action and resistance and strategies to counter resistance. Antimicrob Agents Chemother 44:3249–3256

Kovrigina E, Wesolowski D, Altman S (2003) Coordinate inhibition of expression of several genes for protein subunits of human nuclear RNase P. Proc Natl Acad Sci USA 100:1598–1602

Labuda D, Striker G, Porschke D (1984) Mechanism of codon recognition by transfer RNA and codon-induced tRNA association. J Mol Biol 174:587–604

Lazarus and Kitron (1973) Neomycin inhibition of DNA polymerase. Biochem Pharmacol 22:3115–3117

Litovchick A, Lapidot A, Eisenstein M, Kalinkovich A, Borkow G (2001) Neomycin B-arginine conjugate, a novel HIV-1 Tat antagonist: synthesis and anti-HIV activities. Biochemistry 40:15612–15623

Luedtke NW, Liu Q, Tor Y (2003) RNA-ligand interactions: affinity and specificity of aminoglycoside dimmers and acridine conjugates to the HIV-1 rev response element. Biochemistry 42:11391–11403

Mandal M, Boese B, Barrick JE, Winkler WC, Breaker RR (2003) Riboswitches control fundamental biochemical pathways in Bacillus subtilis and other bacteria. Cell 113:577–586

Mattick JS (2003) Challenging the dogma: the hidden layer of non-protein-coding RNAs in complex organisms. Bioessays 25:930–939

McDonald LJ, Mamrack MD (1995) Phosphoinositide hydrolysis by phospholipase C modulated by multivalent cations La(3+), Al(3+), neomycin, polyamines, and melittin. J Lipid Mediat Cell Signal 11:81–91

McManus MT (2004) Small RNAs and immunity. Immunity 21:747–756

Mei H-Y, Galan AA, Halim NS, Mack DP, Moreland DW, Sanders KB, Truong HN, Czarnik AW (1995) Inhibition of an HIV-1 Tat-derived peptide binding to TAR RNA by aminoglycoside antibiotics. Bioorg Med Chem Lett 5:2755–2760

Mikkelsen NE, Brännvall M, Virtanen A, Kirsebom LA (1999) Inhibition of RNase P RNA cleavage by aminoglycosides. Proc Natl Acad Sci USA 96:6155–6160

Mikkelsen NE, Johansson K, Virtanen A, Kirsebom LA (2001) Aminoglycoside binding displaces a divalent metal ion in a tRNA-neomycin B complex. Nat Struct Biol 8:510–514

Mingeot-Leclercq MP, Tulkens PM (1999) Aminoglycosides: nephrotoxicity. Antimicrob Agents Chemother 43:1003–1012

Mingeot-Leclercq MP, Glupczynski Y, Tulkens PM (1999) Aminoglycosides: activity and resistance. Antimicrob Agents Chemother 43:727–737

Møller T, Franch T, Udesen C, Gerdes K, Valentin-Hansen P (2002) Spot 42 RNA mediates discoordinate expression of the E. coli galactose operon. Genes Dev 16:1696–1706

Nahvi A, Sudarsan N, Ebert MS, Zou X, Brown KL, Breaker RR (2002) Genetic control by a metabolite binding mRNA. Chem Biol 9:1043–1049

Nissen P, Hansen J, Ban N, Moore PB, Steitz TA (2000) The structural basis of ribosome activity in peptide bond synthesis. Science 289:920–930

Ogle JM, Brodersen DE, Clemons Jr WM, Tarry MJ, Carter AP, Ramakrishnan V (2001) Crystal structure of an initiation factor bound to the 30S ribosomal subunit. Science 292:897–902

Patel DJ, Suri AK, Jiang F, Jiang L, Fan P, Kumar RA, Nonin S (1997) Structure, recognition and adaptive binding in RNA aptamer complexes. J Mol Biol 272:645–664

Pedersen LC, Benning MM, Holden HM (1995) Structural investigation of the antibiotic and ATP-binding sites in kanamycin nucleotidyltransferase. Biochemistry 34:13305–13311

Piolleti M, Schlünzen F, Harms J, Zarivach R, Glühmann M, Avila H, Bashan A, Bartels H, Auerbach T, Jacobi C, Hartsch T, Yonath A, Franceschi F (2001) Crystal structures of complexes of the small ribosomal subunits with tetracycline, edeine and IF3. EMBO J 20:1829–1839

Posey JE, Gherardini FC (2000) Lack of a role for iron in the Lyme disease pathogen. Science 288:1651–1653

Recht MI, Douthwaite S, Puglisi JD (1999) Basis for prokaryotic specificity of action of aminoglycoside antibiotics. EMBO J 18:3133–3138

Ren Y-G, Martínez J, Kirsebom LA, Virtanen A (2002) Inhibition of Klenow DNA polymerase and poly(A)-specific ribonuclease by aminoglycosides. RNA 8:1393–1400

Rogers J, Chang AH, von Ahsen U, Schroeder R, Davies J (1996) Inhibition of the self-cleavage reaction of the human hepatitis delta virus ribozyme by antibiotics. J Mol Biol 259:916–925

Sakon J, Liao HH, Kanikula AM, Benning MM, Rayment I, Holden HM (1993) Molecular structure of kanamycin nucleotidyltransferase determined to 3.0-Å resolution. Biochemistry 32:11977–11984

Schlünzen F, Zarivach R, Harms J, Bashan A, Tocilj A, Albrecht R, Yonath A, Franceschi F (2000) Structural basis for the interaction of antibiotics with the peptidyl transferase centre in eubacteria. Nature 413:814–821

Schroeder R, Waldisch C, Wank H (2000) Modulation of RNA function by aminoglycoside antibiotics. EMBO J 19:1–9

Shi H, Moore PB (2000) The crystal structure of yeast phenylalanine tRNA at 1.93 Å resolution: a classic structure revisited. RNA 6:1091–1105

Stage TK, Hertel KJ, Uhlenbeck OC (1995) Inhibition of the hammerhead ribozyme by neomycin. RNA 1:95–101

Stewart PS, Costerton JW (2001) Antibiotic resistance of bacteria in biofilms. Lancet 358:135–138

Sucheck SJ, Shue YK (2001) Combinatorial synthesis of aminoglycoside libraries. Curr Opin Drug Discov Devel 4:462–470

Sullenger BA, Gilboa E (2002) Emerging clinical applications of RNA. Nature 418:252–258

Summers JS, Shimko J, Freedman FL, Badger CT, Sturgess M (2002) Displacement of Mn^{2+} from RNA by K^+, Mg^{2+}, neomycin B, and an arginine-rich peptide: indirect detection of nucleic acid/ligand interactions using phosphorus relaxation enhancement. J Am Chem Soc 124:14934–149339

Sutherland R, Boon RJ, Griffin KE, Masters PJ, Slocombe B, White AR (1985) Antibacterial activity of mupirocin (pseudomonic acid), a new antibiotic for topical use. Antimicrob Agents Chemother 27:495–498

Thompson J, Skeggs PA, Cundliffe E (1985) Methylation of 16S ribosomal RNA and resistance to the aminoglycoside antibiotics gentamicin and kanamycin determined by DNA from the gentamicin-producer, Micromonospora purpurea. Mol Gen Genet 201:168–173

Tijsterman M, Ketting RF, Plasterk RH (2002) The genetics of RNA silencing. Annu Rev Genet 36:489–519

Uhlen P, Laestadius A, Jahnukainen T, Söderblom T, Backhed F, Celsi G, Brismar H, Normark S, Aperia A, Richter-Dahlfors A (2000) Alpha-haemolysin of uropathogenic E. coli induces Ca^{2+} oscillations in renal epithelial cells. Nature 405:694–697

Van Bambeke F, Glupczynski Y, Plésiat P, Pechère JC, Tulkens PM (2003) Antibiotic efflux pumps in prokaryotic cells: occurrence, impact on resistance and strategies for the future of antimicrobial therapy. J Antimicrob Chemother 51:1055–1065

Vicens Q, Westhof E (2001) Crystal structure of paromomycin docked into the eubacterial ribosomal decoding A site. Structure 9:647–658

Vicens Q, Westhof E (2002) Crystal structure of a complex between the aminoglycoside tobramycin and an oligonucleotide containing the ribosomal decoding A site. Chem Biol 9:747–756

Vicens Q, Westhof E (2003a) RNA as a drug target: the case of aminoglycosides. Chembiochem 4:1018–1023

Vicens Q, Westhof E (2003b) Crystal structure of geneticin bound to a bacterial 16S ribosomal RNA A site oligonucleotide. J Mol Biol 326:1175–1188

Vioque A, Arnez J, Altman S (1988) Protein-RNA interactions in the RNase P holoenzyme from Escherichia coli. J Mol Biol 1988 202:835–848

Vitreschak AG, Rodionov DA, Mironov AA, Gelfand MS (2004) Riboswitches: the oldest mechanism for the regulation of gene expression? Trends Genet 20:44–50

Vogel J, Argaman L, Wagner EGH, Altuvia S (2005) The small RNA IstR inhibits synthesis of an SOS-induced toxic response. Curr Biol 14:2271–2276

von Ahsen U, Davies J, Schroeder R (1991) Antibiotic inhibition of group I ribozyme function. Nature 353:368–370

von Ahsen U, Davies J, Schroeder R (1992) Non-competitive inhibition of group I intron RNA self-splicing by aminoglycoside antibiotics. J Mol Biol 226:935–941

Walsh C (2003) Antibiotics: actions, origins, resistance. ASM press, Washington DC

Walter F, Vicens Q, Westhof E (1999) Aminoglycoside-RNA interactions. Curr Opin Chem Biol 3:694–704

Walter F, Pütz J, Giegé R, Westhof E (2002) Binding of tobramycin leads to conformational changes in yeast tRNAAsp and inhibition of aminoacylation. EMBO J 21:760–768

Ward A, Campoli-Richards DM (1986) Mupirocin. A review of its antibacterial activity, pharmacokinetic properties and therapeutic use. Drugs 32:425–444

Wassarman KM (2003) Diverse regulators of gene expression in response to environmental changes. Cell 109:141–144

Wassarman KM, Storz G (2000) 6S RNA regulates E. coli RNA polymerase activity. Cell 101:613–623

Winkler W, Nahvi A, Breaker RR (2002) Thiamine derivatives bind messenger RNAs directly to regulate bacterial gene expression. Nature 419:952–956

Withey JH, Friedman DI (2003) A salvage pathway for protein structures: tmRNA and trans-translation. Annu Rev Microbiol 57:101–123

Woegerbauer M, Burgmann H, Davies J, Graninger W (2000) DNase I induced DNA degradation is inhibited by neomycin. J Antibiot (Tokyo) 53:129–137

Yao S, Sgarbi PW, Marby KA, Rabuka D, O'Hare SM, Cheng ML, Bairi M, Hu C, Hwang S-B, Hwang C-K, Ichikawa Y, Sears P, Sucheck SJ (2004) Glyco-optimization of aminoglycosides: new aminoglycosides as novel anti-infective agents. Bioorg Med Chem Lett 14:3733–3738

Yonath A, Bashan A (2004) Ribosomal crystallography: initiation, peptide bond formation, and amino acid polymerization are hampered by antibiotics. Annu Rev Microbiol 58:233–251

Zapp ML, Stern S, Green MR (1993) Small molecules that selectively block RNA binding of HIV-1 Rev protein inhibit Rev function and viral production. Cell 74:969–978

Zembower TR, Noskin GA, Postelnick MJ, Nguyen C, Peterson LR (1998) The utility of aminoglycosides in an era of emerging drug resistance. Int J Antimicrob Agents 10:95–105

High-Throughput RNA Interference in Functional Genomics

M. Janitz (✉) · D. Vanhecke · H. Lehrach

Department Vertebrate Genomics, Max Planck Institute for Molecular Genetics,
Fabeckstr. 60–62, 14195 Berlin, Germany
janitz@molgen.mpg.de

1	Introduction	97
2	Transfected Cell Arrays	98
3	High-Throughput RNA Interference	100
3.1	Silencing Endogenous Genes	101
3.2	Multiplexing of siRNA	102
4	Outlook	102
	References	103

Abstract RNA interference (RNAi) refers to post-transcriptional silencing of gene expression as a result of the introduction of double-stranded RNA into cells. The application of RNAi in experimental systems has significantly accelerated elucidation of gene functions. In order to facilitate large-scale functional genomics studies using RNAi, several high-throughput approaches have been developed based on microarray or microwell assays. The recent establishment of large libraries of RNAi reagents combined with a variety of detection assays has further improved the performance of functional genome-wide screens in mammalian cells.

Keywords RNA interference · Microarray · Reverse transfection · Cell array

1
Introduction

Microarray technology enables investigators to take a comprehensive approach in the analysis of gene expression. Messenger (m)RNA profiles of cells and tissues are identified by using DNA arrays and provide important information on differential gene expression in health and disease (Chittur 2004; Kaynak et al. 2003; Grzeskowiak et al. 2003; Boer et al. 2001). Likewise, protein arrays are being developed to identify patterns of protein abundance, protein modifications, and protein–protein and protein–DNA interactions (Gutjahr et al. 2005; Stoll et al. 2005; Walter et al. 2000). The combination of gene expression profiling studies using genome-wide DNA microarrays and the annotation of thousands of expressed sequence tags (ESTs) to concrete genes resulted in the unprecedented acceleration of expression data acquisition and interpretation.

However, in vivo functional analysis of thousands of novel genes still presents a challenge toward our understanding of cellular processes on a global scale. Functional analysis—which aims to discern the causal role of newly discovered candidate genes and proteins in cellular processes and their potential clinical impact or suitability as drug targets—is being revealed as slow and rate limiting. Functional validation is usually accomplished in molecular- and cell-based assays on a gene-by-gene basis, creating a bottleneck effect for the characterization of the huge numbers of targets arising from genomic and proteomic surveys.

2
Transfected Cell Arrays

If microarray technology works for DNA and proteins it may work for cells as well. Researchers analyze cells not only in the petri dish or multi-well plate but are starting to adapt cell culture and analysis to the microscale. The advantages of miniaturization are obvious: smaller amounts of reagents, less storage space and handling time, and fewer cells are required to perform an experiment when compared to conventional culture systems. Especially when only small numbers of cells are available, a miniaturized format may even provide the basis to address questions that have not been possible to address before in the conventional format. After the first arrays were developed, many ideas followed on how and why cells should be analyzed in arrays. Cellular responses to overexpression or knockdown of genes can be investigated, the impact of many different molecular environments on cell function can be assessed, and cell surface markers can be determined in a rapid way that, in the future, may serve for diagnosis and prognosis of diseases.

Overexpression of genes or knockdown of mRNA using cell transfection methods has become a common approach to analyze the effects of gene expression in cells. Since the human genome and genomes of other species have been sequenced, collections of full-length complementary (c)DNAs and small interfering (si)RNA probes are constantly growing.

Efficient methods are now needed to cope with the number of genetic probes and cellular assays for the functional analysis of the transcriptome. The cell microarray developed by Ziauddin and Sabatini constituted the first approach to address this challenge by miniaturizing and parallelizing the transfection of cells with plasmids containing cDNA sequences (Ziauddin and Sabatini 2001). cDNA plasmids are mixed with gelatin and printed in arrays on glass slides. The arrays are covered with transfection reagent and a monolayer of cells is subsequently added. During an incubation period of 40 h, the cells become transfected by the underlying plasmid resulting in an array of clusters of cells expressing the genes encoded by the respective cDNAs.

Since the identity of the genes is known from the array coordinates, the technology allows a rapid identification of gene products that alter cellular physiology or induce a desired phenotype. These changes can be detected using commonly used methods, like immunostaining of proteins or tagging proteins with autofluorescent proteins to follow their cellular localization. Fluorescent labeling techniques can be used to visualize other physiological events in the cell, such as transient increases of intracellular calcium concentrations with calcium-sensitive fluorophores (Mishina et al. 2004).

Conrad et al. (2004) have recently applied machine learning-based classification methods to the microarray to classify cell phenotypes in response to overexpression or knockdown of genes. These authors also improved the automation of the technology by spotting all transfection components at once. The use of siRNAs to knock down expression of selected genes extends the usage of the transfected cell arrays significantly (Erfle et al. 2004).

Moreover, combination screens, in which two siRNAs or expression vectors and siRNA were transfected into cells of the same cell cluster, have been performed (Wheeler et al. 2004; Baghdoyan et al. 2004). With a capacity of 6,000–10,000 cell clusters per standard microscope slide, a small number of slides would suffice to overexpress or knock down the entire set of human genes in cell arrays. The combining of large collections of full-length gene clones [the FLEXGene (the acronym comes from "full-length expression") repository, from the Harvard Institute of Proteomics; the IMAGE (this acronym derives from "integrated molecular analysis of genomes and their expression") cDNA collection; the German cDNA Network (Wiemann et al. 2003)] or collections of effective siRNA sequences could make the system extremely productive for future global characterizations of gene function analyses (Carpenter and Sabatini 2004).

Transfected cell arrays offer a number of advantages when used to study gene function. One of the most important features is that this assay provides a unique opportunity to study protein function in the context of the living cell. Proteins expressed in the cell array can undergo post-translational processing such as glycosylation and folding, which can be very distinct in different organisms and different cell types (Colosimo et al. 2000). Functional analysis of proteins in their natural environment guarantees that additional molecules involved in the functional process are present during the assay (e.g., cofactors, proteins involved in large complex formation, etc.).

There are a number of advantages in using transfected-cell array (TCA) as compared to functional assays performed on mammalian cells in microwell plate format. The array approach requires fewer cells per number of genes tested. This is especially relevant for human primary cells since (1) only small numbers of these can be isolated out of a tissue and (2) the in vitro expansion of these cells is rather limited. Although primary cells are notorious for their capacity to be transfected, reverse transfection of these cells is possible, but requires careful optimization of the experimental conditions for each cell type tested (Yoshikawa et al. 2004). Transfected cell arrays also require far less DNA/RNA as well as transfection and signal development reagents compared to assays performed in microwell plate format. Since, in the case of high-throughput technologies, the cost of the single-sample analysis must be reduced to absolute minimum, cell arrays are, at the moment, the most cost-effective functional genomic tool available.

3
High-Throughput RNA Interference

RNA interference (RNAi) has been employed in genome-wide phenotype screens in *Caenorhabditis elegans* (Ashrafi et al. 2003) and *Drosophila melanogaster* (Lum et al. 2003; Wheeler et al. 2004). Although siRNAs synthesized chemically or enzymatically have mainly been used to study the function of individual mammalian genes, RNAi-based genome-wide functional analysis for mammalian cells is also well underway. Indeed, several laboratories have already created and successfully applied siRNA-based libraries targeting large groups of human genes (Zheng et al. 2004; Berns et al. 2004; Kittler et al. 2004). It soon became clear that cell arrays provided a suitable platform for high-throughput cellular assays using RNAi. Several groups, including this author's laboratory, focused on adaptation of the TCA platform for the purpose of large-scale loss-of-function studies (Mousses et al. 2003; Kumar et al. 2003; Silva et al. 2004). The basic protocol for preparation of RNAi cell arrays includes the spotting of siRNAs in gelatin solution onto a modified glass surface using a standard robotic arrayer. The RNA array is subsequently treated with transfection reagent and covered with a monolayer of recipient cells. Transfection efficiency can be monitored by using fluorochrome conjugated siRNA (Mousses et al. 2003). Alternatively, the siRNA are cotransfected with a reporter plasmid, and gene silencing can be monitored in cells expressing the reporter gene (Silva et al. 2004).

The availability of vector-based RNAi systems (Brummelkamp et al. 2002) offers the possibility of applying cell array technology using DNA-based reverse transfection protocols. The potential drawback of this approach is a substantial delay in exertion of the silencing effect due to the de novo siRNA synthesis within the cell. Nevertheless, Silva and coworkers recently showed effective silencing of the reporter protein using co-transfection of short hairpin (sh)RNA-expressing plasmid (Silva et al. 2004).

Silencing of targeted genes is monitored using, for example, fluorescent-labeled monoclonal antibodies or autofluorescent reporter proteins. Targeted genes are either exogenous genes, introduced by transient transfection or stable integration of gene expression plasmids, or endogenous genes. The former approach allows for the evaluation of silencing capacity of different siRNAs specific for the same target gene and can be used as a robust validation platform for functional siRNA selection (Kumar et al. 2003; Silva et al. 2004). Functional validation of potential siRNA molecules is still required despite a number of algorithms that exist for sequence design (Paddison and Hannon 2003). Indeed, for most algorithms, it is estimated that only one out of two or even one out of five designed siRNAs will efficiently knock down a target gene.

3.1
Silencing Endogenous Genes

A more challenging application is the targeting of endogenous genes by reverse transfection of siRNA molecules. Endogenous gene silencing on cell arrays can be monitored by means of specific fluorescent-labeled monoclonal antibodies. This results in clusters of cells that are negative for the protein, thus creating arrays of black holes in monolayers of cells that are expressing the protein (Fig. 1). However, detection with gene-specific antibodies puts a limit to the number of genes that can be assayed in the cell array. For the functional evaluation of a genome-wide collection of siRNA molecules using cell arrays, other detection methods have to be used. For example, downstream cellular processes that are affected by silencing endogenous genes can be monitored by specific antibodies that can detect changes in the phosphorylation state of cell membrane-bound receptors or transcription factors. This would allow for the identification of essential or even novel cell signaling molecules using genome-

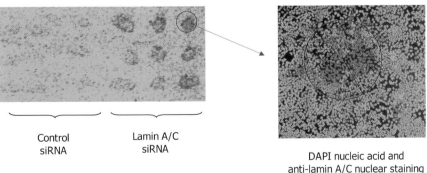

Fig. 1 Downregulation of lamin A/C gene expression using the cell array platform. Silencing of the lamin A/C gene in cells covering siRNA spots can be visualized with an array scanner (*left*) as black holes in the monolayer of cells stained with anti-lamin A/C fluorescein isothiocyanate (FITC)-labeled antibody. The *red circle* indicates a single reverse transfected cell cluster shown with 40× magnification using a Zeiss Axiocam microscope (*right*). Cellular nuclei were counterstained with 4′-6′-diamidino-2-phenylindole (*DAPI*)

wide siRNA libraries. Alternatively, the effects of silencing genes can result in cellular processes such as apoptosis or changes in intercellular adhesion or morphological changes of the transfected cells (Ziauddin and Sabatini 2001).

3.2
Multiplexing of siRNA

Since transfected cell arrays allow the analysis of many genes in parallel and furthermore can be adopted to many existing cell-based assays, these arrays could become an efficient molecular profiling tool for disease-related proteins or to study protein–protein and protein–drug interactions. Interesting perspectives arise when double knockdowns are considered. Two different siRNA or vectors expressing shRNAs could be co-transfected together or a single siRNA could be transfected to the cell line stably transfected with siRNA-vector (van de Wetering et al. 2003). Cell arrays would have enough spotting capacity to test hundreds of different siRNA combinations in a single experiment. This would open the possibility of screening synthetic lethal interactions, which occur by appearance of two non-lethal mutations that, acting together, result in cell death. Such a functional interaction of two corresponding gene products greatly helps the in identification of novel drug targets, e.g., against cancer. Most of the synthetic lethal interactions have been determined in lower eukaryotes such as *C. elegans* (Wang et al. 2004). In contrast, in mammalian systems synthetic lethality has been detected only in few studies, mainly because efficient molecular tools for identification of such reactions has been lacking so far. Application of high-throughput RNAi-based cell arrays could substantially accelerate progress in this field.

4
Outlook

The combination of cell arrays and cell lines stably transfected with inducible shRNA plasmid constructs is also well suited for screening of small molecules for their ability to specifically interact with protein targets such as G protein-coupled receptors (GPCRs) (Mishina et al. 2004). Also, in this case the cell microarrays present a more advanced alternative to standard protein arrays, since in TCA the proteins are synthesized in situ in the physiological environment of the cells, thus enabling proper post-translational protein folding and glycosylation. Taken together, cell arrays represent an emerging high-throughput platform in RNAi research, especially in applications where cell numbers are a limiting factor. Further optimization of the reverse transfection protocol for different cell types should open the possibility to perform loss-of-function studies in a tissue- or organ-specific environment.

References

Ashrafi K, Chang FY, Watts JL, Fraser AG, Kamath RS, Ahringer J, Ruvkun G (2003) Genome-wide RNAi analysis of Caenorhabditis elegans fat regulatory genes. Nature 421:268–272

Baghdoyan S, Roupioz Y, Pitaval A, Castel D, Khomyakova E, Papine A, Soussaline F, Gidrol X (2004) Quantitative analysis of highly parallel transfection in cell microarrays. Nucleic Acids Res 32:e77

Berns K, Hijmans EM, Mullenders J, Brummelkamp TR, Velds A, Heimerikx M, Kerkhoven RM, Madiredjo M, Nijkamp W, Weigelt B, Agami R, Ge W, Cavet G, Linsley PS, Beijersbergen RL, Bernards R (2004) A large-scale RNAi screen in human cells identifies new components of the p53 pathway. Nature 428:431–437

Boer JM, Huber WK, Sultmann H, Wilmer F, von Heydebreck A, Haas S, Korn B, Gunawan B, Vente A, Fuzesi L, Vingron M, Poustka A (2001) Identification and classification of differentially expressed genes in renal cell carcinoma by expression profiling on a global human 31,500-element cDNA array. Genome Res 11:1861–1870

Brummelkamp TR, Bernards R, Agami R (2002) A system for stable expression of short interfering RNAs in mammalian cells. Science 296:550–553

Carpenter AE, Sabatini DM (2004) Systematic genome-wide screens of gene function. Nat Rev Genet 5:11–22

Chittur SV (2004) DNA microarrays: tools for the 21st century. Comb Chem High Throughput Screen 7:531–537

Colosimo A, Goncz KK, Holmes AR, Kunzelmann K, Novelli G, Malone RW, Bennett MJ, Gruenert DC (2000) Transfer and expression of foreign genes in mammalian cells. Biotechniques 29:314–318

Conrad C, Erfle H, Warnat P, Daigle N, Lorch T, Ellenberg J, Pepperkok R, Eils R (2004) Automatic identification of subcellular phenotypes on human cell arrays. Genome Res 14:1130–1136

Erfle H, Simpson JC, Bastiaens PI, Pepperkok R (2004) siRNA cell arrays for high-content screening microscopy. Biotechniques 37:454–462

Grzeskowiak R, Witt H, Drungowski M, Thermann R, Hennig S, Perrot A, Osterziel KJ, Klingbiel D, Scheid S, Spang R, Lehrach H, Ruiz P (2003) Expression profiling of human idiopathic dilated cardiomyopathy. Cardiovasc Res 59:400–411

Gutjahr C, Murphy D, Lueking A, Koenig A, Janitz M, O'Brien J, Korn B, Horn S, Lehrach H, Cahill DJ (2005) Mouse protein arrays from a TH1 cell cDNA library for antibody screening and serum profiling. Genomics 85:285–296

Kaynak B, von Heydebreck A, Mebus S, Seelow D, Hennig S, Vogel J, Sperling HP, Pregla R, Alexi-Meskishvili V, Hetzer R, Lange PE, Vingron M, Lehrach H, Sperling S (2003) Genome-wide array analysis of normal and malformed human hearts. Circulation 107:2467–2474

Kittler R, Putz G, Pelletier L, Poser I, Heninger AK, Drechsel D, Fischer S, Konstantinova I, Habermann B, Grabner H, Yaspo ML, Himmelbauer H, Korn B, Neugebauer K, Pisabarro MT, Buchholz F (2004) An endoribonuclease-prepared siRNA screen in human cells identifies genes essential for cell division. Nature 432:1036–1040

Kumar R, Conklin DS, Mittal V (2003) High-throughput selection of effective RNAi probes for gene silencing. Genome Res 13:2333–2340

Lum L, Yao S, Mozer B, Rovescalli A, Von Kessler D, Nirenberg M, Beachy PA (2003) Identification of Hedgehog pathway components by RNAi in Drosophila cultured cells. Science 299:2039–2045

Mishina YM, Wilson CJ, Bruett L, Smith JJ, Stoop-Myer C, Jong S, Amaral LP, Pedersen R, Lyman SK, Myer VE, Kreider BL, Thompson CM (2004) Multiplex GPCR assay in reverse transfection cell microarrays. J Biomol Screen 9:196–207

Mousses S, Caplen NJ, Cornelison R, Weaver D, Basik M, Hautaniemi S, Elkahloun AG, Lotufo RA, Choudary A, Dougherty ER, Suh E, Kallioniemi O (2003) RNAi microarray analysis in cultured mammalian cells. Genome Res 13:2341–2347

Paddison PJ, Hannon GJ (2003) siRNAs and shRNAs: skeleton keys to the human genome. Curr Opin Mol Ther 5:217–224

Silva JM, Mizuno H, Brady A, Lucito R, Hannon GJ (2004) RNA interference microarrays: high-throughput loss-of-function genetics in mammalian cells. Proc Natl Acad Sci USA 101:6548–6552

Stoll D, Templin MF, Bachmann J, Joos TO (2005) Protein microarrays: applications and future challenges. Curr Opin Drug Discov Devel 8:239–252

van de Wetering M, Oving I, Muncan V, Pon Fong MT, Brantjes H, van Leenen D, Holstege FC, Brummelkamp TR, Agami R, Clevers H (2003) Specific inhibition of gene expression using a stably integrated, inducible small-interfering-RNA vector. EMBO Rep 4:609–615

Walter G, Bussow K, Cahill D, Lueking A, Lehrach H (2000) Protein arrays for gene expression and molecular interaction screening. Curr Opin Microbiol 3:298–302

Wang JC, Walker A, Blackwell TK, Yamamoto KR (2004) The Caenorhabditis elegans ortholog of TRAP240, CeTRAP240/let-19, selectively modulates gene expression and is essential for embryogenesis. J Biol Chem 279:29270–29277

Wheeler DB, Bailey SN, Guertin DA, Carpenter AE, Higgins CO, Sabatini DM (2004) RNAi living-cell microarrays for loss-of-function screens in Drosophila melanogaster cells. Nat Methods 1:127–132

Wiemann S, Bechtel S, Bannasch D, Pepperkok R, Poustka A, German cDNA Network (2003) The German cDNA network: cDNAs, functional genomics and proteomics. J Struct Funct Genomics 4:87–96

Yoshikawa T, Uchimura E, Kishi M, Funeriu DP, Miyake M, Miyake J (2004) Transfection microarray of human mesenchymal stem cells and on-chip siRNA gene knockdown. J Control Release 96:227–232

Zheng L, Liu J, Batalov S, Zhou D, Orth A, Ding S, Schultz PG (2004) An approach to genomewide screens of expressed small interfering RNAs in mammalian cells. Proc Natl Acad Sci USA 101:135–140

Ziauddin J, Sabatini DM (2001) Microarrays of cells expressing defined cDNAs. Nature 411:107–110

Antiviral Applications of RNAi

K. V. Morris[1] · J. J. Rossi[2] (✉)

[1]Division of Rheumatology, Department of Molecular and Experimental Medicine, The Scripps Research Institute, 10550 N. Torrey Pines Road, La Jolla CA, 92037, USA

[2]Division of Molecular Biology, Beckman Research Institute of the City of Hope, 1450 E. Duarte Rd., Duarte CA, 91010, USA
JRossi@coh.org

1	RNA Interference	105
2	Diversity of Viral Targets	108
3	siRNA Selection	108
4	Delivery of siRNAs to Target Cells	109
5	siRNA Challenges	112
	References	112

Abstract RNA interference is a natural mechanism by which small interfering (si)RNA operates to specifically and potently down-regulate the expression of a target gene. This down-regulation has been thought to predominantly function at the level of the messenger (m)RNA, post-transcriptional gene silencing (PTGS). Recently, the discovery that siRNAs can function to suppress a gene's expression at the level of transcription, i.e., transcriptional gene silencing (TGS), has created a major paradigm shift in mammalian RNAi. These recent findings significantly broaden the role RNA, specifically siRNAs and potentially microRNAs, plays in the regulation of gene expression as well as the breadth of potential siRNA target sites. Indeed, the specificity and simplicity of design makes the use of siRNAs to target and suppress virtually any gene or gene promoter of interest a realized technology. Furthermore, since siRNAs are a small nucleic acid reagent, they are unlikely to elicit an immune response, making them a theoretically good future therapeutic. This review will focus on the development, delivery, and potential therapeutic use of antiviral siRNAs in treating viral infections as well as emerging viral threats.

Keywords RNAi · siRNA · PTGS · TGS · HIV-1

1
RNA Interference

RNA interference (RNAi), first described in plants and termed cosuppression (reviewed in Tijsterman et al. 2002), is a process in which double-stranded (ds)RNA induces homology-dependent degradation of mRNA (Montgomery

1998; Nishikura 2001; Sharp 2001). RNAi is a process involving small interfering double-stranded (si)RNAs 21–22 bp in length, with 3′ overhanging ends that can induce a homology-dependent degradation of cognate messenger (m)RNA (Nishikura 2001). The generation of siRNA is the result of a multistep process that involves the action of RNase III endonuclease Dicer (Bernstein et al. 2001, 2003; Sui 2002; Fig. 1). The approximately 22-bp siRNAs that is processed by Dicer provide much of the specificity in the silencing process. However, the necessity for an exact sequence match in the sense strand of siRNA duplexes has been questioned, as single stranded antisense siRNAs can guide target RNA cleavage (Martinez et al. 2002) and as many as five mismatches in the sense strand RNA may be tolerated (Sumimoto 2003). In contrast, a single base pair mismatch relative to the target RNA on the antisense strand has been shown to significantly reduce siRNA-mediated message degradation (Hamada et al. 2002). Following the action of Dicer, the ∼21-bp siRNAs are incorporated into the RNA-induced silencing complex (RISC), which identifies and silences by slicing the mRNAs complementary to the 21-bp siRNA through interactions with Argonaute 2 (Liu et al. 2004; Fig. 1). The specificity juxtaposed with potent suppression of target genes by siRNA has truly adopted RNAi as a standard methodology for gene specific silencing in mammalian cells.

Mechanistically, RNAi can suppress gene expression via two distinct pathways: transcriptional (TGS) and post-transcriptional (PTGS) gene silencing. PTGS involves siRNAs targeting of either mRNA or pre-mRNA, including intronic sequences in *Caenorhabditis elegans* and yeast (Bosher 1999). TGS involves silencing at the chromatin and was first observed when doubly transformed tobacco plants exhibited a suppressed phenotype of the transformed transgene. Careful analysis indicated that methylation of the targeted gene was involved in the suppression (Matzke 1989). TGS mediated by dsRNAs was further substantiated in viroid-infected plants and was shown to be due to RNA-dependent methylation of DNA (RdDM) (Wassenegger 1994). The observed TGS in viroid-infected plants contained viral promoters expressing integrated transgenes. Interestingly, these promoters became methylated at sites matching the small double stranded viral RNAs, and transcription of the viral promoters was suppressed as a result of these homologous viral RNAs entering the nucleus and inducing TGS (Wassenegger 1994, 2000), i.e., RNA directed suppression of gene expression at the promoter. In human cells, gene silencing induced by RNAi was initially thought to be restricted to action on cytoplasmic mRNA or RNA at the nuclear pore (Zeng 2002), similar to most reports in *C. elegans* and *Trypanosoma brucei* (Fire 1998; Montgomery 1998; Ngo 1998). To date, TGS has been found to occur in plants, *Drosophila*, and in *Schizosaccharomyces pombe* in centromeric regulation (Volpe 2002). Recently, TGS was reported to be operable in mammalian cells and appeared to rely on the delivery of the siRNA to the nucleus (Kawasaki et al. 2005; Morris et al. 2004a). However, the strict requirements of nuclear delivery may not be necessary if temporal factors are included in the analysis (Kawasaki and Taira

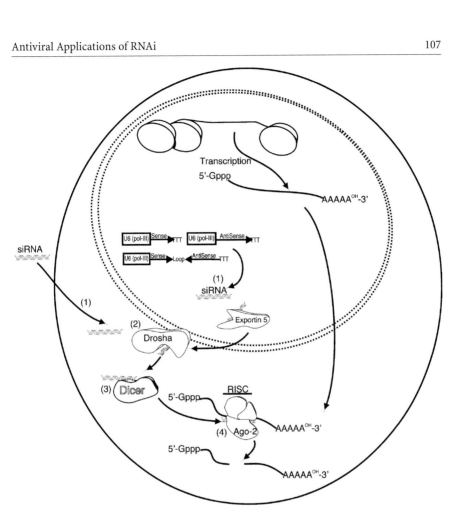

Fig. 1 Post-transcriptional RNAi in mammalian cells. Synthetic siRNAs or those generated by Dicer ex vivo can be transfected directly into cells using lipid-based transfection reagents or with siRNAs expressed from within the cell from lentiviral or other gene therapy-based vector systems (*1*). A cell can be stably transduced with a lentiviral vector that expresses siRNAs either from two independent promoters (U6, Pol III) or a single promoter driving the expression of a hairpin shRNA targeting a particular gene of interest (*1*). The vector-expressed siRNAs are probably bound by Exportin 5 and Drosha (*2*; Lee et al. 2003; Lund et al. 2004), and then get shuttled out of the nucleus and handed off to Dicer, which then cleaves the loop from the hairpin (*3*) producing the siRNA that is then loaded into RISC, ultimately leading to slicing of the target mRNA (*4*), essentially driving post-transcriptional gene silencing (PTGS)

2004). The observed TGS in mammalian cells appears to involve DNA methylation, specifically DNMT1, DNMT3b (Kawasaki and Taira 2004), and DNMT3a (Jeffery and Nakielny 2004), as well as histone deacetylation, as the observed inhibition of gene expression was reversible with the addition of 5-azacytidine (5′Aza-C, 4 µM) and trichostatin A (TSA, 0.05 mM; Morris et al. 2004a).

2
Diversity of Viral Targets

Targeted suppression of human immunodeficiency virus (HIV)-1 has been achieved through siRNAs directed against HIV-1 tat and rev (Coburn 2002; Lee 2002; Novina 2002; Surabhi and Gaynor 2002), reverse transcriptase (Morris 2004; Surabhi and Gaynor 2002), *trans*-activating response region (TAR), and the 3'-untranslated region (UTR), Vif (Jacque 2002), as well as gag and the HIV-1 co-receptor CD4 (Novina 2002) and co-receptor CCR5 (Qin 2002; reviewed in Lee and Rossi 2004).

Viruses other than HIV-1 have also been successfully targeted by siRNAs in vitro with some success, including Semliki forest virus (SFV), poliovirus, dengue virus, influenza virus, hepatitis C virus, and many others (reviewed in Radhakrishnan et al. 2004). The fact that such a wide berth of varying viruses can be successfully targeted by siRNAs suggests that these nucleic acid molecules can be used to theoretically target virtually any emerging or present-day infectious agent. However, despite the excitement and the early proofs-of-principle in the literature, there are important issues and concerns about therapeutic application of this technology, including difficulties with efficient delivery, uncertainty about potential toxicity, and the emergence of siRNA-resistant viruses. In particular, certain viruses encode proteins that block one or more steps in the RNAi pathway (Bennasser et al. 2005; Hamilton et al. 2002; Johansen and Carrington 2001; Li et al. 2002; Llave et al. 2000; Mallory et al. 2001, 2002). Indeed resistance to siRNA occurs rather rapidly and is only contingent on a single nucleotide substitution (Gitlin 2002), and recently HIV-1 was shown to elude siRNA targeting by the evolution of alternative splice variants for the siRNA-targeted transcripts (Westerhout et al. 2005). A possible way to circumvent such a conclusion in siRNA-mediated therapies for human viral infections could be to (1) design siRNAs to best fit targets from an extensive database of the variants in the particular target virus (Morris 2004) and (2) incorporate these best-fit siRNAs into a multiple anti-viral siRNA-expressing transgene vector. Undeniably, the multiplexing of several different siRNAs targeting different sites in the HIV genome along with non-essential cellular targets such as CCR5 should be utilized to harness the full potential of this mechanism in treating HIV-1 with siRNA technology. Alternatively, siRNAs designed to more conserved regions, such as to target viral intron/exon splice junctions, might also prove more resistant to the emergence of variant viral strains as the result of siRNA-mediated targeting.

3
siRNA Selection

There are many commercially available reagents as well as PCR-based methodologies (Castanotto and Rossi 2004) for use in the generation of synthetic siRNAs. The usefulness of first generating and testing siRNA on a particular target prior to construction and generation of a vector system for the delivery and expression of a particular siRNA species (Morris 2004) cannot be overstated. Specific targeting of siRNAs is extremely important, as slight positional changes in the siRNA relative to the mRNA can have drastic effects on silencing (Holen 2002), indicating that the target mRNA secondary structure plays a role in the siRNA accessibility. Indeed not all siRNAs are functional, and a computational design or algorithm that provides 100% successful selection of efficacious siRNAs has not, to our knowledge, been developed. However, a set of common rules has begun to emerge from many of the studies done. SiRNAs in which the helix at the 5'-end of the antisense strand has a lower stability than the 3'-end of the siRNA are generally more effective than those with the opposite arrangement. A biochemical basis for the thermodynamic arrangement of effective siRNAs was provided by biochemical studies of the mRNA cleavage complex RISC in *Drosophila* embryo extracts, which showed unequal incorporation of the two strands of the siRNA into RISC (Schwarz et al. 2003). Strand biases could be manipulated by altering the thermodynamic stability of the terminal nucleotides in a way that precisely matched the rules that were derived from empirical studies. Finally, an examination of microRNAs (miRNAs), most of which produce RISC-like complexes containing only one strand of the precursor, showed the same pattern of thermodynamic asymmetry as did effective siRNAs (reviewed in Meissner 2001).

Another important factor in siRNA-mediated RNAi is based on cell type. siRNA-transfected cells that are actively dividing lose transcriptional silencing over roughly 96 h (Novina 2002; Tuschl 2002), possibly due to the cell division and subsequent loss of the required template mRNA (Holen 2002). In non-dividing cells, siRNA silencing has been retained long-term and correlates well with the presence of the mRNA target (Song 2003). Consequently, successful targeting of a desired transcript should involve prior attempts to model the siRNA accessibility to the template mRNA, similar to approaches employed with ribozyme and antisense RNA targeting (Scherr 1998). Furthermore, when targeting the RNA of a virus, conserved regions that cannot accommodate evolved point mutations should be preferentially selected. Certainly the sequence-specific ability of siRNA to inhibit gene expression suggests broad applications, including targeting of viral infections such as HIV-1. However, the sensitivity of siRNA to single base pair mismatches, coupled with extant data on the rapidity of evolution of drug resistance (Richman 1994) in the face of selective pressure, may limit the overall target selection in some viral infections.

4
Delivery of siRNAs to Target Cells

Once an siRNA or multiple siRNAs targeting a particular viral RNA have been designed and tested in vitro with transient-based transfection assays (Fig. 1), it may prove necessary to express the siRNA from the context of the cell. The introduction of siRNAs into mammalian cells can be achieved through a variety of standard transfection methods (Fig. 1). The strength and duration of the silencing response delivered in the context of such transfection methods, however, is determined or limited (or both determined and limited) by several factors. On a population basis, the overall efficiency of transfection is a major determinant, which must be addressed by optimizing conditions. In each individual cell, silencing depends upon a combination of the amount of siRNA that is delivered and the potential of the siRNA to suppress its target (the potency). Even a relatively poor siRNA can silence its target provided that sufficient quantities are delivered. However, overloading the system with a high-concentration of siRNAs is likely to lead to undesired effects, including off-target suppression as well as the induction of a PKR response (Sledz et al. 2003). Indeed, there are innumerable methodologies available for expressing siRNAs from the context of the cell, including transient transfection of the synthesized or plasmid-expressed siRNA and stable expression of the particular siRNA by lentiviral vector delivery (Banerjea et al. 2003; Fig. 1).

Lentiviral vectors are emerging as one of the best candidates currently available for delivering and stably expressing short hairpin (sh)RNAs or siRNAs in target cells (Fig. 2). Lentiviruses, unlike retroviruses such as Moloney murine leukemia virus (MoMuLV), tend to preferentially integrate downstream of active promoters within the active transcriptional unit, potentially limiting their overall oncogenicity (Wu et al. 2003). Moreover, lentiviral-based vectors are capable of transducing non-dividing cells (Buchschacher 2000) and specifically targeting the nucleus. HIV-1-, HIV-2/SIV-, and feline immunodeficiency virus (FIV)-based lentiviral vectors are produced by co-transfecting vector, packaging, and envelope into producer cells, and collecting the resultant supernatants that contain the packaged vector 48 h later (Fig. 2). Lentiviral vectors are capable of stably transducing many cell types, including hematopoietic stem cells (Gervaix et al. 1997), integrating into the target genome, and expressing desired transgenes (Poeschla 1996; Price 2002; Quinonez 2002; Yam 2002). Lentiviruses have also been shown to cross-package one another (Browning 2001; Goujon 2003; White 1999). This observation has been carried over experimentally with HIV-1 and HIV-2 vectors being cross-packaged by FIV and capable of stably transducing and protecting human primary blood mononuclear cells from HIV-1 infection (Morris et al. 2004b). The cross-packaging of lentiviral vectors such as HIV-1 with an FIV packaging system offers a unique and possibly safer method for delivering anti-viral vectors to target cells in HIV-1-infected individuals. For instance, FIV-packaged HIV-1 or HIV-2 vectors reduce the likelihood of immune recognition, or seroconversion, due to exposure to HIV-1 structural proteins. Finally, lentiviral vectors can be specifically pseudotyped (Kobinger 2001; Sandrin 2003) or designed with a receptor-ligand bridge to target specific cell types (Boerger 1999).

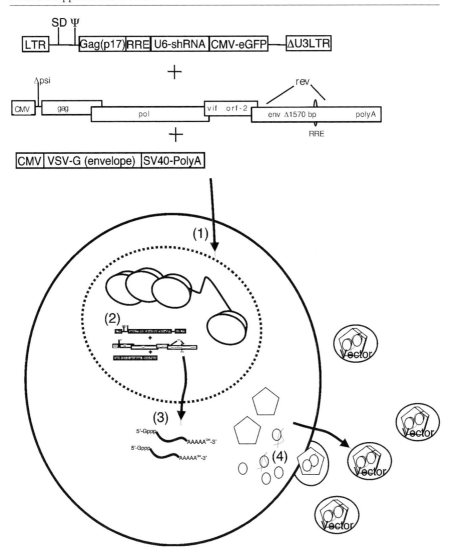

Fig. 2 Production of lentiviral vectors. Lentiviral vectors are produced by (*1*) transfecting 293T producer cells with the lentiviral vector, packaging, and envelope plasmids. Next, the transfected cell transcribes the respective plasmids (*2* and *3*) subsequently producing the packaging co-factors (*4*) and vector RNA which is then packaged into the budding particles (*4*). The culture supernatants are collected 48–72 h later, and vector concentration is determined by titering on target cells

Therapeutically, the use of lentiviral or other stable integrating vector systems may not prove useful in the application of siRNAs in treating transient infections such as influenza or severe acute respiratory syndrome (SARS). One alternative is the use of cationic lipid complexes to systemically or locally deliver the viral or disease-specific shRNA or siRNA to the infected individual. Systemic delivery of siRNAs have been shown in mice and could be used to aid or augment the immune response during times of duress (Sioud and Sorensen 2003; Sorensen et al. 2003).

5
siRNA Challenges

Indubitably, one of the advantages of using siRNAs to treat emerging infectious agents such as viral infections is the relative ease of design, construction, and testing. The emerging field of RNAi—and siRNAs in particular—provides a potentially cost-effective and relatively quick methodology to treating some of the worlds most deadly emerging viral infections, such as Ebola and SARS, or to even deal with theoretical threats of smallpox or other viruses. Moreover, RNAi technology can also be used beyond the scope of human disease to treat agricultural, horticultural, and wildlife diseases. However, there are two important issues currently facing RNAi-mediated technologies that must be circumvented prior to the realization of RNAi in human therapeutics. These two constraints are the avoidance of off-target effects and the delivery of the siRNA to the target cell.

Steady progress has been made with regards to gene therapy-based delivery systems, specifically lentiviral-based vector systems. Regarding off-target effects, the use of siRNAs to target specific cellular or viral transcripts relies essentially on hijacking the endogenous RNAi machinery, of which we know very little, i.e., what is the potential for saturating the RNAi pathway. Indeed there is evidence that RISC can be saturated at least in the context of cultured cells (Pasquinelli 2002; Pasquinelli and Ruvkun 2002). Consequently, endogenous RNAi pathways appear to be susceptible to high concentrations of exogenous siRNA, suggesting that it will probably be imperative to not only quantitate siRNA-mediated silencing but to also monitor other genes in siRNA-treated cultures for untoward off-target effects. Indeed a thorough understanding of the mechanism(s) leading to nonspecific off-target effects as the result of siRNA treatment is essential before siRNAs or shRNAs can become realized in human therapies to treat viral infections.

Acknowledgements This work was supported by funds from NIH NIAID and HLB to J.J.R. and the NCI Comprehensive Cancer Center Grant to both K.V.M. and J.J.R.

References

Banerjea A, Li MJ, Bauer G, Remling L, Lee NS, Rossi J, Akkina R (2003) Inhibition of HIV-1 by lentiviral vector-transduced siRNAs in T lymphocytes differentiated in SCID-hu mice and CD34+ progenitor cell-derived macrophages. Mol Ther 8:62–71

Bennasser Y, Le SY, Benkirane M, Jeang KT (2005) Evidence that HIV-1 encodes an siRNA and a suppressor of RNA silencing. Immunology 22:607–619

Bernstein E, Caudy AA, Hammond SM, Hannon GJ (2001) Role for a bidentate ribonuclease in the initiation step of RNA interference. Nature 409:363–366

Bernstein E, Kim SY, Carmell MA, Murchison EP, Alcorn H, Li MZ, Mills AA, Elledge SJ, Anderson KV, Hannon GJ (2003) Dicer is essential for mouse development. Nat Genet 35:215–217

Boerger AL, Snitkovsky S, Young JAT (1999) Retroviral vectors preloaded with a viral receptor-ligand bridge protein are targeted to specific cell types. Proc Natl Acad Sci USA 96:9867–9872

Bosher JM, Dugourcq P, Sookhareea S, Labouesse M (1999) RNA interference can target pre-mRNA: consequences for gene expression in a Caenorhabditis elegans operon. Genetics 153:1245–1256

Browning MT, Schmidt RD, Lew KA, Rizvi TA (2001) Primate and feline lentivirus vector RNA packaging and propagation by heterologous lentivirus virions. J Virol 75:5129–5140

Buchschacher GL, Wong-Staal F (2000) Development of lentiviral vectors for gene therapy for human diseases. Blood 95:2499–2504

Castanotto D, Rossi JJ (2004) Construction and transfection of PCR products expressing siRNAs or shRNAs in mammalian cells. Methods Mol Biol 252:509–514

Coburn GA, Cullen BR (2002) Potent and specific inhibition of human immunodeficiency virus type 1 replication by RNA interference. J Virol 76:9225–9231

Fire A, Xu S, Montgomery MK, Kostas SA, Driver SE, Mello CC (1998) Potent and specific genetic interference by double stranded RNA in Caenorhabditis elegans. Nature 391:806–811

Gervaix A, West D, Leoni LM, Richman DD, Wong-Staal F, Corbeil J (1997) A new reporter cell line to monitor HIV infection and drug susceptibility in vitro. Proc Natl Acad Sci U S A 94:4653–4658

Gitlin L, Karelsky S, Andino R (2002) Short interfering RNA confers intracellular antiviral immunity in human cells. Nature 26:1–5

Goujon C, Jarrosson-Wuilleme L, Bernaud J, Rigal D, Darlix J, Cimarelli A (2003) Heterologous human immunodeficiency virus type 1 lentiviral vectors packaging a simian immunodeficiency virus-derived genome display a specific postentry transduction defect in dendritic cells. J Virol 787:9295–9304

Hamada M, Ohtsuka T, Kawaida R, Koizumi M, Morita K, Furukawa H, Imanishi T, Miyagishi M, Taira K (2002) Effects on RNA interference in gene expression (RNAi) in cultured mammalian cells of mismatches and the introduction of chemical modifications at the 3'-ends of siRNAs. Antisense Nucleic Acid Drug Dev 12:301–309

Hamilton A, Voinnet O, Chappell L, Baulcombe D (2002) Two classes of short interfering RNA in RNA silencing. EMBO J 21:4671–4679

Holen T, Amarzguioui M, Wiiger MT, Babaie E, Prydz H (2002) Positional effects of short interfering RNAs targeting the human coagulation trigger tissue factor. Nucleic Acids Res 30:1757–1766

Jacque J, Triques K, Stevenson M (2002) Modulation of HIV-1 replication by RNA interference. Nature 26:1–4

Jeffery L, Nakielny S (2004) Components of the DNA methylation system of chromatin control are RNA-binding proteins. J Biol Chem 279:49479–49487

Johansen LK, Carrington JC (2001) Silencing on the spot. Induction and suppression of RNA silencing in the Agrobacterium-mediated transient expression system. Plant Physiol 126:930–938

Kawasaki H, Taira K (2004) Induction of DNA methylation and gene silencing by short interfering RNAs in human cells. Nature 9:211–217

Kawasaki H, Taira K, Morris KV (2005) siRNA induced transcriptional gene silencing in mammalian cells. Cell Cycle 3:442–448

Kobinger GP, Weiner DJ, Yu Q, Wilson JM (2001) Filovirus-pseudotyped lentiviral vector can efficiently and stably transduce airway epithelia in vivo. Nat Biotechnol 19:225–230

Lee NS, Rossi JJ (2004) Control of HIV-1 replication by RNA interference. Virus Res 102:53–58

Lee NS, Dohjima T, Bauer G, Li H, Li M, Ehsani A, Salvaterra P, Rossi J (2002) Expression of small interfering RNAs targeted against HIV-1 rev transcripts in human cells. Nat Biotechnol 19:500–505

Lee Y, Ahn C, Han J, Choi H, Kim J, Yim J, Lee J, Provost P, Radmark O, Kim S, Kim VN (2003) The nuclear RNase III Drosha initiates microRNA processing. Nature 425:415–419

Li H, Li WX, Ding SW (2002) Induction and suppression of RNA silencing by an animal virus. Science 296:1319–1321

Liu J, Carmell MA, Rivas FV, Marsden CG, Thomson JM, Song JJ, Hammond SM, Joshua-Tor L, Hannon GJ (2004) Argonaute2 is the catalytic engine of mammalian RNAi. Science 305:1437–1441

Llave C, Kasschau KD, Carrington JC (2000) Virus-encoded suppressor of posttranscriptional gene silencing targets a maintenance step in the silencing pathway. Proc Natl Acad Sci USA 97:13401–13406

Lund E, Guttinger S, Calado A, Dahlberg JE, Kutay U (2004) Nuclear export of microRNA precursors. Science 303:95–98

Mallory AC, Ely L, Smith TH, Marathe R, Anandalakshmi R, Fagard M, Vaucheret H, Pruss G, Bowman L, Vance VB (2001) HC-Pro suppression of transgene silencing eliminates the small RNAs but not transgene methylation or the mobile signal. Plant Cell 13:571–583

Mallory AC, Reinhart BJ, Bartel D, Vance VB, Bowman LH (2002) A viral suppressor of RNA silencing differentially regulates the accumulation of short interfering RNAs and micro-RNAs in tobacco. Proc Natl Acad Sci U S A 99:15228–15233

Martinez J, Patkaniowska A, Urlaub H, Luhrmann R, Tuschl T (2002) Single-stranded antisense siRNAs guide target RNA cleavage in RNAi. Cell 110:563–574

Matzke MA, Primig M, Trnovsky J, Matzke AJM (1989) Reversible methylation and inactivation of marker genes in sequentially transformed tobacco plants. EMBO J 8:643–649

Meissner W, Rothfels H, Schafer B, Seifart K (2001) Development of an inducible pol III transcription system essentially requiring a mutated form of the TAT-binding protein. Nucleic Acids Res 29:1672–1682

Montgomery MK, Xu S, Fire A (1998) RNA as a target of double-stranded RNA-mediated genetic interference in Caenorhabditis elegans. Proc Natl Acad Sci USA 95:15502–15507

Morris KV, Chung C, Witke W, Looney DJ (2004) Inhibition of HIV-1 replication by siRNA targeting conserved regions of gag/pol. RNA Biol 1:114–117

Morris KV, Chan SW, Jacobsen SE, Looney DJ (2004a) Small interfering RNA-induced transcriptional gene silencing in human cells. Science 305:1289–1292

Morris KV, Gilbert J, Wong-Staal F, Gasmi M, Looney DJ (2004b) Transduction of cell lines and primary cells by FIV-packaged HIV vectors. Mol Ther 10:181–190

Ngo H, Tschudi C, Gull K, Ullu E (1998) Double-stranded RNA induces mRNA degradation in Trypanosoma brucei. Proc Natl Acad Sci USA 95:14687–14692

Nishikura K (2001) A short primer on RNAi: RNA-directed RNA polymerase acts as a key catalyst. Cell 107:415–418

Novina CD, Murray MF, Dykxhoorn DM, Beresford PJ, Riess J, Lee S, Collman RG, Leiberman J, Shankar P, Sharp PA (2002) siRNA-directed inhibition of HIV-1 infection. Nat Med 8:681–686

Pasquinelli AE (2002) MicroRNAs: deviants no longer. Trends Genet 18:171–173

Pasquinelli AE, Ruvkun G (2002) Control of developmental timing by micrornas and their targets. Annu Rev Cell Dev Biol 18:495–513

Poeschla E, Corbeau P, Wong-Staal F (1996) Development of HIV vectors for anti-HIV gene therapy. Proc Natl Acad Sci U S A 93:11395–11399

Price MA, Case SS, Carbonaro DA, Yu XJ, Petersen D, Sabo KM, Curran MA, Engel BC, Margarian H, Abkowitz JL, Nolan GP, Kohn DB (2002) Expression from second-generation feline immunodeficiency virus vectors is impaired in human hematopoietic cells. Mol Ther 6:645–652

Qin X, An D, Chen ISY, Baltimore D (2002) Inhibiting HIV-1 infection in human T cells by lentiviral mediated delivery of small interfering RNA against CCR5. Proc Natl Acad Sci USA 100:183–188

Quinonez R, Sutton RE (2002) Lentiviral Vectors for gene delivery into cells. DNA Cell Biol 12:937–951

Radhakrishnan S, Gordon J, Del Valle L, Cui J, Khalili K (2004) Intracellular approach for blocking JC virus gene expression by using RNA interference during viral infection. J Virol 78:7264–7269

Richman DD, Havlir D, Corbeil J, Looney D, Ignacio C, Spector SA, Sullivan J, Cheeseman S, Barringer K, Pauletti D (1994) Nevirapine resistance mutations of human immunodeficiency virus type 1 selected during therapy. J Virol 68:1660–1666

Sandrin V, Russell SJ, Cosset FL (2003) Targeting retroviral and lentiviral vectors. Curr Top Microbiol Immunol 281:137–178

Scherr M, Rossi JJ (1998) Rapid determination and quantitation of the accessibility to native RNAs by antisense oligodeoxynucleotides in murine cell extracts. Nucleic Acids Res 26:5079–5085

Schwarz DS, Hutvagner G, Du T, Xu Z, Aronin N, Zamore PD (2003) Asymmetry in the assembly of the RNAi enzyme complex. Cell 115:199–208

Sharp PA (2001) RNA interference. Genes Dev 15:485–490

Sioud M, Sorensen DR (2003) Cationic liposome-mediated delivery of siRNAs in adult mice. Biochem Biophys Res Commun 312:1220–1225

Sledz CA, Holko M, de Veer MJ, Silverman RH, Williams BR (2003) Activation of the interferon system by short-interfering RNAs. Nat Cell Biol 5:834–839

Song E, S Lee, Dykxhoorn DM, Novina C, Zhang D, Crawford K, Cerny J, Sharp PA, Leiberman J, Manjunath N, Shankar P (2003) Sustained small interfering RNA-mediated human immunodeficiency virus type 1 inhibition in primary macrophages. J Virol 77:7174–7181

Sorensen DR, Leirdal M, Sioud M (2003) Gene silencing by systemic delivery of synthetic siRNAs in adult mice. J Mol Biol 327:761–766

Sui G, Soohoo C, Affar E, Gay F, Shi Y, Forrester WC, Shi Y (2002) A DNA vector-based RNAi technology to suppress gene expression in mammalian cells. Proc Natl Acad Sci USA 99:5515–5520

Sumimoto H, Miyagishi M, Miyoshi H, Taira K, Kawakami Y (2003) Development of an efficient small interfering RNA (siRNA) expression system with a lentiviral vector. Mol Ther 7:S34

Surabhi RM, Gaynor RB (2002) RNA interference directed against viral and cellular targets inhibits human immunodeficiency virus type 1 replication. J Virol 76:12963–12973

Tijsterman M, Ketting RF, Plasterk RH (2002) The genetics of RNA silencing. Annu Rev Genet 36:489–519

Tuschl T (2002) Expanding small RNA interference. Nat Biotechnol 20:446–448
Volpe TA, Kidner C, Hall IM, Teng G, Grewal SIS, Martienssen RA (2002) Regulation of heterochromatic silencing and histone H3 lysine-9 methylation by RNAi. Science 297:1833–1837
Wassenegger M (2000) RNA-directed DNA methylation. Plant Mol Biol 43:203–220
Wassenegger M, Graham MW, Wang MD (1994) RNA-directed de novo methylation of genomic sequences in plants. Cell 76:567–576
Westerhout EM, Ooms M, Vink M, Das AT, Berkhout B (2005) HIV-1 can escape from RNA interference by evolving an alternative structure in its RNA genome. Nucleic Acids Res 33:796–804
White SM, Renda M, Nam NY, Klimatcheva E, Zhu Y, Fisk J, Halterman M, Rimel BJ, Federoff H, Pandya S, Rosenblatt JR, Planelles V (1999) Lentivirus vectors using human and simian immunodeficiency virus elements. J Virol 73:2832–2840
Wu X, Li Y, Crise B, Burgess SM (2003) Transcription start regions in the human genome are favored targets for MLV integration. Science 300:1749–1751
Yam PY, Li S, Wu JU, Hu J, Zaia JA, Yee J (2002) Design of HIV vectors for efficient gene delivery into human hematopoietic cells. Mol Ther 5:479–484
Zeng Y, Cullen BR (2002) RNA interference in human cells is restricted to the cytoplasm. RNA 8:855–860

RNA Interference: Its Use as Antiviral Therapy

J. Haasnoot · B. Berkhout (✉)

Department of Human Retrovirology, Academic Medical Centre, University of Amsterdam, Meibergdreef 15, 1105 AZ Amsterdam, The Netherlands
b.berkhout@amc.uva.nl

1	Introduction	118
2	The Biology of RNA Interference	118
2.1	miRNA Biosynthesis and DICER-Mediated Processing	119
2.2	RISC-Mediated Target RNA Cleavage	120
2.3	RNAi and the Impact on Chromosome Structure and Transcription	121
3	Optimizing the siRNA Inhibitor	122
4	Stable Intracellular Expression of Short Hairpin RNA	123
5	Inhibition of Viruses by RNAi	124
5.1	HIV-1 Inhibition by RNAi, Viral Escape and Human Countermeasures	125
5.2	In Vivo Evidence for Inhibition of Respiratory Viruses	132
5.3	Inhibition of HCV by RNAi	133
5.4	Inhibition of Human Coronaviruses by RNAi	134
5.5	Inhibition of HBV by RNAi	134
5.6	Inhibition of DNA Viruses by RNAi	135
6	Off-Target Effects	136
7	Viruses Utilize RNAi in Their Replication Strategy	137
7.1	Viroids Also Utilize the RNAi Mechanism	138
8	RNAi Versus the Antiviral Interferon System	138
9	Adenovirus VA RNAs as Suppressor of the RNAi and Interferon Systems	139
10	The Future of RNAi Therapeutics	140
	References	140

Abstract RNA interference (RNAi) is a sequence-specific gene-silencing mechanism that has been proposed to function as a defence mechanism of eukaryotic cells against viruses and transposons. RNAi was first observed in plants in the form of a mysterious immune response to viral pathogens. But RNAi is more than just a response to exogenous genetic material. Small RNAs termed microRNA (miRNA) regulate cellular gene expression programs to control diverse steps in cell development and physiology. The discovery that exogenously delivered short interfering RNA (siRNA) can trigger RNAi in mammalian cells has made it into a powerful technique for generating genetic knock-outs. It also raises the possibility to use RNAi technology as a therapeutic tool against pathogenic viruses. Indeed, inhibition of virus replication has been reported for several human pathogens including human immunodeficiency virus, the hepatitis B and C viruses and influenza virus. We reviewed the field of antiviral RNAi research in 2003 (Haasnoot et al. 2003), but many new studies have recently been published. In this review, we present a complete listing of all antiviral strategies published up to and including December 2004. The latest developments in the RNAi field and their antiviral application are described.

Keywords RNA interference · siRNA · shRNA · RISC · Virus · Antiviral therapy

1
Introduction

RNA interference (RNAi) is certainly not the first mechanism with therapeutic potential that is based on nucleic acid. The idea of using nucleic acid-based inhibitors of gene expression stems from research performed in the 1980s when antisense DNA oligonucleotides could be chemically synthesized for the first time. Scientists had high hopes for antisense, but thus far only a single drug, called Vitravene, cleared the Food and Drug Administration (FDA) in 1998 and is currently used to treat cytomegalovirus-induced eye infections in acquired immunodeficiency syndrome (AIDS) patients. The exploitation of DNA-based therapeutic agents remains limited because of delivery problems and unpredictable effectiveness. Similarly, ribozymes once held promise, but not one ribozyme-based drug has come to the market. RNAi is about to change the momentum. Acuity Pharmaceuticals in Philadelphia filed an investigational new drug application with the FDA in August 2003 to begin testing small interfering RNA (siRNA) treatment for the wet form of age-related macular degeneration (AMD). Still, it could be years before the first RNAi drug is approved for consumer use; but the field is moving at an unprecedented speed. The RNAi pathway is commonly considered a cellular defence mechanism against viruses, and it can be used as a therapeutic tool against pathogenic viruses. We will summarize the current status of RNAi-mediated inhibition of virus replication and discuss the possibilities for the development of RNAi-based antiviral therapeutics.

2
The Biology of RNA Interference

RNAi is a cellular regulatory pathway triggered in response to double-stranded RNA (dsRNA) (Hannon 2002). Since its first description in *Caenorhabditis elegans*, RNAi has been found to exist in many eukaryotic organisms and to be involved in an extraordinary number of gene-silencing phenomena (Fire et al.

1998; Hannon 2002). The RNAi machinery consists of a conserved core of factors with roles in recognizing, processing and effecting the responses to dsRNA. MicroRNAs (miRNAs) are a growing family of small non-protein-coding regulatory transcripts found in many eukaryotic organisms. miRNAs are processed via the RNAi machinery, and some have been shown to regulate the expression of homologous target-gene transcripts. miRNAs were first described in *C. elegans*; the lin-4 and let-7 transcripts. Both target the 3' untranslated regions (3'-UTRs) of developmental transcription factors and suppress their translation (Grishok et al. 2001; Reinhart et al. 2000; Vella et al. 2004). Current estimates for the number of miRNAs represent 1% of the predicted number of genes in a mammalian genome, similar to the proportion represented by large gene families such as transcription factors. Computational methods are being used to identify targets with the expectation that the results will provide clues as to the mechanism of action of a particular miRNA (Lai 2004). As more miRNA:target pairs were described, it became apparent that regulation of development might be a common theme in miRNA biology. miRNAs and their targets have been found to affect diverse processes, including flowering time and leaf patterning in *Arabidopsis*, neuronal asymmetry in *C. elegans*, and developmentally regulated cell proliferation in *Drosophila* (Aukerman and Sakai 2003; Brennecke et al. 2003; Kidner and Martienssen 2004; Palatnik et al. 2003). Little is known about the roles of miRNAs in mammals, but the field is moving fast (Miska et al. 2004).

2.1
miRNA Biosynthesis and DICER-Mediated Processing

miRNA genes are often located in clusters that appear to be transcribed as a polycistronic transcript (Lee et al. 2002b). Perhaps surprisingly, neither the miRNA promoters nor the RNA polymerase responsible for miRNA transcription has been characterized in detail thus far. However, recent evidence indicates that human miRNAs are processed from capped, polyadenylated transcripts that can also function as messenger (m)RNAs (Cai et al. 2004). It is also clear that miRNA genes are often under strict developmental- and tissue-specific control (Dostie et al. 2003; Houbaviy et al. 2003; Krichevsky et al. 2003; Lagos-Quintana et al. 2002; Moss and Tang 2003). Many primary miRNA transcripts are predicted by computer algorithms to adopt elaborate stem-loop structures (Lau et al. 2001; Lee and Ambros 2001; Mourelatos et al. 2002). Cleavage of the stem-loops by the RNase III enzyme Drosha liberates approximately 70-nucleotide (nt) precursor miRNAs (pre-miRNAs) (Lee et al. 2003b). These RNAs have the characteristic two-nt 3' overhang end structure left by the staggered cut of RNase III enzymes. Recent studies have shown that Exportin-5 mediates the nuclear export of pre-miRNAs in a Ran–guanosine triphosphate (GTP)-dependent manner (Bohnsack et al. 2004; Lund et al. 2004; Yi et al. 2003). Interestingly, Exportin-5-mediated nuclear export of another cargo RNA, the adenovirus VA1 noncoding RNA, also requires a two- to three-nt 3' overhang (Gwizdek et al. 2004), suggesting that the structural determinants of Exportin-5 recognition may be implicit in Drosha processing. Entry into the cytoplasm brings pre-miRNAs into contact with Dicer, a predominantly cytoplasmic enzyme (Billy et al. 2001; Findley et al. 2003; Provost et al. 2002). Dicer cleaves pre-miRNAs into mature ~22mer miRNAs, but it can also cleave dsRNA into ~22mer siRNAs (Bernstein et al. 2001; Grishok et al. 2001; Hutvagner et al. 2001; Ketting et al. 2001; Knight and Bass 2001). Dicer is a modular enzyme composed of two RNaseIII domains, a DExH/DEAH box RNA helicase domain and a PAZ domain, as well as a domain of unknown function (DUF283) and a double-stranded RNA-binding motif (Bernstein et al. 2001).

2.2
RISC-Mediated Target RNA Cleavage

Dicer cleavage is followed by release of the mature miRNA or siRNA, and its incorporation into a RISC (RNA-induced silencing complex) effector complex whose diverse functions include mRNA cleavage, translational suppression, transcriptional silencing and heterochromatin formation. The *C. elegans* double-stranded RNA-binding protein RDE-4 and its *Drosophila* homologue R2D2 facilitate the transfer of siRNAs to RISC (Liu et al. 2003; Tabara et al. 2002). There is also likely to be a number of additional components that aid in RISC assembly (Tomari et al. 2004a), particularly those that unwind siRNAs or miRNAs and through this action determine which RNA strand is chosen for silencing (Schwarz et al. 2003). Which strand of the miRNA or siRNA duplex is preferentially incorporated into RISC is also determined by the thermodynamic properties of the nucleic acid duplex (Martinez et al. 2002a; Schwarz et al. 2003).

RISC ribonucleoprotein complexes contain members of the PAZ–Piwi domain Argonaute family of proteins, siRNAs or miRNAs, and miRNA/siRNA complementary mRNAs. The Argonaute protein recognizes the 3' overhang of the single-stranded (ss)RNA and is a crucial component of RISC. The loaded ssRNA guides the search for mRNA with complementary sequences and defines the actual site of cleavage (~10 bases from the 3' overhang). The first crystal structure of an archaeal Argonaute protein revealed an RNaseH-like fold with a catalytic Asp-Asp-Glu (DDE) motif for RNA cleavage, suggesting that this protein is the actual RNA slicer (Parker et al. 2004; Song et al. 2004). Mammalian cells have four Argonaute homologues, but recent evidence suggests that Ago2 has the RNA-cleaving activity (Liu et al. 2004a; Rand et al. 2004). There may be a division of labour between Argonautes. For instance, *Drosophila melanogaster* was recently shown to have Ago2 for unwinding of siRNA, which is a prerequisite for siRNA-mediated cleavage, and Ago1 interacts directly with Dicer, leading to stabilization of mature miRNAs (Okamura et al. 2004). The endonucleolytic cleavage by RISC generates two RNA fragments with a 3′ hydroxyl and 5′ phosphate (Schwarz et al. 2004). These mRNA fragments must be eliminated because they may encode unwanted polypeptide fragments when translated. The fate of the mRNA cleavage products remained unclear for some time, but recent evidence has implicated a 5′ to 3′ exonuclease, the cytoplasmic AtXRN4 enzyme in *Arabidopsis*, in removal of 3′ fragments (Gazzani et al. 2004; Souret et al. 2004).

RISC contains a number of accessory factors, some of which have activities necessary for effector function. The precise biochemical mechanisms whereby RISCs carry out their functions are unknown. Two well-characterized RISC activities are mRNA cleavage and translational suppression. RISC is a multiple-turnover enzyme complex, and, once incorporated, an siRNA or miRNA can direct multiple rounds of target cleavage (Hutvagner and Zamore 2002). Whether RISC can also act in a catalytic manner in mediating translational suppression is not clear. It is also not known whether miRNA–mRNA duplexes require specific features (e.g. a mismatch) to be recognized by factors that mediate the translational repression. Recent evidence indicates that the primary function of the miRNAs is to guide their associated proteins to the mRNA. Tethering Ago proteins to the 3′-UTR of an mRNA by other means also resulted in translational repression (Pillai et al. 2004).

2.3
RNAi and the Impact on Chromosome Structure and Transcription

Although miRNA molecules are thought to act at the post-transcriptional level by interfering with mRNAs, there is accumulating evidence that some of them, and other non-coding RNA molecules, are involved in transcriptional silencing and heterochromatin formation as well. RNAi is known to work transcriptionally in plants by methylating gene promoters with sequences complementing the RNAs. De novo cytosine methylation of genomic DNA was shown to occur in plants infected with RNA viroids whose sequences were homologous to the methylated genomic sequences (Wassenegger et al. 1994). This process is referred to as RNA-directed DNA methylation (RdDM). Subsequently, dsRNA targeting a promoter region was shown to induce RdDM and to trigger transcriptional silencing. This silencing was accompanied by siRNA production, pointing to an RNAi-like mechanism for gene silencing at the transcriptional level (Mette et al. 1999). There is recent evidence that DNA methylation and gene silencing can also be induced by siRNA in human cells (Kawasaki and Taira 2004; Morris et al. 2004). Cells typically pass this DNA-regulating modification on to daughter cells, possibly permitting more lasting inhibitory effects.

RNAi has been linked with heterochromatin formation, gene silencing and chromosome segregation in fission yeast. Transcripts derived from the outer centromere repeats of the yeast chromosomes are chopped up by the RNAi machinery, thus forming siRNAs that are required for the formation of heterochromatin over the outer repeats. Chromosome analysis in a mammalian cell with a conditional knockout of Dicer suggested that RNAi does also play a role in heterochromatin formation at centromeres in vertebrates (Fukagawa et al. 2004).

This second DNA pathway could provide an alternative means to inhibit certain viruses by RNAi, in particular viruses that utilize DNA transcription as part of their replication strategy, which includes all DNA viruses and retroviruses. Compared to the RNA phase, the existence of a DNA phase of the RNAi machinery could provide longer-lasting effects and thus open new therapeutic strategies. On the other hand, a broader mechanism of action does also increase the likelihood of unintended consequences through off-target silencing.

3
Optimizing the siRNA Inhibitor

The results of detailed biochemical studies suggest a need to revise the current design rules for the construction of siRNAs (Khvorova et al. 2003; Schwarz et al. 2003). One could select RNA targets for which the corresponding siRNA molecule has an optimal thermodynamic signature. Naturally occurring miRNAs show a strong bias for accumulating only one strand into functional RISC complexes (Bartel 2004). In the course of identifying more active and more specific siRNAs, it was noticed that the sequence composition of the siRNA duplex influences the ratio of the "sense" and "antisense" (complementary to the target gene) siRNAs entering the RISC complex (Khvorova et al. 2003; Schwarz et al. 2003). Particularly important is a low base-pairing stability at the 5′ end of the antisense strand relative to the 3′ end, which increases the chance that this antisense strand enters the RISC complex. An additional advantage of such biased uptake in RISC is that there will be less off-target effects through RISC complexes that are loaded with the sense strand. In *Drosophila*, the orientation of the Dicer-2/R2D2 protein heterodimer determines which siRNA strand associates with the core RISC protein Argonaute 2 (Tomari et al. 2004b). R2D2 binds the siRNA end with the greatest double-stranded character, thereby orienting the heterodimer on the siRNA duplex. Strong R2D2 binding requires a 5′-phosphate on the RNAi strand that is subsequently excluded from RISC. This explains why relatively low base-pairing at the 5′ end of the antisense strand is beneficial.

In case the selected siRNA is still sub-optimal, one could improve its thermodynamic signature by altering the structure of the siRNA duplex. This can be done by introduction of weak G-U base-pairs, internal loops or bulges at distinct positions within the duplex, obviously without changing the antisense strand that mediates subsequent mRNA-cleavage (Miyagishi et al. 2004). In cells, it may even be feasible to convert adenosine (A) to inosine (I) within dsRNA by RNA editing enzymes (Tonkin and Bass 2003). In case of synthetic siRNA, one could introduce chemical modifications at specific positions within the sense strand of the siRNA molecule to reach the optimal thermodynamic signature. Chemical modification of one of the strands may also be used to either block RISC incorporation of the sense strand or trigger incorporation of the antisense strand. This optimization will also reduce the likelihood of sense-strand-directed off-target effects.

Other aspects may also influence the choice of optimal antivirals, including the effect of RNA secondary structure in the target and the presence of target RNA-associated proteins. When targeting a viral pathogen, it is also particularly important to consider the degree in which the target sequence is conserved among virus isolates. For instance, we selected eight siRNAs against well-conserved and highly accessible domains of the human immunodeficiency virus (HIV)-1 RNA genome, but the majority of these molecules were ineffective, possibly because they did not obey these new rules for optimal siRNA-design (Das et al. 2004).

4
Stable Intracellular Expression of Short Hairpin RNA

The addition of chemically or enzymatically synthesized siRNA to cells is the most convenient way to induce RNAi in the laboratory setting, but stable intracellular expression may be required for several therapeutic applications. The first vectors for the expression of functional siRNA [mostly short hairpin (sh)RNA] were described in 2002 (Brummelkamp et al. 2002; Paddison et al. 2002). Several improvements have been reported, but much optimization is still to be expected. RNA polymerase III (Pol III) promoters have been widely used to express shRNA for silencing a variety of target genes (Brummelkamp et al. 2002), and several modifications of this system were recently published. It was reported that the enhancer from the cytomegalovirus immediate-early promoter (a Pol II unit) can enhance the synthesis of siRNA from a Pol III unit (Xia et al. 2003).

The two different human Pol III promoters that have been widely used naturally encode either a small nuclear (sn)RNA (U6 unit) or part of the RNase P molecule (H1 unit), but more efficient siRNA expression was recently reported for a modified transfer (t)RNA^{met} promoter (MTD unit) (Boden et al. 2003b). Another major improvement is the design of a doxycycline-regulated H1 promoter that allows the inducible knockdown of gene expression by siRNAs (van de Wetering et al. 2003). Similarly, a doxycycline-inducible siRNA expression cassette was inserted in a lentivirus vector (Wiznerowicz and Trono 2003). Alternatively, the ecdysone system has been transplanted onto the U6 promoter for inducible shRNA expression (Gupta et al. 2004). The Cre–loxP recombination system has also been used to switch on shRNA expression, which would allow regulation in a spatially, temporally or cell- or tissue-specific manner (Kasim et al. 2004). These inducible systems should be particularly useful for the study of proteins that have an impact on cell growth or cell differentiation (e.g. oncogenes and tumour-suppressor genes). The presence of an inverted repeat to encode the hairpin RNA may affect the stability of some vectors, and a convergent transcription unit with two opposing U6 promoters has therefore been designed (Tran et al. 2003). Strategies to express multivalent shRNA constructs have also been described (Anderson et al. 2003), which seems very important to avoid the danger of viral escape (Berkhout 2004). Much remains to be learnt about the shRNA design rules and features that promote efficient shRNA expression modification, intracellular transport, processing, etc. For instance, a detailed mutational analysis elegantly demonstrated that it is possible to introduce multiple G-U base-pairs by mutation of the sense strand (Miyagishi et al. 2004).

5
Inhibition of Viruses by RNAi

Viruses are both inducers and targets of RNA silencing in plants (Vance and Vaucheret 2001). The antiviral capacity of RNA silencing has been used as a tool to generate virus resistance in plants (Lindbo and Dougherty 1992; Smith et al. 2000; Waterhouse et al. 1998). RNAi technology is currently being used to inhibit viral replication in animal cells. Promising results have been obtained with RNAi against several animal viruses both in in vitro and in vivo settings.

The first demonstration of RNAi-mediated inhibition of a human pathogenic virus was reported by Bitko and Barik in 2001 (Bitko and Barik 2001). These authors reported a tenfold inhibition of human respiratory syncytial virus (HRSV) replication in vitro using nanomolar concentrations of synthetic siRNAs that targeted the viral polymerase subunit P and the fusion protein F. Currently, many other studies have described RNAi-mediated inhibition of a large variety of viruses. RNAi-mediated inhibition of HIV-1 has received much attention (see below, Tables 1 and 2). In addition, 17 different RNA viruses, and 10 different DNA viruses have been efficiently targeted by RNAi (Tables 3 and 4). These include important human pathogens such as hepatitis C virus (HCV), dengue (DEN) virus, severe acute respiratory syndrome (SARS) coronavirus, poliovirus, influenza A virus, hepatitis D virus (HDV), human rhinovirus-16 (HRV-16), hepatitis B virus (HBV), herpes simplex virus type-1 (HSV-1), human papillomavirus (HPV), JC virus (JCV), Epstein-Barr virus (EBV), and human cytomegalovirus (HCMV). Other viruses listed in Tables 1 and 4 are: enterovirus 71 (EV71), Semliki Forest virus (SFV), rhesus rotavirus (RRV), flock house virus (FHV), Rous sarcoma virus (RSV), porcine endogenous retrovirus (PERV), foot-and-mouth disease virus (FMDV), murine herpesvirus 68

(MHV68), *Orgyia pseudotsugata* M nucleopolyhedrovirus (OpMNPV), *Autographa californica* nucleopolyhedrovirus (AcNPV), *Microplitis demolitor* bracovirus (MdBV).

Initially, the standard method to induce RNAi towards viruses in mammalian cells was transfection of synthetic siRNAs corresponding to viral sequences shortly before or after a viral challenge. Currently, transient transfection of plasmids that express antiviral shRNAs is also commonly used. Both strategies can result in potent, albeit temporary inhibition of virus replication. In order to obtain long-term virus resistance, researchers have turned to a combined RNAi/gene therapy approach. In this approach, lenti-, retro- or adeno-associated virus (AAV) vectors are used to stably transduce cells with constructs expressing shRNA, resulting in viral resistance. We provide a complete overview of these antiviral studies in Tables 1–4 and discuss the possibility to develop RNAi-based antiviral therapies. The focus will be on HIV-1, and a few RNA and DNA viruses.

5.1
HIV-1 Inhibition by RNAi, Viral Escape and Human Countermeasures

Several studies reported that siRNA can suppress HIV-1 (Capodici et al. 2002; Coburn and Cullen 2002; Hu et al. 2002; Jacque et al. 2002; Lee et al. 2002a; Martinez et al. 2002b; Novina et al. 2002; Park et al. 2002; Qin et al. 2003; Surabhi and Gaynor 2002). There is some evidence suggesting that the genomic RNA present within an infecting virion particle is targeted for destruction, but it appears that new viral transcripts, synthesized from the integrated provirus, are more efficient targets. Most studies used chemically synthesized siRNAs that were transfected into cells either shortly before or after challenge with HIV-1. Despite the transient nature of such a transfection experiment, a single siRNA application is able to achieve relatively long-lasting suppression (Song et al. 2003). Other studies used transient transfection of siRNA-expression vectors. However, the development of efficient vector delivery systems capable of mediating stable siRNA expression in mature T lymphocytes or progenitor stem cells will be a minimal requirement for RNAi to be used as a therapeutic modality against HIV-1. Lentiviral vectors with a Pol III expression cassette are an efficient means to deliver anti-HIV siRNAs into haematopoietic precursor cells. In a recent report, the transduced human cells were allowed to differentiate in vivo in the SCID-hu thymopoiesis mouse model (Banerjea et al. 2003), and the mature T lymphocytes derived from this model resisted HIV-1 infection ex vivo.

Two studies addressed the potency and durability of anti-HIV RNAi approaches. Boden et al. expressed an shRNA against the tat gene in an AAV vector with an H1-promoter (Boden et al. 2003a). Potent inhibition was scored, but an escape virus variant appeared in prolonged cultures. Similar results were described by Das et al. using a lentiviral vector with an H1 unit expressing

Table 1 Inhibition of HIV-1 by RNAi

Target gene	RNAi inducer	Cell type	Fold inhibition of virus replication	Reference
Tat, Rev	Intracellular siRNA	293/EcR	10,000	Lee 2002a
LTR, Vif, Nef	siRNA, shRNA	Magi, PBLs	>20	Jacque 2002
Gag	siRNA	Magi-CCR5, Hela-CD4	>4	Novina 2002
Gag, Pol	siRNA	Hos.T4.CXCR4	>10	Hu 2002
Gag, LTR	siRNA	U87-CD4$^+$/CXCR4$^+$, CCR5$^+$, 293T	~4	Capodici 2002
Tat+Rev	siRNA	Jurkat, HPBLs	>10	Coburn 2002
Gag, Env	500-nt dsRNAs	PBMCs, COS, Hela-CD4$^+$	70	Park 2002
Nef	556-nt dsRNA	MT-4 T, U937	2	Yamamoto 2002
Tat, Rt	siRNA	Magi	5–100	Surabhi 2002
Env	siRNA	Cos, Hela-CD4$^+$, PBMCs	~10	Park 2003
Tat	Stable shRNA[a]	macrophages	12	Lee 2003a
Rev	Stable shRNA[a]	CD34$^+$ derived macrophages, T cells in SCID mice	8–16	Banerjea 2003
Tat/Rev, Rev	Stable shRNA[a]	PBMC	~1,000	Li 2003
Tat	Stable shRNA[b], siRNA and shRNA	293T, H9	33	Boden 2003a
	shRNA, siRNA	293T	14	Boden 2003b
	Stable shRNA[b]	293T, H9	1,200	Boden 2004a
	Intracellular pre-miRNA*	293	25–45	Boden 2004b
Gag	shRNA	293T	275	Pusch 2003
Pol	shRNA	293, Hela	10	Paul 2003
p24	siRNA	MDM	~5	Song 2003
Luc, GFP	Stable shRNA[a]	293T, PBMCs	10	Nishitsuji 2004
Nef	Stable shRNA[c]	SupT1	>10	Das 2004
Env, Tat/Rev, Rev, Nef, Pol	shRNA	293T, CEM	1,000	Scherer 2004
gp41, Nef, Tat, Rev	siRNA	HelaCD4-LTR-β-Gal, HelaCD4, 293T	~200–1,000	Dave 2004

Table 1 (continued)

Target gene	RNAi inducer	Cell type	Fold inhibition of virus replication	Reference
PBS	siRNAs, shRNAs, stable shRNA[b] and HIV vectors[d]	SupT1	~76	Han 2004
Rev	shRNA, stable shRNA[e]	EcR-293, HT1080, 293, CEM	~10	Unwalla 2004
Nef	shRNA	Hela, BHK, Jurkat, MT-4, CRFK	~50	Omoto 2004

All siRNAs were chemically synthesized and transfected into cells unless indicated otherwise. ShRNAs were intracellularly expressed from transfected plasmids under the control of a Pol III promoter (H1 or U6). The fold inhibition of virus production represents the result obtained with the most efficient siRNA or shRNA. [a-e] Stable expression of shRNAs was obtained using: [a] a lentiviral, [b] AAV vector, [c] a retroviral vector, [d] an HIV-vector containing expression cassette for a shRNA against the heterologous PBS sequence, or [e] a lentiviral vector containing a HIV-1 Tat inducible shRNA expression cassette.
*Pre-miRNA are shRNAs against HIV made to resemble miRNAs

an siRNA against sequences in the nef gene (Das et al. 2004). The latter study described seven independent HIV-1 escape variants. The combined results convincingly demonstrate that inhibition was potent and sequence-specific, but also that HIV-1 is able to escape from the inhibitory action of a single siRNA. Boden et al. described a single revertant with a point mutation in the target sequence, and Das et al. described a large variety of escape routes (point mutation, double point mutation, partial or complete deletion of the target sequence). A deletion-based resistance mechanism seems impossible in case essential HIV-1 genes or critical sequence motifs are targeted. Thus, one should preferentially target essential sequences that are well conserved among HIV-1 isolates. Interesting targets with relatively little mutational freedom are the multiple overlaps in reading frames within the HIV-1 genome, including a triple overlap (*tag–rev–env*). Ideally, one should target more than one of these essential and well-conserved viral sequences. Such combination siRNA-therapy mimics the successful strategy to combat HIV-1 with multiple antiviral drugs, and should avoid the evolution of escape variants.

We recently discovered an alternative resistance mechanism that is not triggered by mutation of the target sequence. Instead, a mutation in the flanking sequences was selected, which was subsequently shown to induce a conformational change within the target sequence such that it is protected from RISC attack (Westerhout et al. 2005). This finding indicates that it will not be very straightforward to predict viral escape routes. Nevertheless, one could make a first estimation of the chance of viral escape in a therapeutic setting

Table 2 Inhibition of HIV by RNAi-mediated silencing of essential host genes

Target gene	RNAi inducer	Cell type	Fold inhibition of virus replication	Reference
CXCR4/CCR5	siRNA	U87-CD4$^+$/CXCR4$^+$, CCR5$^+$	2–3	Martinez 2002b
NF-κB	siRNA	Magi, 293T	5	Surabhi 2002
CD4	siRNA	Magi-CCR5, Hela-CD4	4	Novina 2002
CCR5	siRNA	MDM	∼3	Song 2003
CCR5	shRNA	Magi-CCR5, PBLs	3–7	Qin 2003
CCR5	Stable shRNA[a]	macrophages	∼6	Lee 2003
PARP-1	siRNA	Hela, J111, Magic-5A	∼16	Kameoka 2004
CDK9/CyclinT1	siRNA	Hela, Magi	∼12	Chiu 2004
CXCR4	siRNA	293, HosCD4CXCR4, HosCD4CCR5	2	Zhou 2004
DC-SIGN	Stable shRNA[a]	DCs, Raji B, Hela	6–32	Arrighi 2004
CyPA	snRNA[b] and shRNA	293T, Hela, Jurkat, CEM-SS	∼6	Liu 2004b
SPT5	siRNA	Magi	∼32	Ping 2004

All siRNAs were chemically synthesized and transfected into cells unless indicated otherwise. ShRNAs were intracellularly expressed from transfected plasmids under the control a Pol III promoter (H1 or U6). The fold inhibition of virus production represents the result obtained with the most efficient siRNA or shRNA. [a]Stable expression of shRNAs was obtained using a lentiviral vector. [b]snRNA stands for antisense U7 small nuclear RNAs that disrupt CyPA splicing

with one or multiple siRNAs (Berkhout 2004). If we assume that an essential viral sequence is targeted, deletion is no option. Therefore, one and more likely two nt substitutions are required per 19-nt target sequence to obtain a fair level of resistance (Das et al. 2004). Assuming that 2-point mutations are needed to obtain complete resistance, and further assuming an error rate of the reverse transcriptase polymerase of 2×10^{-5}, the chance of viral escape in a single replication cycle is $19 \times [(2 \times 10^{-5})]^2 = 1.44 \times 10^{-7}$. Studies in the field of drug-resistance indicate that an untreated HIV-infected individual contains an effective viral population size of 10^4–10^5 (Rouzine and Coffin 1999), which means that most 1-nt substitutions will already be present within the viral population. Starting in an untreated patient with a moderate viral load, this means that resistance is likely to occur. Thus, it may indeed be important to consider siRNA combination therapy (SIRCT) (Berkhout 2004). With four effective siRNAs, the chance of viral escape drops to 2.1×10^{-14}. In practice, this

Table 3 Inhibition of RNA viruses by RNAi

Virus	Target gene	RNAi inducer	Cell type	Fold inhibition of virus replication	Reference
HRSV	P, F	siRNA	A549	~10	Bitko 2001
	P	siRNA	A549, Mice	~5,000	Bitko 2005
	NS1	shRNA	A549, Vero, Mice	~100	Zhang 2005
HPIV	P	siRNA	A549, Mice	~100	Bitko 2005
Dengue virus	Capsid, PrM, NS5	~250-nt ssRNA[a]	Mosquitoes, BHK, C6/36	>50	Adelman 2001
	PrM	290-nt dsRNA[b]	C6/36	100	Adelman 2002
	PrM, E, NS1, NS5	77-nt dsRNA	C6/36	10	Caplen 2002
HCV	NS3, NS5B	siRNA	Huh-7	10	Kapadia 2003
	Capsid, NS4B	siRNA	Huh-7.5	~100	Randall 2003
	5'-UTR	siRNA	Huh-7	7	Seo 2003
	5'-UTR, NS3, NS5b	siRNA, intracell. siRNA	Huh-7	~10	Wilson 2003
	5'-UTR	siRNA, shRNA, intracell. siRNA	Huh-7	~5	Yokota 2003
	5'-UTR, NS4B, NS5A, NS5B	15–40-nt siRNAs, stable shRNA[c]	Huh-7, 9B	20–100	Krönke 2004
	NS5A	siRNA	HepG2, Hep5A	4–5	Sen 2003
	5'-UTR, core, NS3, NS5B	siRNA, shRNA, stable shRNA[d]	Huh-7	~7	Takigawa 2004
Poliovirus	Capsid, 3Dpol	siRNA	Hela S3, mouse fibroblasts	100	Gitlin 2002
EV71	VP1, 3D	shRNA	Hela, Vero	5–10	Lu 2004b
SFV	Nsp-1,-2,-4	77-nt dsRNA	C6/36	2	Caplen 2002
RRV	VP4	siRNA	MA104	~4	Dector 2002
	NSP4, VP7	siRNA	MA104	4–5	Lopez 2005

Table 3 (continued)

Virus	Target gene	RNAi inducer	Cell type	Fold inhibition of virus replication	Reference
FHV	3'-UTR	500-nt dsRNA	S2	>100	Li 2002
RSV	Gag	siRNA	Chicken embryos, DF-1	5–10	Hu 2002
PERV	Gag, Pol, Env	siRNA, shRNA	293	5–10	Karlas 2004
FMDV	VP1	shRNA	BHK-21, Mice	5–10	Chen 2004
	3B, 3D	siRNA	BHK-21	>10	Kahana 2004
HDV	δAg	siRNA	Huh-7	5–20	Chang 2003
SARS-CoV	Pol	shRNA	Vero	5	Wang 2004b
	Pol	shRNA	Vero-E6, 293, Hela	16	Lu 2004a
	Pol	siRNA	FRhk-4	7–14	He 2003
	Spike	shRNA	293-T, Vero E6	~10	Zhang 2004b
HCoV-NL63	Spike	siRNA	LLC-Mk2	>10	Pyrc in prep
HRV-16	5'-UTR, vp4, vp2, vp3, vp1, 2A, 2C, 3A, 3C, 3D	siRNA	Hela	~10–20	Phipps 2004
Influenza A virus	PB1, PB2, PA, NP, M, NS	siRNA	MDCK, chicken embryos	200	Ge 2003
	NP, PA	siRNA	Mice	56	Tompkins 2004
	NP, PA, PB1	siRNA, stable shRNA[d]	Mice, Vero	~10	Ge 2004

All siRNAs were chemically synthesized and transfected into cells unless indicated otherwise. ShRNAs were intracellularly expressed from transfected plasmids under the control a Pol III promoter (H1 or U6). The fold inhibition of virus production represents the result obtained with the most efficient siRNA or shRNA. [a]Intracellular expressed dengue virus ssRNA using Sindbis virus as a vector. [b]290-bp hairpin RNA expressed from a transfected plasmid under the control of hsp 70 promoter. [c,d]Stable expression of shRNAs was obtained using [c] a retroviral vector or [d] a lentiviral vector

Table 4 Inhibition of DNA viruses by RNAi

Virus	Target	RNAi inducer	Cell type	Fold inhibition of virus replication	Reference
HPV-16	E6, E7	siRNA	CASKi, SiHa	No[a]	Jiang 2002
HBV	X, core	shRNA	Huh-7, HepG2	20	Shlomai 2003
	Core, HbsAg/Pol, X	shRNA	Huh-7, Mice	>6	McCaffrey 2003
	Core	siRNA	Huh-7, HepG2	~5	Hamasaki 2003
	Core	siRNA	HepAD38, HepAD79	4–50	Ying 2003
	HbsAg	siRNA	HepG2.2.15, Mice	5–100	Giladi 2003
	Core, S	siRNA	Mice	3	Klein 2003
	PA, PreC, S	siRNA	HepG2, 2.2.15	1.7–4.5	Konishii 2004
	hLa[b]	shRNA	HepG2, 2.2.15	19	Ni 2004
	X, core, Pol, S	shRNA	Huh-7	2.5–7	Zhang 2004a
MHV-68	Rta, ORF 45	siRNA	293T	>43	Jia 2003
OpMNPV	Op-iap3	511-nt dsRNA	Sf21, Ld652Y	No[a]	Means 2003
AcNPV	gp64, ie1	619, 451-nt dsRNA	Sf21, *T. mollitor* larvae	>20	Valdes 2003
MdBV	Glc1.8, egf1.0	289, 359-nt dsRNA	High Five	No[c]	Beck 2003
HSV-1	gE	siRNA	HaCaT	~4	Bhuyan 2004
JCV	VP1, Agno, T-Ag	siRNA	SVG-A	10–24	Orba 2004
	Agno, T-Ag	siRNA	phFA	~12	Radhakrishnan 2004
EBV	Zta	shRNA	NPC-TW01, 293A	11–16	Chang 2004
	LMP-1	shRNA	C666	No[d]	Li 2004a
HCMV	UL54	siRNA	U373	~2,000	Wiebusch 2004

All siRNAs were chemically synthesized and transfected into cells unless indicated otherwise. ShRNAs were intracellularly expressed from transfected plasmids under the control a Pol III promoter (H1 or U6). The fold inhibition of virus production represents the result obtained with the most efficient siRNA or shRNA. [a]*E6, E7* and *Op-iap3* are nonessential viral genes, but virus production was negatively affected through apoptosis of the host cell. [b]hLa is a host factor that is required for HBV replication. [c]*Glc1.8* and *egf1.0* are nonessential viral genes. [d]*LMP-1* is a non-essential viral gene that plays a role in cell transformation means that viral escape is impossible as long as viral suppression is complete. Even if several assumptions are wrong, the prospects are favourable that one can achieve effective and long-term viral suppression. An alternative strategy is to target unmutable host-encoded functions that are important for viral replication, but not essential for survival of the host cell (Table 2; Haasnoot et al. 2003).

Effective RNAi-based antiviral therapy is still facing serious technical hurdles, the major one being the delivery of siRNAs into the right cells. Some recent progress has been achieved in this field. Simple conjugates of siRNA and cholesterol, which was chemically linked to the terminal hydrozyl group of the sense RNA strand, were recently reported to trigger tissue delivery (Soutschek et al. 2004). Intravenous injections of the conjugate in mice resulted in uptake into several tissues, including the liver, jejunum, heart, kidneys, lungs and fat tissue. However, many questions remain before this method sees application in humans. For instance, the treatment might require the lifetime use of cholesterol-lowering compounds. Thus, the research on improved delivery systems based on proteins, liposomes or other molecules should continue apace.

Alternatively, RNAi-triggering genes could be transferred into the appropriate target cells. Such a gene therapy protocol seems ideally suited for the treatment of individuals that are chronically infected with HIV-1 and that fail on standard antiretroviral therapy. In chronically infected individuals, HIV-1 infects a significant fraction of the mature T cells each day, leading to cell killing either directly by HIV-1 or indirectly by the HIV-induced immune system. Thus, the preferential survival of even a minority of siRNA-expressing cells will result in their outgrowth over time. One could treat either the mature immune cells from the blood or haematopoietic stem cells from a patient's bone marrow, and put them back into the patient. The latter cells will proliferate into mature T cells and move to the periphery, thus forming a constant supply of cells that resist HIV-1 infection. This means that even a relatively inefficient ex vivo gene therapy protocol could be beneficial. Retroviral and especially lentiviral vectors are frequently used to deliver the siRNA-expression cassette in mammalian cells, although there is concern because the former vector triggered leukaemia in two children in a gene therapy trial (Check 2002). Results are expected soon from the first human trial using lentiviral vectors (Lu et al. 2004d; Lu et al. 2004c).

5.2
In Vivo Evidence for Inhibition of Respiratory Viruses

RNAi-mediated virus inhibition has been studied for a large group of RNA viruses (see Table 3). Recent studies in mice suggest that RNAi holds great promise for the prevention and treatment of infection of the respiratory viruses with an RNA genome such as influenza A virus, human parainfluenza virus

(HPIV) and HRSV (Bitko et al. 2005; Ge et al. 2004; Tompkins et al. 2004; Zhang et al. 2005). Due to limitations of anti-influenza vaccines and drugs, there is a real need for novel strategies to inhibit influenza virus. Worldwide, an estimated half million deaths per year are attributed to influenza virus, and there is the continuous threat of the emergence of a novel pandemic strain. To use siRNA as an in vivo therapeutic, it must be delivered efficiently to the appropriate tissue(s), in this case the lungs. Lungs are perhaps the most readily transfectable organs because they are likely the most vascularized tissue in the body. Furthermore, injected materials will first traverse the capillary beds of the lungs upon intravenous administration. Researchers have used polyethyleneimine (PEI) injected intravenously or intratracheally to deliver siRNAs and a lentiviral DNA vector expressing shRNAs (Ge et al. 2004). PEI is a cationic polymer that has been used to deliver DNA into lung cells. Others have delivered anti-influenza siRNAs intranasally with the cationic transfection reagent Oligofectamine (Tompkins et al. 2004). Reduction of the virus titre in the lungs and lethality was observed when the antivirals were administered either prior or subsequent to virus challenge.

Similarly, replication of HRSV and HPIV in mice could be blocked by intranasal delivery of synthetic siRNAs (Bitko et al. 2005). The authors show that this approach is effective both with and without the use of transfection reagents. Besides synthetic siRNAs, also intranasal administration of plasmids expressing shRNA against HRSV results in a significant decrease of viral titres (Zhang et al. 2005). These findings suggest that low dosages of inhaled or intravenously administered siRNAs/shRNAs might provide an easy and efficient basis for prophylaxis and antiviral therapy against respiratory viruses in human populations.

5.3
Inhibition of HCV by RNAi

HCV is a major cause of chronic hepatitis and hepatocellular carcinoma. Currently, no vaccines are available for HCV, and several groups have used RNAi to target HCV replication. HCV belongs to the family Flaviviridae, and its genome is encoded by a 9.6-kb RNA of positive polarity. Because there is no cell culture system for HCV replication, all studies have used replicon systems in Huh-7 cells as a model for HCV replication. These replicons support HCV RNA transcription and protein synthesis, but do not produce infectious virus. Regions that have been targeted include conserved sequences in the capsid, NS3, NS4A/B, NS5B, and the 5'-UTR (see Table 3). The 5'-UTR of the HCV RNA is a good potential target because it is the most conserved part of the HCV genome that harbours the internal ribosomal entry site (IRES), which is required for translation. Researchers have used siRNAs and shRNAs to block HCV replication (Takigawa et al. 2004). In addition, they looked at virus replication in cells that were transduced with a lentiviral vector to stably express

HCV specific shRNAs. They obtained good inhibition of HCV with shRNAs targeting both *NS3* and *NS5B* with shRNA and in the transduced cells. Poor inhibition was found with shRNA against the 5'-UTR. However, other groups have shown that the 5'-UTR can indeed be a good target (Kronke et al. 2004; Seo et al. 2003; Wilson et al. 2003; Yokota et al. 2003). This again shows that the effectiveness of siRNAs and shRNAs is still difficult to predict.

5.4
Inhibition of Human Coronaviruses by RNAi

Since it became clear that the outbreak of SARS beginning 2003 is caused by the virus currently known as SARS-coronavirus (SARS-CoV), researchers have tried to find cures for this new virus. It was shown that both shRNAs and siRNAs could efficiently block SARS-CoV replication in tissue culture systems (He et al. 2003; Lu et al. 2004a; Wang et al. 2004b; Zhang et al. 2004b). In these studies, the main target was the polymerase gene, whereas one paper describes inhibition of SARS-CoV by targeting the spike protein, which is essential for particle formation and entry (Zhang et al. 2004b). Additionally, replication of the newly discovered human coronavirus NL63, HCoV-NL63 (Van der Hoek et al. 2004), could also be inhibited by siRNAs that target the spike gene (Pyrc et al. in preparation). Because SARS is a disease of the upper airways and lungs, it could be relatively easy to administer therapeutic siRNAs. For influenza virus, HRSV and HPIV, it has been shown that virus replication in the lungs of mice can be inhibited by intravenous or intranasal administration of siRNAs/shRNAs (Bitko et al. 2005; Ge et al. 2004; Tompkins et al. 2004; Zhang et al. 2005). Following a similar route, siRNAs against SARS-CoV might be effective as a new antiviral therapeutic.

5.5
Inhibition of HBV by RNAi

In contrast to RNA viruses, RNAi against DNA viruses targets only the viral mRNA transcripts, but not the viral genome. This suggest that RNAi against DNA viruses might be less effective than RNAi against RNA viruses. However, the published data on RNAi mediated inhibition of DNA viruses indicate that this is not the case. HBV is a member of the Hepadnaviridae and its genome is a 3.2-kb double-stranded circular DNA. HBV infection can cause liver cirrhosis, which may ultimately lead to hepatocellular carcinoma. Although vaccines have been developed that can prevent infection, HBV remains a serious health problem in many countries. A number of studies show that RNAi induced by synthetic siRNAs can block HBV both in cell culture and in mouse model systems (Hamasaki et al. 2003; Klein et al. 2003; Konishi et al. 2003; Ying et al. 2003). For instance, Giladi et al. showed a 5–12-fold inhibition of HBV replication in mice that have been treated with synthetic siRNAs targeting the small HBV surface antigen, HbsAg (Giladi et al. 2003). Earlier, a 6-fold inhibition of HBV in the liver of mice was obtained by intravenous injection of plasmids expressing shRNAs against the core, HbsAg/Pol and X gene (McCaffrey et al. 2003). Possibly, a similar approach could be used in infected patients to lower virus titres. Besides targeting the viral RNA, it is also possible to inhibit HBV replication by targeting the mRNAs of cellular factors that are required for virus replication. Ni et al. have shown that targeting the hLa protein results in a decrease in virus replication (Ni et al. 2004). Additionally, it has been shown that RNAi can be used to prevent HCV- or HBV-induced disease of the liver by silencing the expression of the cellular Fas gene. During HCV or HBV infection Fas-mediated apoptosis of hepatocytes is triggered as a self-destructive inflammatory response of the liver. Silencing Fas expression with synthetic siRNAs blocks this reponse (Li et al. 2004a).

5.6
Inhibition of DNA Viruses by RNAi

HSV-1 is a large DNA virus that infects epithelial and neuronal cells. Bhuyan and co-workers used siRNAs against glycoprotein E (gE) to inhibit HSV-1 replication. The gE is important for cell-to-cell spread and evasion from complement and antibody responses, and silencing the expression of gE resulted in a fourfold inhibition of HSV-1 replication (Bhuyan et al. 2004). An 11- to 16-fold inhibition of the gamma herpesvirus EBV was obtained with shRNA against the essential viral gene Zta (Chang et al. 2004). Zta is involved in the reactivation of EBV and important for expression of lytic genes and viral DNA replication. In addition to targeting EBV replication, one report describes inhibition of an EBV oncogene to inhibit the pathogenic effects of EBV infection (Li et al. 2004a). EBV is associated with the development of highly metastatic nasopharyngeal carcinoma (NPC). Important in the development of NPC is the viral latent membrane protein-1 (LMP-1), which is involved in cell transformation and tumour metastasis. Suppression of LMP-1 expression by RNAi resulted in altered cell motility, surface adhesion and transmembrane invasion ability, suggesting that RNAi can be used to inhibit the metastatic potential of the EBV-positive carcinoma cells.

Two studies have used RNAi to inhibit the small DNA virus human polyomavirus JCV (Orba et al. 2004; Radhakrishnan et al. 2004). JCV can cause progressive multifocal leukoencephalopathy (PML) in patients with impaired immune systems. As such, PML has become a major neurologic problem among patients with AIDS. In both studies, synthetic siRNAs were used to target the VP1, Agno and T-Ag genes, resulting in 10- to 24-fold inhibition of virus replication in vitro.

6
Off-Target Effects

An important question mark concerning RNAi-therapy is its specificity. The introduced siRNA may negatively affect the production or activity of endogenous RNAi pathways that are involved in the regulation of cellular gene expression. This is particularly important because there is growing evidence for the biological significance of RNAi in development (Bernstein et al. 2003; Wienholds et al. 2003). It also needs to be proved that RNAi will not cross-silence cellular mRNAs with a sequence motif that resembles the actual viral target. Genome-wide expression profiling yielded promising results (Chi et al. 2003; Li et al. 2004b; Semizarov et al. 2003), but other studies reported that siRNA-treatment can cause changes in the expression of dozens of cellular genes that are not directly targeted (Jackson et al. 2003; Persengiev et al. 2004; Scacheri et al. 2004). However, microarray expression profiling may provide a very sensitive readout compared to a typical functional screen. With shRNA molecules it seems critical to use hairpins with a stem of 19 base-pairs or less in order to avoid induction of the interferon response and cytotoxic effects (Fish and Kruithof 2004).

siRNAs with imperfect complementarity can silence genes by repressing mRNA translation while not affecting the levels of mRNA, and up to three or four mismatches are tolerated in this system (Saxena et al. 2003). This finding, combined with the observation that the G-U wobble is recognized as a regular base-pair, emphasizes the possibility of off-target silencing by siRNAs. Most of these studies were performed with synthetic siRNA, of which the concentration added to cells may also be a critical factor (Persengiev et al. 2004). Intracellularly expressed shRNA may affect cellular genes by other means (Bridge et al. 2003). It is therefore obvious that the use of RNAi-based therapeutic agents requires a further demonstration of the absence of adverse effects on cell physiology.

Another issue is whether siRNA will induce the interferon system, which causes cells to shut themselves down in response to invading RNA viruses. In contrast to plants and *C. elegans*, transfection of dsRNA longer than 30 base-pairs into mammalian cells induces this interferon pathway via dsRNA-dependent protein kinase (PKR). The induced antiviral response involves non-specific degradation of RNA and generalized inhibition of translation. siRNA was supposed not to be detected by the radar of the interferon system, but this may not be completely true. Two recent papers examined the effect of siRNAs on the interferon system (Bridge et al. 2003; Sledz et al. 2003). Both studies reported nonspecific changes in expression of interferon-stimulated genes (Moss and Taylor 2003). One study reported activation of the PKR by a synthetic 21-bp siRNA (Sledz et al. 2003), but this finding is not consistent with previous results (Manche et al. 1992). A detailed characterization of the minimal RNA motif for PKR activation revealed the following characteristics:

a hairpin with a (possibly imperfect) 16-bp stem flanked by 10- to 15-nt single-stranded tail (Zheng and Bevilacqua 2000). These results indicate that siRNA is not a PKR-activator, but does not rule out a role for PKR in RNAi biology.

It was reported that certain genes of the interferon system are activated upon siRNA-introduction into cells, thus providing another serious warning that off-target effects may be more common than initially anticipated (Sledz et al. 2003). Furthermore, dsRNA can activate several protein kinases such as p38, c-Jun N-terminal kinase (JNK)2 and I kappa B kinase (IKK) in addition to PKR. Induction of these signalling pathways can alter gene expression by regulating the activity of transcription factors such as nuclear factor (NF)-κB, interferon regulatory factor (IRF)-3, and ATF-1 (Williams 1999). Therefore, caution must be exerted in the interpretation of data from experiments using RNAi technology for suppression of specific genes. It will be important to first add to our basic understanding of how siRNA exactly works before rushing into clinical trials. Among other things, these experiments need to be repeated in animals, and that is exactly what was done very recently (Heidel et al. 2004). Regardless of the injection method into mice, siRNA failed to trigger a strong type I interferon response, unlike the long double-stranded control RNA poly(I:C). The siRNAs were shown to silence genes in a sequence-specific manner, demonstrating that they reached the intracellular target. The absence of both an interferon and inflammatory response in vivo is good news for using siRNAs therapeutically.

7
Viruses Utilize RNAi in Their Replication Strategy

Alternatively, viruses could exploit RNA silencing to control the expression of genes of viral or host origin. The first example was recently provided for EBV, a large DNA virus of the herpes family that preferentially infects human B cells (Pfeffer et al. 2004). When the small RNAs from a latently EBV-infected Burkitt's lymphoma cell line were cloned, 4% of them originated from two regions of the EBV genome. A computational method was used to identify potential targets of these EBV-encoded miRNAs. Among the predicted targets were regulators of cell proliferation, apoptosis, transcriptional regulators and components of signal transduction pathways. Although these targets should be verified experimentally, it is striking that several of these genes have more than one binding site for a particular EBV-miRNA. Degradation of a cellular DNA polymerase was demonstrated experimentally (Pfeffer et al. 2004). Furthermore, the expression of the EBV-miRNAs was shown to differ in the lytic versus latent stage, suggesting tight regulation during viral infection. The viral miRNAs could be involved in tumour formation and may explain how EBV hides so well. Other members of the herpesvirus family and other viruses with a large DNA genome could encode miRNAs in order to exploit RNA silencing for the regulation of host and viral gene expression.

Both a theoretical and an experimental study recently identified several potential miRNAs encoded by the HIV-1 RNA genome (Bennasser et al. 2004; Omoto et al. 2004). In addition, Bennasser et al. (2004) identified several corresponding cellular RNAs that could potentially be targeted by these viral miRNAs.

7.1
Viroids Also Utilize the RNAi Mechanism

Viroids are single RNA molecules that have no protective protein coat and that do not encode protein. Viroids can cause severe disease in plants, and several hypotheses have been proposed to explain disease induction in the absence of any viral protein. Viroids could interrupt the function of an unknown host cell factor or use small regulatory RNAs to influence host gene expression. Now it appears that viroid pathogenicity may involve the RNA silencing pathway. Wang et al. reported that engineered tomato plants, which express virion-derived non-infectious hairpin RNA, had symptoms mimicking those of viroid infection (Wang et al. 2004a). Much remains to be learned, for instance how nuclear-replicating viroids can exploit the RNAi machinery that is located in the cytoplasm, and which cellular mRNA is targeted. Nevertheless, the EBV and viroid examples underscore the putative pathogenic function of virus-derived small RNA molecules.

8
RNAi Versus the Antiviral Interferon System

Assuming that RNAi acts as an antiviral mechanism in humans, one would predict that human viruses have developed countermeasures, although it has been questioned whether RNAi plays a major role in the antiviral defence in vertebrates (Saksela 2003). However, unlike plants and invertebrates, vertebrates also have the interferon system that responds to dsRNA by inducing the synthesis of a large group of proteins that have a general inhibitory effect on virus multiplication. The best-characterized interferon-induced genes encode PKR kinase and the $2'$-$5'$oligo A synthetase enzymes, both of which are activated in response to dsRNA (Goodbourn et al. 2000). Activated PKR causes an inhibition of protein synthesis by phosphorylation of eIF2, and $2'$-$5'$oligoA synthetase induces general RNA degradation via activation of RNase L, leading to ultimate cell death via apoptosis. Thus, mammals with their adaptive immune system have already defence mechanisms that respond to dsRNA. However, the discovery that RNAi is triggered by siRNAs (Caplen et al. 2001; Elbashir et al. 2001), which are too short to efficiently activate the interferon response pathway (Moss and Taylor 2003), suggested that RNAi may also play a role in the cellular defence against infection by human viruses. Thus, human cells may have two alternative pathways to combat dsRNA; long dsRNA activates the interferon response pathway, whereas short dsRNA (< 30 bp) activates RNAi.

9
Adenovirus VA RNAs as Suppressor of the RNAi and Interferon Systems

It is well established that most mammalian viruses have evolved defence strategies to suppress the negative effects that the interferon response pathway has on virus multiplication. This is of vital importance for the capacity of a virus to multiply successfully. Numerous viruses encode proteins or decoy RNAs that inhibit the activity of PKR by a surprisingly large range of different strategies (Gale and Katze 1998). For example, human adenovirus type 5 (ad5) encodes two approximately 160-nt non-translated RNA polymerase III transcripts; the highly structured VA RNAI and VA RNAII (Mathews 1995). VA RNAI has been shown to stimulate protein synthesis in infected cells and in transient transfection assays by blocking activation of the interferon-induced antiviral defence system (Kitajewski et al. 1986; Svensson and Akusjarvi 1984). VA RNAI binds to PKR and acts as a competitive inhibitor (Gale and Katze 1998), thus preventing viral dsRNA that is produced by symmetrical transcription of the viral DNA from activating PKR.

We recently demonstrated that human adenovirus also inhibits the RNAi machinery at late times of infection (Andersson et al. 2005). The suppression of RNAi results from a virus-induced block of the two key enzymatic activities in RNAi, Dicer and RISC. We further showed that VA RNAI and VA RNAII have the capacity to suppress RNAi in transient transfection experiments. Mechanistically, the VA RNAs appear to block RNAi by acting as competitive substrates that squelch Dicer. Since VA RNAs are expressed at copious amounts at late times of infection [up to 10^8 copies/cell, (Soderlund et al. 1976)], one would expect that they are produced in great excess over any aberrantly formed dsRNA. Therefore, a simple competitive inhibition for binding to Dicer would be sufficient to explain the inhibitory effect of the VA RNAs on RNAi. VA RNAs might function as suppressors of RNAi because they form highly structured motifs with imperfect stems that resemble precursors to miRNA and therefore might sequester Dicer by acting as competing substrates, or pseudo-substrates. The finding that VA RNAI and VA RNAII are indeed processed by Dicer into siRNA both in vitro and during a lytic infection supports this model and shows that the VA RNAs can interact with Dicer. Unlike Dicer inhibition, the VA RNAs are not required for the observed inhibition of RISC during infection. Most likely the inhibition of RISC requires another, yet-to-be-identified, viral factor. Recently, Lu et al. (2004) published similar results for adenovirus VA RNAI (Lu and Cullen 2004). Collectively, these results suggest that the adenovirus VA RNAs antagonize the cellular defence pathways directed against both long (interferon-induced) and short (RNAi-induced) dsRNA by inactivating two key enzymes, PKR and Dicer.

These findings have an impact on strategies that use viral vectors for RNA silencing purposes. The possible existence of viral suppressors of RNAi should have consequences on how adenoviral vectors, and potentially other viral vectors, are designed to create optimal vectors for siRNA delivery to target cells. For instance, the VA RNAs are expressed not only from replicating adenoviral vectors, but also from non-replicating adenoviral vectors. Thus, one would expect that VA RNA might negatively affect the efficiency of adenovirus-delivered shRNA. Further, it is possible that an adenoviral vector may alter cellular gene expression as a result of competition between VA RNA and cellular miRNAs, as has been seen in virus-infected plants (Kasschau et al. 2003).

10
The Future of RNAi Therapeutics

One of the most important consequences of the RNA silencing revolution is the ability to use the RNAi pathways to determine gene function and to apply RNA silencing as a tool in agriculture and medicine, e.g. to protect against viral infections. Indeed, RNAi therapy is expected to make its way towards clinical trials in the near future. The interplay between RNAi and viruses is very complex as viruses can be inducers, suppressors and actual targets of the RNAi silencing mechanism. There is an interesting and strong argument that RNA was the primordial biopolymer of life because of its multi-functionality (both enzyme and replicon). A relatively small cadre of scientists has always been dedicated to studying RNA, but RNAi has brought about a real RNA revolution that seems to have converted all scientists.

Acknowledgements RNA research in the Berkhout laboratory is sponsored by Senter (grant with Viruvation BV), NWO-CW (TOP grant) and ZonMw (VICI grant).

References

Adelman ZN, Blair CD, Carlson JO, Beaty BJ, Olson KE (2001) Sindbis virus-induced silencing of dengue viruses in mosquitoes. Insect Mol Biol 10:265–273

Adelman ZN, Sanchez-Vargas I, Travanty EA, Carlson JO, Beaty BJ, Blair CD, Olson KE (2002) RNA silencing of dengue virus type 2 replication in transformed C6/36 mosquito cells transcribing an inverted-repeat RNA derived from the virus genome. J Virol 76:12925–12933

Anderson J, Banerjea A, Akkina R (2003) Bispecific short hairpin siRNA constructs targeted to CD4, CXCR4, and CCR5 confer HIV-1 resistance. Oligonucleotides 13:303–312

Andersson MG, Haasnoot PCJ, Xu N, Berenjian S, Berkhout B, Akusjärvi G (2005) Suppression of RNA interference by adenovirus VA RNA. J Virol 79:9556–9565

Arrighi JF, Pion M, Wiznerowicz M, Geijtenbeek TB, Garcia E, Abraham S, Leuba F, Dutoit V, Ducrey-Rundquist O, van Kooyk Y, Trono D, Piguet V (2004) Lentivirus-mediated RNA interference of DC-SIGN expression inhibits human immunodeficiency virus transmission from dendritic cells to T cells. J Virol 78:10848–10855

Aukerman MJ, Sakai H (2003) Regulation of flowering time and floral organ identity by a MicroRNA and its APETALA2-like target genes. Plant Cell 15:2730–2741

Banerjea A, Li MJ, Bauer G, Remling L, Lee NS, Rossi J, Akkina R (2003) Inhibition of HIV-1 by lentiviral vector-transduced siRNAs in T lymphocytes differentiated in SCID-hu mice and CD34+ progenitor cell-derived macrophages. Mol Ther 8:62–71

Bartel DP (2004) MicroRNAs: genomics, biogenesis, mechanism, and function. Cell 116:281–297

Beck M, Strand MR (2003) RNA interference silences Microplitis demolitor bracovirus genes and implicates glc1.8 in disruption of adhesion in infected host cells. Virology 314:521–535

Bennasser Y, Le SY, Yeung ML, Jeang KT (2004) HIV-1 encoded candidate micro-RNAs and their cellular targets. Retrovirology 1:43

Berkhout B (2004) RNA interference as an antiviral approach: targeting HIV-1. Curr Opin Mol Ther 6:141–145

Bernstein E, Caudy AA, Hammond SM, Hannon GJ (2001) Role for a bidentate ribonuclease in the initiation step of RNA interference. Nature 409:363–366

Bernstein E, Kim SY, Carmell MA, Murchison EP, Alcorn H, Li MZ, Mills AA, Elledge SJ, Anderson KV, Hannon GJ (2003) Dicer is essential for mouse development. Nat Genet 35:215–217

Bhuyan PK, Kariko K, Capodici J, Lubinski J, Hook LM, Friedman HM, Weissman D (2004) Short interfering RNA-mediated inhibition of herpes simplex virus type 1 gene expression and function during infection of human keratinocytes. J Virol 78:10276–10281

Billy E, Brondani V, Zhang H, Muller U, Filipowicz W (2001) Specific interference with gene expression induced by long, double-stranded RNA in mouse embryonal teratocarcinoma cell lines. Proc Natl Acad Sci U S A 98:14428–14433

Bitko V, Barik S (2001) Phenotypic silencing of cytoplasmic genes using sequence-specific double-stranded short interfering RNA and its application in the reverse genetics of wild type negative-strand RNA viruses. BMC Microbiol 1:34

Bitko V, Musiyenko A, Shulyayeva O, Barik S (2005) Inhibition of respiratory viruses by nasally administered siRNA. Nat Med 11:50–55

Boden D, Pusch O, Lee F, Tucker L, Ramratnam B (2003a) Human immunodeficiency virus type 1 escape from RNA interference. J Virol 77:11531–11535

Boden D, Pusch O, Lee F, Tucker L, Shank PR, Ramratnam B (2003b) Promoter choice affects the potency of HIV-1 specific RNA interference. Nucleic Acids Res 31:5033–5038

Boden D, Pusch O, Lee F, Tucker L, Ramratnam B (2004a) Efficient gene transfer of HIV-1-specific short hairpin RNA into human lymphocytic cells using recombinant adeno-associated virus vectors. Mol Ther 9:396–402

Boden D, Pusch O, Silbermann R, Lee F, Tucker L, Ramratnam B (2004b) Enhanced gene silencing of HIV-1 specific siRNA using microRNA designed hairpins. Nucleic Acids Res 32:1154–1158

Bohnsack MT, Czaplinski K, Gorlich D (2004) Exportin 5 is a RanGTP-dependent dsRNA-binding protein that mediates nuclear export of pre-miRNAs. RNA 10:185–191

Brennecke J, Hipfner DR, Stark A, Russell RB, Cohen SM (2003) Bantam encodes a developmentally regulated microRNA that controls cell proliferation and regulates the proapoptotic gene hid in Drosophila. Cell 113:25–36

Bridge AJ, Pebernard S, Ducraux A, Nicoulaz AL, Iggo R (2003) Induction of an interferon response by RNAi vectors in mammalian cells. Nat Genet 34:263–264

Brummelkamp TR, Bernards R, Agami R (2002) A system for stable expression of short interfering RNAs in mammalian cells. Science 296:550–553

Cai X, Hagedorn CH, Cullen BR (2004) Human microRNAs are processed from capped, polyadenylated transcripts that can also function as mRNAs. RNA 10:1957–1966

Caplen NJ, Parrish S, Imani F, Fire A, Morgan RA (2001) Specific inhibition of gene expression by small double-stranded RNAs in invertebrate and vertebrate systems. Proc Natl Acad Sci U S A 98:9742–9747

Caplen NJ, Zheng Z, Falgout B, Morgan RA (2002) Inhibition of viral gene expression and replication in mosquito cells by dsRNA-triggered RNA interference. Mol Ther 6:243–251

Capodici J, Kariko K, Weissman D (2002) Inhibition of HIV-1 infection by small interfering RNA-mediated RNA interference. J Immunol 169:5196–5201

Chang J, Taylor JM (2003) Susceptibility of human hepatitis delta virus RNAs to small interfering RNA action. J Virol 77:9728–9731

Chang Y, Chang SS, Lee HH, Doong SL, Takada K, Tsai CH (2004) Inhibition of the Epstein-Barr virus lytic cycle by Zta-targeted RNA interference. J Gen Virol 85:1371–1379

Check E (2002) A tragic setback. Nature 420:116–118

Chen W, Yan W, Du Q, Fei L, Liu M, Ni Z, Sheng Z, Zheng Z (2004) RNA interference targeting VP1 inhibits foot-and-mouth disease virus replication in BHK-21 cells and suckling mice. J Virol 78:6900–6907

Chi JT, Chang HY, Wang NN, Chang DS, Dunphy N, Brown PO (2003) Genomewide view of gene silencing by small interfering RNAs. Proc Natl Acad Sci U S A 100:6343–6346

Chiu YL, Cao H, Jacque JM, Stevenson M, Rana TM (2004) Inhibition of human immunodeficiency virus type 1 replication by RNA interference directed against human transcription elongation factor P-TEFb (CDK9/CyclinT1). J Virol 78:2517–2529

Coburn GA, Cullen BR (2002) Potent and specific inhibition of human immunodeficiency virus type 1 replication by RNA interference. J Virol 76:9225–9231

Das AT, Brummelkamp TR, Westerhout EM, Vink M, Madiredjo M, Bernards R, Berkhout B (2004) Human immunodeficiency virus type 1 escapes from RNA interference-mediated inhibition. J Virol 78:2601–2605

Dave RS, Pomerantz RJ (2004) Antiviral effects of human immunodeficiency virus type 1-specific small interfering RNAs against targets conserved in select neurotropic viral strains. J Virol 78:13687–13696

Dector MA, Romero P, Lopez S, Arias CF (2002) Rotavirus gene silencing by small interfering RNAs. EMBO Rep 3:1175–1180

Dostie J, Mourelatos Z, Yang M, Sharma A, Dreyfuss G (2003) Numerous microRNPs in neuronal cells containing novel microRNAs. RNA 9:180–186

Elbashir SM, Lendeckel W, Tuschl T (2001) RNA interference is mediated by 21- and 22-nucleotide RNAs. Genes Dev 15:188–200

Findley SD, Tamanaha M, Clegg NJ, Ruohola-Baker H (2003) Maelstrom, a Drosophila spindle-class gene, encodes a protein that colocalizes with Vasa and RDE1/AGO1 homolog, Aubergine, in nuage. Development 130:859–871

Fire A, Xu S, Montgomery MK, Kostas SA, Driver SE, Mello CC (1998) Potent and specific genetic interference by double-stranded RNA in Caenorhabditis elegans. Nature 391:806–811

Fish RJ, Kruithof EK (2004) Short-term cytotoxic effects and long-term instability of RNAi delivered using lentiviral vectors. BMC Mol Biol 5:9

Fukagawa T, Nogami M, Yoshikawa M, Ikeno M, Okazaki T, Takami Y, Nakayama T, Oshimura M (2004) Dicer is essential for formation of the heterochromatin structure in vertebrate cells. Nat Cell Biol 6:784–791

Gale M, Katze MG (1998) Molecular mechanisms of interferon resistance mediated by viral-directed inhibition of PKR, the interferon-induced protein kinase. Pharmacol Ther 78:29–46

Gazzani S, Lawrenson T, Woodward C, Headon D, Sablowski R (2004) A link between mRNA turnover and RNA interference in Arabidopsis. Science 306:1046–1048

Ge Q, McManus MT, Nguyen T, Shen CH, Sharp PA, Eisen HN, Chen J (2003) RNA interference of influenza virus production by directly targeting mRNA for degradation and indirectly inhibiting all viral RNA transcription. Proc Natl Acad Sci USA 100:2718–2723

Ge Q, Filip L, Bai A, Nguyen T, Eisen HN, Chen J (2004) Inhibition of influenza virus production in virus-infected mice by RNA interference. Proc Natl Acad Sci U S A 101:8676–8681

Giladi H, Ketzinel-Gilad M, Rivkin L, Felig Y, Nussbaum O, Galun E (2003) Small interfering RNA inhibits hepatitis B virus replication in mice. Mol Ther 8:769–776

Gitlin L, Karelsky S, Andino R (2002) Short interfering RNA confers intracellular antiviral immunity in human cells. Nature 418:430–434

Goodbourn S, Didcock L, Randall RE (2000) Interferons: cell signalling, immune modulation, antiviral response and virus countermeasures. J Gen Virol 81:2341–2364

Grishok A, Pasquinelli AE, Conte D, Li N, Parrish S, Ha I, Baillie DL, Fire A, Ruvkun G, Mello CC (2001) Genes and mechanisms related to RNA interference regulate expression of the small temporal RNAs that control C. elegans developmental timing. Cell 106:23–34

Gupta S, Schoer RA, Egan JE, Hannon GJ, Mittal V (2004) Inducible, reversible, and stable RNA interference in mammalian cells. Proc Natl Acad Sci U S A 101:1927–1932

Gwizdek C, Ossareh-Nazari B, Brownawell AM, Evers S, Macara IG, Dargemont C (2004) Minihelix-containing RNAs mediate exportin-5-dependent nuclear export of the double-stranded RNA-binding protein ILF3. J Biol Chem 279:884–891

Haasnoot PC, Cupac D, Berkhout B (2003) Inhibition of virus replication by RNA interference. J Biomed Sci 10:607–616

Hamasaki K, Nakao K, Matsumoto K, Ichikawa T, Ishikawa H, Eguchi K (2003) Short interfering RNA-directed inhibition of hepatitis B virus replication. FEBS Lett 543:51–54

Han W, Wind-Rotolo M, Kirkman RL, Morrow CD (2004) Inhibition of human immunodeficiency virus type 1 replication by siRNA targeted to the highly conserved primer binding site. Virology 330:221–232

Hannon GJ (2002) RNA interference. Nature 418:244–251

He ML, Zheng B, Peng Y, Peiris JS, Poon LL, Yuen KY, Lin MC, Kung HF, Guan Y (2003) Inhibition of SARS-associated coronavirus infection and replication by RNA interference. JAMA 290:2665–2666

Heidel JD, Hu S, Liu XF, Triche TJ, Davis ME (2004) Lack of interferon response in animals to naked siRNAs. Nat Biotechnol 22:1579–1582

Houbaviy HB, Murray MF, Sharp PA (2003) Embryonic stem cell-specific microRNAs. Dev Cell 5:351–358

Hu WY, Myers CP, Kilzer JM, Pfaff SL, Bushman FD (2002) Inhibition of retroviral pathogenesis by RNA interference. Curr Biol 12:1301–1311

Hutvagner G, Zamore PD (2002) A microRNA in a multiple-turnover RNAi enzyme complex. Science 297:2056–2060

Hutvagner G, McLachlan J, Pasquinelli AE, Balint E, Tuschl T, Zamore PD (2001) A cellular function for the RNA-interference enzyme Dicer in the maturation of the let-7 small temporal RNA. Science 293:834–838

Jackson AL, Bartz SR, Schelter J, Kobayashi SV, Burchard J, Mao M, Li B, Cavet G, Linsley PS (2003) Expression profiling reveals off-target gene regulation by RNAi. Nat Biotechnol 21:635–637

Jacque JM, Triques K, Stevenson M (2002) Modulation of HIV-1 replication by RNA interference. Nature 418:435–438

Jia Q, Sun R (2003) Inhibition of gamma herpesvirus replication by RNA interference. J Virol 77:3301–3306

Jiang M, Milner J (2002) Selective silencing of viral gene expression in HPV-positive human cervical carcinoma cells treated with siRNA, a primer of RNA interference. Oncogene 21:6041–6048

Kahana R, Kuznetzova L, Rogel A, Shemesh M, Hai D, Yadin H, Stram Y (2004) Inhibition of foot-and-mouth disease virus replication by small interfering RNA. J Gen Virol 85:3213–3217

Kameoka M, Nukuzuma S, Itaya A, Tanaka Y, Ota K, Ikuta K, Yoshihara K (2004) RNA interference directed against Poly(ADP-Ribose) polymerase 1 efficiently suppresses human immunodeficiency virus type 1 replication in human cells. J Virol 78:8931–8934

Kapadia SB, Brideau-Andersen A, Chisari FV (2003) Interference of hepatitis C virus RNA replication by short interfering RNAs. Proc Natl Acad Sci USA 100:2014–2018

Karlas A, Kurth R, Denner J (2004) Inhibition of porcine endogenous retroviruses by RNA interference: increasing the safety of xenotransplantation. Virology 325:18–23

Kasim V, Miyagishi M, Taira K (2004) Control of siRNA expression using the Cre-loxP recombination system. Nucleic Acids Res 32:e66

Kasschau KD, Xie Z, Allen E, Llave C, Chapman EJ, Krizan KA, Carrington JC (2003) P1/HC-Pro, a viral suppressor of RNA silencing, interferes with Arabidopsis development and miRNA unction. Dev Cell 4:205–217

Kawasaki H, Taira K (2004) Induction of DNA methylation and gene silencing by short interfering RNAs in human cells. Nature 431:211–217

Ketting RF, Fischer SE, Bernstein E, Sijen T, Hannon GJ, Plasterk RH (2001) Dicer functions in RNA interference and in synthesis of small RNA involved in developmental timing in C. elegans. Genes Dev 15:2654–2659

Khvorova A, Reynolds A, Jayasena SD (2003) Functional siRNAs and miRNAs exhibit strand bias. Cell 115:209–216

Kidner CA, Martienssen RA (2004) Spatially restricted microRNA directs leaf polarity through ARGONAUTE1. Nature 428:81–84

Kitajewski J, Schneider RJ, Safer B, Munemitsu SM, Samuel CE, Thimmappaya B, Shenk T (1986) Adenovirus VAI RNA antagonizes the antiviral action of interferon by preventing activation of the interferon-induced eIF-2 alpha kinase. Cell 45:195–200

Klein C, Bock CT, Wedemeyer H, Wustefeld T, Locarnini S, Dienes HP, Kubicka S, Manns MP, Trautwein C (2003) Inhibition of hepatitis B virus replication in vivo by nucleoside analogues and siRNA. Gastroenterology 125:9–18

Knight SW, Bass BL (2001) A role for the RNase III enzyme DCR-1 in RNA interference and germ line development in Caenorhabditis elegans. Science 293:2269–2271

Konishi M, Wu CH, Wu GY (2003) Inhibition of HBV replication by siRNA in a stable HBV-producing cell line. Hepatology 38:842–850

Krichevsky AM, King KS, Donahue CP, Khrapko K, Kosik KS (2003) A microRNA array reveals extensive regulation of microRNAs during brain development. RNA 9:1274–1281

Kronke J, Kittler R, Buchholz F, Windisch MP, Pietschmann T, Bartenschlager R, Frese M (2004) Alternative approaches for efficient inhibition of hepatitis C virus RNA replication by small interfering RNAs. J Virol 78:3436–3446

Lagos-Quintana M, Rauhut R, Yalcin A, Meyer J, Lendeckel W, Tuschl T (2002) Identification of tissue-specific microRNAs from mouse. Curr Biol 12:735–739

Lai EC (2004) Predicting and validating microRNA targets. Genome Biol 5:115

Lau NC, Lim LP, Weinstein EG, Bartel DP (2001) An abundant class of tiny RNAs with probable regulatory roles in Caenorhabditis elegans. Science 294:858–862

Lee MT, Coburn GA, McClure MO, Cullen BR (2003a) Inhibition of human immunodeficiency virus type 1 replication in primary macrophages by using Tat- or CCR5-specific small interfering RNAs expressed from a lentivirus vector. J Virol 77:11964–11972

Lee NS, Dohjima T, Bauer G, Li H, Li MJ, Ehsani A, Salvaterra P, Rossi J (2002a) Expression of small interfering RNAs targeted against HIV-1 rev transcripts in human cells. Nat Biotechnol 20:500–505

Lee RC, Ambros V (2001) An extensive class of small RNAs in Caenorhabditis elegans. Science 294:862–864

Lee Y, Jeon K, Lee JT, Kim S, Kim VN (2002b) MicroRNA maturation: stepwise processing and subcellular localization. EMBO J 21:4663–4670

Lee Y, Ahn C, Han J, Choi H, Kim J, Yim J, Lee J, Provost P, Radmark O, Kim S, Kim VN (2003b) The nuclear RNase III Drosha initiates microRNA processing. Nature 425:415–419

Li H, Li WX, Ding SW (2002) Induction and suppression of RNA silencing by an animal virus. Science 296:1319–1321

Li MJ, Bauer G, Michienzi A, Yee JK, Lee NS, Kim J, Li S, Castanotto D, Zaia J, Rossi JJ (2003) Inhibition of HIV-1 infection by lentiviral vectors expressing Pol III-promoted anti-HIV RNAs. Mol Ther 8:196–206

Li XP, Li G, Peng Y, Kung HF, Lin MC (2004a) Suppression of Epstein-Barr virus-encoded latent membrane protein-1 by RNA interference inhibits the metastatic potential of nasopharyngeal carcinoma cells. Biochem Biophys Res Commun 315:212–218

Li Y, Wasser S, Lim SG, Tan TM (2004b) Genome-wide expression profiling of RNA interference of hepatitis B virus gene expression and replication. Cell Mol Life Sci 61:2113–2124

Lindbo JA, Dougherty WG (1992) Untranslatable transcripts of the tobacco etch virus coat protein gene sequence can interfere with tobacco etch virus replication in transgenic plants and protoplasts. Virology 189:725–733

Liu J, Carmell MA, Rivas FV, Marsden CG, Thomson JM, Song JJ, Hammond SM, Joshua-Tor L, Hannon GJ (2004a) Argonaute2 Is the Catalytic Engine of Mammalian RNAi. Science 305:1437–1441

Liu Q, Rand TA, Kalidas S, Du F, Kim HE, Smith DP, Wang X (2003) R2D2, a bridge between the initiation and effector steps of the Drosophila RNAi pathway. Science 301:1921–1925

Liu S, Asparuhova M, Brondani V, Ziekau I, Klimkait T, Schumperli D (2004b) Inhibition of HIV-1 multiplication by antisense U7 snRNAs and siRNAs targeting cyclophilin A. Nucleic Acids Res 32:3752–3759

Lopez T, Camacho M, Zayas M, Najera R, Sanchez R, Arias CF, Lopez S (2005) Silencing the morphogenesis of rotavirus. J Virol 79:184–192

Lu A, Zhang H, Zhang X, Wang H, Hu Q, Shen L, Schaffhausen BS, Hou W, Li L (2004a) Attenuation of SARS coronavirus by a short hairpin RNA expression plasmid targeting RNA-dependent RNA polymerase. Virology 324:84–89

Lu S, Cullen BR (2004) Adenovirus VA1 noncoding RNA can inhibit small interfering RNA and MicroRNA biogenesis. J Virol 78:12868–12876

Lu WW, Hsu YY, Yang JY, Kung SH (2004b) Selective inhibition of enterovirus 71 replication by short hairpin RNAs. Biochem Biophys Res Commun 325:494–499

Lu X, Humeau L, Slepushkin V, Binder G, Yu Q, Slepushkina T, Chen Z, Merling R, Davis B, Chang YN, Dropulic B (2004c) Safe two-plasmid production for the first clinical lentivirus vector that achieves >99% transduction in primary cells using a one-step protocol. J Gene Med 6:963–973

Lu X, Yu Q, Binder GK, Chen Z, Slepushkina T, Rossi J, Dropulic B (2004d) Antisense-mediated inhibition of human immunodeficiency virus (HIV) replication by use of an HIV type 1-based vector results in severely attenuated mutants incapable of developing resistance. J Virol 78:7079–7088

Lund E, Guttinger S, Calado A, Dahlberg JE, Kutay U (2004) Nuclear export of microRNA precursors. Science 303:95–98

Manche L, Green SR, Schmedt C, Mathews MB (1992) Interactions between double-stranded RNA regulators and the protein kinase DAI. Mol Cell Biol 12:5238–5248

Martinez J, Patkaniowska A, Urlaub H, Luhrmann R, Tuschl T (2002a) Single-stranded antisense siRNAs guide target RNA cleavage in RNAi. Cell 110:563–574

Martinez MA, Gutierrez A, Armand-Ugon M, Blanco J, Parera M, Gomez J, Clotet B, Este JA (2002b) Suppression of chemokine receptor expression by RNA interference allows for inhibition of HIV-1 replication. AIDS 16:2385–2390

Mathews MB (1995) Structure, function, and evolution of adenovirus virus-associated RNAs. Curr Top Microbiol Immunol 199:173–187

McCaffrey AP, Nakai H, Pandey K, Huang Z, Salazar FH, Xu H, Wieland SF, Marion PL, Kay MA (2003) Inhibition of hepatitis B virus in mice by RNA interference. Nat Biotechnol 21:639–644

Means JC, Muro I, Clem RJ (2003) Silencing of the baculovirus Op-iap3 gene by RNA interference reveals that it is required for prevention of apoptosis during Orgyia pseudotsugata M nucleopolyhedrovirus infection of Ld652Y cells. J Virol 77:4481–4488

Mette MF, van der Winden J, Matzke MA, Matzke AJ (1999) Production of aberrant promoter transcripts contributes to methylation and silencing of unlinked homologous promoters in trans. EMBO J 18:241–248

Miska EA, Alvarez-Saavedra E, Townsend M, Yoshii A, Sestan N, Rakic P, Constantine-Paton M, Horvitz HR (2004) Microarray analysis of microRNA expression in the developing mammalian brain. Genome Biol 5:R68

Miyagishi M, Sumimoto H, Miyoshi H, Kawakami Y, Taira K (2004) Optimization of an siRNA-expression system with an improved hairpin and its significant suppressive effects in mammalian cells. J Gene Med 6:715–723

Morris KV, Chan SW, Jacobsen SE, Looney DJ (2004) Small interfering RNA-induced transcriptional gene silencing in human cells. Science 305:1289–1292

Moss EG, Tang L (2003) Conservation of the heterochronic regulator Lin-28, its developmental expression and microRNA complementary sites. Dev Biol 258:432–442

Moss EG, Taylor JM (2003) Small-interfering RNAs in the radar of the interferon system. Nat Cell Biol 5:771–772

Mourelatos Z, Dostie J, Paushkin S, Sharma A, Charroux B, Abel L, Rappsilber J, Mann M, Dreyfuss G (2002) miRNPs: a novel class of ribonucleoproteins containing numerous microRNAs. Genes Dev 16:720–728

Ni Q, Chen Z, Yao HP, Yang ZG, Liu KZ, Wu LL (2004) Inhibition of human La protein by RNA interference downregulates hepatitis B virus mRNA in 2.2.15 cells. World J Gastroenterol 10:2050–2054

Nishitsuji H, Ikeda T, Miyoshi H, Ohashi T, Kannagi M, Masuda T (2004) Expression of small hairpin RNA by lentivirus-based vector confers efficient and stable gene-suppression of HIV-1 on human cells including primary non-dividing cells. Microbes Infect 6:76–85

Novina CD, Murray MF, Dykxhoorn DM, Beresford PJ, Riess J, Lee SK, Collman RG, Lieberman J, Shankar P, Sharp PA (2002) siRNA-directed inhibition of HIV-1 infection. Nat Med 8:681–686

Okamura K, Ishizuka A, Siomi H, Siomi MC (2004) Distinct roles for Argonaute proteins in small RNA-directed RNA cleavage pathways. Genes Dev 18:1655–1666

Omoto S, Ito M, Tsutsumi Y, Ichikawa Y, Okuyama H, Andi BE, Saksena NK, Fuji Y (2004) HIV-1 nef suppression by virally encoded microRNA. Retrovirology 1:44

Orba Y, Sawa H, Iwata H, Tanaka S, Nagashima K (2004) Inhibition of virus production in JC virus-infected cells by postinfection RNA interference. J Virol 78:7270–7273

Paddison PJ, Caudy AA, Bernstein E, Hannon GJ, Conklin DS (2002) Short hairpin RNAs (shRNAs) induce sequence-specific silencing in mammalian cells. Genes Dev 16:948–958

Palatnik JF, Allen E, Wu X, Schommer C, Schwab R, Carrington JC, Weigel D (2003) Control of leaf morphogenesis by microRNAs. Nature 425:257–263

Park WS, Miyano-Kurosaki N, Hayafune M, Nakajima E, Matsuzaki T, Shimada F, Takaku H (2002) Prevention of HIV-1 infection in human peripheral blood mononuclear cells by specific RNA interference. Nucleic Acids Res 30:4830–4835

Park WS, Hayafune M, Miyano-Kurosaki N, Takaku H (2003) Specific HIV-1 env gene silencing by small interfering RNAs in human peripheral blood mononuclear cells. Gene Ther 10:2046–2050

Parker JS, Roe SM, Barford D (2004) Crystal structure of a PIWI protein suggests mechanisms for siRNA recognition and slicer activity. EMBO J 23:4727–4737

Paul CP, Good PD, Winer I, Engelke DR (2002) Effective expression of small interfering RNA in human cells. Nat Biotechnol 20:505–508

Persengiev SP, Zhu X, Green MR (2004) Nonspecific, concentration-dependent stimulation and repression of mammalian gene expression by small interfering RNAs (siRNAs). RNA 10:12–18

Pfeffer S, Zavolan M, Grasser FA, Chien M, Russo JJ, Ju J, John B, Enright AJ, Marks D, Sander C, Tuschl T (2004) Identification of virus-encoded microRNAs. Science 304:734–736

Phipps KM, Martinez A, Lu J, Heinz BA, Zhao G (2004) Small interfering RNA molecules as potential anti-human rhinovirus agents: in vitro potency, specificity, and mechanism. Antiviral Res 61:49–55

Pillai RS, Artus CG, Filipowicz W (2004) Tethering of human Ago proteins to mRNA mimics the miRNA-mediated repression of protein synthesis. RNA 10:1518–1525

Ping YH, Chu CY, Cao H, Jacque JM, Stevenson M, Rana TM (2004) Modulating HIV-1 replication by RNA interference directed against human transcription elongation factor SPT5. Retrovirology 1:46

Provost P, Dishart D, Doucet J, Frendewey D, Samuelsson B, Radmark O (2002) Ribonuclease activity and RNA binding of recombinant human Dicer. EMBO J 21:5864–5874

Pusch O, Boden D, Silbermann R, Lee F, Tucker L, Ramratnam B (2003) Nucleotide sequence homology requirements of HIV-1-specific short hairpin RNA. Nucleic Acids Res 31:6444–6449

Qin XF, An DS, Chen ISY, Baltimore D (2003) Inhibiting HIV-1 infection in human T cells by lentiviral-mediated delivery of small interfering RNA against CCR5. Proc Natl Acad Sci USA 100:183–188

Radhakrishnan S, Gordon J, Del Valle L, Cui J, Khalili K (2004) Intracellular approach for blocking JC virus gene expression by using RNA interference during viral infection. J Virol 78:7264–7269

Rand TA, Ginalski K, Grishin NV, Wang X (2004) Biochemical identification of Argonaute 2 as the sole protein required for RNA-induced silencing complex activity. Proc Natl Acad Sci U S A 101:14385–14389

Randall G, Grakoui A, Rice CM (2003) Clearance of replicating hepatitis C virus replicon RNAs in cell culture by small interfering RNAs. Proc Natl Acad Sci USA 100:235–240

Reinhart BJ, Slack FJ, Basson M, Pasquinelli AE, Bettinger JC, Rougvie AE, Horvitz HR, Ruvkun G (2000) The 21-nucleotide let-7 RNA regulates developmental timing in Caenorhabditis elegans. Nature 403:901–906

Rouzine IM, Coffin JM (1999) Linkage disequilibrium test implies a large effective population number for HIV in vivo. Proc Natl Acad Sci U S A 96:10758–10763

Saksela K (2003) Human viruses under attack by small inhibitory RNA. Trends Microbiol 11:345–347

Saxena S, Jonsson ZO, Dutta A (2003) Small RNAs with imperfect match to endogenous mRNA repress translation. Implications for off-target activity of small inhibitory RNA in mammalian cells. J Biol Chem 278:44312–44319

Scacheri PC, Rozenblatt-Rosen O, Caplen NJ, Wolfsberg TG, Umayam L, Lee JC, Hughes CM, Shanmugam KS, Bhattacharjee A, Meyerson M, Collins FS (2004) Short interfering RNAs can induce unexpected and divergent changes in the levels of untargeted proteins in mammalian cells. Proc Natl Acad Sci U S A 101:1892–1897

Scherer LJ, Yildiz Y, Kim J, Cagnon L, Heale B, Rossi JJ (2004) Rapid assessment of anti-HIV siRNA efficacy using PCR-derived Pol III shRNA cassettes. Mol Ther 10:597–603

Schwarz DS, Hutvagner G, Du T, Xu Z, Aronin N, Zamore PD (2003) Asymmetry in the assembly of the RNAi enzyme complex. Cell 115:199–208

Schwarz DS, Tomari Y, Zamore PD (2004) The RNA-induced silencing complex is a Mg^{2+}-dependent endonuclease. Curr Biol 14:787–791

Semizarov D, Frost L, Sarthy A, Kroeger P, Halbert DN, Fesik SW (2003) Specificity of short interfering RNA determined through gene expression signatures. Proc Natl Acad Sci U S A 100:6347–6352

Sen A, Steele R, Ghosh AK, Basu A, Ray R, Ray RB (2003) Inhibition of hepatitis C virus protein expression by RNA interference. Virus Res 96:27–35

Seo MY, Abrignani S, Houghton M, Han JH (2003) Small interfering RNA-mediated inhibition of hepatitis C virus replication in the human hepatoma cell line Huh-7. J Virol 77:810–812

Shlomai A, Shaul Y (2003) Inhibition of hepatitis B virus expression and replication by RNA interference. Hepatology 37:764–770

Sledz CA, Holko M, de Veer MJ, Silverman RH, Williams BR (2003) Activation of the interferon system by short-interfering RNAs. Nat Cell Biol 5:834–839

Smith NA, Singh SP, Wang MB, Stoutjesdijk PA, Green AG, Waterhouse PM (2000) Total silencing by intron-spliced hairpin RNAs. Nature 407:319–320

Soderlund H, Pettersson U, Vennstrom B, Philipson L, Mathews MB (1976) A new species of virus-coded low molecular weight RNA from cells infected with adenovirus type 2. Cell 7:585–593

Song E, Lee SK, Dykxhoorn DM, Novina C, Zhang D, Crawford K, Cerny J, Sharp PA, Lieberman J, Manjunath N, Shankar P (2003) Sustained small interfering RNA-mediated human immunodeficiency virus type 1 inhibition in primary macrophages. J Virol 77:7174–7181

Song JJ, Smith SK, Hannon GJ, Joshua-Tor L (2004) Crystal structure of Argonaute and its implications for RISC slicer activity. Science 305:1434–1437

Souret FF, Kastenmayer JP, Green PJ (2004) AtXRN4 degrades mRNA in Arabidopsis and its substrates include selected miRNA targets. Mol Cell 15:173–183

Soutschek J, Akinc A, Bramlage B, Charisse K, Constien R, Donoghue M, Elbashir S, Geick A, Hadwiger P, Harborth J, John M, Kesavan V, Lavine G, Pandey RK, Racie T, Rajeev KG, Rohl I, Toudjarska I, Wang G, Wuschko S, Bumcrot D, Koteliansky V, Limmer S, Manoharan M, Vornlocher HP (2004) Therapeutic silencing of an endogenous gene by systemic administration of modified siRNAs. Nature 432:173–178

Surabhi RM, Gaynor RB (2002) RNA interference directed against viral and cellular targets inhibits human immunodeficiency virus type 1 replication. J Virol 76:12963–12973

Svensson C, Akusjarvi G (1984) Adenovirus VA RNAI: a positive regulator of mRNA translation. Mol Cell Biol 4:736–742

Tabara H, Yigit E, Siomi H, Mello CC (2002) The dsRNA binding protein RDE-4 interacts with RDE-1, DCR-1, and a DExH-box helicase to direct RNAi in C. elegans. Cell 109:861–871

Takigawa Y, Nagano-Fujii M, Deng L, Hidajat R, Tanaka M, Mizuta H, Hotta H (2004) Suppression of hepatitis C virus replicon by RNA interference directed against the NS3 and NS5B regions of the viral genome. Microbiol Immunol 48:591–598

Tomari Y, Du T, Haley B, Schwarz DS, Bennett R, Cook HA, Koppetsch BS, Theurkauf WE, Zamore PD (2004a) RISC assembly defects in the Drosophila RNAi mutant armitage. Cell 116:831–841

Tomari Y, Matranga C, Haley B, Martinez N, Zamore PD (2004b) A protein sensor for siRNA asymmetry. Science 306:1377–1380

Tompkins SM, Lo CY, Tumpey TM, Epstein SL (2004) Protection against lethal influenza virus challenge by RNA interference in vivo. Proc Natl Acad Sci U S A 101:8682–8686

Tonkin LA, Bass BL (2003) Mutations in RNAi rescue aberrant chemotaxis of ADAR mutants. Science 302:1725

Tran N, Cairns MJ, Dawes IW, Arndt GM (2003) Expressing functional siRNAs in mammalian cells using convergent transcription. BMC Biotechnol 3:21

Unwalla HJ, Li MJ, Kim JD, Li HT, Ehsani A, Alluin J, Rossi JJ (2004) Negative feedback inhibition of HIV-1 by TAT-inducible expression of siRNA. Nat Biotechnol 22:1573–1578

Valdes VJ, Sampieri A, Sepulveda J, Vaca L (2003) Using double-stranded RNA to prevent in vitro and in vivo viral infections by recombinant baculovirus. J Biol Chem 278:19317–19324

van de Wetering M, Oving I, Muncan V, Pon Fong MT, Brantjes H, van Leenen D, Holstege FC, Brummelkamp TR, Agami R, Clevers H (2003) Specific inhibition of gene expression using a stably integrated, inducible small-interfering-RNA vector. EMBO Rep 4:609–615

van der Hoek L, Pyrc K, Jebbink MF, Vermeulen-Oost W, Berkhout RJ, Wolthers KC, Wertheim-van Dillen PM, Kaandorp J, Spaargaren J, Berkhout B (2004) Identification of a new human coronavirus. Nat Med 10:368–373

Vance V, Vaucheret H (2001) RNA silencing in plants-defense and counterdefense. Science 292:2277–2280

Vella MC, Choi EY, Lin SY, Reinert K, Slack FJ (2004) The C. elegans microRNA let-7 binds to imperfect let-7 complementary sites from the lin-41 3′ UTR. Genes Dev 18:132–137

Wang MB, Bian XY, Wu LM, Liu LX, Smith NA, Isenegger D, Wu RM, Masuta C, Vance VB, Watson JM, Rezaian A, Dennis ES, Waterhouse PM (2004a) On the role of RNA silencing in the pathogenicity and evolution of viroids and viral satellites. Proc Natl Acad Sci U S A 101:3275–3280

Wang Z, Ren L, Zhao X, Hung T, Meng A, Wang J, Chen YG (2004b) Inhibition of severe acute respiratory syndrome virus replication by small interfering RNAs in mammalian cells. J Virol 78:7523–7527

Wassenegger M, Heimes S, Riedel L, Sanger HL (1994) RNA-directed de novo methylation of genomic sequences in plants. Cell 76:567–576

Waterhouse PM, Graham MW, Wang MB (1998) Virus resistance and gene silencing in plants can be induced by simultaneous expression of sense and antisense RNA. Proc Natl Acad Sci USA 95:13959–13964

Westerhout EM, Ooms M, Vink M, Das AT, Berkhout B (2005) HIV-1 can escape from RNA interference by evolving an alternative structure in its RNA genome. Nucleic Acids Res 33:796–804

Wiebusch L, Truss M, Hagemeier C (2004) Inhibition of human cytomegalovirus replication by small interfering RNAs. J Gen Virol 85:179–184

Wienholds E, Koudijs MJ, van Eeden FJ, Cuppen E, Plasterk RH (2003) The microRNA-producing enzyme Dicer1 is essential for zebrafish development. Nat Genet 35:217–218

Williams BR (1999) PKR; a sentinel kinase for cellular stress. Oncogene 18:6112–6120

Wilson JA, Jayasena S, Khvorova A, Sabatinos S, Rodrigue-Gervais IG, Arya S, Sarangi F, Harris-Brandts M, Beaulieu S, Richardson CD (2003) RNA interference blocks gene expression and RNA synthesis from hepatitis C replicons propagated in human liver cells. Proc Natl Acad Sci USA 100:2783–2788

Wiznerowicz M, Trono D (2003) Conditional suppression of cellular genes: lentivirus vector-mediated drug-inducible RNA interference. J Virol 77:8957–8961

Xia XG, Zhou H, Ding H, Affar eB, Shi Y, Xu Z (2003) An enhanced U6 promoter for synthesis of short hairpin RNA. Nucleic Acids Res 31:e100

Yamamoto T, Omoto S, Mizuguchi M, Mizukami H, Okuyama H, Okada N, Saksena NK, Brisibe EA, Otake K, Fuji YR (2002) Double-stranded nef RNA interferes with human immunodeficiency virus type 1 replication. Microbiol Immunol 46:809–817

Yi R, Qin Y, Macara IG, Cullen BR (2003) Exportin-5 mediates the nuclear export of pre-microRNAs and short hairpin RNAs. Genes Dev 17:3011–3016

Ying C, De Clercq E, Neyts J (2003) Selective inhibition of hepatitis B virus replication by RNA interference. Biochem Biophys Res Commun 309:482–484

Yokota T, Sakamoto N, Enomoto N, Tanabe Y, Miyagishi M, Maekawa S, Yi L, Kurosaki M, Taira K, Watanabe M, Mizusawa H (2003) Inhibition of intracellular hepatitis C virus replication by synthetic and vector-derived small interfering RNAs. EMBO Rep 4:1–7

Zhang W, Yang H, Kong X, Mohapatra S, Juan-Vergara HS, Hellermann G, Behera S, Singam R, Lockey RF, Mohapatra SS (2005) Inhibition of respiratory syncytial virus infection with intranasal siRNA nanoparticles targeting the viral NS1 gene. Nat Med 11:56–62

Zhang XN, Xiong W, Wang JD, Hu YW, Xiang L, Yuan ZH (2004a) siRNA-mediated inhibition of HBV replication and expression. World J Gastroenterol 10:2967–2971

Zhang Y, Li T, Fu L, Yu C, Li Y, Xu X, Wang Y, Ning H, Zhang S, Chen W, Babiuk LA, Chang Z (2004b) Silencing SARS-CoV Spike protein expression in cultured cells by RNA interference. FEBS Lett 560:141–146

Zheng X, Bevilacqua PC (2000) Straightening of bulged RNA by the double-stranded RNA-binding domain from the protein kinase PKR. Proc Natl Acad Sci U S A 97:14162–14167

Gene Silencing of Virus Replication by RNA Interference

N. Miyano-Kurosaki · H. Takaku (✉)

Department of Life and Environmental Sciences and High Technology Research Center, Chiba Institute of Technology, 2-17-1 Narashino, Tsudanuma, 275-0016 Chiba, Japan
hiroshi.takaku@it-chiba.ac.jp

1	General Mechanism of RNAi	151
2	Synthetic siRNAs	155
3	DNA Vector-Mediated siRNAs or Short Hairpin RNAs	156
4	Inhibition of HIV-1 Replication by RNAi	158
5	siRNA Agents Work as Ligands for Toll-Like Receptors	161
6	Virus Escape from RNAi	162
7	Using RNAi to Treat Other Viruses	163
8	Concluding Remarks	165
	References	166

Abstract Small interfering RNAs (siRNAs) are as effective as long double-stranded RNAs (dsRNAs) at targeting and silencing genes by RNA interference (RNAi). siRNAs are widely used for assessing gene function in cultured mammalian cells or early developing vertebrate embryos. They are also promising reagents for developing gene-specific therapeutics. The specific inhibition of viral replication is particularly well suited to RNAi, as several stages of the viral life cycle and many viral and cellular genes can be targeted. The future success of this approach will depend on the recent advances in siRNA-based clinical trials.

Keywords RNA interference · Gene silencing · Virus · Toll-like receptors · Virus escape

1
General Mechanism of RNAi

Eukaryotes have evolved a cellular defense system that responds to double-stranded (ds)RNAs and protects their genomes against these invading foreign elements. dsRNA delivery into cells has been used to elucidate the role of cellular genes that are homologous in sequence to the introduced dsRNAs by means of sequence-specific gene silencing (Fire et al. 1998). RNA interference (RNAi)-based reverse genetic analysis now provides a rapid link between sequence data and biological function. RNAi is particularly useful for the analysis of gene function in *Caenorhabditis elegans* (for reviews, see Hope 2001; Kim 2001). Effective gene silencing typically requires long dsRNAs (Parrish et al. 2000; Elbashir et al. 2001c). However, its application in vertebrates, including mammals, has proved to be difficult because of the presence of additional dsRNA-triggered pathways that mediate the non-specific suppression of gene expression (Caplen et al. 2000; Nakano et al. 2000; Oates et al. 2000; Zhao et al. 2001). These non-specific responses to long dsRNAs are not, however, triggered by small interfering RNAs (siRNAs) (Bitko and Barik 2001; Caplen et al. 2001; Elbashir et al. 2001a; Zhou et al. 2002). siRNAs can target genes as effectively as long dsRNAs (Elbashir et al. 2001c) and are widely used for assessing gene function in cultured mammalian cells or early developing vertebrate embryos (Harborth et al. 2001; Elbashir et al. 2002; Zhou et al. 2002). siRNAs are also promising reagents for developing gene-specific therapeutics (Tuschl and Borkhardt 2002). However, another major problem for using RNAi as a tool to inhibit viral replication is predicting the effectiveness of a specific siRNA. The difficulty lies in making siRNAs trigger silencing in a gene-specific manner without causing non-target-related biological effects or the emergence of escape variants by foreign siRNAs. Work over the past 2 years has allowed investigators to meet this challenge, and the siRNA approach has now been adopted as a standard methodology for sequence-specific silencing in mammalian cells. This review focuses on RNAi as it relates to mammalian systems and the application of siRNAs for targeting genes that are expressed in virus-infected cell lines.

Studies in plants and *Drosophila* have provided fundamental insights into the mechanism of RNAi, following the demonstration that RNAi was activated by dsRNAs and the suggestion that it might involve a derivative of dsRNAs (Fire et al. 1998; Fig. 1).

Biochemical characterization has shown that siRNAs are 21- to 23-nt dsRNA duplexes with symmetric 2- to 3-nt $3'$ overhangs, and $5'$-phosphate and $3'$-hydroxyl groups (Elbashir et al. 2001b; Fig. 1). This structure is characteristic of an RNase III-like enzymatic cleavage pattern, which led to the identification of the highly conserved Dicer family of RNase III enzymes as the mediators of dsRNA cleavage (Bernstein et al. 2001; Billy et al. 2001; Ketting et al. 2001).

Extensive biochemical and genetic evidence has allowed a better understanding of how long dsRNAs trigger degradation of the target messenger RNAs (mRNAs) (Fig. 1; for recent reviews, see Sharp 2001; Hannon 2002; McManus and Sharp 2002; Zamore 2002). Several studies have shown that this process is restricted to the cytoplasm (Hutvagner and Zamore 2002; Zeng and Cullen 2002; Kawasaki and Taira 2003). In the first step, Dicer cleaves long dsRNAs to produce siRNAs, which are incorporated into a multiprotein RNA-inducing silencing complex (RISC). There is a strict requirement for the siRNAs to be $5'$ phosphorylated in order to enter into the RISC (Nykanen et al. 2001; Schwarz et al. 2002). siRNAs that lack a $5'$ phosphate are rapidly phosphorylated by

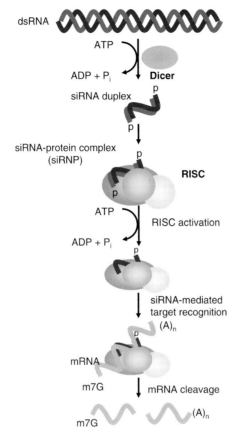

Fig. 1 Model for RNA-mediated interference and silencing. The cellular RNase III enzyme Dicer processes double-stranded RNA (*dsRNA*) to 21- to 23-nt short-interfering RNA (*siRNA*) duplexes in an ATP-dependent manner. The siRNAs are incorporated into a siRNA–ribonucleoprotein complex (*siRNP*), which uses ATP to rearrange itself into the RNA-induced silencing complex (*RISC*) by unwinding the siRNA duplex. Once unwound, the single-stranded antisense siRNA guides the RISC to mRNA with a complementary sequence, causing endonucleolytic cleavage of the target mRNA. The mRNA-cleavage products are then released and the RISC can be reactivated for another round of catalytic target RNA cleavage

an endogenous kinase (Schwarz et al. 2002). The duplex siRNA is unwound, leaving the antisense strand to guide the RISC to its homologous target mRNA for endonucleolytic cleavage. The target mRNA is cleaved at a single site in the center of the duplex region between the guide siRNA and the target mRNA, 10 nt from the 5′ end of the siRNA (Elbashir et al. 2001a,b).

Interestingly, endogenously expressed siRNAs have not been found in mammals. However, related microRNAs (miRNAs) have been cloned from various organisms and cell types (Pasquinelli 2002; Fig. 2). These short (22 nt) RNA

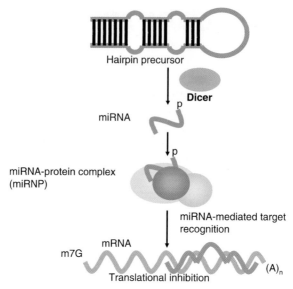

Fig. 2 MicroRNAs might use the same RNA-processing complex to direct silencing. Processing of the microRNA (*miRNA*) precursor hairpin (~70 nt) or long double-stranded RNA (dsRNA) would lead to single-stranded 21- to 23-nt RNA that is associated with the miRNA-protein complex (*miRNP*). This complex might direct either mRNA translation repression or mRNA target cleavage, depending on the degree of complementarity between the 21- to 23-nt RNA and the mRNA

species are produced by Dicer cleavage of longer (70 nt) endogenous precursors with imperfect hairpin RNA structures (Fig. 2). The miRNAs are believed to bind to sites that have partial sequence complementarity in the 3′ untranslated region (UTR) of their target mRNAs, causing the repression of translation and the inhibition of protein synthesis (Pasquinelli and Ruvken 2002). More recently, Zeng et al. (2003) demonstrated that an endogenously encoded human miRNA was able to cleave an mRNA bearing fully complementary target sites, whereas an exogenously supplied siRNA could inhibit the expression of an mRNA bearing partially complementary sequences without inducing detectable RNA cleavage. These data suggest that miRNAs and siRNAs can use similar mechanisms to repress mRNA expression, and that the choice of mechanism might be largely, or entirely, determined by the degree of complementary of the RNA target. In addition to Dicer, other PAZ/PIWI domain proteins (PPD), including eukaryotic translation-initiation factor 2C2 (eIF2C2), are likely to function in both pathways (Grishok et al. 2001; Hutvagner and Zamore 2002; Mourelatos et al. 2002).

Although the apparent lack of RNAi in mammalian cell culture was unexpected, yet RNAi has been found in mouse oocytes and early embryos (Svoboda et al. 2000; Wianny and Zernicka-Goetz 2000). Similarly, RNAi-related transgenic-mediated co-suppression has been observed in cultured Rat-1 fibroblasts (Bahramian and Zarbl 1999). Notably, dsRNAs in the cytoplasm of mammalian cells have been reported to trigger profound physiological reactions that lead to the induction of interferon (IFN) synthesis (Lengyel 1987; Stark et al. 1998; Barber 2001). In the IFN response, dsRNAs that are longer than 30 bp bind and activate the protein kinase PKR and 2′,5′-oligoadenylate synthetase (2′,5′-AS) (Minks et al. 1979; Manche et al. 1992). Activated PKR suppresses translation by phosphorylating the translation-initiation factor eIF2α, while activated 2′,5′-AS causes mRNA degradation by stimulating RNase L. These responses are intrinsically sequence-non-specific with respect to the inducing dsRNAs.

2
Synthetic siRNAs

RNAi mediated by siRNAs is a powerful tool for dissecting gene function and drug-target validation. siRNAs can be synthesized in large quantities and thus can be used to analyze large numbers of sequences emerging from genome projects in a cost-effective manner. However, the phenomenon might reflect an incorrect sequence of RNAi, poor penetration of the mammalian cells by the nucleotides, or insufficient knowledge of the protein in question. siRNAs for gene-targeting experiments have only been introduced into cells via classic gene-transfer methods, such as liposome-mediated transfection, electroporation, and microinjection, all of which require the chemical or enzymatic synthesis of siRNAs (Donze and Picard 2002). Synthetic siRNA duplexes can be incubated with lipid formulations to generate liposomes containing siRNAs. In such formulations, cationic lipids bind to oligoribonucleotides through anion–cation and hydrophobic interactions. The efficiency of siRNA uptake is dependent upon the cell type. As high concentrations of cationic liposomes can be toxic, their application must be optimized for each type of target cell (Felgner et al. 1994). In particular, the transfection efficiency into suspension cells, such as T-cell lines and primary cells, using cationic liposomes is often less than 10%. Most of the cationic lipid reagents that are currently used for siRNAs are formulated as liposomes (lipofectamine) containing two lipid species: the polycationic lipid 2,3-diolexyoloxy-*N*-[2(spermine-carboxamido)ethyl] *N,N*-dimethyl-1-propanaminium trifluoroacetate (DOSPA) and the neutral lipid dioleoylphosphatidylethanolamine (DOPE) (3:1 w/w). The efficient delivery of siRNAs by the lipofectamine reagent has been reported for siRNA-mediated RNAi in cultured mammalian cells (Caplen et al. 2001; Elbashir et al. 2001a; Garrus et al. 2001; Paul et al. 2002) and siRNA-mediated anti-acquired immunodeficiency syndrome (AIDS) therapeutics (Gitlin et al. 2002; Jacque et al. 2002; Novina et al. 2002; Park et al. 2002, 2003). The further development of liposomes might enhance their ability to deliver siRNAs to a broader range of target cells.

3
DNA Vector-Mediated siRNAs or Short Hairpin RNAs

siRNA-directed silencing by transfection is limited in mammals by its transient nature. To overcome some of the shortcomings of transfecting chemically synthesized siRNAs into cells, several groups have developed DNA vector-mediated mechanisms to express substrates that can be converted into siRNAs in vivo (Kennerdell and Carthew 2000; Tavernarakis et al. 2000; Svoboda et al. 2001; Brummelkamp et al. 2002a; Lee et al. 2002; McManus et al. 2002; Miyagishi and Taira 2002; Paddison et al. 2002a,b; Paul et al. 2002; Sui et al. 2002; Yu et al. 2002; Kawasaki and Taira 2003) (Fig. 3). Alternatively, small RNA molecules might also be expressed in cells following the cloning of siRNA templates into RNA polymerase III (pol III) transcription units, which are based on the sequences of the natural transcription units of the small-nuclear RNA (snRNA) U6 or the human RNase P RNA H1 (Medina and Joshi 1999; Paule and White 2000; Myslinski et al. 2001).

Two approaches have been used to express siRNA species through constructs that are driven by RNA pol III. In the first approach, the sense and antisense strands of the siRNA duplex are expressed from different, usually tandem, promoters (Fig. 3) (Lee et al. 2002; Miyagishi and Taira, 2002; Yu et al. 2002). In vivo, these strands come together to form a 19-nt duplex with 4-nt overhangs from the pol III-termination signal. The second approach uses Dicer to express and process short hairpin (sh)RNAs into siRNAs (Fig. 1). Dicer is required for the processing of pre-*let7* RNA—which is a structured, approximately 70-nt hairpin—into the mature, 22-nt active miRNA species (Reinhart et al. 2000; Grishok et al. 2001; Hutvagner et al. 2001; Knight and Bass 2001; Hutvagner and Zamore 2002). H1 RNA–pol III-based shRNA expression vector has been used to produce hairpin RNA with a 19-nt stem and a short loop (Brummelkamp et al. 2002a; Fig. 3). This system was used to inhibit the expression of E-cadherin (CDH1) and p53 with a comparable efficiency to siRNA transfection. Using RNA based on the *let7* precursor, *luciferase* mRNA has been targeted for degradation by including a 32-nt *luciferase*-complementary sequence in the stem of the hairpin (Paddison et al. 2002b). When transfected into *Drosophila* S2 cells, they found that although *let7*-basedpre-*let7* RNA structures could target the *luciferase* mRNA, the most effective inhibitors had a simple hairpin structure with full complementarity in the stem. To express hairpin RNA in mammalian cells, they developed a U6 RNA–pol III-based expression system, which used a 29-nt sequence complementary to the luciferase gene and an 8-nt loop.

Although most expression systems use either the U6 or H1 promoter, an expression system that uses the transfer (t)RNA^{Val} promoter was described. shRNAs that have been generated using this expression system show a strong cytoplasmic localization and are efficiently processed by Dicer into siRNAs (Kawasaki and Taira 2003).

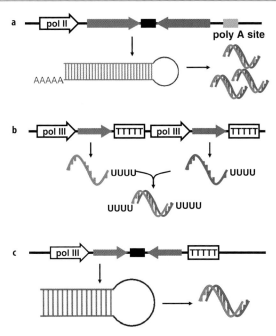

Fig. 3a–c Endogenous expression of short-interfering RNAs (siRNAs). **a** Long hairpin RNA expressed from an RNA polymerase (pol II) promoter yields a population of siRNAs with various sequence specificities. **b** An expression cassette for sense and antisense siRNAs using the tandem pol III small-nuclear RNA (snRNA) promoter. The preferred target site, which has been selected for optimal vector design, is indicated at the *bottom*. **c** A single pol III cassette for expressing hairpin RNAs that are subsequently processed to siRNAs. In this case, transcript synthesis is initiated with a +1 guanosine, and the 3′ end of the sense strand is joined by short oligonucleotide loops with the antisense strand

Some investigators have employed viral vectors in order to facilitate the introduction of siRNA-expressing cassettes into cells. Human immunodeficiency virus type-1 (HIV-1)-based lentivirus vectors have attracted particular attention in this regard. These vectors exploit the ability of HIV-1 to infect non-dividing cells (Weinberg et al. 1991; Bukrinsky et al. 1992; Lewis et al. 2002). HIV-1-based lentivirus vectors retain this central characteristic and, as such, are particularly suitable for the transduction of non-dividing cells, such as neurons and hematopoietic progenitor cells (Naldini et al. 1996).

Lentivirus vectors expressing shRNAs have been shown to promote specific gene silencing in primary dendritic cells (Stewart et al. 2003), while CD8-specific shRNAs expressed from an HIV-1-based vector were capable of silencing CD8 expression both in vitro and in vivo (Rubinson et al. 2003). Collectively, these studies illustrate the broad utility of RNAi for the silencing of viral and cellular processes in vitro and in vivo.

4
Inhibition of HIV-1 Replication by RNAi

The introduction of combination antiretroviral therapy has resulted in a remarkable improvement of the life expectancy of individuals infected with HIV and has significantly reduced their likelihood of developing AIDS. However, despite this progress, HIV infection remains incurable. Toxicity problems associated with current drug therapies and the emergence of drug resistance clearly indicate the need for alternative therapeutic approaches. Retroviral infection with HIV results in the stable integration of proviral DNA into the genome of target cells, and can therefore be viewed as an acquired genetic disease. Thus, the modulation of HIV replication by the expression of antiviral genes might be a therapeutic option for HIV infection. Baltimore (1988) was the first to suggest the concept of gene therapy as an intracellular immunization against HIV. Recently, numerous anti-HIV gene-therapy approaches have been developed and tested in clinical trials. These strategies can be divided into two main categories: first, the genetic modification of HIV target cells or their progeny in order to inhibit HIV replication and second, the genetic modification of cells in order to generate an immune response against HIV or HIV-infected cells. The latter category can be viewed as gene therapy-based immunotherapy and will not be discussed further in this review.

The inhibition of HIV replication involves the transfer of genetic material into HIV-1 target cells or their progenitors (CD4$^+$ T cells or hematopoietic stem cells). A typical gene-therapy approach for HIV-1 infection is schematically depicted in Fig. 4. HIV-1 is well suited for target RNAi because dsRNAs act at multiple steps during the HIV-1 replication cycle (Fig. 4). The inhibitory proteins that are used against HIV act intracellularly and include antibody fragments, single-chain variable fragments, transdominant negative HIV, and cellular proteins. Most of these approaches target viral RNA or proteins. Additional cellular factors that are prerequisites for HIV infection or replication are also potential targets for anti-HIV gene therapy. A number of studies have reported that the transient transfection of siRNAs directed to several HIV-1 genes (*HIV-1 LTR*, *gag*, *vif*, *nef*, *tat* and *rev*) induced pre-integrated HIV-1 RNA degradation and consequently reduced HIV-1 antigen production by infected cells (Brummelkamp et al. 2002a; Capodici et al. 2002; Coburn and Cullen 2002; Jacque et al. 2002; Lee et al. 2002; Lewis et al. 2002; Novina et al. 2002; Paul et al. 2002; Surabhi and Gaynor 2002; Yamamoto et al. 2002; Yu et al. 2002; Song et al. 2003a). Lee and colleagues, and Banerjea and co-workers, demonstrated that a psiRNA approach can be used to inhibit the expression of HIV-1 *rev* and/or *tat* transcripts in transient transfections (Lee et al. 2002) or from lentiviral-transduced hematopoietic progenitor cells (Banerjea et al. 2003). In this approach, the vectors contain two tandem human U6 snRNA promoters followed by 21-mers encoding sense and antisense siRNAs. In co-transfection experiments, psiRNAs that were co-transfected with the HIV-1 pNL4-3 provi-

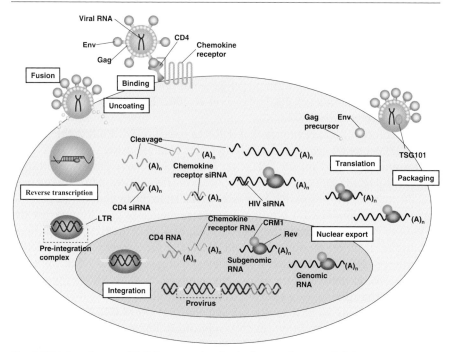

Fig. 4 RNA interference (RNAi) target sites in the human immunodeficiency virus type-1 (HIV-1) replication cycle. Short-interfering RNAs (siRNAs) that target HIV-1 RNA might induce the cleavage of pre-integrated RNA or interfere with post-integration HIV-1 RNA transcripts and block progeny virus production. siRNAs targeting CD4, CXCR4, or CCR5 RNA transcripts inhibit virus attachment to the CD4 receptor or chemokine receptor-mediated HIV-1 fusion and entry. As cleavage of the messenger RNA (mRNA) target requires a high degree of complementarity between the siRNA and its target sequence, heterogeneity in the virus population might prevent efficient silencing of some virus variants by specific siRNAs

ral DNA inhibited HIV-1 p24 antigen expression by up to 4 logs (Lee et al. 2002). This strong inhibition was achieved by simultaneously targeting two essential sites (*rev* and *tat*). Synthetic siRNAs targeted to HIV-1 *rev* and *tat* mRNAs were also shown to inhibit HIV-1 gene expression and replication in both human T cell lines and primary lymphocytes (Coburn and Cullen 2002).

Additional studies have demonstrated that siRNAs act at a later stage of the HIV-1 life cycle, causing post-integration degradation of HIV-1 RNA transcripts (Jacque et al. 2002; Lee et al. 2002; Novina et al. 2002).

The HIV-1 gag gene is expressed during the later steps of HIV-1 replication and encodes the gag-precursor protein, which is proteolytically cleaved into p24 and other polypeptides. p24 forms the HIV-1 core and functions by uncoating and packaging viral RNA. Novina et al. (2002) transfected cells with anti-gag siRNAs, exposed them to HIV-1, and observed a decrease in the in vitro production of p24. Co-transfections by Jacque et al. (2002) of a proviral

HIV-1 clone, 19-bp stem siRNAs directed against other HIV-1 genes (such as *vif* and LTR-*TAR*), and insertational mutagenesis of *nef* by a green fluorescent protein (GFP) gene, showed a significant suppression of virus production compared with non-transfected cells. Such siRNA- or shRNA-expression systems, if stable, might allow long-term target-gene suppression in cells (Naldini et al. 1996; Brummelkamp et al. 2002a; Lewis et al. 2002; Paddison et al. 2002b; Paul et al. 2002; Yu et al. 2002).

We demonstrated previously that dsRNAs specifically suppress the expression of HIV-1 genes (Park et al. 2002). In order to study dsRNA-mediated gene interference in HIV-1 infected cells, six long dsRNAs were designed to target the HIV-1 *gag* and *env* genes. HIV-1 replication was suppressed in a sequence-specific manner by these dsRNAs in infected cells. In particular, the E2 dsRNA, containing the major CD4 binding-domain sequence of gp 120 to target the HIV-1 env gene, dramatically inhibited the expression of the HIV-1 p24 antigen in peripheral blood mononuclear cells (PBMCs) for 2 weeks. More effective inhibition of HIV-1 replication was achieved using four siRNAs that were targeted to several regions of the HIV-1 env genes (Park et al. 2003). The mRNA targets for the siRNAs were selected from the middle of the *env* regions in the HIV-1 genome, as we previously showed that 531-bp (7.070–7.600) E2 dsRNAs complementary to the *env* mRNA-containing V3 loop and the major CD4 binding-domain sequence of gp 120 were more effective inhibitors than those targeted to the *gag* gene. Furthermore, the envelope protein (Env) of HIV-1 mediates functions that are critical to the viral life cycle, including viral attachment to target cells, and fusion of the viral and cellular membranes. We also showed the inhibition of HIV-1 replication in T cells using E2 shRNA directly from a lentivirus vector (Hayafune et al. 2005). On the other hand, we have shown that vif shRNA specifically suppresses the expression of HIV-1 (Barnor et al. in press). The HIV-1-encoded vif protein is essential for viral replication, virion production, and pathogenicity. HIV-1 vif interacts with the endogenous human APOBEC3G protein (an mRNA editor) in target cells to prevent its virions from encapsidation. Previous studies have established targets within the HIV-1 vif gene that are important for its biological function; however, it is important to determine effective therapeutic targets within *vif* because of its critical role in HIV-1 vif-dependent infectivity and pathogenicity. vif shRNAs increased the inhibition of HIV-1 replication in a long-term culture assay.

Rather than targeting the viral RNA, an alternative way of inhibiting virus replication by RNAi is to silence the expression of cellular genes that are critically involved in viral replication. For HIV-1, these targets include the mRNAs encoding the CD4 receptor and the CCR5 or CXCR4 co-receptors. These receptors are essential for attachment of the HIV-1 particle to the cell and for subsequent viral entry. RNAi against the viral RNA does not protect the cell against viral entry. By silencing these receptors, the HIV-1 particle will be unable to attach to, and enter, the cell, thus yielding a form of HIV-1 resistance.

Novina et al. (2002) showed that specific siRNAs that were directed against either CD4 or gag genes were able to prevent HIV-1 replication in MAGI and H9 cells. siRNA targeting rendered the receptor unavailable for virus attachment, thereby inhibiting HIV-1 entry and virus production. However, CD4 targeting might not be a feasible therapeutic approach because of its importance in immune function. By contrast, CCR5, which is the major HIV-1 co-receptor for viral entry into macrophages, might be a potentially useful cellular target, as a 32-bp homozygous deletion of the gene abolishes its function without deleterious immunological consequences and provides protection from HIV-1 infection (Martinez et al. 2002; Qin et al. 2003; Song et al. 2003a).

In this regard, RNAi is a powerful tool with which to determine the role of cellular co-factors in HIV-1 replication. Indeed, the first study to use RNAi in HIV-1 research silenced the expression of TSG-101, which is a component of the class E vacuolar protein-sorting pathway, by means of siRNAs (Garrus et al. 2001). This revealed a critical role for TSG-101 in the budding of HIV-1 virions. Moreover, when a lentivirus-based vector system was used to introduce shRNAs against CCR5 into peripheral blood T lymphocytes, the expression of CCR5 on the cell surface was reduced tenfold, resulting in a three- to sevenfold decrease in the number of infected cells (Qin et al. 2003). Lee et al. (2003) also showed inhibition of HIV-1 replication in macrophages using tat or CCR5 directly from a lentivirus vector. Similarly, siRNAs directed against CXCR4 co-receptors blocked HIV-1 entry, and protected cells from infection and delayed virus replication (Anderson et al. 2003). Another host factor that is important for HIV-1 replication is the transcription factor nuclear factor (NF)-κB to motifs in the long terminal repeat (LTR) promoter of the integrated provirus is required for viral transcription (Surabhi and Gaynor 2002). However, targeting NF-κB is not an appropriate therapeutic option, owing to the important role of NF-κB in cells.

5
siRNA Agents Work as Ligands for Toll-Like Receptors

Surprisingly, recent studies have indicated that siRNAs can induce global up-regulation of the expression of IFN-stimulated genes (Bridge et al. 2003; Jackson et al. 2003; Sledz et al. 2003; Kariko et al. 2004; Persengiev et al. 2004; Fig. 5). This effect was detected with synthetic siRNAs that were transfected into cells, and with siRNAs that were produced within cells by the expression of shRNAs. Both of these papers documented significant non-specific changes in gene expression as a consequence of the delivery of siRNAs. Sledz et al. (2003) observed a 2-fold induction of 52 out of 850 putative IFN-stimulated genes using synthetic siRNAs. By contrast, Bridge et al. (2003) observed a 50-fold induction of the IFN-stimulated gene *OAS1* with one siRNA vector alone, and a 500-fold induction when two vectors were used simultaneously. These results suggest

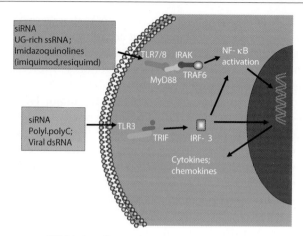

Fig. 5 Toll-like receptor (TLR) signaling pathways. TLRs recognize molecular patterns associated with bacterial pathogens, double-stranded RNA and siRNA for TLR3; siRNA, UG-rich ssRNA, imidazoquinoline and its derivatives for TLR7/8. TLR signaling pathways are separated into two groups: (1) A MyD88-dependent pathway that leads to the production of pro-inflammatory cytokines with quick activation of NF-κB and MAPK; and (2) TRIF, which may exist downstream of TLR3/4, and IKKε and TBK1 to mediate the MyD88-independent pathways leading to production of IFN-β and IFN-inducible genes

that the ability to induce the IFN system depends on both the siRNA sequence and the method of delivery. Both groups pointed out that increasing the quantity of the siRNAs enhanced the effect. Furthermore, two recent studies have indicated that the mechanism of the IFN response might include recognition of the siRNAs by Toll-like receptor 3 (TLR3) (Heidel et al. 2004; Kariko 2004). One simple method for limiting the risk of inducing an IFN response is to use the lowest effective dose of shRNA vector, as advocated by Bridge et al. (2003).

Recently, Kim et al. (2004) showed that siRNAs synthesized using the T7 RNA polymerase system can trigger the apotent induction of IFN-α and -β in a variety of cells. The mediators of this response revealed that an initiating 5′-triphosphate was required for IFN induction. These findings have led to the development of an improved method for bacteriophage polymerase-mediated siRNA synthesis that incorporates two 3′ adenosines in order to prevent base-pairing with the initiating Gs, thereby allowing RNase T1 and calf intestine alkaline phosphatase (CIP) to remove the initiating 5′ nucleotides and triphosphates of the transcripts.

6
Virus Escape from RNAi

When profound inhibition of virus replication is obtained by means of RNAi technology, the possibility of viral escape must be considered. This potential problem is particularly relevant for viruses that exhibit significant genetic variation due to an error-prone replication machinery.

This risk might be more severe for RNA viruses and retroviruses than for DNA viruses. The variability of HIV caused by its error-prone reverse transcriptase has been shown to generate mutations in the gene being targeted, thus allowing it to rapidly evade siRNAs (Boden et al. 2003). Similar results were observed for RNAi of a poliovirus infection (Gitlin et al. 2002). Synthetic siRNAs against poliovirus inhibited virus production 100-fold; however, the virus titer increased to high levels upon prolonged incubation. Sequence analysis of the progeny virus demonstrated a single escape mutation in the center of the siRNA target sequence. These findings indicate that the point mutation occurred in the middle of the signal target sequences. Two recent studies have demonstrated that transfection with siRNAs containing mismatches to the target sequence in the middle of the siRNA molecules reduces the efficiency of gene silencing (Brummelkamp et al. 2002b; Amarzguioui et al. 2003). These findings suggest that in order for RNAi to durably suppress HIV-1 replication, more potent shRNAs will need to be designed that can target highly conserved regions of the viral genome (for example, *gag* and *pol*) that are essential for the viral life cycle. Alternatively, RNAi constructs co-expressing multiple shRNAs could be developed that simultaneously target different regions of the viral genome, thereby reducing the probability of generating shRNA escape mutants.

7
Using RNAi to Treat Other Viruses

Although many previous studies on RNAi-mediated inhibition have focused on HIV-1, there is a growing body of data addressing the inhibition of other animal and human viruses. These include RNA viruses such as hepatitis C virus (HCV), poliovirus, Semliki Forest virus (SFV), influenza virus A, rhesus rotavirus (RRV), and Rous sarcoma virus (RSV), and DNA viruses such as human papillomavirus type 16 (HPV-16) and hepatitis B virus (HBV). In most of these studies, the RNAi machinery was directly targeted towards the viral RNA using synthetic siRNAs.

Hepatitis induced by HBV or HCV is a major health problem. At present, hundreds of millions of individuals are infected worldwide. Although there is an effective vaccine against HBV, it is only useful for the prevention of viral infection. There is no vaccine for HCV. Hepatitis caused by these two viruses has therefore been an important target for potential RNAi therapy.

The first demonstration of RNAi efficacy against a virus in vivo involved the hydrodynamic co-delivery of an HBV replicon and an expression unit encoding an anti-HBV shRNA in mice. HBV is a member of the family Hepadnaviridae and has a 3.2-kb circular dsDNA genome. During infection, four RNAs are transcribed, which encode the coat protein (CP), polymerase (P), surface antigen (S), and transactivator of transcription (X). HBV production in Huh-7 cells was shown to be reduced by up to 20-fold through the transfection of a vector-expressing shRNA against the X mRNA (Shlomai and Shaul 2003).

Inhibition of HBV in the liver of mice was achieved through the co-transfection of HBV DNA and shRNA-expressing plasmids (McCaffrey et al. 2003), which resulted in a six fold decrease in the amount of secreted HBV surface antigen in the serum. This small-animal model of human infectious disease shows that it is possible to use RNAi as a potent antiviral therapy in mammals.

HCV is a major cause of chronic liver disease, which can lead to liver cirrhosis and hepatocellular carcinoma (Reed and Rice 2000). The HCV genome is a positive-strand RNA molecule with a single open reading frame encoding a polyprotein that is processed post-translationally to produce at least 10 proteins. HCV is a member of the family Flaviviridae and has a (+) single-stranded (ss)RNA genome.

Subgenomic and full-length HCV replicons that replicate and express HCV proteins in stably transfected human hepatoma-derived Huh-7 cells have been used to study the effects of various antiviral drugs (Lohmann, et al. 1999; Pietschmann, et al. 2001; Ikeda et al. 2002). Several groups have now tested the efficacy of the siRNA-mediated inhibition of replicon function using these systems (Kapadia et al. 2003; Randall et al. 2003; Wilson et al. 2003). These replicons support HCV RNA transcription and protein synthesis, but do not produce infectious viruses.

siRNAs targeted against sequences in the viral non-structural proteins NS3 and NS5B have been shown to cause profound (up to 100-fold) inhibition of HCV replicon function in cell cultures (Kapadia et al. 2003; Randall et al. 2003; Seo et al. 2003; Wilson et al. 2003). Furthermore, the internal ribosomal-entry site (IRES) in the well-conserved 5′ UTR of the HCV RNA has also been a good target. Both siRNAs and shRNAs have been reported to inhibit HCV replicon function in cells (Seo et al. 2003; Wilson et al. 2003; Yokota et al. 2003; Hamazaki et al. 2005).

In another in vivo study, siRNAs were used to treat fulminant hepatitis induced by an agonistic Fas-specific antibody in mice (Song et al. 2003b). Fas-mediated apoptosis of hepatocytes can be triggered by HBV and HCV infection. Infusing siRNAs targeting Fas mRNAs into the tails of the mice blocked this self-destructive inflammatory response of the liver. These findings indicate that major hurdles remain before this therapy can be applied to humans.

As with HIV therapeutics, delivery of the siRNAs or shRNA vectors is the main challenge for the successful treatment of HCV.

Generally, human influenzal lesions are local infections that remain in the upper portion of the respiratory tract and do not proceed to pneumonia. Nevertheless, in high-risk patients, cases of influenzal pneumonia have been reported in which expansion of the virally infectious focus is observed in a pulmonary lesion. Furthermore, such cases are often accompanied by a secondary bacterial pneumonia. The influenza virus belongs to the family Orthomyxoviridae and has a (−) ssRNA genome. Its genome is composed of eight separate segments. The proteins encoded by the eight segmented genes include HA and NA, as well as the M1 and M2 membrane proteins, which are located on the surface of the envelope. Furthermore, a nucleoprotein complex (RNP) is located at the center of the virus and is composed of the gene RNA, three RNA polymerase subunits (PB1, PB2, and PA) and a nucleoprotein (NP). A non-structural protein (NS) is synthesized from the eighth segmented gene. Amantadine and rimantadine are known antiviral agents for the influenza A virus; however, neither drug can cope with mutants and both have strong side effects (Atmar et al. 1990; Wang et al. 1993). Treatment by an inactivated vaccine has also been attempted; however, the vaccine cannot sustain antibody productivity for a long period and, thus, cannot completely prevent the spread of infection (Hirota et al. 1996).

Ge et al. (2003) showed that siRNAs targeting conserved regions (PA and NP) of the influenza genome inhibited virus production in cell culture and in embryonated chicken eggs. Furthermore, RNAi mediated by PA-, NP-, and PB1-specific siRNAs or shRNAs expressed from DNA vectors prevented and treated influenza A virus infection in mice (Ge et al. 2004). In addition, Tompkins et al. (2004) showed that the administration of influenza-specific siRNAs decreased lung virus titers and protected mice from lethal challenge by a variety of influenza A viruses, including the potential pandemic subtypes H5 and H7. This specific inhibition of influenza virus replication requires homology between the siRNAs and gene targets, and is not the result of IFN induction by dsRNAs. For therapeutic applications against the influenza A virus, the siRNAs can be administered via intranasal or pulmonary routes. RNAi is more potent than the antisense approach (Mizuta et al. 1999), and the evaluation of this technology as a treatment for the influenza virus through human clinical trials is expected to take place in the near future.

8
Concluding Remarks

While the results obtained to date should be considered preliminary in terms of their application to humans, they do provide strong justification for further investigations into the use of RNAi for the treatment of viruses in a clinical setting. The major problem with using RNAi as a tool to inhibit viral replication is the fact that it is still difficult to predict the effectiveness of specific siRNAs. It is clear from numerous studies that not all siRNAs are equally effective at generating an RNAi response. It has generally been assumed that siRNAs are under the control of the host interference-response mechanism. Recently, some expressed shRNAs have been shown to activate at least one of the arms of the human IFN-response mechanism. In addition, viral escape from RNA silencing is clearly a problem for developing effective RNAi-based antiviral therapy. Furthermore, some viral RNA sequences might be buried within secondary structures or highly folded regions. Clearly, difficulties concerning the delivery, specificity, and effectiveness of siRNAs remain. However, once these fundamental questions have been addressed, it seems likely that RNAi therapy against viral infections will progress towards clinical trials.

Acknowledgements This work was supported in part by a Grant-in-Aid for High Technology Research from the Ministry of Education, Science, Sports, and Culture of Japan, a Grant from the Japan Society for the Promotion of Science as part of in the Research for the Future program (JSPS-RFTF97L00593) and a Research Grant from the Human Science Foundation (HIV-SA-14719). We thank Drs. Habu Y, Park W-S, Kusunoki A, Takahashi H, Barnor J, Hayafune M, and Hamazaki H for their contributions to the RNAi studies described here.

References

Amarzguioui M, Holen T, Babaie E, Prydz H (2003) Tolerance for mutations and chemical modifications in a siRNA. Nucleic Acids Res 31:589–595
Anderson J, Banerjea A, Planelles V, Akkina R (2003) Potent suppression of HIV type 1 infection by a short hairpin anti-CXCR4 siRNA. AIDS Res Hum Retroviruses 19:699–706
Atmar RL, Greenberg SB, Quarles JM, Wilson SZ, Tyler B, Feldman S, Couch RB (1990) Safety and pharmacokinetics of rimantadine small-particle aerosol. Antimicrob Agents Chemother 34:2228–2233
Bahramian MB, Zarbl H (1999) Transcriptional and posttranscriptional silencing of rodent alpha1(I) collagen by a homologous transcriptionally self-silenced transgene. Mol Cell Biol 19:274–283
Baltimore D (1998) Gene therapy. Intracellular immunization [news]. Nature 335:395–396
Banerjea A, Li M-J, Bauer G, Remling L, Lee N-S, Rossi J, Akkina R (2003) Inhibition of HIV-1 by lentiviral vector-transduced siRNAs in T lymphocytes differentiated in SCID-hu mice and CD34+ progenitor cell-derived macrophages. Mol Ther 8:62–71
Barber GN (2001) Host defense, viruses and apoptosis. Cell Death Differ 2:113–126
Barnor J, Miyano-Kurosaki N, Abuni Y, Yamaguchi K, Shiina H, Ishikawa K, Yamamoto N, Takaku H (2005) Lentiviral-mediated delivery of combined HIV-1 decoy TAR and vif-siRNAs as a single RNA molecule that cleaves to inhibit HIV-1 in transduced cells. Nucleosides Nucleotides Nucleic Acids 24 (in press)
Bernstein E, Caudy AA, Hammond SM, Hannon GJ (2001) Role for a bidentate ribonuclease in the initiation step of RNA interference. Nature 409:363–366
Billy E, Brondani V, Zhang H, Muller U, Filipowicz W (2001) Specific interference with gene expression induced by long, double-stranded RNA in mouse embryonal teratocarcinoma cell lines. Proc Natl Acad Sci USA 98:14428–14433
Bitko V, Barik S (2001) Phenotypic silencing of cytoplasmic genes using sequence-specific double-stranded short interfering RNA and its application in the reverse genetics of wild type negative-strand RNA viruses. BMC Microbiol 1:34
Boden D, Pusch O, Lee F, Tucker L, Ramratnam B (2003) Human immunodeficiency virus type 1 escape from RNA interference. J Virol 77:11531–11535
Bridge AJ, Pebernated S, Ducraux A, Nicoulaz A-L, Iggo R (2003) Induction of an interferon response by RNAi vectors in mammalian cells. Nat Genet 34:263–264
Brummelkamp TR, Bernards R, Agami R (2002a) A system for stable expression of short interfering RNAs in mammalian cells. Science 296:550–553
Brummelkamp TR, Bernards R, Agami R (2002b) Stable suppression of tumorigenicity by virus-mediated RNA interference. Cancer Cell 2:243–247

Bukrinsky MI, Sharova N, Dempsey MP, Stanwick TL, Bukrinskaya AG, Haggerty S, Stevenson M (1992) Active nuclear import of human immunodeficiency virus type 1 preintegration complexes. Proc Natl Acad Sci USA 89:6580–6584

Caplen NJ, Fleenor J, Fire A, Morgan RA (2000) dsRNA-mediated gene silencing in cultured Drosophila cells: a tissue culture model for the analysis of RNA interference. Gene 252:95–105

Caplen NJ, Parrish S, Imani F, Fire A, Morgan RA (2001) Specific inhibition of gene expression by small double-stranded RNAs in invertebrate and vertebrate systems. Proc Natl Acad Sci USA 98:9742–9747

Capodici J, Kariko K, Weissman D (2002) Inhibition of HIV-1 infection by small interfering RNA-mediated RNA interference. J Immunol 169:5196–5201

Coburn G, Cullen BR (2002) Potent and specific inhibition of human immunodeficiency virus type 1 replication by RNA interference. J Virol 76:9225–9231

Donze O, Picard D (2002) RNA interference in mammalian cells using siRNAs synthesized with T7 RNA polymerase. Nucleic Acids Res 30:e46

Elbashir SM, Harborth J, Lendeckel W, Yalcin A, Weber K, Tuschl T (2001a) Duplexes of 21-nucleotide RNAs mediate RNA interference in cultured mammalian cells. Nature 411:494–498

Elbashir SM, Lendeckel W, Tuschl T (2001b) RNA interference is mediated by 21- and 22-nucleotide RNAs. Genes Dev 15:188–200

Elbashir SM, Martinez J, Patkaniowska A, Lendeckel W, Tuschl T (2001c) Functional anatomy of siRNAs for mediating efficient RNAi in Drosophila melanogaster embryo lysate. EMBO J 20:6877–6888

Elbashir SM, Harborth J, Weber K, Tuschl T (2002) Analysis of gene function in somatic mammalian cells using small interfering RNAs. Methods 2:199–213

Felgner JH, Kumar R, Sridhar CN, Wheeler CJ, Tsai YJ, Border R, Ramsey P, Martin M, Felgner PL (1994) Enhanced gene delivery and mechanism studies with a novel series of cationic lipid formulations. J Biol Chem 269:2550–2561

Fire A, Xu S, Montgomery MK, Kostas SA, Driver SE, Mello CC (1998) Potent and specific genetic interference by double-stranded RNA in Caenorhabditis elegans. Nature 391:806–811

Garrus JE, von Schwedler UK, Pornillos OW, Morham SG, Zavitz KH, Wang HE, Wettstein DA, Stray KM, Cote M, Rich RL, Myszka DG, Sundquist WI (2001) Tsg101 and the vacuolar protein sorting pathway are essential for HIV-1 budding. Cell 107:55–65

Ge Q, McManus MT, Nguyen T, Shen CH, Sharp PA, Eisen HN, Chen J (2003) RNA interference of influenza virus production by directly targeting mRNA for degradation and indirectly inhibiting all viral RNA transcription. Proc Natl Acad Sci USA 100:2718–2723

Ge Q, Filip L, Bai A, Nguyen T, Eisen HN, Chen J (2004) Inhibition of influenza virus production in virus-infected mice by RNA interference. Proc Natl Acad Sci USA 101:8676–8681

Gitlin L, Karelsky S, Andino R (2002) Short interfering RNA confers intracellular antiviral immunity in human cells. Nature 418:430–434

Grishok A, Pasquinelli AE, Conte D, Li N, Parrish S, Ha I, Baillie DL, Fire A, Ruvkun G, Mello CC (2001) Genes and mechanisms related to RNA interference regulate expression of the small temporal RNAs that control C. elegans developmental timing. Cell 106:23–34

Hamazaki H, Takahashi H, Miyano-kurosaki N, Shimotohno K, Takkau H (2005) Inhibition of HCV replication in HCV replicon cells by siRNAs. Nucleosides Nucleotides Nucleic Acids 24 (in press)

Hannon GJ (2002) RNA interference. Nature 418:244–251

Harborth J, Elbashir SM, Bechert K, Tuschl T, Weber K (2001) Identification of essential genes in cultured mammalian cells using small interfering RNAs. J Cell Sci 114:4557–4565

Hayafune M, Miyano-Kurosaki N, Park W-S, Takaku H (2005) Silencing of HIV-1 gene expression by siRNAs in transduced cells. Nucleosides Nucleotides Nucleic Acids 24 (in press)

Heidel JD, Hu S, Liu XF, Triche TJ, Davis ME (2004) Lack of interferon response in animals to naked siRNAs. Nat Biotechnol 22:1579–1582

Hirota Y, Fedson DS, Kaji M (1996) Japan lagging in influenza jabs. Nature 380:18

Hope IA (2001) Broadcast interference-functional genomics. Trends Genet 6:297–299

Hutvagner G, Zamore PD (2002) A microRNA in a multiple-turnover RNAi enzyme complex. Science 297:2056–2060

Hutvagner G, McLachlan J, Pasquinelli AE, Balint E, Tuschl T, Zamore PD (2001) A cellular function for the RNA-interference enzyme Dicer in the maturation of the let-7 small temporal RNA. Science 293:834–838

Ikeda M, Yi M, Li K, Lemon SM (2002) Selectable subgenomic and genome-length dicistronic RNAs derived from an infectious molecular clone of the HCV-N strain of hepatitis C virus replicate efficiently in cultured Huh7 cells. J Virol 76:2997–3006

Jackson AL, Bartz SR, Schelter J, Kobayashi SV, Burchard J, Mao M, Li B, Cavet G, Linsley PS (2003) Expression profiling reveals off-target gene regulation by RNAi. Nat Biotechnol 21:635–637

Jacque JM, Triques K, Stevenson M (2002) Modulation of HIV-1 replication by RNA interference. Nature 418:435–438

Kapadia SB, Brideau-Andersen A, Chisari FV (2003) Interference of hepatitis C virus RNA replication by short interfering RNAs. Proc Natl Acad Sci USA 100:2014–2018

Kariko K, Bhuyan P, Capodici J, Ni H, Lubinski J, Friedman H, Weissman D (2004) Exogenous siRNA mediates sequence-independent gene suppression by signaling through toll-like receptor 3. Cells Tissues Organs 177:132–138

Kawasaki H, Taira K (2003) Short hairpin type of dsRNAs that are controlled by tRNA(Val) promoter significantly induce RNAi-mediated gene silencing in the cytoplasm of human cells. Nucleic Acids Res 31:700–707

Kennerdell JR, Carthew RW (2000) Heritable gene silencing in Drosophila using double-stranded RNA. Nat Biotechnol 18:896–898

Ketting RF, Fischer SE, Bernstein E, Sijen T, Hannon GJ, Plasterk RH (2001) Dicer functions in RNA interference and in synthesis of small RNA involved in developmental timing in C. elegans. Genes Dev 15:2654–2659

Kim DH, Longo M, Han Y, Lundberg P, Cantin E, Rossi JJ (2004) Interferon induction by siRNAs and ssRNAs synthesized by phage polymerase. Nat Biotechnol 22:321–325

Kim SK (2001) Functional genomics: the worm scores a knockout. Curr Biol 11:R85–87

Knight SW, Bass BL (2001) A role for the RNase III enzyme DCR-1 in RNA interference and germ line development in Caenorhabditis elegans. Science 293:2269–2271

Lee MT, Coburn GA, McClure MO, Cullen BR (2003) Inhibition of human immunodeficiency virus type 1 replication in primary macrophages by using Tat- or CCR5-specific small interfering RNAs expressed from a lentivirus vector. J Virol 77:11964–11972

Lee NS, Dohjima T, Bauer G, Li H, Li MJ, Ehsani A, Salvaterra P, Rossi J (2002) Expression of small interfering RNAs targeted against HIV-1 rev transcripts in human cells. Nat Biotechnol 20:500–505

Lengyel P (1987) Double-stranded RNA and interferon action. J Interferon Res 5:511–519

Lewis D, Hagstrom J, Loomis A, Wolff J, Herweijer H (2002) Efficient delivery of siRNA for inhibition of gene expression in postnatal mice. Nat Genet 32:107–108

Lohmann V, Korner F, Koch J, Herian U, Theilmann L, Bartenschlager R (1999) Replication of subgenomic hepatitis C virus RNAs in a hepatoma cell line. Science 285:110–113

Manche L, Green SR, Schmedt C, Mathews MB (1992) Interactions between double-stranded RNA regulators and the protein kinase DAI. Mol Cell Biol 12:5238–5248

Martinez MA, Gutierrez A, Armand-Ugon M, Blanco J, Parera M, Gomez J, Clotet B, Este JA (2002) Suppression of chemokine receptor expression by RNA interference allows for inhibition of HIV-1 replication. AIDS 16:2385–2390

McCaffrey AP, Nakai H, Pandey K, Huang Z, Salazar FH, Xu H, Wieland SF, Marion PL, Kay MA (2003) Inhibition of hepatitis B virus in mice by RNA interference. Nat Biotechnol 21:639–644

McManus MT, Sharp PA (2002) Gene silencing in mammals by small interfering RNAs. Nat Rev Genet 3:737–747

McManus MT, Petersen CP, Haines BB, Chen J, Sharp PA (2002) Gene silencing using microRNA designed hairpins. RNA 8:842–850

Medina MF, Joshi S (1999) RNA-polymerase III-driven expression cassettes in human gene therapy. Curr Opin Mol Ther 1:580–594

Minks MA, West DK, Benvin S, Baglioni C (1979) Structural requirements of double-stranded RNA for the activation of $2',5'$-oligo(A) polymerase and protein kinase of interferon-treated HeLa cells. J Biol Chem 254:10180–10183

Miyagishi M, Taira K (2002) U6 promoter-driven siRNAs with four uridine $3'$ overhangs efficiently suppress targeted gene expression in mammalian cells. Nat Biotechnol 20:497–500

Mizuta T, Fujiwara M, Hatta T, Abe T, Kurosaki N, Shigeta S, Yokota T, Takaku H (1999) Antisense oligonucleotides directed against the viral RNA polymerase gene enhance survival of mice infected with influenza A. Nat Biotechnol 17:583–587

Mourelatos Z, Dostie J, Paushkin S, Sharma A, Charroux B, Abel L, Rappsilber J, Mann M, Dreyfuss G (2002) miRNPs: a novel class of ribonucleoproteins containing numerous microRNAs. Genes Dev 16:720–728

Myslinski E, Ame JC, Krol A, Carbon P (2001) An unusually compact external promoter for RNA polymerase III transcription of the human H1RNA gene. Nucleic Acids Res 29:2502–2509

Nakano H, Amemiya S, Shiokawa K, Taira M (2000) RNA interference for the organizer-specific gene Xlim-1 in Xenopus embryos. Biochem Biophys Res Commun 274:434–439

Naldini L, Blomer U, Gallay P, Ory D, Mulligan R, Gage FH, Verma IM, Trono D (1996) In vivo gene delivery and stable transduction of nondividing cells by a lentiviral vector. Science 272:263–267

Novina CD, Murray MF, Dykxhoorn DM, Beresford PJ, Riess J, Lee SK, Collman RG, Lieberman J, Shankar P, Sharp PA (2002) siRNA-directed inhibition of HIV-1 infection. Nat Med 8:681–686

Nykanen A, Haley B, Zamore PD (2001) ATP requirements and small interfering RNA structure in the RNA interference pathway. Cell 107:309–321

Oates AC, Bruce AE, Ho RK (2000) Too much interference: injection of double-stranded RNA has nonspecific effects in the zebrafish embryo. Dev Biol 224:20–28

Paddison PJ, Caudy AA, Hannon GJ (2002a) Stable suppression of gene expression by RNAi in mammalian cells. Proc Natl Acad Sci USA 99:1443–1448

Paddison PJ, Caudy AA, Bernstein E, Hannon GJ, Conklin DS (2002b) Short hairpin RNAs (shRNAs) induce sequence-specific silencing in mammalian cells. Genes Dev 16:948–958

Park WS, Miyano-Kurosaki N, Hayafune M, Nakajima E, Matsuzaki T, Shimada F, Takaku H (2002) Prevention of HIV-1 infection in human peripheral blood mononuclear cells by specific RNA interference. Nucleic Acids Res 30:4830–4835

Park WS, Hayafune M, Miyano-Kurosaki N, Takaku H (2003) Specific HIV-1 env gene silencing by small interfering RNAs in human peripheral blood mononuclear cells. Gene Ther 10:2046–2050

Parrish S, Fleenor J, Xu S, Mello C, Fire A (2000) Functional anatomy of a dsRNA trigger: differential requirement for the two trigger strands in RNA interference. Mol Cell 5:1077–1087

Pasquinelli AE (2002) MicroRNAs: deviants no longer. Trends Genet 18:171–173

Pasquinelli AE, Ruvkun G (2002) Control of developmental timing by microRNAs and their targets. Annu Rev Cell Dev Biol 18:495–513

Paul CP, Good PD, Winer I, Engelke DR (2002) Effective expression of small interfering RNA in human cells. Nat Biotechnol 20:505–508

Paule MR, White RJ (2000) Survey and summary: transcription by RNA polymerases I and III. Nucleic Acids Res 28:1283–1298

Persengiev SP, Zhu X, Green MR (2004) Nonspecific concentration-dependent stimulation and repression of mammalian gene expression by small interfering RNAs (siRNA). RNA 10:12–18

Pietschmann T, Lohmann V, Rutter G, Kurpanek K, Bartenschlager R (2001) Characterization of cell lines carrying self-replicating hepatitis C virus RNAs. J Virol 75:1252–1264

Qin XF, An DS, Chen IS, Baltimore D (2003) Inhibiting HIV-1 infection in human T cells by lentiviral-mediated delivery of small interfering RNA against CCR5. Proc Natl Acad Sci USA 100:183–188

Randall G, Grakoui A, Rice CM (2003) Clearance of replicating hepatitis C virus replicon RNAs in cell culture by small interfering RNAs. Proc Natl Acad Sci USA 100:235–240

Reed KE, Rice CM (2000) Overview of hepatitis C virus genome structure, polyprotein processing and protein properties. Curr Top Microbiol Immunol 242:55–84

Reinhart BJ, Slack FJ, Basson M, Pasquinelli AE, Bettinger JC, Rougvie AE, Horvitz HR, Ruvkun G (2000) The 21-nucleotide let-7 RNA regulates developmental timing in Caenorhabditis elegans. Nature 403:901–906

Rubinson D, Dillon CP, Kwiatkowski AV, Sievers C, Yang L, Kopinja J, Rooney DL, Ihrig MM, McManus MT, Gertler FB, Scott ML, Van Parijs L (2003) A lentivirus-based system to functionally silence genes in primary mammalian cells, stem cells and transgenic mice by RNA interference. Nat Genet 33:401–406

Schwarz DS, Hutvagner G, Haley B, Zamore PD (2002) Evidence that siRNAs function as guides, not primers, in the Drosophila and human RNAi pathways. Mol Cell 10:537–548

Seo MY, Abrignani S, Houghton M, Han JH (2003) Small interfering RNA-mediated inhibition of hepatitis C virus replication in the human hepatoma cell line Huh-7. J Virol 77:810–812

Sharp PA (2001) RNA interference. Genes Dev 15:485–490

Shlomai A, Shaul Y (2003) Inhibition of hepatitis B virus expression and replication by RNA interference. Hepatology 37:764–770

Sledz CA, Holko M, de Veer MJ, Silverman RH, Williams BR (2003) Activation of the interferon system by short-interfering RNAs. Nat Cell Biol 9:834–839

Song E, Lee SK, Dykxhoorn DM, Novina C, Zhang D, Crawford K, Cerny J, Sharp PA, Lieberman J, Manjunath N, Shankar P (2003a) Sustained small interfering RNA-mediated human immunodeficiency virus type 1 inhibition in primary macrophages. J Virol 77:7174–7181

Song E, Lee SK, Wang J, Ince N, Ouyang N, Min J, Chen J, Shankar P, Lieberman J (2003b) RNA interference targeting Fas protects mice from fulminant hepatitis. Nat Med 9:347–351

Stark GR, Kerr IM, Williams BR, Silverman RH, Schreiber RD (1998) How cells respond to interferons. Annu Rev Biochem 67:227–264

Stewart S, Dykxhoorn DM, Palliser D, Mizuno H, Yu EY, An DS, Sabatini DM, Chen IS, Hahn WC, Sharp PA, Weinberg RA, Novina CD (2003) Lentivirus-delivered stable gene silencing by RNAi in primary cells. RNA 9:493–501

Sui G, Soohoo C, Affar el B, Gay F, Shi Y, Forrester WC, Shi Y (2002) A DNA vector-based RNAi technology to suppress gene expression in mammalian cells. Proc Natl Acad Sci USA 99:5515–5520

Surabhi R, Gaynor R (2002) RNA interference directed against viral and cellular targets inhibits human immunodeficiency virus type 1 replication. J Virol 76:12963–12973

Svoboda P, Stein P, Hayashi H, Schultz RM (2000) Selective reduction of dormant maternal mRNAs in mouse oocytes by RNA interference. Development 127:4147–4156

Svoboda P, Stein P, Schultz RM (2001) RNAi in mouse oocytes and preimplantation embryos: effectiveness of hairpin dsRNA. Biochem Biophys Res Commun 287:1099–1104

Tavernarakis N, Wang SL, Dorovkov M, Ryazanov A, Driscoll M (2000) Heritable and inducible genetic interference by double-stranded RNA encoded by transgenes. Nat Genet 24:180–183

Tompkins SM, Lo CY, Tumpey TM, Epstein SL (2004) Protection against lethal influenza virus challenge by RNA interference in vivo. Proc Natl Acad Sci USA 101:8682–8686

Tuschl T, Borkhardt A (2002) Small interfering RNAs—a revolutionary tool for analysis of gene function and gene therapy. Mol Interv 2:42–51

Wang C, Takeuchi K, Pinto LH, Lamb RA (1993) On channel activity of influenza A virus M2 protein: characterization of the amantadine block. J Virol 67:5585–5594

Weinberg JB, Matthews TJ, Cullen BR, Malim MH (1991) Productive human immunodeficiency virus type 1 (HIV-1) infection of non-proliferating human monocytes. J Exp Med 174:1477–1482

Wianny F, Zernicka-Goetz M (2000) Specific interference with gene function by double-stranded RNA in early mouse development. Nat Cell Biol 2:70–75

Wilson JA, Jayasena S, Khvorova A, Sabatinos S, Rodrigue-Gervais IG, Arya S, Sarangi F, Harris-Brandts M, Beaulieu S, Richardson CD (2003) RNA interference blocks gene expression and RNA synthesis from hepatitis C replicons propagated in human liver cells. Proc Natl Acad Sci USA 100:2783–2788

Yamamoto T, Omoto S, Mizuguchi M, Mizukami H, Okuyama H, Okada N, Saksena NK, Brisibe EA, Otake K, Fuji YR (2002) Double-stranded nef RNA interferes with human immunodeficiency virus type 1 replication. Microbiol Immunol 46:809–817

Yokota T, Sakamoto N, Enomoto N, Tanabe Y, Miyagishi M, Maekawa S, Yi L, Kurosaki M, Taira K, Watanabe M, Mizusawa H (2003) Inhibition of intracellular hepatitis C virus replication by synthetic and vector-derived small interfering RNAs. EMBO Rep 4:1–7

Yu JY, DeRuiter SL, Turner DL (2002) RNA interference by expression of short-interfering RNAs and hairpin RNAs in mammalian cells. Proc Natl Acad Sci USA 99:6047–6052

Zamore PD (2002) Ancient pathways programmed by small RNAs. Science 296:1265–1269

Zeng Y, Cullen BR (2002) RNA interference in human cells is restricted to the cytoplasm. RNA 8:855–860

Zeng Y, Yi R, Cullen BR (2003) MicroRNAs and small interfering RNAs can inhibit mRNA expression by similar mechanisms. Proc Natl Acad Sci USA 100:9779–9784

Zhao Z, Cao Y, Li M, Meng A (2001) Double-stranded RNA injection produces nonspecific defects in zebrafish. Dev Biol 229:215–223

Zhou Y, Ching YP, Kok KH, Kung HF, Jin DY (2002) Post-transcriptional suppression of gene expression in Xenopus embryos by small interfering RNA. Nucleic Acids Res 30:1664–1669

Progress in the Development of Nucleic Acid Therapeutics

A. Kalota · V. R. Dondeti · A. M. Gewirtz (✉)

Division of Hematology/Oncology, Department of Medicine, University of Pennsylvania,
421 Curie Blvd, Philadelphia PA, 19104, USA
gewirtz@mail.med.upenn.edu

1	Introduction	174
2	Gene Silencing Strategies for Targeting mRNA	174
2.1	Antisense Oligonucleotides	174
2.2	Ribozymes and DNAzymes	175
2.3	RNA Interference	177
3	Improving the Effectiveness of Antisense Molecules	178
3.1	Structural Modifications of Antisense Oligonucleotides	178
3.2	Selecting Target mRNA	181
3.3	Selecting Accessible Sites Within the Targeted mRNA	181
3.4	Delivery	183
4	Clinical Applications	184
4.1	Inflammatory Diseases	185
4.2	Cardiovascular Diseases	186
4.3	Cancer Therapy	186
4.4	Immunotherapy	189
5	Conclusion	189
	References	190

Abstract Abnormal gene expression is a hallmark of many diseases. Gene-specific down-regulation of aberrant genes could be useful therapeutically and potentially less toxic than conventional therapies due its specificity. Over the years, many strategies have been proposed for silencing gene expression in a gene-specific manner. Three major approaches are antisense oligonucleotides (AS-ONs), ribozymes/DNAzymes, and RNA interference (RNAi). In this brief review, we will discuss the successes and shortcomings of these three gene-silencing methods, and the approaches being taken to improve the effectiveness of antisense molecules. We will also provide an overview of some of the clinical applications of antisense therapy.

Keywords Antisense · Oligonucleotides · Modifications · Accessible sites · Antisense therapy

1
Introduction

Abnormalities of gene expression are associated with numerous diseases. Gene-specific therapies that can target the abnormal genes are much sought after by scientists, physicians, and pharmaceutical companies because of the promise of high specificity and minimal side effects (Vile et al. 2000). Several gene-silencing strategies that target messenger (m)RNA using nucleic acids have been developed over the years. These strategies include antisense oligonucleotides (AS-ONs), ribozymes/DNAzymes, and RNA interference (RNAi). Paterson et al. (1977) were the first to use complementary exogenous nucleic acids to inhibit translation of mRNA. Subsequently, it was demonstrated that a 13mer antisense DNA oligonucleotide can inhibit Rous sarcoma virus replication and cell transformation in culture (Zamecnik and Stephenson 1978). Since then, considerable progress has been made using antisense approaches, but many obstacles need to be overcome before such therapies can be taken from the bench to the bedside (Opalinska and Gewirtz 2002). Rational selection of mRNA target sites, intracellular stability, and delivery into cells of interest are among the problems that need to be resolved. In this review, we will discuss current and newly developed gene silencing approaches that target mRNA and examine some clinical applications.

2
Gene Silencing Strategies for Targeting mRNA

There are three major approaches for targeting mRNA for the purpose of gene silencing: AS-ONs, ribozymes and DNAzymes, and RNAi.

2.1
Antisense Oligonucleotides

AS-ONs can be used to silence gene expression either by degradation or inhibition of translation of the target mRNA (Fig. 1). Short fragments of single-stranded DNA or RNA that are complementary to the target mRNA can be used as AS-ONs. After delivery of AS-ON into cells, the AS-ONs hybridize to the complementary fragment of the endogenously expressed target mRNA. Binding of the AS-ON to the target mRNA can result in cleavage of the mRNA or disruption of translation, depending on the class of AS-ON (Kalota et al. 2004). Two classes of oligonucleotides have been investigated in considerable detail. The first class of AS-ON facilitates RNase H-mediated cleavage of the target mRNA, and the second class of AS-ON disrupts translation by blocking the ribosome. The modification of the oligonucleotide determines the class to which it belongs. The applicability of AS-ONs for gene silencing has been extensively studied in

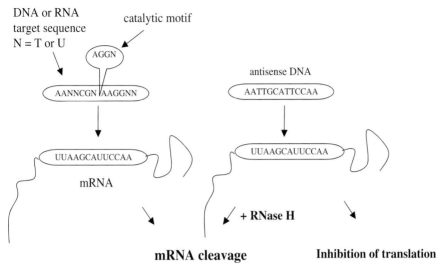

Fig. 1 Strategies to inhibit gene expression. DNAzymes and ribozymes are (DNA or RNA) oligonucleotides with catalytic motifs capable of cleaving the mRNA of interest. Antisense oligonucleotides bind to the complementary target mRNA sequence and inhibit translation either by activating RNase H to cleave the target mRNA or by blocking the ribosome

vitro and in vivo (Zamecnik and Stephenson 1978; Anfossi et al. 1989; Gewirtz 2000; Agrawal and Kandimalla 2001; Dias and Stein 2002; Crooke 2004), and AS-ONs have also been tested for use as cancer therapy (Table 1).

2.2
Ribozymes and DNAzymes

Ribozymes and DNAzymes are oligonucleotides with catalytic activity. Ribozymes are made of ribonucleotides, whereas DNAzymes are composed of deoxyribonucleotides, and both can be designed to target specific transcripts. Upon hybridizing to the target mRNA, they cleave the transcript in a sequence-specific fashion (Fig. 1). Ribozymes have been more extensively investigated than DNAzymes. The results of these studies show that ribozymes can cleave phosphodiester bonds without the aid of protein enzymes (Jen and Gewirtz 2000), and that the catalytic moiety of ribozymes recognizes a specific nucleotide sequence, GUX, where X=C, U, or A (Ruffner et al. 1989) or in some cases, NUX, where N is any nucleotide (Xing and Whitton 1992). Hammerhead, hairpin, group I intron, ribonuclease P, and the hepatitis delta virus ribozymes are five naturally occurring ribozymes that are used as the basis for the catalytic motifs of engineered ribozymes. A greater emphasis has been placed on studying hammerhead and hairpin ribozymes due their simplicity, relatively small size, and their ability to be incorporated into a variety of flanking

Table 1 Antisense oligonucleotide targets in oncology tested in vitro and in animals

Target	Cell type analyzed	Biological endpoints
BCL2	B cell lymphoma, melanoma, lung tumor	Apoptosis
Survivin	Cervical tumor, lung cancer	Apoptosis
MDM 2	Multiple tumors	p53 activation
BCLXL	Endothelial cells, lung cancer cells	Apoptosis
RelA	Fibrosarcoma cell line	Cell adhesion, tumorigenicity
RAS	Endothelial cells, bladder cancer	CAM expression, proliferation
RAF	Endothelial cells, smooth muscle cells	CAM expression, proliferation
BCR-ABL	Primary progenitor bone marrow cells	Adhesion, proliferation
Jun N-terminal kinase 1 and Jun N-terminal kinase 2	Renal epithelial cells	Apoptosis
Telomerase	Prostate cell lines	Cell death
c-MYC	Leukemia cell lines	Proliferation, apoptosis
c-MYB	Leukemia cell lines	Proliferation

Table reprinted with permission from Elsevier (Lancet 2001; 358:489–497)

sequence motifs without changing site-specific cleavage capacities (Sun et al. 2000).

The hammerhead motif consists of a highly conserved catalytic core, which will cleave substrate RNA at NUH triplets 3′, surrounded by a flanking sequence that is responsible for specifically binding to the target. Hammerhead ribozymes require the presence of divalent metal ions, of which magnesium is the most often used in vitro, for carrying out the catalytic cleavage reaction. In contrast to the hammerhead motif, the hairpin model contains four base-paired helices and two unpaired loops, with the reactive phosphodiester located within one of the loops. The effectiveness of ribozymes has been proved both in vitro and in vivo (Bramlage et al. 1999; Giannini et al. 1999; Parry et al. 1999). Though ribozymes have been shown to be effective, their effects are usually short lived due the high susceptibility of ribozymes to endogenous nucleases. In order to increase the intracellular stability of ribozymes and make them more resistant to endogenous nucleases, many modifications have been introduced into ribozyme backbones. The most common modification is at the 2′-position of the ribosyl moiety (Pieken et al. 1991). In contrast to AS-ONs, ribozymes are RNase H independent, so 2′ modifications to these nucleic acids generally increase their stability without diminishing the antisense effect.

Breaker and Joyce (1994) demonstrated that deoxyribonucleic acid oligomers can also be designed to have enzymatic activity and can be used to downregulate gene expression. The first DNAzyme named "10–23 DNA enzyme" possessed a catalytic domain of approximately 15 nt and two substrate-recognition domains of approximately 8 nt each. The advantage of DNAzymes over ribozymes is that since they are made up of deoxyribonucleotides, they are cheaper to synthesize and more resistant to nuclease degradation compared to ribozymes. However, issues regarding delivery into target cells, optimal hybridization with target mRNA, and toxicity need to be overcome before DNAzymes can be used as drugs in antisense therapy.

2.3
RNA Interference

RNAi, also known as post-transcriptional gene silencing (PTGS), is the most recently discovered method for gene silencing. RNA interference is a phenomenon in which introduced double-stranded (ds)RNAs cause the degradation of their cognate mRNAs, thus silencing gene expression. RNAi is an evolutionary conserved mechanism found in fungi, plants, and animals, and is involved in genome defense against transposons and repetitive elements, antiviral response, and developmental regulation (Bernstein et al. 2001). A key hallmark of RNAi is the cleavage of long dsRNA by the Dicer enzyme into short duplexes of 21–25 nt called small interfering (si)RNAs (Elbashir et al. 2001b). Dicer is an RNase III protein believed to act as a dimer that cleaves dsRNA into siRNAs with a 2-nt overhang at the 3′-end. These siRNAs are incorporated into the RNA-induced silencing complex (RISC) and guide a nuclease to the target mRNA, and then the mRNA is degraded (Fig. 2).

Gene silencing using long dsRNA has been successfully demonstrated in many model organisms. However, when dsRNA longer than 30 bp is used, it typically results in the activation of the interferon response in differentiated cells, which eventually leads to a global shutdown of protein synthesis and nonspecific mRNA degradation (Elbashir et al. 2001b; Yang et al. 2001). As a consequence of this interferon response, RNAi in mammalian cells had remained elusive for a while. However, it has been shown that if 21-nt siRNA duplexes are used instead of long dsRNA, gene-specific silencing occurs without the nonspecific RNA degradation pathway being turned on (Elbashir et al. 2001a). The siRNA duplex consists of 21 nt of the mRNA target and 21-nt antisense siRNA, where the siRNA hybridizes with 19 nt of the target mRNA and has a 2-nt overhang at the 3′-end. The use of siRNAs has become the most common approach used for downregulating gene expression. The effectiveness of this gene-silencing modality has been evaluated extensively in vitro and in vivo. siRNAs have been successfully utilized to silence the expression of BCR-ABL (breakpoint cluster region-Abelson murine leukemia viral oncogene homolog), erbB1, Bcl-2, Raf-1, androgen receptor, and Lyn in vitro (Futami

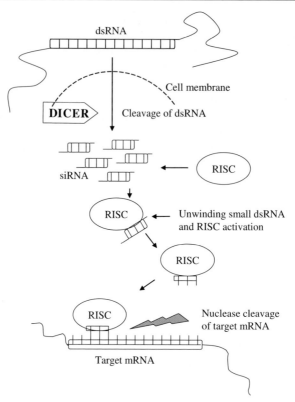

Fig. 2 Schematic mechanism of RNA interference. Long dsRNA delivered into the cell are cleaved by a ribonuclease (Dicer) into short 21–25 nt RNAs (siRNAs). The resultant siRNAs are incorporated into the RISC complex and guide the nuclease to the complementary target mRNA to degrade it

et al. 2002; Aoki et al. 2003; Nagy et al. 2003; Wohlbold et al. 2003; Yin et al. 2003; Ptasznik et al. 2004). In vivo, siRNA has been used successfully to diminish Fas expression in mice (Song et al. 2003). The results of these studies are very promising and suggest an important role for siRNA in treating disease.

3
Improving the Effectiveness of Antisense Molecules

3.1
Structural Modifications of Antisense Oligonucleotides

The success of AS-ONs as gene-silencing agents is dependent on sequence-specific hybridization to target mRNA, stability in biological fluids, and transport through the cell membrane. Initial work employing AS-ONs was carried out using unmodified DNA molecules, but rapid degradation of these molecules by endonucleases as well as 3′ exonucleases became a major obstacle (Gewirtz and Calabretta 1988). In order to improve the stability of AS-ONs, as well as increase the strength of hybridization with mRNA, investigators began to focus on chemical modifications of DNA molecules. A broad range of modifications has been introduced to the nitrogen base, sugar, and phosphate moieties (Fig. 3). The most widely studied DNA analog is the phosphorothioate (PS) oligodeoxynucleotide (ODN)—the so-called first-generation analog—in which the non-bridging oxygen is replaced with sulfur. Molecules with this type of backbone are relatively nuclease resistant, water soluble, and elicit an RNase H response. Phosphorothioate ODNs have been shown to be highly successful and have been employed in many laboratory experiments and clinical trials to date (Ratajczak et al. 1992b; Tolcher 2001; Marshall et al. 2004a; Marshall et al. 2004b). However, a major disadvantage of PS molecules is that they interact with many proteins, which results in nonspecific activity and cell toxicity.

With the goal of overcoming the disadvantages of phosphorothioate ODNs, a "second generation" of DNA analogs has been introduced. The major focus within this group was on variety of modifications to the sugar moiety, especially the 2′-position like the 2′-O-methyl, the 2′-O-methoxy-ethyl, or the 2′-O-alkyl modifications. These modifications make DNA molecules more RNA-like, and as a result increase their ability to hybridize with RNA, and the modifications

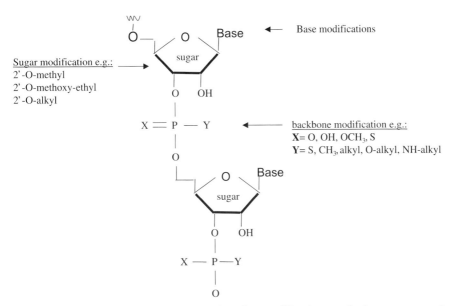

Fig. 3 Chemical modifications of oligonucleotides. Modifications at the base, sugar, and phosphodiester linkage of oligonucleotides have been prepared

also improve nuclease resistance. Nevertheless, these backbone modifications do not recruit RNase H, and thus their efficacy becomes significantly reduced. Developing DNA analogs to meet all the criteria required for effective antisense-based gene silencing remains a big challenge.

Several new modifications collectively called the "third generation" have been proposed. The most significant among them are peptide nucleic acids (PNAs), morpholino oligonucleotides (PMOs), locked nucleic acids (LNAs), and oxetane-modified deoxyribonucleotides (OXE) (Fig. 4). Peptide nucleic acids are oligonucleotides in which the phosphodiester backbone is replaced with a pseudo-peptide polymer [N-(2-aminoethyl) glycine] to which the nitrogen bases are attached via a methyl carbonyl linker (Paulasova and Pellestor 2004). PNA modifications result in an enhanced ability to hybridize with DNA and RNA, and thus increase the stability of PNA–DNA and PNA–RNA duplexes. In addition, PNAs can form triplexes with dsDNA, as well displace a DNA strand in a duplex (Opalinska and Gewirtz 2002). Furthermore, the unnatural backbone of PNAs makes them almost completely resistant to nucleases.

Morpholino oligonucleotides have each ribonucleoside moiety converted to morpholino moiety (C_4H_9NO) and the phosphodiester linkage is replaced with a phosphorodiamidate linkage (Heasman 2002). They have proved to be very stable under physiological conditions and show sequence specificity with very little or no off-target activity (Summerton and Weller 1997). However, both PMOs and PNAs do not recruit RNase H due to their lack of charge. The lack of charge also complicates their delivery into cells.

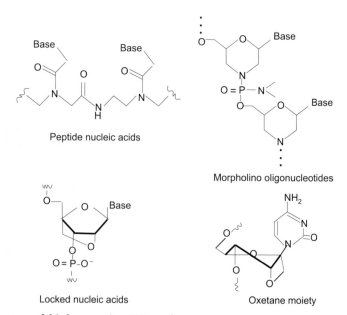

Fig. 4 Structure of third-generation DNA analogs

Another "third-generation" analog, locked nucleic acids contain one or more LNA monomers in which the sugar has a linkage between the 2′-oxygen and 4′-carbon via a methylene unit (Vester and Wengel 2004). This type of chemical structure greatly increases affinity towards RNA/DNA targets and reduces susceptibility to nucleases. AS-ONs made entirely of LNA monomers are not recognized by nucleases as a substrate, and so are quite stable. However, one major disadvantage is that RNase H does not recognize them either. This fact prompted the investigation of mixed backbone oligonucleotides containing LNA monomers with a gap of a few DNA nucleotides. The results from a number of studies show that a gap size of 7–10 unmodified oligonucleotides is optimal for achieving complete RNase H activity (Kurreck et al. 2002; Frieden et al. 2003). However, when the DNA gap is present the susceptibility to endonucleases increases.

Recently introduced oligonucleotides with an oxetane modification [oxetane, 1-(1′, 3′-O-anhydro-β-D-psicofuranosyl) nucleosides] appear to be very promising (Pradeepkumar et al. 2003). This type of backbone enhances nuclease resistance, forms stable hybrids with target RNA with melting temperatures similar to those predicted for RNA/RNA hybrids, and maintains the ability to activate RNase H. Opalinska et al. (2004) demonstrated that oxetane modified AS-ONs are very effective in inhibiting gene expression in cell culture. Nonetheless, the delivery of oxetane ONs to cells remains a challenge.

3.2
Selecting Target mRNA

When selecting a target, several things must be considered in order to obtain a successful outcome. First of all, the gene must be specific to the condition under study, but this is not sufficient. For antisense therapy to be successful, the target mRNA and the protein it codes for should have short half-lives, so that when the target mRNA is degraded, the cell will not be able to recover the function of the target gene quickly. In addition, if multiple sites within a given target mRNA or if multiple genes are targeted simultaneously, the chances of success can increase. This concept was successfully demonstrated in our lab by showing that simultaneously targeting *Myb* and *Vav* (both important hematopoietic genes) significantly diminishes colony formation of primary chronic myelogenous leukemia (CML) cells when compared to targeting *Myb* or *Vav* alone (Opalinska et al. 2005).

3.3
Selecting Accessible Sites Within the Targeted mRNA

A major obstacle in employing antisense nucleic acids for post-transcriptional gene silencing is the inability to readily identify hybridization accessible sites within the target mRNA. Target site accessibility is influenced by RNA secondary and tertiary structure, and the proteins associated with RNA.

A number of experimental and computational approaches have been developed to identify accessible regions within a given mRNA molecule. One of the first strategies that addressed this problem was the so-called "walk along" method. A series of AS-ONs complementary to various regions of the target mRNA were synthesized and tested for their ability to inhibit mRNA expression (Bacon and Wickstrom 1991). This method is effective, but also potentially expensive and time consuming. Many sequences have to be tested to find the one that shows the optimal antisense effect.

Another experimental approach for mapping RNA employs a library of semi-random oligonucleotides (ODNs) to probe the target mRNA. The accessible region of the RNA should hybridize to complementary sequences found within the library. These regions are then identified using RNase H to cleave the RNA strand of the RNA/DNA hybrid, followed by gel sequencing (Ho et al. 1996) or matrix-assisted laser desorption ionization-time of flight (MALDI-TOF) spectrometry (Gabler et al. 2003) of the cleaved fragments. An oligonucleotide library has also been used in conjunction with a reverse transcription assay to identify accessible sites of the target mRNA (Allawi et al. 2001). The random oligonucleotides were used to prime complementary (c)DNA synthesis in the presence of reverse transcriptase, in order to generate a series of cDNA fragments. The initiation of cDNA synthesis occurs only at RNA sites that are available for ODN binding. The products were then analyzed by gel sequencing. Another method proposed for selecting effective target sites exploited arrays of oligonucleotides complementary to the target mRNA. In this method, potential sites for antisense are identified as ODNs that form stable hybrids with the target mRNA (Milner et al. 1997; Mir and Southern 1999).

In addition to the many experimental approaches for mapping RNA, investigators have also attempted to identify accessible sites using computer-aided predictions of secondary structure of the target mRNA. The folding potential of RNA is calculated using structure prediction algorithms (Zuker 1989, 2003; Stull et al. 1992; Sczakiel et al. 1993). However, the reliability of the secondary structure prediction programs remains to be proved. Computer methods alone seem to be insufficient for designing effective antisense.

Recent work from our laboratory introduces a new RNA mapping strategy, combining computational and experimental approaches (Gifford et al. 2005). Fluorescent self-quenching reporter molecules (SQRM) are used to identify regions within the target mRNA that are available for hybridization with AS-ON. SQRMs are 20–30 bases long DNA hairpin molecules with 4–5 base complementary ends. The 5'-end of the sequence is linked to a fluorophore (fluorescein) and 3'-end to the quencher (DABCYL) (Fig. 5). In the absence of the target mRNA, the SQRM folds such that the fluorescein is quenched by DABCYL and no signal is detected. Once the SQRM is mixed with the target and hybridization takes place, the fluorophore and quencher are separated and the fluorescence

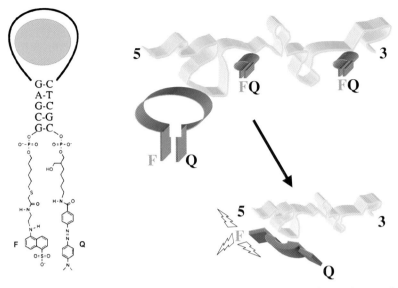

Fig. 5 Diagram of a self-quenching reporter molecule (SQRM). F is a fluorophore and Q is a quencher

signal can be detected. In order to design SQRM sequences, a computer algorithm was developed to scan the sequence of the gene of interest for inverted repeats of 4–5 bases separated by an intervening sequence of 18–20 bases. Antisense sequences selected based on this strategy were successfully tested in cell culture (Opalinska et al. 2004).

The studies described here have made some progress in identifying accessible sites, but questions remain as to whether the same target sites are suitable for both antisense and RNAi. Reports discussing this issue are contradictory. Some investigators have suggested that effective AS-ONs and siRNA have different preferences for target sites within mRNA (Xu et al. 2003), while others show that there is some correlation between successful antisense and siRNA sites (Kretschmer-Kazemi Far and Sczakiel 2003). Nevertheless, the rules governing the relationship between RNA structure and effectiveness of siRNA remain unknown.

3.4
Delivery

Despite the encouraging prospects of nucleotide chemistry and the successful identification of accessible sites within mRNA described in the previous sections, a significant obstacle to be overcome is the delivery of antisense molecules into cells. Oligonucleotide molecules possess very little or no ability to diffuse across cell membranes, therefore they are mostly taken up by cells through energy-dependent mechanisms, typically endocytosis. As a result, oligonucleotides become trapped in endosomes and/or lysosomes, where they usually become degraded (Gewirtz et al. 1996; Gewirtz 2000). A number of methods have been developed for delivery of AS-ONs. Some of them are based on physical disruption of the cell membrane using electroporation or permeabilization reagents like streptolysin O (Giles et al. 1995; Flanagan and Wagner 1997) or digitonin (Adam et al. 1990).

An alternative strategy is to use delivery vehicles that have the ability to form stable complexes with oligonucleotides through electrostatic interactions between the negatively charged DNA and positively charged vehicle. Many classes of compounds have been used as delivery vehicles for ONs, including cationic liposomes, cationic porphyrins, fusogenic peptides, and artificial virosomes. Though these methods have achieved some degree of success (Bennett et al. 1992; Capaccioli et al. 1993), their effectiveness is mostly limited to adherent cells. In our laboratory, we have been very successful in using nucleoporation for delivery of antisense ONs as well as siRNAs into many types of suspended cells (Opalinska et al. 2004; Ptasznik et al. 2004). Although these methods have been effectively used in tissue-culture experiments, their usefulness in vivo in animal systems as well as for clinical applications is limited. To overcome these limitations, polyethylenimine (PEI) complexes have been intensively studied. These cationic polymers have been successfully used in cell culture, as well in animal experiments (Koping-Hoggard et al. 2001; De Rosa et al. 2003). Positively charged PEI polymers help neutralize the negative charges of oligonucleotides and help in intracellular delivery. Additionally, they have the capacity to condense ONs for delivery and prolong the half-life of the incorporated material in the circulation of the recipient, which is an important feature in animal experiments and clinical applications. In spite of these useful properties of PEI complexes, delivery to specific target sites has been problematic due to the non-specific interactions of polyplexes with blood components and non-target cells (Kichler 2004).

Another approach proposed for delivery of nucleic acid molecules employs viral vectors, in which genes essential for virus replications have been replaced with sequences of interest (Hoeller et al. 2002; Thomas et al. 2003). The main classes of viral vectors that have been utilized include retroviruses, adenoviruses, lentiviruses, adeno-associated viruses, and herpes simplex viruses. Although the retroviral strategy appears to be very effective, there are some safety issues that need to be dealt with (Kalota et al. 2004).

4
Clinical Applications

All the work described above was focused on designing and synthesizing the most effective antisense molecules. Some of the AS-ONs tested had very encouraging results in laboratory experiments and have been evaluated in the clinic. One of them, Vitravene, developed by Isis Pharmaceuticals, has been approved by the US Food and Drug Administration (FDA) in 1998. Vitravene (sodium fomivirsen) is used to treat an inflammatory viral infection of the eye, caused by cytomegalovirus (CMV) in immunocompromised patients and in patients with acquired immunodeficiency syndrome (AIDS) (Opalinska and Gewirtz 2002). Another representative of antiviral agents is ISIS 14803—a phosphorothioated ODN directed against the hepatitis C virus (HCV). The results of phase I clinical trials have been reported (Soler et al. 2004). The study was performed on 24 patients with chronic hepatitis C. Two of the patients showed significant viral load reduction and nine of them minor viral load reduction. AS-ONs have also been used in combating inflammation, cardiovascular diseases, and cancer. Oligonucleotides could also play a role in immunotherapy, because of their ability to mimic bacterial DNA.

4.1
Inflammatory Diseases

Alicaforsen is example of an AS-ON used as an anti-inflammatory agent, and it is directed against intracellular adhesion molecule (ICAM)-1. ICAM-1 is a cell surface protein that plays an important role in the recruitment of leukocytes to sites of inflammation, and its expression is upregulated in the intestinal mucosa of inflammatory bowel disease (Crohn's disease). Administration of alicaforsen (ISIS 2302) against ICAM-1 results in inhibition of ICAM-1 receptor expression and reduces inflammation (Bennett et al. 1994; Nestle et al. 1994). A few clinical trials using alicaforsen have been reported. In a study reported by Yacyshyn et al. (1998), patients were randomized to receive placebo or alicaforsen. The drug was well tolerated and at the end of the study, 47% of patients in the antisense group went into remission compared to 20% in the placebo group. More recent trials evaluated alicaforsen in patients with active steroid-dependent Crohn's disease (Yacyshyn et al. 2002). A total of 299 patients were enrolled in this study and treated with placebo or ISIS 2302 (at dose of 2 mg/kg three times per week) for 2–4 weeks. Unfortunately, this study failed to demonstrate the efficacy of ISIS 2302, most likely due to insufficient dosage of the drug. The most significant difference in steroid-free remission between the antisense and placebo groups was observed at week 14 and was 78% versus 64%, respectively. These data resulted in examining of higher doses in phase III clinical trials. Primary results from this study have been reported by Yu et al. (2003) and indicate that a fixed dose regiment (250–400 mg) given three times per week provides a desirable clinical response. However, a larger number of patients need to be tested to obtain a statistically meaningful difference between placebo and active treatment groups.

4.2
Cardiovascular Diseases

Restenosis in coronary vessels after angioplasty remains an ongoing problem. In one-third of the cases, arterial blockage returns, usually within 6 months of the procedure. There are two major mechanisms causing this restenosis. The first is thrombosis, blood clotting at the site of treatment, and the second is the proliferation of endothelial cells that normally line blood vessels. The myelocytomatosis viral oncogene homolog (c-MYC) has been identified as an important mediator in this process through its effect on regulating the growth of vascular cells in atherosclerotic lesions. This makes c-MYC an attractive target for antisense therapy to prevent post-angioplasty complications. At least two clinical trials using antisense ODN against c-MYC have been reported (Roque et al. 2001; Kutryk et al. 2002). Both of the studies showed the safety of intracoronary application of the drug, but no significant clinical response (Opalinska and Gewirtz 2002).

4.3
Cancer Therapy

Cancers, in particular, are attractive candidates for antisense therapy. The problem with conventional therapies is that they are highly toxic. Since antisense strategies are directed against specific genes that are aberrantly expressed in diseased cells, it is expected that this approach will cause fewer and less serious side effects. Several genes implicated in many cancers have been intensively studied as potential targets for antisense therapy and these studies are discussed below.

B-Cell Lymphoma Protein 2 (*Bcl-2*) *Bcl-2* is an apoptosis-regulating oncogene and its overexpression has been reported in most follicular non-Hodgkin's lymphoma (NHL), other lymphomas and leukemias, as well in other kinds of cancers like lung, breast, colorectal, gastric, prostate, renal, and neuroblastoma (Kalota et al. 2004). This overexpression is usually associated with an aggressive malignant clone characterized by resistance to standard chemotherapeutic agents, early relapse, and poor survival outcome. Laboratory studies have shown that exposing cells to the oligonucleotides directed against *Bcl-2* will specifically and significantly decrease mRNA and protein expression (Reed et al. 1990). For all of these reasons, there is a great deal of interest in targeting *Bcl-2* for therapeutic purposes. The results of several phase I and II clinical trials that employed an 18-base phosphorothioate oligonucleotide (G3139) complementary to the first 6 codons of Bcl-2 mRNA have been reported (Webb et al. 1997; Jansen et al. 2000; Waters et al. 2000; Tolcher 2001). The studies revealed that G3139 (Oblimersen) alone does not consistently produce a strong antitumor response. Therefore, subsequent trials investigated

Oblimersen combined with chemotherapy. G3139, when combined with paclitaxel in a cytotoxic dose range, has been well tolerated by patients with chemorefractory small-cell lung cancer (Rudin et al. 2002). Furthermore, it can be safely administered with fludarabine (FL), cytarabine (ARA-C), and granulocyte colony stimulating factor (G-CSF) salvage chemotherapy in patients with refractory or relapsed leukemia (Marcucci et al. 2003). In spite of the promising early results, Oblimersen has not achieved the desired outcome in several trials. The FDA's Oncology Drug Advisory Committee advised against approving it for metastatic melanoma due to the lack of effectiveness as measured by response rate and progression-free survival in relation to its toxicity (low-grade fever, usually resolving within 1–2 days, nausea but no vomiting, and thrombocytopenia). In phase III trials, Oblimersen failed to reach the primary endpoint in the treatment of multiple myeloma. However, it did meet the primary endpoint for the treatment of chronic lymphocytic leukemia, though it failed to meet the secondary endpoints of time-to-progression and overall survival. In order to understand why Oblimersen failed to achieve the desired result, we need to examine its pharmacodynamics in detail to see if it got into cells and hit the target. This information will help us decide whether Bcl-2 is a good target or not. Only by understanding the pharmacodynamics in detail can we hope to achieve the desired outcome in this case and others like it.

c-myb The c-myb proto-oncogene is a normal homolog of the avian myeloblastosis viral oncogene (*v-myb*). *c-myb* encodes MYB, a protein that is responsible for regulation of cell cycle transition and cellular maturation, primarily in hematopoietic cells. AS-ONs targeting *c-myb* have been shown to be successful in decreasing mRNA and protein levels of MYB, as well as inhibit proliferation of acute myeloid leukemia (AML), chronic myeloid leukemia (CML), and T cell leukemic cells in culture by inducing apoptosis (Ratajczak et al. 1992a). In a phase I clinical pilot study, LR3001 has been used to purge marrow autografts administrated to allograft-ineligible chronic myelogenous leukemia patients. $CD34^+$ marrow cells were purged by being exposed to c-myb AS-ON for either 24 or 72 h. Patients received busulfan and cyclophosphamide chemotherapy, followed by reinfusion of previously cryopreserved and purged marrow cells. Post-purging, c-myb mRNA levels declined substantially in roughly 50% of patients. Analysis of BCR-ABL expression in a surrogate stem-cell assay indicated that purging has been accomplished at a primitive cell level in more than 50% of patients. Cytogenetic analysis was performed at day 100 on patients who engrafted without the need for an administration of unmanipulated "backup" marrow ($n = 14$), and 6 out of 14 patients demonstrated a major cytogenetic response. However, conclusions regarding the clinical efficacy of ODN marrow purging cannot be drawn from this small pilot study (Luger et al. 2002).

Protein Kinase C-α (*PKC*-α) PKC-α is a cytoplasmic serine/threonine kinase that belongs to the family of isoenzymes involved in signal transduction in response to growth factors, hormones, and neurotransmitters. It regulates cell proliferation and differentiation. Overexpression of this kinase causes uncontrolled proliferation of cancer cells in several tumors. The antitumor activity of antisense inhibitors of PKC-α has been illustrated in cell culture studies (Dean et al. 1994; Nemunaitis et al. 1999) and in animal models (Dean and McKay 1994). A number of clinical trials employing AS-ONs designed to target protein kinase C-α have been carried out. Results from clinical studies reported by Nemunaitis et al. (1999) and Yuen et al. (1999) revealed that the antisense therapy is well tolerated, but the antitumor effect is poor. Subsequent data from phase II studies in patients with colorectal (Cripps et al. 2002) and prostate (Tolcher et al. 2002) cancer confirmed the results—no significant antitumor activity was observed. Anti-PKC-α ODNs have also been studied in combination with 5-fluorouracil and leucovorin in patients with advanced cancer (Mani et al. 2002). This combination therapy has shown antitumor activity, but also some toxicity including alopecia, fatigue, mucositis, diarrhea, anorexia, nausea, and tumor pain. In addition, a phase III study comparing LY900003 (Affinitac) plus gemcitabine and cisplatin versus gemcitabine and cisplatin in patients with advanced, previously untreated non-small cell lung cancer has been suspended due to lack of efficacy. Before antisense therapy to PKC-α succeeds, the pharmacodynamics of LY900003 needs to be thoroughly investigated to ensure that it is hitting its target.

H-Ras Pathway Another attractive target for antisense therapy is *H-Ras*, a regulator of several interconnected signaling pathways. Constitutive activation of this gene promotes uncontrolled proliferation and malignant transformation in many human tumors. Several clinical trials have used AS-ONs directed against *H-Ras* (ISIS 2503). Results from a phase I study in patients with advanced carcinoma presented by Cunningham et al. (2001) have shown that ISIS 2503 gives only a partial response in some individuals and is associated with mild toxicity. However, stabilization of disease has been reported in several patients. ISIS 2503 tested in combination with gemcitabine has been well tolerated, and a partial response in patients with metastatic breast cancer has been documented (Adjei et al. 2003). The results of a more recent trial using ISIS 2503 in combination with gemcitabine to treat patients with pancreatic adenocarcinoma are promising (Alberts et al. 2004). A median overall survival of 6.6 months and a response rate of 10% were achieved. However, further work needs to be done to clarify tumor characteristics that may help better predict response to the therapy. Downstream effectors of the RAS signal transduction pathway like c-Raf have also has been investigated as potential targets for antisense therapy. Based on the encouraging results of in vitro experiments (Brennscheidt et al. 1994) and in vivo tumor xenograft mouse models (Monia et al. 1996), several clinical trails have been conducted. Although no major tumor response has been reported, some patients have shown a stabilization of their disease (Coudert et al. 2001; Rudin et al. 2001; Oza et al. 2003).

4.4
Immunotherapy

Oligonucleotides with unmethylated deoxycytidyl-deoxyguanosine (CpG) dinucleotides have been intensively studied over the past few years as potential immunological adjuvants. CpG-ODNs are able to mimic the immunostimulatory activity of bacterial DNA, which makes them recognizable by several types of immune cells. As a consequence, they can activate the immune response and lead to the activation of natural killer (NK) cells, dendritic cells, macrophages, and B cells. This characteristic property of CpG-ODNs suggests that CpG-containing oligonucleotides might be effective adjuvants for the immunotherapy of cancer, and for enhancing the immune responses to antigens that are less efficient. These ideas were proved to be correct and tested successfully in cell culture and in animal studies (Brazolot Millan et al. 1998; Krug et al. 2001; Wooldridge and Weiner 2003; Alignani et al. 2005). Recent data reported by Friedberg et al. (2005) presents the results from a phase I clinical study involving immunotherapy with the CpG oligonucleotide 1018 ISS in combination with rituximab in patients with non-Hodgkin's lymphoma. CpG-ODN treatment resulted in no significant toxicity and a dose-dependent increase in the expression of several interferon-α/β-inducible genes, providing rationale for further testing of this combination immunotherapy approach.

5
Conclusion

Gene silencing strategies have a long history and have been evolving very rapidly over the past few years. Many new techniques and molecules have been developed with the goal of improving the effectiveness of antisense therapy, but some obstacles still remain. The development of effectively targeted and efficiently delivered nucleic acid molecules will lead to a transformation in the way human disease is treated. As was true for the field of monoclonal antibody-based therapies, where hype was followed by disappointment and then finally genuine triumph of the concept, we strongly believe that breakthroughs in the area of nucleic acid-mediated gene silencing will shortly be forthcoming and will more than justify the time and resources expended in developing the therapeutic use of these molecules.

Acknowledgements V.R.D. is supported by a grant from the NIH (T32 GM07170). A.M.G. is supported by grants from the Doris Duke Charitable Foundation and the NIH (RO1-CA101859).

References

Adam SA, Marr RS, Gerace L (1990) Nuclear protein import in permeabilized mammalian cells requires soluble cytoplasmic factors. J Cell Biol 111:807–816

Adjei AA, Dy GK, Erlichman C, Reid JM, Sloan JA, Pitot HC, Alberts SR, Goldberg RM, Hanson LJ, Atherton PJ, Watanabe T, Geary RS, Holmlund J, Dorr FA (2003) A phase I trial of ISIS 2503, an antisense inhibitor of H-ras, in combination with gemcitabine in patients with advanced cancer. Clin Cancer Res 9:115–123

Agrawal S, Kandimalla ER (2001) Antisense and/or immunostimulatory oligonucleotide therapeutics. Curr Cancer Drug Targets 1:197–209

Alberts SR, Schroeder M, Erlichman C, Steen PD, Foster NR, Moore DF Jr, Rowland KM Jr, Nair S, Tschetter LK, Fitch TR (2004) Gemcitabine and ISIS-2503 for patients with locally advanced or metastatic pancreatic adenocarcinoma: a North Central Cancer Treatment Group phase II trial. J Clin Oncol 22:4944–4950

Alignani D, Maletto B, Liscovsky M, Ropolo A, Moron G, Pistoresi-Palencia MC (2005) Orally administered OVA/CpG-ODN induces specific mucosal and systemic immune response in young and aged mice. J Leukoc Biol 77:898–905

Allawi HT, Dong F, Ip HS, Neri BP, Lyamichev VI (2001) Mapping of RNA accessible sites by extension of random oligonucleotide libraries with reverse transcriptase. RNA 7:314–327

Anfossi G, Gewirtz AM, Calabretta B (1989) An oligomer complementary to c-myb-encoded mRNA inhibits proliferation of human myeloid leukemia cell lines. Proc Natl Acad Sci U S A 86:3379–3383

Aoki Y, Cioca DP, Oidaira H, Kamiya J, Kiyosawa K (2003) RNA interference may be more potent than antisense RNA in human cancer cell lines. Clin Exp Pharmacol Physiol 30:96–102

Bacon TA, Wickstrom E (1991) Walking along human c-myc mRNA with antisense oligodeoxynucleotides: maximum efficacy at the $5'$ cap region. Oncogene Res 6:13–19

Bennett CF, Chiang MY, Chan H, Shoemaker JE, Mirabelli CK (1992) Cationic lipids enhance cellular uptake and activity of phosphorothioate antisense oligonucleotides. Mol Pharmacol 41:1023–1033

Bennett CF, Condon TP, Grimm S, Chan H, Chiang MY (1994) Inhibition of endothelial cell adhesion molecule expression with antisense oligonucleotides. J Immunol 152:3530–3540

Bernstein E, Denli AM, Hannon GJ (2001) The rest is silence. RNA 7:1509–1521

Bramlage B, Alefelder S, Marschall P, Eckstein F (1999) Inhibition of luciferase expression by synthetic hammerhead ribozymes and their cellular uptake. Nucleic Acids Res 27:3159–3167

Brazolot Millan CL, Weeratna R, Krieg AM, Siegrist CA, Davis HL (1998) CpG DNA can induce strong Th1 humoral and cell-mediated immune responses against hepatitis B surface antigen in young mice. Proc Natl Acad Sci USA 95:15553–15558

Breaker RR, Joyce GF (1994) A DNA enzyme that cleaves RNA. Chem Biol 1:223–229

Brennscheidt U, Riedel D, Kolch W, Bonifer R, Brach MA, Ahlers A, Mertelsmann RH, Herrmann F (1994) Raf-1 is a necessary component of the mitogenic response of the human megakaryoblastic leukemia cell line MO7 to human stem cell factor, granulocyte-macrophage colony-stimulating factor, interleukin 3, and interleukin 9. Cell Growth Differ 5:367–372

Capaccioli S, Di Pasquale G, Mini E, Mazzei T, Quattrone A (1993) Cationic lipids improve antisense oligonucleotide uptake and prevent degradation in cultured cells and in human serum. Biochem Biophys Res Commun 197:818–825

Chanan-Khan A (2004) Bcl-2 antisense therapy in hematologic malignancies. Curr Opin Oncol 16:581–585

Coudert B, Anthoney A, Fiedler W, Droz JP, Dieras V, Borner M, Smyth JF, Morant R, de Vries MJ, Roelvink M, Fumoleau P (2001) Phase II trial with ISIS 5132 in patients with small-cell (SCLC) and non-small cell (NSCLC) lung cancer. A European Organization for Research and Treatment of Cancer (EORTC) Early Clinical Studies Group report. Eur J Cancer 37:2194–2198

Cripps MC, Figueredo AT, Oza AM, Taylor MJ, Fields AL, Holmlund JT, McIntosh LW, Geary RS, Eisenhauer EA (2002) Phase II randomized study of ISIS 3521 and ISIS 5132 in patients with locally advanced or metastatic colorectal cancer: a National Cancer Institute of Canada clinical trials group study. Clin Cancer Res 8:2188–2192

Crooke ST (2004) Antisense strategies. Curr Mol Med 4:465–487

Cunningham CC, Holmlund JT, Geary RS, Kwoh TJ, Dorr A, Johnston JF, Monia B, Nemunaitis J (2001) A Phase I trial of H-ras antisense oligonucleotide ISIS 2503 administered as a continuous intravenous infusion in patients with advanced carcinoma. Cancer 92:1265–1271

De Rosa G, Bochot A, Quaglia F, Besnard M, Fattal E (2003) A new delivery system for antisense therapy: PLGA microspheres encapsulating oligonucleotide/polyethyleneimine solid complexes. Int J Pharm 254:89–93

Dean NM, McKay R (1994) Inhibition of protein kinase C-alpha expression in mice after systemic administration of phosphorothioate antisense oligodeoxynucleotides. Proc Natl Acad Sci U S A 91:11762–11766

Dean NM, McKay R, Condon TP, Bennett CF (1994) Inhibition of protein kinase C-alpha expression in human A549 cells by antisense oligonucleotides inhibits induction of intercellular adhesion molecule 1 (ICAM-1) mRNA by phorbol esters. J Biol Chem 269:16416–16424

Dias N, Stein CA (2002) Antisense oligonucleotides: basic concepts and mechanisms. Mol Cancer Ther 1:347–355

Elbashir SM, Harborth J, Lendeckel W, Yalcin A, Weber K, Tuschl T (2001a) Duplexes of 21-nucleotide RNAs mediate RNA interference in cultured mammalian cells. Nature 411:494–498

Elbashir SM, Martinez J, Patkaniowska A, Lendeckel W, Tuschl T (2001b) Functional anatomy of siRNAs for mediating efficient RNAi in Drosophila melanogaster embryo lysate. EMBO J 20:6877–6888

Flanagan WM, Wagner RW (1997) Potent and selective gene inhibition using antisense oligodeoxynucleotides. Mol Cell Biochem 172:213–225

Friedberg JW, Kim H, McCauley M, Hessel EM, Sims P, Fisher DC, Nadler LM, Coffman RL, Freedman AS (2005) Combination immunotherapy with a CpG oligonucleotide (1018 ISS) and rituximab in patients with non-Hodgkin lymphoma: increased interferon-alpha/beta-inducible gene expression, without significant toxicity. Blood 105:489–495

Frieden M, Christensen SM, Mikkelsen ND, Rosenbohm C, Thrue CA, Westergaard M, Hansen HF, Orum H, Koch T (2003) Expanding the design horizon of antisense oligonucleotides with alpha-L-LNA. Nucleic Acids Res 31:6365–6372

Futami T, Miyagishi M, Seki M, Taira K (2002) Induction of apoptosis in HeLa cells with siRNA expression vector targeted against bcl-2. Nucleic Acids Res Suppl 251–252

Gabler A, Krebs S, Seichter D, Forster M (2003) Fast and accurate determination of sites along the FUT2 in vitro transcript that are accessible to antisense oligonucleotides by application of secondary structure predictions and RNase H in combination with MALDI-TOF mass spectrometry. Nucleic Acids Res 31:e79

Gewirtz AM (2000) Oligonucleotide therapeutics: a step forward. J Clin Oncol 18:1809–1811

Gewirtz AM, Calabretta B (1988) A c-myb antisense oligodeoxynucleotide inhibits normal human hematopoiesis in vitro. Science 242:1303–1306

Gewirtz AM, Stein CA, Glazer PM (1996) Facilitating oligonucleotide delivery: helping antisense deliver on its promise. Proc Natl Acad Sci U S A 93:3161–3163

Giannini CD, Roth WK, Piiper A, Zeuzem S (1999) Enzymatic and antisense effects of a specific anti-Ki-ras ribozyme in vitro and in cell culture. Nucleic Acids Res 27:2737–2744

Gifford LK, Opalinska JB, Jordan D, Pattanayak V, Greenham P, Kalota A, Robbins M, Vernovsky K, Rodriguez LC, Do BT, Lu P, Gewirtz AM (2005) Identification of antisense nucleic acid hybridization sites in mRNA molecules with self-quenching fluorescent reporter molecules. Nucleic Acids Res 33:e28

Giles RV, Ruddell CJ, Spiller DG, Green JA, Tidd DM (1995) Single base discrimination for ribonuclease H-dependent antisense effects within intact human leukaemia cells. Nucleic Acids Res 23:954–961

Heasman J (2002) Morpholino oligos: making sense of antisense? Dev Biol 243:209–214

Ho SP, Britton DH, Stone BA, Behrens DL, Leffet LM, Hobbs FW, Miller JA, Trainor GL (1996) Potent antisense oligonucleotides to the human multidrug resistance-1 mRNA are rationally selected by mapping RNA-accessible sites with oligonucleotide libraries. Nucleic Acids Res 24:1901–1907

Hoeller D, Petrie N, Yao F, Eriksson E (2002) Gene therapy in soft tissue reconstruction. Cells Tissues Organs 172:118–125

Jansen B, Wacheck V, Heere-Ress E, Schlagbauer-Wadl H, Hoeller C, Lucas T, Hoermann M, Hollenstein U, Wolff K, Pehamberger H (2000) Chemosensitisation of malignant melanoma by BCL2 antisense therapy. Lancet 356:1728–1733

Jen KY, Gewirtz AM (2000) Suppression of gene expression by targeted disruption of messenger RNA: available options and current strategies. Stem Cells 18:307–319

Kalota A, Shetzline SE, Gewirtz AM (2004) Progress in the development of nucleic acid therapeutics for cancer. Cancer Biol Ther 3:4–12

Kichler A (2004) Gene transfer with modified polyethylenimines. J Gene Med 6 Suppl 1:S3–10

Koping-Hoggard M, Tubulekas I, Guan H, Edwards K, Nilsson M, Varum KM, Artursson P (2001) Chitosan as a nonviral gene delivery system. Structure-property relationships and characteristics compared with polyethylenimine in vitro and after lung administration in vivo. Gene Ther 8:1108–1121

Kretschmer-Kazemi Far R, Sczakiel G (2003) The activity of siRNA in mammalian cells is related to structural target accessibility: a comparison with antisense oligonucleotides. Nucleic Acids Res 31:4417–4424

Krug A, Towarowski A, Britsch S, Rothenfusser S, Hornung V, Bals R, Giese T, Engelmann H, Endres S, Krieg AM, Hartmann G (2001) Toll-like receptor expression reveals CpG DNA as a unique microbial stimulus for plasmacytoid dendritic cells which synergizes with CD40 ligand to induce high amounts of IL-12. Eur J Immunol 31:3026–3037

Kurreck J, Wyszko E, Gillen C, Erdmann VA (2002) Design of antisense oligonucleotides stabilized by locked nucleic acids. Nucleic Acids Res 30:1911–1918

Kutryk MJ, Foley DP, van den Brand M, Hamburger JN, van der Giessen WJ, deFeyter PJ, Bruining N, Sabate M, Serruys PW (2002) Local intracoronary administration of antisense oligonucleotide against c-myc for the prevention of in-stent restenosis: results of the randomized investigation by the Thoraxcenter of antisense DNA using local delivery and IVUS after coronary stenting (ITALICS) trial. J Am Coll Cardiol 39:281–287

Luger SM, O'Brien SG, Ratajczak J, Ratajczak MZ, Mick R, Stadtmauer EA, Nowell PC, Goldman JM, Gewirtz AM (2002) Oligodeoxynucleotide-mediated inhibition of c-myb gene expression in autografted bone marrow: a pilot study. Blood 99:1150–1158

Mani S, Rudin CM, Kunkel K, Holmlund JT, Geary RS, Kindler HL, Dorr FA, Ratain MJ (2002) Phase I clinical and pharmacokinetic study of protein kinase C-alpha antisense oligonucleotide ISIS 3521 administered in combination with 5-fluorouracil and leucovorin in patients with advanced cancer. Clin Cancer Res 8:1042–1048

Marcucci G, Byrd JC, Dai G, Klisovic MI, Kourlas PJ, Young DC, Cataland SR, Fisher DB, Lucas D, Chan KK, Porcu P, Lin ZP, Farag SF, Frankel SR, Zwiebel JA, Kraut EH, Balcerzak SP, Bloomfield CD, Grever MR, Caligiuri MA (2003) Phase 1 and pharmacodynamic studies of G3139, a Bcl-2 antisense oligonucleotide, in combination with chemotherapy in refractory or relapsed acute leukemia. Blood 101:425–432

Marshall J, Chen H, Yang D, Figueira M, Bouker KB, Ling Y, Lippman M, Frankel SR, Hayes DF (2004a) A phase I trial of a Bcl-2 antisense (G3139) and weekly docetaxel in patients with advanced breast cancer and other solid tumors. Ann Oncol 15:1274–1283

Marshall JL, Eisenberg SG, Johnson MD, Hanfelt J, Dorr FA, El-Ashry D, Oberst M, Fuxman Y, Holmlund J, Malik S (2004b) A phase II trial of ISIS 3521 in patients with metastatic colorectal cancer. Clin Colorectal Cancer 4:268–274

Milner N, Mir KU, Southern EM (1997) Selecting effective antisense reagents on combinatorial oligonucleotide arrays. Nat Biotechnol 15:537–541

Mir KU, Southern EM (1999) Determining the influence of structure on hybridization using oligonucleotide arrays. Nat Biotechnol 17:788–792

Monia BP, Johnston JF, Geiger T, Muller M, Fabbro D (1996) Antitumor activity of a phosphorothioate antisense oligodeoxynucleotide targeted against C-raf kinase. Nat Med 2:668–675

Nagy P, Arndt-Jovin DJ, Jovin TM (2003) Small interfering RNAs suppress the expression of endogenous and GFP-fused epidermal growth factor receptor (erbB1) and induce apoptosis in erbB1-overexpressing cells. Exp Cell Res 285:39–49

Nemunaitis J, Holmlund JT, Kraynak M, Richards D, Bruce J, Ognoskie N, Kwoh TJ, Geary R, Dorr A, Von Hoff D, Eckhardt SG (1999) Phase I evaluation of ISIS 3521, an antisense oligodeoxynucleotide to protein kinase C-alpha, in patients with advanced cancer. J Clin Oncol 17:3586–3595

Nestle FO, Mitra RS, Bennett CF, Chan H, Nickoloff BJ (1994) Cationic lipid is not required for uptake and selective inhibitory activity of ICAM-1 phosphorothioate antisense oligonucleotides in keratinocytes. J Invest Dermatol 103:569–575

Opalinska JB, Gewirtz AM (2002) Nucleic-acid therapeutics: basic principles and recent applications. Nat Rev Drug Discov 1:503–514

Opalinska JB, Kalota A, Gifford LK, Lu P, Jen KY, Pradeepkumar PI, Barman J, Kim TK, Swider CR, Chattopadhyaya J, Gewirtz AM (2004) Oxetane modified, conformationally constrained, antisense oligodeoxyribonucleotides function efficiently as gene silencing molecules. Nucleic Acids Res 32:5791–5799

Opalinska JB, Machalinski B, Ratajczak J, Ratajczak MZ, Gewirtz AM (2005) Multi-gene targeting with antisense oligodeoxynucleotides: an exploratory study employing primary human leukemia cells. Clin Cancer Res 11:4948–4954

Oza AM, Elit L, Swenerton K, Faught W, Ghatage P, Carey M, McIntosh L, Dorr A, Holmlund JT, Eisenhauer E (2003) Phase II study of CGP 69846A (ISIS 5132) in recurrent epithelial ovarian cancer: an NCIC clinical trials group study (NCIC IND.116). Gynecol Oncol 89:129–133

Parry TJ, Cushman C, Gallegos AM, Agrawal AB, Richardson M, Andrews LE, Maloney L, Mokler VR, Wincott FE, Pavco PA (1999) Bioactivity of anti-angiogenic ribozymes targeting Flt-1 and KDR mRNA. Nucleic Acids Res 27:2569–2577

Paterson BM, Roberts BE, Kuff EL (1977) Structural gene identification and mapping by DNA-mRNA hybrid-arrested cell-free translation. Proc Natl Acad Sci U S A 74:4370–4374

Paulasova P, Pellestor F (2004) The peptide nucleic acids (PNAs): a new generation of probes for genetic and cytogenetic analyses. Ann Genet 47:349–358

Pieken WA, Olsen DB, Benseler F, Aurup H, Eckstein F (1991) Kinetic characterization of ribonuclease-resistant 2′-modified hammerhead ribozymes. Science 253:314–317

Pradeepkumar PI, Amirkhanov NV, Chattopadhyaya J (2003) Antisense oligonucleotides with oxetane-constrained cytidine enhance heteroduplex stability, and elicit satisfactory RNase H response as well as showing improved resistance to both exo and endonucleases. Org Biomol Chem 1:81–92

Ptasznik A, Nakata Y, Kalota A, Emerson SG, Gewirtz AM (2004) Short interfering RNA (siRNA) targeting the Lyn kinase induces apoptosis in primary, and drug-resistant, BCR-ABL1(+) leukemia cells. Nat Med 10:1187–1189

Ratajczak MZ, Hijiya N, Catani L, DeRiel K, Luger SM, McGlave P, Gewirtz AM (1992a) Acute- and chronic-phase chronic myelogenous leukemia colony-forming units are highly sensitive to the growth inhibitory effects of c-myb antisense oligodeoxynucleotides. Blood 79:1956–1961

Ratajczak MZ, Kant JA, Luger SM, Hijiya N, Zhang J, Zon G, Gewirtz AM (1992b) In vivo treatment of human leukemia in a scid mouse model with c-myb antisense oligodeoxynucleotides. Proc Natl Acad Sci U S A 89:11823–11827

Reed JC, Stein C, Subasinghe C, Haldar S, Croce CM, Yum S, Cohen J (1990) Antisense-mediated inhibition of BCL2 protooncogene expression and leukemic cell growth and survival: comparisons of phosphodiester and phosphorothioate oligodeoxynucleotides. Cancer Res 50:6565–6570

Roque F, Mon G, Belardi J, Rodriguez A, Grinfeld L, Long R, Grossman S, Malcolm A, Zon G, Ormont ML, Fischman DL, Shi Y, Zalewski A (2001) Safety of intracoronary administration of c-myc antisense oligomers after percutaneous transluminal coronary angioplasty (PTCA). Antisense Nucleic Acid Drug Dev 11:99–106

Rudin CM, Holmlund J, Fleming GF, Mani S, Stadler WM, Schumm P, Monia BP, Johnston JF, Geary R, Yu RZ, Kwoh TJ, Dorr FA, Ratain MJ (2001) Phase I trial of ISIS 5132, an antisense oligonucleotide inhibitor of c-raf-1, administered by 24-hour weekly infusion to patients with advanced cancer. Clin Cancer Res 7:1214–1220

Rudin CM, Otterson GA, Mauer AM, Villalona-Calero MA, Tomek R, Prange B, George CM, Szeto L, Vokes EE (2002) A pilot trial of G3139, a bcl-2 antisense oligonucleotide, and paclitaxel in patients with chemorefractory small-cell lung cancer. Ann Oncol 13:539–545

Ruffner DE, Dahm SC, Uhlenbeck OC (1989) Studies on the hammerhead RNA self-cleaving domain. Gene 82:31–41

Sczakiel G, Homann M, Rittner K (1993) Computer-aided search for effective antisense RNA target sequences of the human immunodeficiency virus type 1. Antisense Res Dev 3:45–52

Soler M, McHutchison JG, Kwoh TJ, Dorr FA, Pawlotsky JM (2004) Virological effects of ISIS 14803, an antisense oligonucleotide inhibitor of hepatitis C virus (HCV) internal ribosome entry site (IRES), on HCV IRES in chronic hepatitis C patients and examination of the potential role of primary and secondary HCV resistance in the outcome of treatment. Antivir Ther 9:953–968

Song E, Lee SK, Wang J, Ince N, Ouyang N, Min J, Chen J, Shankar P, Lieberman J (2003) RNA interference targeting Fas protects mice from fulminant hepatitis. Nat Med 9:347–351

Stull RA, Taylor LA, Szoka FC Jr (1992) Predicting antisense oligonucleotide inhibitory efficacy: a computational approach using histograms and thermodynamic indices. Nucleic Acids Res 20:3501–3508

Summerton J, Weller D (1997) Morpholino antisense oligomers: design, preparation, and properties. Antisense Nucleic Acid Drug Dev 7:187–195

Sun LQ, Cairns MJ, Saravolac EG, Baker A, Gerlach WL (2000) Catalytic nucleic acids: from lab to applications. Pharmacol Rev 52:325–347

Thomas CE, Ehrhardt A, Kay MA (2003) Progress and problems with the use of viral vectors for gene therapy. Nat Rev Genet 4:346–358

Tolcher AW (2001) Preliminary phase I results of G3139 (bcl-2 antisense oligonucleotide) therapy in combination with docetaxel in hormone-refractory prostate cancer. Semin Oncol 28:67–70

Tolcher AW, Reyno L, Venner PM, Ernst SD, Moore M, Geary RS, Chi K, Hall S, Walsh W, Dorr A, Eisenhauer E (2002) A randomized phase II and pharmacokinetic study of the antisense oligonucleotides ISIS 3521 and ISIS 5132 in patients with hormone-refractory prostate cancer. Clin Cancer Res 8:2530–2535

Vester B, Wengel J (2004) LNA (locked nucleic acid): high-affinity targeting of complementary RNA and DNA. Biochemistry 43:13233–13241

Vile RG, Russell SJ, Lemoine NR (2000) Cancer gene therapy: hard lessons and new courses. Gene Ther 7:2–8

Waters JS, Webb A, Cunningham D, Clarke PA, Raynaud F, di Stefano F, Cotter FE (2000) Phase I clinical and pharmacokinetic study of bcl-2 antisense oligonucleotide therapy in patients with non-Hodgkin's lymphoma. J Clin Oncol 18:1812–1823

Webb A, Cunningham D, Cotter F, Clarke PA, di Stefano F, Ross P, Corbo M, Dziewanowska Z (1997) BCL-2 antisense therapy in patients with non-Hodgkin lymphoma. Lancet 349:1137–1141

Wohlbold L, van der Kuip H, Miething C, Vornlocher HP, Knabbe C, Duyster J, Aulitzky WE (2003) Inhibition of bcr-abl gene expression by small interfering RNA sensitizes for imatinib mesylate (STI571). Blood 102:2236–2239

Wooldridge JE, Weiner GJ (2003) CpG DNA and cancer immunotherapy: orchestrating the antitumor immune response. Curr Opin Oncol 15:440–445

Xing Z, Whitton JL (1992) Ribozymes which cleave arenavirus RNAs: identification of susceptible target sites and inhibition by target site secondary structure. J Virol 66:1361–1369

Xu Y, Zhang HY, Thormeyer D, Larsson O, Du Q, Elmen J, Wahlestedt C, Liang Z (2003) Effective small interfering RNAs and phosphorothioate antisense DNAs have different preferences for target sites in the luciferase mRNAs. Biochem Biophys Res Commun 306:712–717

Yacyshyn BR, Bowen-Yacyshyn MB, Jewell L, Tami JA, Bennett CF, Kisner DL, Shanahan WR Jr (1998) A placebo-controlled trial of ICAM-1 antisense oligonucleotide in the treatment of Crohn's disease. Gastroenterology 114:1133–1142

Yacyshyn BR, Chey WY, Goff J, Salzberg B, Baerg R, Buchman AL, Tami J, Yu R, Gibiansky E, Shanahan WR (2002) Double blind, placebo controlled trial of the remission inducing and steroid sparing properties of an ICAM-1 antisense oligodeoxynucleotide, alicaforsen (ISIS 2302), in active steroid dependent Crohn's disease. Gut 51:30–36

Yang S, Tutton S, Pierce E, Yoon K (2001) Specific double-stranded RNA interference in undifferentiated mouse embryonic stem cells. Mol Cell Biol 21:7807–7816

Yin JQ, Gao J, Shao R, Tian WN, Wang J, Wan Y (2003) siRNA agents inhibit oncogene expression and attenuate human tumor cell growth. J Exp Ther Oncol 3:194–204

Yu RZ, Su JQ, Grundy JS, Geary RS, Sewell KL, Dorr A, Levin AA (2003) Prediction of clinical responses in a simulated phase III trial of Crohn's patients administered the antisense phosphorothioate oligonucleotide ISIS 2302: comparison of proposed dosing regimens. Antisense Nucleic Acid Drug Dev 13:57–66

Yuen AR, Halsey J, Fisher GA, Holmlund JT, Geary RS, Kwoh TJ, Dorr A, Sikic BI (1999) Phase I study of an antisense oligonucleotide to protein kinase C-alpha (ISIS 3521/CGP 64128A) in patients with cancer. Clin Cancer Res 5:3357–3363

Zamecnik PC, Stephenson ML (1978) Inhibition of Rous sarcoma virus replication and cell transformation by a specific oligodeoxynucleotide. Proc Natl Acad Sci U S A 75:280–284

Zuker M (1989) Computer prediction of RNA structure. Methods Enzymol 180:262–288

Zuker M (2003) Mfold web server for nucleic acid folding and hybridization prediction. Nucleic Acids Res 31:3406–3415

Screening and Determination of Gene Function Using Randomized Ribozyme and siRNA Libraries

S. Matsumoto · H. Akashi · K. Taira (✉)

Department of Chemistry and Biotechnology, School of Engineering, The University of Tokyo, 7-3-1 Hongo, 113-8656 Tokyo, Japan
taira@chembio.t.u-tokyo.ac.jp

The first two authors have contributed equally to this work.

1	Introduction	198
2	Expression of Ribozymes In Vivo	199
2.1	The pol III Expression System	200
2.2	A Strategy for Accelerating of the Rate of Cleavage by a Natural Hammerhead Ribozyme	201
2.3	Creation of Allosteric Ribozymes	203
2.4	RNA-Protein Hybrid Ribozymes	204
3	Gene Discovery Using a Randomized Ribozyme Library	205
3.1	A Case Study: Identification of Genes Involved in Metastasis	207
4	Ribozyme and siRNA Libraries Compared	210
4.1	Efficacy and Target Specificity	211
4.2	Randomized Libraries	214
5	Concluding Remarks	214
	References	215

Abstract Rapid progress in the sequencing of the genomes of model organisms, such as the mouse, rat, nematode, fly, and *Arabidopsis*, as well as the human genome, has provided abundant sequence information, but functions of long stretches of these genomes remain to be determined. RNA-based technologies hold promise as tools that allow us to identify the specific functions of portions of these genomes. In particular, catalytic RNAs, known also as ribozymes, can be engineered for optimization of their activities in the intracellular environment. The introduction of a library of active ribozymes into cells, with subsequent screening for phenotypic changes, can be used for the rapid identification of a gene function. Ribozyme technology complements another RNA-based tool for the determination of gene function, which is based on libraries of small interfering RNAs (siRNAs).

Keywords Ribozyme library · siRNA library · Hammerhead ribozyme · Gene discovery

1
Introduction

The ability to activate or inactivate the expression of a specific gene is crucial for the analysis of gene function. It is relatively easy to activate the expression of a specific gene because methods for introducing vectors that express target genes inside cells have become very efficient and straightforward. However, until recently, no convenient methods for the inactivation of specific genes were available.

For the generation of a loss-of-function phenotype, early methods relied on gene disruption by random mutagenesis and, subsequently, knockout mice were developed (Austin et al. 2004). Although these methods can provide valuable information about gene function, they are laborious and generally involve disruption of the function of only a single gene. Furthermore, the existence of multiple copies of a gene, homologous genes, genes that can compensate for the loss-of-function of a specific gene, or any combination of these makes it very difficult to suppress a particular function when methods that target the DNA are employed. Multiple copies of the target gene or related genes are transcribed as messenger (m)RNAs with sequences homologous to that of the target mRNA, and it would clearly be advantageous to target these mRNAs in addition to the original target when looking for loss-of-function phenotypes. To achieve this goal, we need simple and reliable knockout tools that function at the mRNA level.

Various constructs that include antisense oligonucleotides and ribozymes have been tested for the suppression of gene expression at the RNA level, but the design of such constructs has been based on trial and error, and the effects depend on the properties of the particular target gene (Stein 1999; Scherer and Rossi 2003). Moreover, the desired effects are difficult to predict, and often only weak suppression can be achieved with conventional antisense and ribozyme technologies.

Ribozyme technology has been used successfully to identify genes that play key roles in several signaling pathways (Kruger et al. 2000; Li et al. 2000; Welch et al. 2000; Beger et al. 2001; Kawasaki et al. 2002; Kawasaki and Taira 2002a,b; Suyama et al. 2003a,b, 2004a,b; Kuwabara et al. 2004; Onuki et al. 2004; Wadhwa et al. 2004; Waninger et al. 2004). We shall focus here on the recent progress that has been made in the optimization of a ribozyme activity in the cellular environment, and we discuss the potential utility of ribozyme libraries as powerful gene-discovery tools. We will also compare ribozyme technologies with RNA interference (RNAi)-based methods for the sequence-specific inactivation of genes at the mRNA level. In this review, "ribozyme" refers to hammerhead ribozymes and their derivatives. In addition, the experiments that "we" performed were carried out by many members of our group over the course of several years.

2
Expression of Ribozymes In Vivo

The hammerhead ribozyme, named after its secondary structure, which resembles a hammerhead (Fig. 1a), is a small and versatile catalytic RNA molecule that was originally discovered in infectious plant-satellite RNA. The intramolecular cleavage activity of a self-cleaving (*cis*-acting) hammerhead ribozyme is indispensable for the replication of virusoid molecules. Hammerhead ribozymes have been engineered so that they can cleave any chosen RNA by intermolecular attack *in trans* (*trans*-acting). Such engineering has been achieved by changing the substrate-recognition arms of the hammerhead ribozyme (Fig. 1a). Moreover, the *trans*-acting ribozyme can bind to another RNA molecule through recognition arms that are complementary to the target sequence, and the ribozyme can cleave the target RNA in the presence of Mg^{2+} ions (Uhlenbeck 1987; Haseloff and Gerlach 1988; Zhou and Taira 1998; Suzumura et al. 2004; Takagi et al. 2004; Tanaka et al. 2004; Warashina et al. 2004). The cleavage of a substrate occurs immediately adjacent to the sequence NUX (where N is any base and X is A, C or U) within the target (Fig. 1a) (Sarver et al. 1990; Shimayama et al. 1995; Zhou and Taira 1998).

The availability of *trans*-acting ribozymes and hydrolysis of various mRNAs as a result of changes in the sequences of recognition arms led to extensive studies of the hammerhead and hairpin ribozymes as gene-knockdown tools

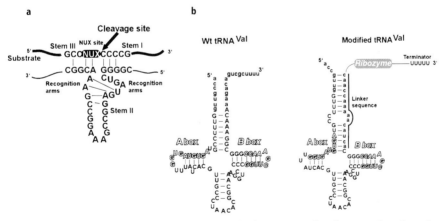

Fig. 1a,b Secondary structure of a hammerhead ribozyme. **a** The cleavage site of a substrate is shown. N represents any base and X can be A, C, or U. **b** Secondary structure of wild-type (*wt*) tRNA^{Val} (*left panel*) and modified tRNA^{Val} (*right panel*) promoters. The *A* and *B* boxes correspond to promoter sequences. The transcripts, which are based on the RNA polymerase III system, include the promoter elements within their transfer (t)RNA sequences. In the modified structure, the 3' half of the wild-type tRNA^{Val} is replaced by a linker sequence to yield a stem/bulge structure. The stem/bulge structure can block the release of the ribozyme from the tRNA^{Val} portion and potential therapeutic agents. Due to their small size, they can be synthesized and easily modified for regulation of the expression of target mRNAs in a strongly sequence-specific manner (Rossi and Sarver 1990; Sarver et al. 1990; Rossi 1995; Sioud 2004). However, the cleavage activity of ribozymes in vitro is not necessarily correlated with their activity in the cellular environments. Therefore, the application of ribozymes to the hydrolysis of specific RNAs in mammalian cells required the development of carefully tailored systems.

2.1
The pol III Expression System

Promoters recognized by RNA polymerase III (pol III promoters) are involved mainly in the transcription of short RNAs, such as transfer (t)RNAs and small nuclear (sn)RNAs (Geiduschek and Tocchini-Valentini 1988). For this reason, pol III expression systems are often employed for the transcription of short RNAs, such as hammerhead and hairpin ribozymes and small interfering (si)RNAs (Cotten and Birnstiel 1989; Shiota et al. 2004). The transcription is two to three orders of magnitude more efficient than that from pol II promoters. In addition, a few extra sequences are added to pol III transcripts, as compared to pol II transcripts, in which a poly(A) tail and a cap structure are added, respectively, at the 3′ and 5′ ends of each transcript. These properties make the pol III system an ideal system for the expression of ribozymes and siRNAs, since high levels of transcripts are required for strong activity, and extra sequences are often inhibitory in this regard. Indeed, the pol III expression system that is based on the promoters of human genes for $tRNA^{Met}$, $tRNA_3^{Lys}$, and $tRNA^{Val}$ and U1, U6, and adenovirus VA1 have been used for the expression of hammerhead and hairpin ribozymes in cells (Good et al. 1997); the $tRNA^{Val}$, H1, $tRNA^{Met}$, U6, and VA1 promoters have similarly been used for the expression of siRNAs (Shiota et al. 2004). With respect to siRNA-expression vectors, the U6 promoter has been used most widely because it is commercially available. However, various promoters seem to confer a variety of properties on the ribozymes and siRNAs transcribed from them, as described below.

As mentioned above, the promoter of the human gene for $tRNA^{Val}$ can be used for the expression of ribozymes (Fig. 1b). Since the tRNA promoter is an internal promoter (the site of initiation of transcription is upstream of the promoter), the transcript contains the tRNA sequence, with parts of the transcript corresponding to internal promoter sequences in its 5′ half and the ribozyme sequence in its 3′ half (hereafter this transcript is referred to as a tRNA-ribozyme). In the presence of the 5′ portion that contains the tRNA sequence, the ribozyme is stabilized and protected from attack by RNases. Such a stabilization of ribozyme transcripts seems to have a favorable effect on the intracellular activity of the tRNA-ribozyme. The $tRNA^{Val}$-based system can also be used for the expression of siRNAs and transcripts composed of $tRNA^{Val}$-

attached short hairpin (sh)RNAs can be processed by Dicer to produce siRNAs in the cytoplasm (Kawasaki and Taira 2003).

In general, tRNA transcripts are processed at their 5′ ends by RNase P and at their 3′ ends by 3′ tRNase, and then are exported from the nucleus to the cytoplasm. If a ribozyme sequence is attached directly downstream of the natural tRNA sequence, 3′ tRNase will cleave the ribozyme portion from tRNA ribozyme transcript. The newly synthesized unprotected ribozyme is then immediately degraded by endogenous RNases. Therefore, to prevent processing at the 3′ end of the tRNA^{Val}-ribozyme, we have modified the sequence at the 3′ end stem by inserting a linker sequence that creates a bulge that blocks the release of the ribozyme from the tRNA^{Val} portion (Fig. 1b). The 5′ tRNA sequence does not hinder the ribozyme's activity, but it appears to have a favorable effect on the ribozyme's intracellular activity, most probably because of the more rapid transport of such tRNA-containing constructs to the cytoplasm and/or their enhanced resistance to RNases.

The subcellular localization of the ribozyme after transcription is another important factor that determines the activity of the ribozyme (Bertrand et al. 1997; Good et al. 1997). We found that tRNA-ribozymes with a high activity were efficiently exported from the nucleus to the cytoplasm, while those with a low activity accumulated in the nucleus (Kato et al. 2001). A modified tRNA^{Val} ribozyme whose secondary structure did not retain the cloverleaf shape remained in the nucleus (Koseki et al. 1999). It appeared that tRNA^{Val} ribozymes with a cloverleaf structure were recognized by exportin-t and transported to the cytoplasm via a nuclear export pathway for tRNAs (Kuwabara et al. 2001), even though these macromolecules could be considered equivalent, in terms of structure, to certain immature tRNAs because of the extra sequence, namely the ribozyme sequence, at their 3′ ends. It has been demonstrated that tRNA^{Val}-attached shRNAs are transported to the cytoplasm by the same mechanism, while U6-driven shRNAs are recognized by exportin-5 and transported to the cytoplasm by a different system (Yi et al. 2003; Lund et al. 2004).

The localization of tRNA-ribozymes is clearly critical to their intracellular activity. A tertiary structure of tRNA-ribozymes, which is influenced by a linker sequence, can affect their export from the nucleus and their intracellular localization. Indeed, a selective choice of linker sequence allows for the engineering of tRNA-ribozymes (Fig. 1b) that are exported to the cytoplasm in mammalian cells and thus exhibit strong intracellular activity.

2.2
A Strategy for Accelerating of the Rate of Cleavage by a Natural Hammerhead Ribozyme

The optimum concentration of Mg^{2+} ions for the reaction catalyzed by the conventional hammerhead ribozyme is above 10 mM and this concentration

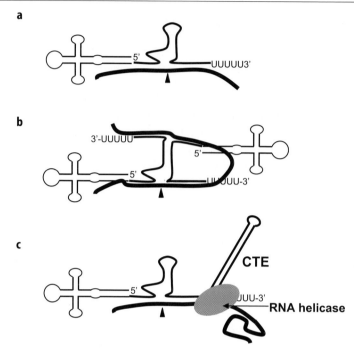

Fig. 2a–c Enhancement of ribozyme activity in the cellular environment. **a** A conventional hammerhead ribozyme. The hammerhead ribozyme is expressed from a tRNA promoter. **b** The heterodimeric maxizyme. The hammerhead ribozyme is shortened in the stem II region, and this shortened ribozyme functions as a dimer with strong suppressive activity. **c** A hybrid ribozyme. This hybrid ribozyme recruits RNA helicase A via the constitutive transport element (CTE), which is an RNA sequence motif that binds RNA helicase. The *ellipse* represents the RNA helicase. The *thick line* represents the target mRNA. The site of cleavage by each ribozyme is indicated by an *arrowhead*

is more than 100-fold higher than that in the cellular environment. Thus, a ribozyme that cleaves its target mRNA efficiently in vitro is not always effective in vivo. Nevertheless, naturally existing satellite RNA appears able to cleave itself quite efficiently via its own ribozyme, at least *in cis*.

A recent report provided one of the reasons why natural ribozymes function so effectively in the cellular environment (Khvorova et al. 2003). Khvorova et al. (2003) demonstrated that the appropriate conformation for cleavage is responsible for the presence of non-conserved nucleotides outside the conventional hammerhead motif. An interaction between the loop of stem I and stem II, through two Watson-Crick base pairs (kissing reactions), is required for the correct folding and strong activity of the ribozyme at physiological concentrations of Mg^{2+} ions. Khvorova et al. (2003) demonstrated this phenomenon in a self-cleaving *cis*-acting ribozyme that mimics satellite RNAs. In the case of a *trans*-acting ribozyme that cleaves RNA molecule *in trans*, the loop of stem I is absent and the rate-limiting step in vivo appears to be the association of the ribozyme with its target RNA (Kato et al. 2001). Further investigations are required to allow application of the kissing strategy to an engineering of *trans*-acting ribozymes that will be effective in vivo.

2.3
Creation of Allosteric Ribozymes

As mentioned above, ribozymes can target RNAs where a specific sequence is present for efficient cleavage Fig. 2a. The cleavage of a substrate occurs immediately after the NUX sequence (where N is any base and X is A, C, or U) within the target (Shimayama et al. 1995; Zhou and Taira 1998). In some cases, a cleavable triplet sequence is not available at a suitable position within the target. For example, in the case of certain abnormal chimeric mRNAs, which are generated by chromosomal translocation, two halves of the fused mRNA are derived from different genes, and an abnormal mRNA must be cleaved while normal transcripts remain untouched. Such chromosomal translocations are often involved in the pathogenesis of disease, and a well-known example is the Philadelphia chromosome, which causes chronic myelogenous leukemia (CML) (Muller et al. 1991). The Philadelphia chromosome occurs as a consequence of a reciprocal translocation that involves the *BCR* and *ABL* genes, which results in a fused mRNA. Since they are tumor-specific and pathogenetically important, fused mRNAs are ideal targets for demonstrations of the potential feasibility of nucleic acid-based therapeutics (Shtivelman et al. 1986). Nevertheless, in the absence of NUX sequences near the junction of the two parental mRNAs, a conventional hammerhead ribozyme cannot target the chimeric mRNA exclusively (Kuwabara et al. 1997, 1998; Warashina et al. 2000). Since both the *BCR* and the *ABL* genes are important for cell survival, ribozyme must cleave only the abnormal chimeric mRNA, without damaging the normal *BCR* and *ABL* mRNAs.

As shown in Fig. 2b, a heterodimeric "maxizyme" that incorporates two different shortened ribozyme monomers (Kuwabara et al. 1996, 1998, 1999) has two substrate-binding arms that are able to recognize two distant sites on the same target RNA simultaneously. Once one arm (the sensor arm) of the maxizyme has recognized and bound to the target sequence, a conformation of the maxizyme changes to generate an active form, which binds Mg^{2+} ions that are essential for hydrolysis reaction. The other arm of the active maxizyme is able to cleave the target mRNA at the NUX sequences that are located at some distance from the junction within the same molecule. These unique features of the maxizyme led to the creation of an allosteric enzyme with a sensor function (that is, an enzyme that is activated *only* in the presence of a junction sequence).

We designed a $tRNA^{Val}$-maxizyme for selective disruption of the abnormal *BCR/ABL* mRNA in cells derived from patients with CML but not the mRNA transcribed from normal *BCR* and *ABL* genes of healthy individuals (Kuwabara et al. 1998). Moreover, we were able to demonstrate the anti-tumor effect of the maxizyme that targeted to the *BCR/ABL* gene fusion in a mouse model of leukemia (Tanabe et al. 2000a). Maxizymes designed to target other chimeric genes and specific splicing variants were also able to suppress the expression of their target genes specifically as a result of their allosteric properties (Tanabe et al. 2000b; Kanamori et al. 2003).

The use of ribozymes derived from hammerheads requires a NUX cleavage site. By contrast, siRNAs can target practically any sequence of interest. Therefore, the transcripts of the above-mentioned chimeric genes can also be destroyed by siRNAs (Oshima et al. 2003; Scherr et al. 2003).

Other types of nucleic-acid-based allosteric aptameric sensors are being developed (Stojanovic and Kolpashchikov 2004). Aptamers are structured nucleic acids that can recognize and bind specific ligands, such as a metabolite, RNA, or protein. In the presence of such specific ligands, an allosteric aptamer changes its structure to allow activation of catalytic functions. Thus, a maxizyme is a type of allosteric sensor that recognizes of a specific abnormal mRNA and cleaves the target. Allosteric aptamers that can be regulated by ATP, theophylline, or flavin mononucleotide (FMN) have been isolated from combinatorial nucleic acid libraries by in vitro evolution and selection methods. Fluorescent aptameric sensors that consist only of a natural RNA have recently been developed (Stojanovic and Kolpashchikov 2004). In the presence of a fluorescent dye, such as malachite green (MG), and target analytes (ATP, theophylline, FMN), an MG aptamer connected to an allosteric aptamer that is controllable by an analyte generates enhanced fluorescence from MG. These observations suggest that it might be possible to generate allosteric aptameric sensors that can be expressed in cells such that minute amounts of specific analytes can be detected by monitoring the fluorescence of MG.

Recent studies in bacteria have revealed that naturally existing RNAs can act as aptameric sensors of vitamin cofactors, glycine, and temperature—which directly regulate the transcription or translation of associated mRNAs—without an involvement of additional proteins (Lai 2003; Mandal and Breaker 2004). Aptameric RNA molecules that act as "riboswitches" play far more important roles in the regulation of gene expression than previously thought. However, natural aptameric sensors in eukaryotic cells have not yet been discovered.

2.4
RNA-Protein Hybrid Ribozymes

The higher-order structure of mRNAs can interfere with the cleavage reaction, since ribozymes cannot access double-stranded stems within a target RNA. Moreover, an exact folding of the target mRNA and the accessibility to the ribozyme cannot easily be predicted and/or determined. Therefore, we postulated that it would be useful to design a ribozyme that recruits a protein that unwinds interfering secondary structures, thereby rendering the target site more accessible to the ribozyme. To create such a ribozyme, we have linked a ribozyme to an RNA-helicase-recruiting motif (Fig. 2c). RNA helicases are known to unwind higher-order structures within RNAs (Luking et al. 1998), and they are involved in post-transcriptional gene silencing (PTGS) in various organisms (Agrawal et al. 2003), as indicated by the observation that PTGS does not occur in helicase-deficient mutant cells. Unwinding of higher-order structures within a target mRNA might be a prerequisite for suppression of gene expression at the mRNA level.

To recruit RNA helicase A to a ribozyme, we exploited the constitutive transport element (CTE), which is an RNA motif that interacts with RNA helicase A both in vitro and in vivo (Tang et al. 1997; Gruter et al. 1998; Braun et al. 1999; Kang and Cullen 1999; Li et al. 1999; Tang and Wong-Staal 2000; Westberg et al. 2000). The CTE-coupled ribozymes had higher activity than non-CTE-coupled ribozymes (Warashina et al. 2001). We also tested an alternative motif, namely the poly(A) sequence (a stretch of 60 As), that recruits a different RNA helicase, eIF4AI, in mammalian cells (Kawasaki et al. 2002; Kawasaki and Taira 2002b). The poly(A) sequence is known to interact with endogenous eIF4AI via interactions with two proteins: poly(A)-binding protein (PABP) and PABP-interacting protein-1 (PAIP). Our poly(A)-ribozymes also had higher activities than the corresponding conventional ribozymes (Kawasaki et al. 2002; Kawasaki and Taira 2002b). Using these improved ribozymes with elevated intracellular activities, we developed a novel gene-discovery system.

3
Gene Discovery Using a Randomized Ribozyme Library

Early attempts at the identification of genes associated with specific cellular phenotypes involved hairpin and hammerhead ribozymes with randomized-substrate recognition arms (Fig. 1a). These randomized ribozymes were constructed by amplification, using polymerase chain reaction (PCR), of mixtures of oligonucleotides with completely randomized recognition arms, each of which included the fixed catalytic core of a ribozyme. PCR products were ligated into an appropriate vector. We used our improved intracellularly active ribozymes with the poly(A) tail to create a randomized ribozyme library (Kawasaki et al. 2002; Kawasaki and Taira 2002b).

Upon the introduction of a library of hybrid ribozymes with randomized binding arms into cells, we might expect some cells to exhibit a change in phenotype (Fig. 3). Although originally ligated products of PCR can be more than 10^{12}, the final number of different ribozymes in the library is between 10^7 and 10^8 because ribozyme-expression constructs are to be amplified in

Escherichia coli cells. Note that after transformation of cells with the ligated ribozyme-containing products of PCR, a diversity decreases during amplification in *E. coli* cells. Depending on the time required for detection of changes in phenotype, the library could be introduced into cells either transiently or stably. The transiently transfected plasmids will be lost with time, and more than one plasmid is likely to enter each cell. Therefore, experiments using transient transfections are informative only if changes in phenotype can be detected within a relatively short period of time (∼60 h). Moreover, repeated transfection is necessary because, in many cases, many ribozyme-bearing plasmids find their way into a single cell. Repeated transfection makes it possible to dilute the selected plasmids to give an average of one expressed plasmid per cell and to increase the ratio of true-positive to false-positive clones upon convergence. If a longer incubation time is required before changes in phenotype are evident, the ribozyme library should be generated in a retrovirus or a lentivirus vector for the establishment of stable transformants. With these viruses as vectors, only a few copies of the ribozyme are introduced into each cell, without a requirement for repeated dilution. Thus, a candidate ribozyme can be isolated after fewer transfections than are required when experiments involve transient

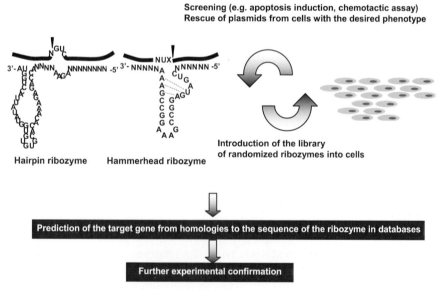

Fig. 3 Strategy for gene discovery with a ribozyme library. Cells that have been transfected with a ribozyme library are subjected to a screening assay, for example, an assay of apoptosis or chemotaxis. Ribozyme sequences are rescued from cells with the phenotype of interest. The target genes of these ribozymes are identified from database searches. The functions of identified genes must then be confirmed in further experiments. *NNNNNN* represents the sequence that is complementary to a randomized target in each recognition arm. *Arrowheads* indicate the sites of cleavage by the ribozyme

transfection. However, more time is required for the establishment of stable transformants than is required for transient transformations.

Nucleotide sequences identified from cells with the desired change in phenotype are considered to be those derived from candidate genes. However, as in any other screening method, it is impossible to avoid false-positive results completely. The background (the ratio of false-positive to true-positive results) of screening methods has a significant effect on an efficiency with which true-positive clones are obtained in cases where many candidate genes are identified or when only a limited number of genes (hidden among a larger number of false-positive genes) are responsible for a specific change in the phenotype. A possible candidate gene must be examined in additional independent experiments, for example, by targeting other sites within the candidate gene with a ribozyme or by RNAi. The advantages of using hammerhead ribozymes include the ease with which an inactive ribozyme can be generated by point mutation at the active site. Thus, only an active ribozyme, and not an inactive counterpart that is targeted to another site within the identified gene of interest, should induce the original change in phenotype. More detailed descriptions of the construction and screening of ribozyme libraries can be found elsewhere (Welch et al. 2000; Kawasaki et al. 2002).

3.1
A Case Study: Identification of Genes Involved in Metastasis

We used a randomized ribozyme library to identify genes that might be involved in the metastasis of cancer cells. In the early stages of cancer, malignant cells are generally localized. As the disease progresses, the expression and/or suppression of various genes allows cells to become invasive and to metastasize. The motility of strongly invasive cancer cells is known to be greater than that of noninvasive or weakly invasive cells, and motility is a prerequisite for metastasis. The mechanisms for metastasis are complicated and remain largely unknown. Therefore, in an attempt to identify individual steps in the metastatic process, we took three different approaches: We screened for an ability to migrate in a (1) chemotactic assay, (2) cell-invasion assay, and (3) assay of metastasis in a mouse model (Suyama et al. 2003a,b, 2004a,b).

In the chemotactic assay, we chose conditions such that 99.9% of cancer cells that were placed in an upper chamber migrated, demonstrating strong invasive potential, toward a lower chamber that contained fibronectin as the chemoattractant (Suyama et al. 2003a; Fig. 4). A ribozyme library was introduced by transient transformation into a line of highly invasive human fibrosarcoma cells. If a specific ribozyme were able to disrupt a gene that is required for invasion of the lower chamber, then a fibrosarcoma cell that harbored this ribozyme would no longer be able to invade the lower chamber. Thus, relevant ribozyme-carrying plasmids could be recovered from cells in the upper chamber. After the third round of transient introduction into cells of the recovered ribozyme

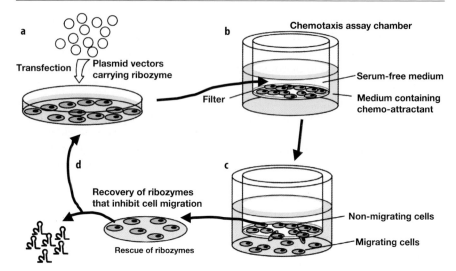

Fig. 4a–d Identification of genes involved in cell migration and thus in metastasis, using a library of randomized ribozymes and an assay of chemotaxis. **a** Cells were transfected with plasmid vectors that carried the ribozyme library. **b** The assay was performed in chambers separated by a porous filter coated with extracellular matrix (ECM) gel. Invasive cells migrated from the upper chamber, through the ECM, to the lower chamber. **c** The cells in the upper chamber, which had failed to migrate, were collected. **d** The ribozymes were rescued and retested

libraries, we identified two independent sequences that corresponded to the ROCK1 gene, which encodes rho-associated, coiled coil-containing protein kinase 1, among other sequences. When we re-introduced two ribozymes that targeted these sequences into cells, both ribozymes efficiently suppressed the expression of *ROCK1* in the strongly invasive fibrosarcoma cells. Moreover, the rates of proliferation of cells treated with either of the two ribozymes and those of non-treated cells were identical. These results suggest that ROCK1 might be an attractive target in efforts to prevent the metastasis of malignant cells, without any negative effects on the viability of normal cells (Suyama et al. 2003a).

In the second approach, we designed a cell-invasion assay (Suyama et al. 2003b). This assay was similar to the chemotactic assay, but the bottom of the top well was coated with an extracellular matrix (ECM) gel. We introduced a retroviral vector that harbored a randomized ribozyme library into mouse NIH3T3 fibroblasts, which were barely able to migrate through the gel-covered filter, and then we isolated RNA from cells that had managed to penetrate the ECM gel after introduction of the library. After reverse transcription of this RNA, we confirmed that eight ribozymes had significantly enhanced the invasive activity of fibroblasts (Suyama et al. 2003b). A database search for candidate genes revealed that one gene corresponded to the Gem gene for a GTPase, which has been reported to be involved in metastasis, while the other seven genes appeared to be novel genes. These genes that we identified should act as suppressors of cell invasion, and the levels of expression of these genes might be expected to provide important information about the malignant potential of cancer cells in a clinical setting.

The random-ribozyme technology allowed us to identify several potentially important genes that might be involved in metastasis, but we were well aware that our screening conditions did not reflect physiological conditions. Therefore, we tried to identify metastasis-related genes using a mouse model of pulmonary tumorigenesis (Suyama et al. 2004a). We injected B16F0 melanoma cells that had been treated with a retroviral vector that harbored a ribozyme library intravenously into mice. The number of pulmonary tumors was significantly higher in mice that had been treated with the ribozyme library as compared with control mice 2 weeks after the injection. We examined sequences of ribozymes that were obtained from the pulmonary tumors and identified eight target genes, three of which are known and five are unknown. We then constructed siRNA vectors targeted to four out of eight candidate genes and introduced them into B16F0 cells. We assayed cells in the wound-scratch assay that is commonly used to study the ability of cells to migrate. Targeting of each of the four genes led, in each case, to significant acceleration of cell migration, which is one of the features of metastasis.

Our experiments resulted in a successful identification of several genes that might be potentially involved in the metastasis of tumors. They also showed that ribozyme technology can be applied even to a mouse model in vivo. In the first experiments, we used poly(A)-ribozyme libraries, which yielded a higher number of potentially informative clones than does a conventional ribozyme library. However, poly(A)-containing plasmids are not very stable (the deletion of As is a problem), and thus, in our next experiment, we used a conventional ribozyme library. Nonetheless, we achieved the successful identification of functional genes in mice. As judged from sequences of those genes that we identified, it might be easier to obtain true-positive clones using hammerhead ribozymes rather than hairpin ribozymes. In the case of hammerhead ribozymes, sequences in all the genes were 100% complementary to the target site of the ribozyme. By contrast, hairpin ribozymes identified some genes with mismatches.

Randomized ribozyme libraries have been used to identify many other kinds of genes, such as genes associated with apoptotic pathways (Kawasaki et al. 2002; Kawasaki and Taira 2002b; Kawasaki and Taira 2002a), Alzheimer's disease (Onuki et al. 2004), muscle differentiation (Wadhwa et al. 2004), and neuronal differentiation (Kuwabara et al. 2004). In particular, we have used a ribozyme library to identify a novel type of non-coding RNA that regulates the differentiation of neuronal stem cells (Fig. 5; Kuwabara et al. 2004). With the availability of the sequence of the entire human genome, our unique gene-discovery technology has become extremely valuable for the identifica-

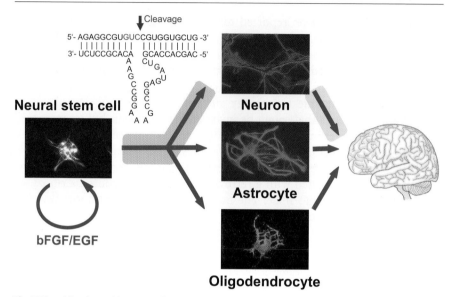

Fig. 5 Identification of functional genes in neuronal development. If neural stem cells that are transfected with a functional ribozyme differentiate in the course as shown by *shaded arrows*, the target gene for the ribozyme would play an important role in neuronal differentiation. Genes responsible for neuronal development can be identified as described in the text

tion of important genes. The successful application of ribozyme libraries to gene-identification studies encouraged us to establish siRNA libraries that are potentially even more effective, as described below (Miyagishi and Taira 2003).

4
Ribozyme and siRNA Libraries Compared

In recent years, RNAi has attracted a great deal of attention as a powerful tool for a gene silencing. After the first report of this phenomenon in response to the introduction of double-stranded (ds)RNA in *Caenorhabditis elegans* (Fire et al. 1998), RNAi was detected in various organisms including plants, *C. elegans*, *Drosophila*, and protozoan and mammalian species (Elbashir et al. 2001; Zamore 2001; Novina and Sharp 2004). In RNAi, exogenous dsRNAs are cleaved into siRNAs, which are subsequently incorporated into the RNA-induced silencing complex (RISC). The siRNA–RISC complex recognizes and cleaves the target mRNA in a sequence-specific manner (Hammond et al. 2001; Nykanen et al. 2001) via a reaction that is similar to the cleavage of target mRNAs by ribozymes. The potential power of RNAi encouraged the scientific community to attempt to exploit this phenomenon as a tool for the analysis of a genome-wide gene function. Such systematic analysis of the genome of *C. elegans* has allowed the identification of the functions of many genes (Fraser et al. 2000; Gonczy et al. 2000). Similar comprehensive analyses of mammalian genomes using libraries of siRNA oligonucleotides and siRNA-expression vectors have been reported (Aza-Blanc et al. 2003; Miyagishi and Taira 2003; Berns et al. 2004; Colland et al. 2004; Hsieh et al. 2004; Kaykas and Moon 2004; Luo et al. 2004; Miyagishi et al. 2004; Paddison et al. 2004; Sen et al. 2004; Shirane et al. 2004; Zheng et al. 2004; Futami et al. 2005).

In the next sections, we compare and contrast the characteristics of ribozyme libraries and siRNA libraries in efforts directed towards the rapid identification of functional genes (Tables 1 and 2).

4.1
Efficacy and Target Specificity

The biggest difference between ribozyme and siRNA technologies is the recruitment of endogenous proteins—applicable in the latter case—which is associated with the maintenance of a high intracellular activity. By contrast, the actions of ribozymes do not depend on intracellular factors (Scherer and Rossi 2003). Thus, siRNAs can exploit many intracellular enzymes, such as helicases and RNases, in cells for cleavage of target mRNAs, and they can be much more effective tools than ribozymes for the suppression of the expression of target mRNAs. However, while in both cases selection of target sites determines activity, it appears that ribozyme activity is probably influenced strongly by a tertiary structure of the target mRNA than is siRNA activity (Yoshinari et al. 2004), and unfortunately it is not easy to predict the best target sites for a ribozyme. By contrast, the suppressive activity of siRNAs depends to a greater extent on the interactions between siRNA and a group of endogenous proteins than it does on the tertiary structure of the target mRNA. Indeed, statistical analysis on the basis of relationships between sequence and activity showed that several important parameters must be considered in attempts to predict the best target sites for siRNAs. In a painstaking analysis, we produced about 1,000 siRNA duplex pairs for a particular gene of approximately 1 kb in length by walking along it and shifting the sequence one base at a time for each pair (Katoh et al. 2003). A complete set of sequence–activity relationships for the 1,000 siRNA pairs enabled us to produce a reliable algorithm by nonlinear regression methods (Miyagishi and Taira 2003). Using that algorithm, we have been able to create libraries of siRNA-expression vectors that circumvent the interferon (IFN) response, targeting individual specific genes in the entire human genome (Matsumoto et al. 2005; H. Akashi, K. Taira, unpublished data).

The most serious disadvantage of siRNA is its non-specific suppressive activity, which is referred to as off-target effect, or an IFN response. Since specificity of the siRNA is not always guaranteed, because off-target effects cannot always be predicted (Bridge et al. 2003; Sledz et al. 2003), ribozymes appear to be superior to siRNAs in this respect, despite their lower activity.

Table 1 The discovery of functional genes using libraries of randomized ribozyme

Selection	Ribozyme	Type*	Identified gene	Cells	Reference
Anchorage-independent cell growth control	Hairpin	Retrovirus	PPAN	HF	Welch et al. 2000
Suppression of fibroblast transformation	Hairpin	Retrovirus	mTERT	NIH3T3	Li et al. 2000
IRES-mediated translation of HCV core protein	Hairpin	Retrovirus	eIF2Bγ, eIF2γ	HeLa	Kruger et al. 2000
Regulation of BRCA1	Hairpin	Retrovirus	Id4		Beger et al. 2001
TNF-α-mediated apoptosis	Hybrid hammerhead	Retrovirus	38 genes	MCF7	Kawasaki et al. 2002
Fas-mediated apoptosis	Hybrid hammerhead	Transient	87 genes	HeLa	Kawasaki et al. 2002b
Fas-mediated apoptosis	Hybrid hammerhead	Transient	7 genes	HeLa	Kawasaki et al. 2002a
Cell migration	Hammerhead	Transient	ROCK1	HT1080	Suyama et al. 2003a
Cell invasion	Hybrid hammerhead	Retrovirus	8 genes	NIH3T3	Suyama et al. 2003b
Metastasis	Hammerhead	Retrovirus	LIMK-2	HT1080	Suyama et al. 2004b
Tunicamycin-induced apoptosis	Hammerhead	Transient EBNA	PKR	SK-N-SH	Onuki et al. 2004
Neuronal differentiation	Hammerhead	Lentivirus	Small modulatory RNA	HCN A94	Kuwabara et al. 2004
Metastasis	Hammerhead	Retrovirus	4 genes	B16F0	Suyama et al. 2004a
Replication of HIV	Hairpin	Retrovirus	Ku80	CEM-GFP	Waninger et al. 2004
Muscle differentiation	Hybrid hammerhead	Retrovirus	$p19^{ARF}$, $p21^{WAF1}$, fem1	C2C12	Wadhwa et al. 2004

*EBNA, semistable lines of cells transfected with Epstein-Barr virus-based vector; lentivirus, stable lines of cells transfected with lentiviral vector; retrovirus, stable lines of cells transfected with retroviral vector; transient, transient transfection

Table 2 Advantages and disadvantages of ribozyme libraries and siRNA libraries for the identification of gene function

	Ribozyme		RNAi	
	Specifically made	Randomized	Specifically made	Randomized/Mixed*
Specificity	High	High	Lower than ribozyme (IFN response/off-target effects)	Lower
Activity	Low	Low	High	High
Target prediction	Difficult	Not required	Possible	Difficult to identify the target gene (randomized)
Performance	Poor	Works well	Good	Might saturate RISC (mixed)
Construction/maintenance	Not feasible because of low activity	Cheap	Expensive	Cheap (randomized)
High-throughput screening	Not feasible	Not feasible	Possible	Not feasible

*The term "mixed" means a case where specifically made siRNA library is used as a mixture

However, improvements in siRNAs are rapidly being made. For example, in mammalian cells, siRNAs of 27 bp in length have been shown to be significantly more effective than the "conventional" 19-bp siRNAs (Kim et al. 2005; Siolas et al. 2005). As a result, picomolar concentrations of 27-bp siRNAs can now be used instead of the nanomolar concentrations of 19-bp siRNAs that have been necessary in the past. The use of picomolar concentrations of 27-bp siRNAs can clearly minimize, if not actually eliminate, any off-target effects. In addition to this improvement at the level of concentration, we have identified some factors that are important in circumventing the interfering activity of short hairpin shRNAs. We found that the introduction of a mutation (either C to U or A to G) in the stem region significantly reduced the interferon response, and this finding was also valid for dsRNAs of more than 100 bp (H. Akashi, K. Taira, unpublished data). Moreover, it is also possible to design siRNA-expression vectors such that only the sense strand of the shRNA is rapidly degraded, and thus off-target effects of the sense strand are minimized (S. Matsumoto, K. Taira, unpublished data). Together, these improvements help to minimize the deleterious, non-specific suppressive activity of siRNA and enhance its potential utility in vivo.

4.2
Randomized Libraries

In our ribozyme libraries, the binding arms were completely randomized. In theory, a similar complete randomization of siRNA duplexes is possible. However, use of such completely randomized siRNA is not feasible in vivo. This problem arises because, in any assay system, it is impossible to remove false positives completely. As mentioned in Sect. 3.1, in the chemotactic assay (Fig. 4), we were able to optimize conditions such that only 0.1% of cancer cells remained in the upper chamber (as false positives). However, if the upper chamber originally contained 10^6 cells (0.1% of 10^6 cells is equal to 10^3 cells), that leaves a large number of false positives. If the number of true positive clones were, for example 100, only 10% of the collected clones would contain useful information. Fortunately, however, ribozymes require a NUX triplet, such as GUC, for cleavage, and thus any sequenced potential targets that lack a cleavable triplet (~90% of collected clones) are easily recognized as false positives and can be discarded. Such discrimination of false positives from true positives is impossible in the case of a completely randomized siRNA library. Moreover, a presence of two potential strands makes it more difficult to identify true positives, because it is not easy to judge which strand has acted as the antisense strand. Therefore, as far as completely randomized libraries are concerned, using ribozymes rather than siRNAs appears more attractive. Nevertheless, as mentioned above, since siRNAs appear to be much more effective in vivo than ribozymes, collection of specifically targeted siRNA clones (libraries of specific siRNAs) should be very useful.

The major advantage of completely randomized libraries is the fact that no sequence information is needed, and thus the discovery of a-yet-unidentified genes is possible. The disadvantage of complete randomization is the fact that a significant portion of the sequences in the library never matches a human gene, and we end up with a huge number of false positives. Nevertheless, with a completely randomized ribozyme library it is possible to identify novel genes, which include genes for small functional RNAs, as long as such small non-coding RNAs contain a GUC triplet (Fig. 5). Thus, despite the popularity of siRNA technology, the ribozymes derived from a hammerhead ribozyme, with their strong specificity and in spite of their lower activity, have proved to be very useful.

5
Concluding Remarks

In this review, we described the use of catalytic RNAs as potential tools for the identification of novel genes and their functions. Since many reports recently have demonstrated the usefulness of siRNAs and siRNA libraries in gene discovery, we compared ribozyme technology and siRNA technology, with special emphasis on the two types of libraries. Even though the design of effective ribozymes for inactivation of a particular gene is generally difficult, we have identified several key parameters, such as strong expression, intracellular stability, efficient export to the cytoplasm, and the ability to access target sequences, that control the activities of ribozymes in vivo. We used our improved ribozymes for the creation of completely randomized libraries that can be used to screen for functional genes that regulate important biological phenomena. In addition, such libraries allow us to identify not only conventional coding genes but also genes for novel non-coding RNAs (Kuwabara et al. 2004).

Recent studies have indicated that non-coding small RNAs, which include microRNAs (miRNAs), silence the expression of certain genes at the level of both transcription (Kawasaki and Taira 2004; Kuwabara et al. 2004; Morris et al. 2004) and translation (Bartel 2004). This process allows levels of expression of specific proteins to be altered very quickly. Analysis in living cells and tissues of the roles of non-coding small RNAs—which appear to play important roles in the differentiation of cells, for example—should help to address questions about their genetics, biogenesis, trafficking, and function. Since non-coding small RNAs or their precursors can be cleaved by ribozymes and siRNAs (Fig. 5; Kuwabara et al. 2004) and since bioinformatics has identified many potential small RNAs within the genomes of higher organisms, future libraries—and, in particular, specifically created (individually tailored) siRNA libraries—should be designed to cover such non-coding small RNAs.

The advent of reliable algorithms for the prediction of targets of siRNAs and the ability to circumvent the IFN response and to avoid off-target effects should enhance the utility of seemingly less-specific siRNAs. Clearly, given that the number of genes in a genome is finite, it should be possible to produce complete siRNA libraries that encompass entire genomes and include genes for potential miRNAs. In fact, such specifically generated siRNA libraries, directed against individual genes, enabled us to identify essential participants in RNAi, including genes that encode a slicer, eIF2C2, and a helicase (M. Miyagishi, K. Taira, unpublished data). In view of the strong specificity of ribozymes and the strong activities of siRNAs, it is clear that the two technologies will play complementary and important roles in the identification of novel genes and their functions.

References

Agrawal N, Dasaradhi PV, Mohmmed A, Malhotra P, Bhatnagar RK, Mukherjee SK (2003) RNA interference: biology, mechanism, and applications. Microbiol Mol Biol Rev 67:657–685

Austin CP, Battey JF, Bradley A, Bucan M, Capecchi M, Collins FS, Dove WF, Duyk G, Dymecki S, Eppig JT, Grieder FB, Heintz N, Hicks G, Insel TR, Joyner A, Koller BH, Lloyd KC, Magnuson T, Moore MW, Nagy A, Pollock JD, Roses AD, Sands AT, Seed B, Skarnes WC, Snoddy J, Soriano P, Stewart DJ, Stewart F, Stillman B, Varmus H, Varticovski L, Verma IM, Vogt TF, von Melchner H, Witkowski J, Woychik RP, Wurst W, Yancopoulos GD, Young SG, Zambrowicz B (2004) The knockout mouse project. Nat Genet 36:921–924

Aza-Blanc P, Cooper CL, Wagner K, Batalov S, Deveraux QL, Cooke MP (2003) Identification of modulators of TRAIL-induced apoptosis via RNAi-based phenotypic screening. Mol Cell 12:627–637

Bartel DP (2004) MicroRNAs: genomics, biogenesis, mechanism, and function. Cell 116:281–297

Beger C, Pierce LN, Kruger M, Marcusson EG, Robbins JM, Welcsh P, Welch PJ, Welte K, King MC, Barber JR, Wong-Staal F (2001) Identification of Id4 as a regulator of BRCA1 expression by using a ribozyme-library-based inverse genomics approach. Proc Natl Acad Sci U S A 98:130–135

Berns K, Hijmans EM, Mullenders J, Brummelkamp TR, Velds A, Heimerikx M, Kerkhoven RM, Madiredjo M, Nijkamp W, Weigelt B, Agami R, Ge W, Cavet G, Linsley PS, Beijersbergen RL, Bernards R (2004) A large-scale RNAi screen in human cells identifies new components of the p53 pathway. Nature 428:431–437

Bertrand E, Castanotto D, Zhou C, Carbonnelle C, Lee NS, Good P, Chatterjee S, Grange T, Pictet R, Kohn D, Engelke D, Rossi JJ (1997) The expression cassette determines the functional activity of ribozymes in mammalian cells by controlling their intracellular localization. RNA 3:75–88

Braun IC, Rohrbach E, Schmitt C, Izaurralde E (1999) TAP binds to the constitutive transport element (CTE) through a novel RNA-binding motif that is sufficient to promote CTE-dependent RNA export from the nucleus. EMBO J 18:1953–1965

Bridge AJ, Pebernard S, Ducraux A, Nicoulaz AL, Iggo R (2003) Induction of an interferon response by RNAi vectors in mammalian cells. Nat Genet 34:263–264

Colland F, Jacq X, Trouplin V, Mougin C, Groizeleau C, Hamburger A, Meil A, Wojcik J, Legrain P, Gauthier JM (2004) Functional proteomics mapping of a human signaling pathway. Genome Res 14:1324–1332

Cotten M, Birnstiel ML (1989) Ribozyme mediated destruction of RNA in vivo. EMBO J 8:3861–3866

Elbashir SM, Harborth J, Lendeckel W, Yalcin A, Weber K, Tuschl T (2001) Duplexes of 21-nucleotide RNAs mediate RNA interference in cultured mammalian cells. Nature 411:494–498

Fire A, Xu S, Montgomery MK, Kostas SA, Driver SE, Mello CC (1998) Potent and specific genetic interference by double-stranded RNA in *Caenorhabditis elegans*. Nature 391:806–811

Fraser AG, Kamath RS, Zipperlen P, Martinez-Campos M, Sohrmann M, Ahringer J (2000) Functional genomic analysis of *C. elegans* chromosome I by systematic RNA interference. Nature 408:325–330

Futami T, Miyagishi M, Taira K (2005) Identification of a network involved in thapsigargin-induced apoptosis using a library of small interfering RNA expression vectors. J Biol Chem 280:826–831

Geiduschek EP, Tocchini-Valentini GP (1988) Transcription by RNA polymerase III. Annu Rev Biochem 57:873–914

Gonczy P, Echeverri C, Oegema K, Coulson A, Jones SJ, Copley RR, Duperon J, Oegema J, Brehm M, Cassin E, Hannak E, Kirkham M, Pichler S, Flohrs K, Goessen A, Leidel S, Alleaume AM, Martin C, Ozlu N, Bork P, Hyman AA (2000) Functional genomic analysis of cell division in C. elegans using RNAi of genes on chromosome III. Nature 408:331–336

Good PD, Krikos AJ, Li SX, Bertrand E, Lee NS, Giver L, Ellington A, Zaia JA, Rossi JJ, Engelke DR (1997) Expression of small, therapeutic RNAs in human cell nuclei. Gene Ther 4:45–54

Gruter P, Tabernero C, von Kobbe C, Schmitt C, Saavedra C, Bachi A, Wilm M, Felber BK, Izaurralde E (1998) TAP, the human homolog of Mex67p, mediates CTE-dependent RNA export from the nucleus. Mol Cell 1:649–659

Hammond SM, Boettcher S, Caudy AA, Kobayashi R, Hannon GJ (2001) Argonaute2, a link between genetic and biochemical analyses of RNAi. Science 293:1146–1150

Haseloff J, Gerlach WL (1988) Simple RNA enzymes with new and highly specific endoribonuclease activities. Nature 334:585–591

Hsieh AC, Bo R, Manola J, Vazquez F, Bare O, Khvorova A, Scaringe S, Sellers WR (2004) A library of siRNA duplexes targeting the phosphoinositide 3-kinase pathway: determinants of gene silencing for use in cell-based screens. Nucleic Acids Res 32:893–901

Kanamori T, Nishimaki K, Asoh S, Ishibashi Y, Takata I, Kuwabara T, Taira K, Yamaguchi H, Sugihara S, Yamazaki T, Ihara Y, Nakano K, Matuda S, Ohta S (2003) Truncated product of the bifunctional *DLST* gene involved in biogenesis of the respiratory chain. EMBO J 22:2913–2923

Kang Y, Cullen BR (1999) The human Tap protein is a nuclear mRNA export factor that contains novel RNA-binding and nucleocytoplasmic transport sequences. Genes Dev 13:1126–1139

Kato Y, Kuwabara T, Warashina M, Toda H, Taira K (2001) Relationships between the activities in vitro and in vivo of various kinds of ribozyme and their intracellular localization in mammalian cells. J Biol Chem 276:15378–15385

Katoh T, Susa M, Suzuki T, Umeda N, Watanabe K (2003) Simple and rapid synthesis of siRNA derived from in vitro transcribed shRNA. Nucleic Acids Res Suppl 249–250

Kawasaki H, Taira K (2002a) A functional gene discovery in the Fas-mediated pathway to apoptosis by analysis of transiently expressed randomized hybrid-ribozyme libraries. Nucleic Acids Res 30:3609–3614

Kawasaki H, Taira K (2002b) Identification of genes by hybrid ribozymes that couple cleavage activity with the unwinding activity of an endogenous RNA helicase. EMBO Rep 3:443–450

Kawasaki H, Taira K (2003) Short hairpin type of dsRNAs that are controlled by tRNA^{Val} promoter significantly induce RNAi-mediated gene silencing in the cytoplasm of human cells. Nucleic Acids Res 31:700–707

Kawasaki H, Taira K (2004) Induction of DNA methylation and gene silencing by short interfering RNAs in human cells. Nature 431:211–217

Kawasaki H, Onuki R, Suyama E, Taira K (2002) Identification of genes that function in the TNF-alpha-mediated apoptotic pathway using randomized hybrid ribozyme libraries. Nat Biotechnol 20:376–380

Kaykas A, Moon RT (2004) A plasmid-based system for expressing small interfering RNA libraries in mammalian cells. BMC Cell Biol 5:16

Khvorova A, Lescoute A, Westhof E, Jayasena SD (2003) Sequence elements outside the hammerhead ribozyme catalytic core enable intracellular activity. Nat Struct Biol 10:708–712

Kim DH, Behlke MA, Rose SD, Chang MS, Choi S, Rossi JJ (2005) Synthetic dsRNA Dicer substrates enhance RNAi potency and efficacy. Nat Biotechnol 23:222–226

Koseki S, Tanabe T, Tani K, Asano S, Shioda T, Nagai Y, Shimada T, Ohkawa J, Taira K (1999) Factors governing the activity in vivo of ribozymes transcribed by RNA polymerase III. J Virol 73:1868–1877

Kruger M, Beger C, Li QX, Welch PJ, Tritz R, Leavitt M, Barber JR, Wong-Staal F (2000) Identification of eIF2Bgamma and eIF2gamma as cofactors of hepatitis C virus internal ribosome entry site-mediated translation using a functional genomics approach. Proc Natl Acad Sci U S A 97:8566–8571

Kuwabara T, Amontov SV, Warashina M, Ohkawa J, Taira K (1996) Characterization of several kinds of dimer minizyme: simultaneous cleavage at two sites in HIV-1 tat mRNA by dimer minizymes. Nucleic Acids Res 24:2302–2310

Kuwabara T, Warashina M, Tanabe T, Tani K, Asano S, Taira K (1997) Comparison of the specificities and catalytic activities of hammerhead ribozymes and DNA enzymes with respect to the cleavage of BCR-ABL chimeric L6 (b2a2) mRNA. Nucleic Acids Res 25:3074–3081

Kuwabara T, Warashina M, Tanabe T, Tani K, Asano S, Taira K (1998) A novel allosterically trans-activated ribozyme, the maxizyme, with exceptional specificity in vitro and in vivo. Mol Cell 2:617–627

Kuwabara T, Warashina M, Nakayama A, Ohkawa J, Taira K (1999) tRNA^{Val}-heterodimeric maxizymes with high potential as geneinactivating agents: simultaneous cleavage at two sites in HIV-1 Tat mRNA in cultured cells. Proc Natl Acad Sci U S A 96:1886–1891

Kuwabara T, Warashina M, Sano M, Tang H, Wong-Staal F, Munekata E, Taira K (2001) Recognition of engineered tRNAs with an extended 3′ end by Exportin-t (Xpo-t) and transport of tRNA-attached ribozymes to the cytoplasm in somatic cells. Biomacromolecules 2:1229–1242

Kuwabara T, Hsieh J, Nakashima K, Taira K, Gage FH (2004) A small modulatory dsRNA specifies the fate of adult neural stem cells. Cell 116:779–793

Lai EC (2003) RNA sensors and riboswitches: self-regulating messages. Curr Biol 13:R285–291

Li J, Tang H, Mullen TM, Westberg C, Reddy TR, Rose DW, Wong-Staal F (1999) A role for RNA helicase A in post-transcriptional regulation of HIV type 1. Proc Natl Acad Sci U S A 96:709–714

Li QX, Robbins JM, Welch PJ, Wong-Staal F, Barber JR (2000) A novel functional genomics approach identifies mTERT as a suppressor of fibroblast transformation. Nucleic Acids Res 28:2605–2612

Luking A, Stahl U, Schmidt U (1998) The protein family of RNA helicases. Crit Rev Biochem Mol Biol 33:259–296

Lund E, Guttinger S, Calado A, Dahlberg JE, Kutay U (2004) Nuclear export of microRNA precursors. Science 303:95–98

Luo B, Heard AD, Lodish HF (2004) Small interfering RNA production by enzymatic engineering of DNA (SPEED). Proc Natl Acad Sci U S A 101:5494–5499

Mandal M, Breaker RR (2004) Gene regulation by riboswitches. Nat Rev Mol Cell Biol 5:451–463

Matsumoto S, Miyagishi M, Akashi H, Nagai R, Taira K (2005) Analysis of Double-stranded RNA-induced apoptosis pathways using interferon-response noninducible small interfering RNA expression vector library. J Biol Chem 280:25687–25696

Miyagishi M, Taira K (2003) Strategies for generation of an siRNA expression library directed against the human genome. Oligonucleotides 13:325–333

Miyagishi M, Matsumoto S, Taira K (2004) Generation of an shRNAi expression library against the whole human transcripts. Virus Res 102:117–124

Morris KV, Chan SW, Jacobsen SE, Looney DJ (2004) Small interfering RNA-induced transcriptional gene silencing in human cells. Science 305:1289–1292

Muller AJ, Young JC, Pendergast AM, Pondel M, Landau NR, Littman DR, Witte ON (1991) BCR first exon sequences specifically activate the BCR/ABL tyrosine kinase oncogene of Philadelphia chromosome-positive human leukemias. Mol Cell Biol 11:1785–1792

Novina CD, Sharp PA (2004) The RNAi revolution. Nature 430:161–164

Nykanen A, Haley B, Zamore PD (2001) ATP requirements and small interfering RNA structure in the RNA interference pathway. Cell 107:309–321

Onuki R, Bando Y, Suyama E, Katayama T, Kawasaki H, Baba T, Tohyama M, Taira K (2004) An RNA-dependent protein kinase is involved in tunicamycin-induced apoptosis and Alzheimer's disease. EMBO J 23:959–968

Oshima K, Kawasaki H, Soda Y, Tani K, Asano S, Taira K (2003) Maxizymes and small hairpin-type RNAs that are driven by a tRNA promoter specifically cleave a chimeric gene associated with leukemia in vitro and in vivo. Cancer Res 63:6809–6814

Paddison PJ, Silva JM, Conklin DS, Schlabach M, Li M, Aruleba S, Balija V, O'Shaughnessy A, Gnoj L, Scobie K, Chang K, Westbrook T, Cleary M, Sachidanandam R, McCombie WR, Elledge SJ, Hannon GJ (2004) A resource for large-scale RNA-interference-based screens in mammals. Nature 428:427–431

Rossi JJ (1995) Controlled, targeted, intracellular expression of ribozymes: progress and problems. Trends Biotechnol 13:301–306

Rossi JJ, Sarver N (1990) RNA enzymes (ribozymes) as antiviral therapeutic agents. Trends Biotechnol 8:179–183

Sarver N, Cantin EM, Chang PS, Zaia JA, Ladne PA, Stephens DA, Rossi JJ (1990) Ribozymes as potential anti-HIV-1 therapeutic agents. Science 247:1222–1225

Scherer LJ, Rossi JJ (2003) Approaches for the sequence-specific knockdown of mRNA. Nat Biotechnol 21:1457–1465

Scherr M, Battmer K, Winkler T, Heidenreich O, Ganser A, Eder M (2003) Specific inhibition of bcr-abl gene expression by small interfering RNA. Blood 101:1566–1569

Sen G, Wehrman TS, Myers JW, Blau HM (2004) Restriction enzyme-generated siRNA (REGS) vectors and libraries. Nat Genet 36:183–189

Shimayama T, Nishikawa S, Taira K (1995) Generality of the NUX rule: kinetic analysis of the results of systematic mutations in the trinucleotide at the cleavage site of hammerhead ribozymes. Biochemistry 34:3649–3654

Shiota M, Sano M, Miyagishi M, Taira K (2004) Ribozymes: applications to functional analysis and gene discovery. J Biochem (Tokyo) 136:133–147

Shirane D, Sugao K, Namiki S, Tanabe M, Iino M, Hirose K (2004) Enzymatic production of RNAi libraries from cDNAs. Nat Genet 36:190–196

Shtivelman E, Lifshitz B, Gale RP, Roe BA, Canaani E (1986) Alternative splicing of RNAs transcribed from the human abl gene and from the bcr-abl fused gene. Cell 47:277–284

Siolas D, Lerner C, Burchard J, Ge W, Linsley PS, Paddison PJ, Hannon GJ, Cleary MA (2005) Synthetic shRNAs as potent RNAi triggers. Nat Biotechnol 23:227–231

Sioud M (2004) Ribozyme- and siRNA-mediated mRNA degradation: a general introduction. Methods Mol Biol 252:1–8

Sledz CA, Holko M, de Veer MJ, Silverman RH, Williams BR (2003) Activation of the interferon system by short-interfering RNAs. Nat Cell Biol 5:834–839

Stein CA (1999) Two problems in antisense biotechnology: in vitro delivery and the design of antisense experiments. Biochim Biophys Acta 1489:45–52

Stojanovic MN, Kolpashchikov DM (2004) Modular aptameric sensors. J Am Chem Soc 126:9266–9270

Suyama E, Kawasaki H, Kasaoka T, Taira K (2003a) Identification of genes responsible for cell migration by a library of randomized ribozymes. Cancer Res 63:119–124

Suyama E, Kawasaki H, Nakajima M, Taira K (2003b) Identification of genes involved in cell invasion by using a library of randomized hybrid ribozymes. Proc Natl Acad Sci U S A 100:5616–5621

Suyama E, Wadhwa R, Kaur K, Miyagishi M, Kaul SC, Kawasaki H, Taira K (2004a) Identification of metastasis-related genes in a mouse model using a library of randomized ribozymes. J Biol Chem 279:38083–38086

Suyama E, Wadhwa R, Kawasaki H, Yaguchi T, Kaul SC, Nakajima M, Taira K (2004b) LIM kinase-2 targeting as a possible anti-metastasis therapy. J Gene Med 6:357–363

Suzumura K, Takagi Y, Orita M, Taira K (2004) NMR-based reappraisal of the coordination of a metal ion at the pro-Rp oxygen of the A9/G10.1 site in a hammerhead ribozyme. J Am Chem Soc 126:15504–15511

Takagi Y, Inoue A, Taira K (2004) Analysis on a cooperative pathway involving multiple cations in hammerhead reactions. J Am Chem Soc 126:12856–12864

Tanabe T, Kuwabara T, Warashina M, Tani K, Taira K, Asano S (2000a) Oncogene inactivation in a mouse model. Nature 406:473–474

Tanabe T, Takata I, Kuwabara T, Warashina M, Kawasaki H, Tani K, Ohta S, Asano S, Taira K (2000b) Maxizymes, novel allosterically controllable ribozymes, can be designed to cleave various substrates. Biomacromolecules 1:108–117

Tanaka Y, Kasai Y, Mochizuki S, Wakisaka A, Morita EH, Kojima C, Toyozawa A, Kondo Y, Taki M, Takagi Y, Inoue A, Yamasaki K, Taira K (2004) Nature of the chemical bond formed with the structural metal ion at the A9/G10.1 motif derived from hammerhead ribozymes. J Am Chem Soc 126:744–752

Tang H, Wong-Staal F (2000) Specific interaction between RNA helicase A and Tap, two cellular proteins that bind to the constitutive transport element of type D retrovirus. J Biol Chem 275:32694–32700

Tang H, Gaietta GM, Fischer WH, Ellisman MH, Wong-Staal F (1997) A cellular cofactor for the constitutive transport element of type D retrovirus. Science 276:1412–1415

Uhlenbeck OC (1987) A small catalytic oligoribonucleotide. Nature 328:596–600

Wadhwa R, Yaguchi T, Kaur K, Suyama E, Kawasaki H, Taira K, Kaul SC (2004) Use of a randomized hybrid ribozyme library for identification of genes involved in muscle differentiation. J Biol Chem 279:51622–51629

Waninger S, Kuhen K, Hu X, Chatterton JE, Wong-Staal F, Tang H (2004) Identification of cellular cofactors for human immunodeficiency virus replication via a ribozyme-based genomics approach. J Virol 78:12829–12837

Warashina M, Takagi Y, Stec WJ, Taira K (2000) Differences among mechanisms of ribozyme-catalyzed reactions. Curr Opin Biotechnol 11:354–362

Warashina M, Kuwabara T, Kato Y, Sano M, Taira K (2001) RNA-protein hybrid ribozymes that efficiently cleave any mRNA independently of the structure of the target RNA. Proc Natl Acad Sci U S A 98:5572–5577

Warashina M, Kuwabara T, Nakamatsu Y, Takagi Y, Kato Y, Taira K (2004) Analysis of the conserved P9-G10.1 metal-binding motif in hammerhead ribozymes with an extra nucleotide inserted between A9 and G10.1 residues. J Am Chem Soc 126:12291–12297

Welch PJ, Marcusson EG, Li QX, Beger C, Kruger M, Zhou C, Leavitt M, Wong-Staal F, Barber JR (2000) Identification and validation of a gene involved in anchorage-independent cell growth control using a library of randomized hairpin ribozymes. Genomics 66:274–283

Westberg C, Yang JP, Tang H, Reddy TR, Wong-Staal F (2000) A novel shuttle protein binds to RNA helicase A and activates the retroviral constitutive transport element. J Biol Chem 275:21396–21401

Yi R, Qin Y, Macara IG, Cullen BR (2003) Exportin-5 mediates the nuclear export of pre-microRNAs and short hairpin RNAs. Genes Dev 17:3011–3016

Yoshinari K, Miyagishi M, Taira K (2004) Effects on RNAi of the tight structure, sequence and position of the targeted region. Nucleic Acids Res 32:691–699

Zamore PD (2001) RNA interference: listening to the sound of silence. Nat Struct Biol 8:746–750

Zheng L, Liu J, Batalov S, Zhou D, Orth A, Ding S, Schultz PG (2004) An approach to genomewide screens of expressed small interfering RNAs in mammalian cells. Proc Natl Acad Sci U S A 101:135–140

Zhou DM, Taira K (1998) The hydrolysis of RNA: from theoretical calculations to the hammerhead ribozyme-mediated cleavage of RNA. Chem Rev 98:991–1026

Ribozymes and siRNAs: From Structure to Preclinical Applications

M. Sioud

Institute for Cancer Research, Department of Immunology, Molecular Medicine Group, The Norwegian Radium Hospital, Montebello, 0310 Oslo, Norway
mosioud@ulrik.uio.no

1	Introduction	224
2	Nuclease-Resistant Ribozymes: Structure–Function Relationships	225
2.1	Effects of Site-Specific Modifications at Position 2.1 upon Ribozyme Cleavage Activity	226
2.1.1	Analysis of the Ribozyme/Substrate Global Structure by Native Gel Electrophoresis	228
3	Small Interfering RNAs: Chemical Modifications	230
4	Hammerhead Ribozyme and siRNA Design: Basic Rules	231
5	Therapeutic Ribozymes and siRNAs	232
5.1	Signalling Pathways as Therapeutic Targets	234
5.1.1	The Ras–Raf Signalling Pathway	234
5.1.2	Protein Kinase C-α	235
5.1.3	Wnt Proteins and Their Receptors	236
6	Tumour–Stroma Interaction: A Novel Target in Cancer	237
7	Conclusions	239
	References	239

Abstract The discovery that nucleic acids mediated the inhibition of gene expression in a sequence-specific manner has provided the scientific community with a potentially important tool to analyse gene function and validate drug targets. Selective inhibition of gene expression by ribozymes and small interfering RNAs (siRNAs) is being explored for potential therapeutics against viral infections, inflammatory disorders, haematological diseases and cancer. In order to be used as pharmaceutical drugs, chemical modifications are necessary to increase their stability in vivo. However, such modifications should not affect either the ribozyme cleavage activity or the incorporation of the siRNAs into the RNA interference (RNAi) targeting complex and subsequent mRNA cleavage. To attain stability, ribozymes and siRNAs must also overcome several other problems, including accessibility to target messenger RNAs (mRNAs), efficient delivery to target cells and unwanted non-specific effects.

Keywords Hammerhead ribozymes · RNA interference · Small interfering RNAs · Tumour cells · Stroma cells

1
Introduction

The ability of RNA to function as a catalyst was first demonstrated for the self-splicing group I intron of *Tetrahymena thermophila* and the RNA moiety for RNAse P (Pan et al. 1993). At present, there are five major RNA catalytic motifs that are derived from naturally occurring ribozymes: hairpin, hammerhead, group I intron, ribonuclease P and hepatitis-δ virus ribozyme. Analysis of these ribozymes indicated that the hammerhead-type ribozyme is the simplest RNA enzyme due to its relatively small size and ability to be incorporated into antisense flanking arms (Haseloff and Gerlach 1988). This enzyme consists of a conserved catalytic core motif that is required for *in trans* cleavage of a phosphodiester bond within an RNA target. In addition to ribozymes, the recently discovered RNA interference (RNAi) pathway is emerging as a powerful genetic tool to silence gene expression in a sequence-specific manner (Fire et al. 1998). In this process, double-stranded (ds)RNA is processed by the cellular enzyme Dicer into small interfering (si)RNAs of 21–23 nucleotides. These siRNAs are then incorporated into a multicomponent nuclease complex known as the RNA-induced silencing complex (RISC), which recognizes the target messenger (m)RNAs for destruction.

In most mammalian systems, introducing longer dsRNAs (>30 bp) induces a potent antiviral response that results in generalized mRNA degradation and inhibition of protein synthesis (Sen 2001). However, recently it has been shown that chemically made siRNAs can induce specific gene silencing in a wide range of mammalian cell lines without causing the non-specific antiviral response (Elbashir et al. 2001). In contrast to siRNAs, hammerhead ribozymes function as molecular scissors to snip an mRNA in the complete absence of proteins. However, we have shown (along with others) that certain cellular proteins can increase hammerhead ribozyme catalysis and stability (Sioud 1994). By linking high-affinity binding RNA sequences for cellular proteins with helix-destabilizing activity, we have found that ribozyme stability and catalysis can be enhanced both in vitro and in vivo (Sioud and Jespersen 1996). The major attraction of this strategy was the use of cell biology. Indeed, proteins and naturally occurring nucleotides are made by the same cell. As mentioned above, siRNAs acquire effector activity only when they are recruited into the RNAi targeting complex RISC. After activation, the antisense strand guides the RISC to target mRNA sequences via base-pair interaction, resulting in mRNA cleavage by, most likely, Argonaute 2 protein (Eister and Tuschl 2004). Initially, RNAi was believed to be exquisitely specific; however, recent reports have indicated that non-specific effects can be induced by siRNAs, both at the level of mRNA and protein. When considering synthetic ribozymes and siRNAs as a novel class of therapeutic agents, the stability and the delivery of these molecules are the major problems. This chapter will focus on structure–function relationship following chemical modifications and the therapeutic applications of ribozymes and siRNAs (see also chapters by Schubert and Kurreck and Miyano-Kurosaki and Takaku).

2
Nuclease-Resistant Ribozymes: Structure–Function Relationships

When RNA enzymes are made in vitro, a variety of modifications can be introduced into the molecule to increase their half-life in biological fluids. Modified nucleotides were very useful for the determination of the functional groups required for the ribozyme cleavage activity (Pieken et al. 1991). Additionally, they were useful for probing nuclease cleavage sites. In this respect, early studies indicated that the 2′-hydroxyl groups of pyrimidines are the primary sites of ribonucleases. Thus, replacement of these nucleotides with their 2′-modified versions should increase the RNA stability. In this connection, the substitution of pyrimidines with 2′-fluoro, 2′-amino, 2′-O-methyl, 2′-O-allyl and/or 2′-C-allyl analogues have provided stable hammerhead ribozymes. However, chemical modifications that improve ribozyme stability without affecting cleavage activity must be chosen carefully (Koizumi and Ohtsuka 1991; Heidenreich et al. 1994). A systemic study of selectively modified hammerhead ribozyme has resulted in the identification of a generic catalytically active and nuclease-resistant ribozyme containing a consensus '5-ribose sequences'. Specifically, ribozymes containing 2′-NH2 substitutions at U_4 and U_7, or 2′-C-allyl substitutions at U4, exhibited enhanced serum stability and retained most of their catalytic activity when compared to the unmodified ribozyme (Beigelman et al. 1995). In contrast to site-specific modifications, uniform modifications abolished ribozyme cleavage activity. Notably, the cleavage activity of ribozyme containing uniform 2′-fluoro pyrimidines was significantly improved by replacing U_4 and U_7 with 2′-amino uridines (Beigelman et al. 1995). Based upon these findings, site-specific modifications at positions U_4 and U_7 combined with other chemical modifications has led to the development of stable and in vivo active ribozymes such as those directed against the vascular endothelial growth factor receptor Flt-1 and stromelysin (Parry et al. 1999; Flory et al. 1996). Indeed, following intraarticular administration, nuclease-resistant ribozymes against stromelysin were taken up by cells in the synovial lining and inhibited the induction of stromelysin gene expression by interleukin-1α.

As mentioned above, uniform modifications (the use of a single type of modification such as amino or fluoro groups) resulted in severe inhibition of the hammerhead ribozyme cleavage activity. Since the cleavage activity of the ribozyme seems to depend largely upon the formation of the correct conformation, we reasoned that 2′-modifications near the cleavage site may perturb the backbone distortions at the three-helix junction necessary for the formation of the active conformation (Murray et al. 1998). Numerous structural studies have been performed with ribozymes modified at specific positions within the catalytic core, because it was believed to be mainly responsible for the formation of the ribozyme active structure. However, we have found that the replacement of the 2′-hydroxyl group at position 2.1, which is not a part of the catalytic core, by an amino group inhibited Mg^{2+}-promoted ribozyme cleavage (Sioud and Leirdal 1999). As illustrated in Fig. 1a, no significant cleavage activity was obtained with the modified ribozyme [2′-NH_2 (2.1)-URz] as compared to its unmodified version (Wt-URz). Interestingly, replacement of the Mg^{2+} with Mn^{2+} significantly restored ribozyme cleavage activity (Fig. 1b). The partial rescue effect of Mn^{2+} on the modified ribozyme may originate from its capacity to co-ordinate to the -NH_2 group or from its ability to alter the global structure of the ribozyme/substrate complexes and therefore the elimination of any potential steric hindrance introduced by the amino group at position 2.1. Additionally, this Mn^{2+} rescue effect would further confirm the integrity of the designed ribozyme.

2.1
Effects of Site-Specific Modifications at Position 2.1 upon Ribozyme Cleavage Activity

To maintain a significant cleavage activity after 2′-amino modification, the hammerhead ribozymes were designed to contain low pyrimidine content (Sioud and Sørensen 1998), especially an unmodified pyrimidine or a purine at position 2.1, because the substitution of the 2′-hydroxyl group at this position by a 2′-amino group inhibited Mg^{2+}-promoted ribozyme cleavage (see Fig. 1). The observed inhibition effect could arise from either the interference with a hydrogen bonding and/or with the local conformation of the ribose at nucleotide 2.1 and thereby ribozyme structure. One method to address this question is to use different 2′-substituents that differ in their polarity. In this respect, three versions of a tumour necrosis factor (TNF)-α ribozyme in which the 2′-hydroxyl group of uridine at position 2.1 (Leirdal and Sioud 1999; Sioud and Leirdal 1999) was substituted with a 2′-amino, 2′-fluoro or 2′-deoxy group were chemically synthesized. These substitutions are expected to be informative for both hydrogen bonding and the conformation of the sugar ring (Guschlbauer and Jankowski 1980). The cleavage activity of the ribozymes was measured under multiple turnover conditions in the presence of 5 mM Mg^{2+}. As summarized in Fig. 1c, only the 2′-amino modification inhibited the ribozyme cleavagecleavage activity activity.

Notably, the analogues used differ significantly in their preference of the 3′-*endo* and the 2′-*endo* conformations. The 2′-amino nucleoside favours the 2′-*endo* conformation more than the 2′-deoxynucleosides. Similar to the 2′-hydroxyl group, the 2′-amino group can play a role as donor or acceptor of

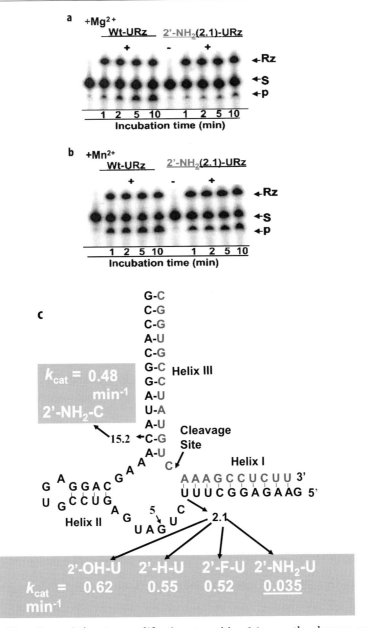

Fig. 1a–c The effects of 2′-amino modification at position 2.1 upon the cleavage activity of a hammerhead ribozyme. **a** A representative example of multiple turnover reactions of the wild-type ribozyme (Wt-URz) and the ribozyme with a single 2′-amino group at position 2.1 [2′-NH$_2$ (2.1)-URz] in the presence of 5 mM Mg^{2+}. **b** The cleavage reaction was performed in the presence of Mn^{2+}. **c** Summary of the ribozyme catalytic activity following various site-specific modifications at position 2.1. Primary and secondary structures of the ribozyme are shown annealed to its substrate. For ribozyme numbering see Hertel (1992)

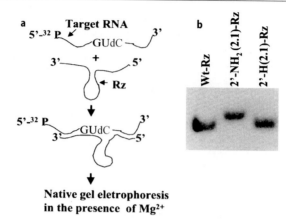

Fig. 2a,b Analysis of the ribozyme/substrate complexes by native gel electrophoresis. **a** Schematic representation of the experimental design. **b** Gel mobility of the ribozyme/substrate complexes on a 10% native gel in the presence of 5 mM Mg^{2+}

hydrogen bonding. In contrast, the 2′-fluoro group can only act as a hydrogen bond acceptor and favours the 3′-*endo* conformation. The deoxy group does not participate in hydrogen bonding (Guschlbauer and Jankowski 1980). From these properties, it would appear that the 2′-hydroxyl group at position 2.1 does not participate in hydrogen bonding since the ribozyme with 2′-deoxy uridine showed a comparable cleavage activity to that of the Wt-URz (Fig. 1c).

To investigate whether the presence of a 2′-amino group at position 15.2 would also affect ribozyme cleavage activity, a TNF-α ribozyme with a single 2′-amino at position 15.2 was chemically synthesized [2′-NH_2(15.2)-URz]. Based upon the ribozyme crystal structure, the 2′-hydroxyl group of position 15.2 forms a hydrogen bonding with the G_5 of the core sequence (Murray et al. 1998; Scott et al. 1995). In principle, the intrinsic role of the amino group as a potential hydrogen bond donor and acceptor would substitute the function of the 2′-hydroxyl group at position 15.2. The 2′-amino modification at position 15.2 did not inhibit ribozyme cleavage as illustrated in Fig. 1c, which may mean that the 2′-amino group also forms a hydrogen bond to G_5.

2.1.1
Analysis of the Ribozyme/Substrate Global Structure by Native Gel Electrophoresis

To probe the global structure of the ribozyme/substrate complex, a number of methods, including crystal structure analysis, examination of distances between helical termini by fluorescence resonance energy transfer and native gel electrophoresis, have been used (Lilley 2004). All crystal structures show that the three helices of the hammerhead ribozyme are arranged in a Y shape, as predicted by fluorescence studies and native gel electrophoresis. As reviewed by Lilley (2004), the ribozyme undergoes a two-stage folding process in the presence of Mg^{2+}. Furthermore, Bassi and colleagues have demonstrated that

$A_{14}G$ and G_8U mutations completely inhibit ion-induced ribozyme folding, whereas modifications at G_5 block the folding at the second stage (Bassi et al. 1999). Mg^{2+}-dependent conformational changes in the hammerhead ribozyme were also reported by the use of a fluorescent 2′-amino purine base as a reporter agent (Menger et al. 1996). From these studies, it appears that the folding process of hammerhead ribozyme/substrate can be blocked at specific stages by 2′-substitutions or mutations at particular positions within the catalytic core. As illustrated in Fig. 2a,b, the folding process can also be altered by 2′-amino modification at position 2.1 within helix I. The substitution of the 2′-fluoro for the 2′-hydroxyl group is a very conservative with regard to the sugar ring conformation (Lilley 2004). The ribozymes with either the 2′-fluoro or hydroxyl group at position 2.1 showed comparable cleavage activity. These results would indicate that the inhibition effect of the 2′-amino group at position 2.1 is related to the conformation of the sugar ring. However, since the energy barrier between 2′-*endo* and 3′-*endo* is rather low, the 2′-amino substituent at position 2.1 may create novel (artificial) interactions in the presence of Mg^{2+} that may cause aberrant folding. These novel interactions seem to be

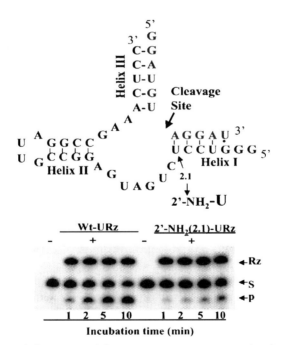

Fig. 3 The effects of 2′-amino modification at position 2.1 upon the cleavage activity of a standard hammerhead ribozyme. Primary and secondary structures of the ribozyme are shown annealed to its substrate (*upper panel*). A representative example of multiple turnover reactions of the wild-type ribozyme (Wt-URz) and the ribozyme with a single 2′-amino group at position 2.1 [2′-NH_2 (2.1)-URz] in the presence of 5 mM Mg^{2+} (*lower panel*)

position-dependent, because no significant effects on ribozyme folding were seen with a single 2′-amino substituent at position 15.2.

To see whether these findings could be applied to other hammerhead ribozymes, a second ribozyme was designed (Fig. 3). This ribozyme has been shown by Fedor and Uhlenbeck (1990) to have a high turnover number and has been further characterized by Eckstein's group (Olsen et al. 1991). Both the unmodified (Wt-URz) and the modified ribozyme [2′-NH$_2$(2.1)-URz] were chemically synthesized and their homogeneity was confirmed by polyacrylamide gel electrophoresis (PAGE) of the 5′-^{32}P-labelled ribozymes. The in vitro cleavage activity of the ribozymes was investigated under multiple turnover reactions and a representative example is shown in Fig. 3. The 2′-NH$_2$(2.1)-URz exhibited a significant reduction in cleavage activity as compared to the Wt-URz. In the presence of 5 mM Mg^{2+}, the Wt-URz and the 2′-NH$_2$(2.1)-URz cleaved the RNA substrate with a K_{cat} of 1.9 min^{-1} (±0.15) and 0.12 (±0.04) min^{-1}, respectively. The inhibition effect seen with this ribozyme is comparable to that obtained with TNF-α-Rz. Similarly to the modified TNF-α ribozyme shown in Fig. 1b, the cleavage activity of the 2′-NH$_2$ (2.1)-URz was substantially rescued when Mg^{2+} was replaced with Mn^{2+}.

Native gel electrophoresis provides a simple yet very powerful method to learn about the global structure of ribozymes, and should facilitate our understanding of the mechanisms of ribozyme folding and activity after chemical modifications. Just as a protein enzyme must fold into the conformation required for activity, ribozymes must also fold to create the local environment that supports catalysis. Therefore, following chemical modifications, the ribozyme/substrate must form a structure compatible with SN$_2$ cleavage reaction. The incorporation of the design rules discussed above is expected to enhance the probability for selecting stable ribozymes with sustained cleavage activity.

3
Small Interfering RNAs: Chemical Modifications

Similar to ribozymes, two main strategies are being explored for delivery (Hannon and Rossi 2004; Sioud 2004). The first strategy focuses on the use of synthetic siRNAs. Although these synthetic siRNAs are expected to be protected from single-stranded specific endonuclease, their therapeutic applications require increased stability in biological fluids. In principle, the variety of chemical modifications that have been developed for antisense oligonucleotides and ribozymes can be incorporated into siRNAs, provided that they do not interfere with the incorporation of the siRNA into the RNAi silencing complex RISC, unwinding of the siRNA duplex by helicases and subsequent cleavage of the target mRNA. Recently, a wide range of chemical modifications that increase stability of siRNA while maintaining an efficient silencing activity has been described (Dorsett and Tuschl 2004). In particular, siRNAs that have been modified with 2′-fluoro (fluoro) pyrimidines exhibited enhanced activity in cell culture as compared to 2′-OH-containing siRNAs. Interestingly, 2′-F-modified siRNA also exhibited activity in mice delivered via the hydrodynamic transfection method (Layzer et al. 2004). Using oligonucleotides containing the T7 promoter sequence, several siRNAs with uniform 2-′modifications have been transcribed in vitro, and a 2′-fluoro pyrimidine-modified siRNA against mouse TNF-α exhibited in vitro and in vivo activity (M. Sioud, unpublished data).

Although some reports suggested modified siRNAs have enhanced activity in cell culture as compared to 2′-OH-containing siRNAs, a recent study indicated that once siRNAs are delivered into cells, modifications that enhance stability had no significant influence on siRNA activity or persistence (Layzer et al. 2004). These results support the notion that siRNA are stabilized by the RISC proteins. Indeed, a recent study revealed that single-stranded siRNAs have an extremely high affinity to Ago proteins even at high salt concentration (Eister and Tuschl 2004). Additionally, it has been shown that Fas gene expression can be inhibited by unmodified siRNAs for several days in vivo. Although further studies are needed, these observations indicate that, once in the cell, the unmodified siRNAs are stable and are not rapidly degraded.

4
Hammerhead Ribozyme and siRNA Design: Basic Rules

Hammerhead ribozymes and siRNAs represent a relatively new addition to antisense technology. These molecules interact with target RNAs via Watson-Crick base pairing of complementary sequences, and suppress expression of a target protein by initiating the specific cleavage of the complementary target mRNA. In principle, hammerhead ribozymes and siRNAs can be designed to cleave any mRNA whose sequences are known. The hammerhead ribozyme cleavage specificity is determined by its hybridizing antisense arms, which anneal to target in a complementary fashion. The cleavage site is a 5′-UH-3′ sequence where H is any nucleotide except G. A detailed kinetic analysis indicated that RNA sequences with GUC cleavage sites were cleaved most efficiently, with CUC and UUC coming next. Therefore, when an mRNA target site is chosen, GUC or CUC may be preferred. Another important feature of a hammerhead ribozyme is the length and base composition of its recognition sequence. This length is usually chosen to be 6 to 8 nucleotides on either side of the cleavage site to facilitate a complete cycle of recognition, catalysis and dissociation in the intracellular environment. This contrasts with in vitro functioning, where the cleavage step was found to be the rate-limiting step. However, within cells the binding step was found to be rate-limiting step (Bertrand et al. 1994; Sioud 1997). In principle, hammerhead ribozymes with longer antisense arms should function properly in vivo.

Regarding siRNA design, the examination of duplex length requirements in mammalian cells indicates that siRNA duplexes must have a minimal length of 19 nucleotides for good silencing efficiency (Eister and Tuschl 2004). The cleavage occurs 10 nucleotides downstream of the 5'-end of the siRNA antisense strand. Notably, 3' overhangs on synthetic siRNAs are not essential for RNAi in HeLa cells. Several groups reported on the issue of sequence specificity, and early reports have shown that single mutations within the centre of the siRNA duplex are more discriminating than mutations located at the 5'- and 3'-ends. Recent studies indicated that some mismatches can be tolerated, but these are dependent on their positions within the duplex and on siRNA base composition (Eister and Tuschl 2004).

Regarding target mRNA recognition, several factors have been suggested to play an important role, including mRNA structure and siRNA internal thermodynamic stability (Raynolds 2004). Based on the thermodynamic stability model, only the strand with the most loosely base paired 5'-end (A:U base pairs) is preferably incorporated into the RISC. Thus, position 19 of the sense and antisense strands of effective siRNAs must be A/U and G/C, respectively. This would facilitate unwinding and incorporation of the antisense strand into the RISC. In addition, a long G/C stretch extending from the 5' sense strand has been shown to prevent complete strand separation (Ui-Tei et al. 2004). Consequently, long G/C stretch should be avoided. Based upon several reports, the criteria for effective RNA interference are the following: low G+C content (30%–50%), absence of internal repeats or palindromes, low internal stability at the 5'-end antisense strand, presence of U at position 10 and high internal stability at the 5' sense strand. Although these rules are promising, we have found that there are several exceptions. For example, a siRNA targeting the mRNA sequence 5'-GAGAUGAUACCACC**UGAAA**-3' (low stability at the 3'-end) exhibited nearly threefold lower activity than a siRNA targeting the same mRNA but a different site sequence 5'-GAAGAUUUGCGCA**GUGGAC**-3' (high stability at the 3'-end; Patzke et al. 2004). A comparison between the results for oligonucleotides and siRNAs showed that targets for RNase H-dependent oligonucleotides are also targets for siRNAs (Vickers et al. 2003; Beale et al. 2003). Notably, placing the recognition site of an active siRNA into a structured mRNA region can abrogate the siRNA activity (Vickers et al. 2003). Collectively, these observations indicate that during siRNA design, several factors should be taken into consideration, including mRNA structure and siRNA base composition.

5
Therapeutic Ribozymes and siRNAs

Ribozymes and siRNA are being used increasingly in various in vitro and in vivo models, and are being explored as potential therapeutics against viral infections, cardiovascular diseases, inflammatory disorders, haematological diseases and cancer (Sioud 2004; Hannon and Rossi 2004). Additionally, a phase I gene transfer study for human immunodeficiency virus (HIV) has been conducted. This study employed a retroviral vector to deliver an anti-HIV ribozyme to $CD34^+$ haematopoietic progenitor cells (Boyd et al. 2004).

With their promise and high specificity, mRNA-targeted therapies are widely expected to make a significant contribution in cancer therapeutics (Fig. 4). Cancer is a complex disease that involves both genetic and epigenetic events. Initiated cells require alterations rendering then self-sufficient for growth, insensitive to growth-inhibitory signals and resistant to programs of terminal differentiation, senescence or apoptosis. Although various abnormalities were detected in cancer cells several years ago, functional genomics studies have now enormously increased our understanding of gene function (Carr et al. 2004). Based on these studies, it has become possible to interfere with a specific component of a particular signalling pathway with therapeutic intent. In addition to genetic changes, several studies underscore the importance of epi-

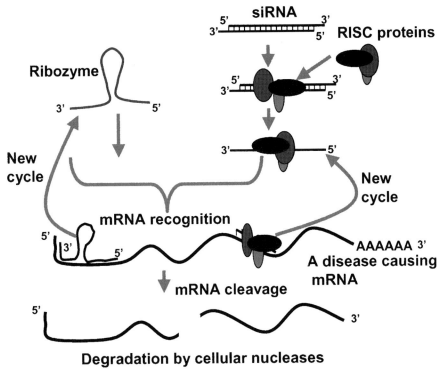

Fig. 4 A schematic of the mode of action of hammerhead ribozymes and siRNAs. Both molecules pair with mRNA that encodes the gene target of interest, leading to the cleavage of the mRNA and thereby preventing its translation into protein

genetic changes leading to altered intracellular signalling in cancer cells (Bode and Dong 2004). These findings led to the conclusion that cancer can now be considered as a disease of altered signalling pathways, which are attracting interest as novel therapeutic targets. The interplay between cancer and stroma cells represents a therapeutic target that must be explored as well (Mueller and Fusenig 2004).

5.1
Signalling Pathways as Therapeutic Targets

5.1.1
The Ras–Raf Signalling Pathway

The four G protein members of the *ras* family of oncogenes (H-*ras*, R-*ras*, K-*ras* 4a and K-*ras* 4b) are central players in the signal transduction pathways controlling cell growth. Ras proteins are posttranslationally modified by isoprenoid lipids (Rebollo and Martinez 1999). For this, farnesyltransferase and geranylgeranyltransferase catalyse the covalent attachment of the farnesyl and geranylgeranyl groups, respectively, to the carboxyl-terminal cysteine of prenylated proteins. Prenylation is essential not only for membrane association but also for biologic activity of Ras proteins. Mutations in the genes encoding the different Ras proteins result in constitutive activation of the proteins, leading to abnormal cell growth and malignant transformation, and have been identified in more than 30% of human tumours (Pruit and Der 2001). Although a Ras mutation has been observed in only 5% of cases, overexpression of Ras proteins has been associated with a more aggressive type of breast cancer. Usually a single base change at codon 12 GGC to GUC or GUU creates a cleavage site for the hammerhead ribozyme. Since the genetic alterations of Ras proteins are known, it is possible to interfere with their constitutive activation. In this regard, a ribozyme against activated H-*ras* targeting codon 12 has been shown to discriminate between both the constitutively activated oncogene and its normal counterpart in transformed cells (Kashani-Sabet and Scanlon 1995). Similarly, targeting the mutated K-*ras* by a ribozyme reversed the tumour phenotype (Kijima et al. 1996). These examples illustrate the capacity of ribozymes to eliminate mutant oncogenes and spare their normal counterparts.

Notably, one of the main downstream targets for Ras proteins are Raf proteins, a family of serine threonine protein kinases. Of the three different Raf proteins (A-Raf, B-Raf and C-Raf), Raf-1 (A), a cytosolic protein serine-threonine kinase, plays a central role in the mitogen-activated protein kinase (MAPK) signalling cascade that has been implicated in many cancers. Impaired Raf-1 expression has been observed in a variety of neoplasia including breast, cervical, hepatocellular and small cell lung carcinomas (Pritchart and McMahon 1997). In addition to its activation by Ras, Raf-1 is also activated independently of Ras by Bcl-2 and protein kinase C (PKC)-α.

Juvenile myelomonocytic leukaemia (JMML) is an aggressive childhood disorder with few therapeutic options (Arico et al. 1997). A major feature of this disease is that the cytokines granulocyte-macrophage colony-stimulating factor (GM-CSF) and TNF-α promote proliferation and viability of JMML cells in vitro via the activation of Raf-1 kinase. Indeed, depletion of Raf-1 protein with a ribozyme induced substantial inhibition of JMML cell-colony formation (Iversen et al. 2002). When immunodeficient mice engrafted with JMML cells were treated continuously with the ribozyme via a peritoneal osmotic minipump for 4 weeks, a profound reduction in the JMML cell numbers in the recipient murine bone marrow was found (Iversen et al. 2002). Although in these experiments we have used a DNAzyme, the data indicated that cleaving Raf-1 mRNA by nucleic acids might hold promise as a clinical therapeutic in several diseases where this kinase is activated.

5.1.2
Protein Kinase C-α

PKC-α is a member of a family of cytoplasmic serine-threonine protein kinases involved signal transduction pathways that control cell proliferation. This family contains at least 12 isoenzymes that have been subdivided into three groups based upon their biochemical properties and sequence homologies. The classical isoenzymes (α, βI, βII, γ) require calcium, diacylglycerol and phosphatidylserine for activation, whereas the novel isoenzymes (δ, ε, η, θ, μ) are regulated by diacylglycerol and phosphatidylserine. The atypical isoenzymes (ζ, τ, λ) are regulated by phosphatidylserine independent of calcium or diacylglycerol (Parekh et al. 2000). Increased PKC-α levels and activation have been associated with a spectrum of malignancies, including, breast, brain, colon, lung, ovarian and melanoma.

A tumour of the central nervous system (CNS) is the most prevalent solid neoplasm of childhood and the second leading cancer-related cause of death in adults between the ages 15–34 years. The most frequent brain tumours are the astrocytomas, which can be divided into low grade, anaplastic and glioblastoma (Rao 2003). The infiltrative growth pattern of these tumours prevents complete curative neurosurgery. In addition, the traditional therapeutic modalities are ultimately ineffective at curing these tumours, because gliomas are resistant to irradiation, chemotherapy and immunotherapy (Avgeropoulos and Batchelor 1999). The difficulty in targeting this type of tumour prompted us to search for novel therapeutic concepts. Given the specificity of Watson-Crick base pairing, nucleic acid enzymes have the capacity to inhibit individual genes that are structurally related, such as isoenzymes. By introducing 2′-amino modification into a RNA ribozyme directed against the PKCα isoform, a nuclease-resistant PKC-α ribozyme was designed. Ribozyme treatment blocked glioma cell proliferation in vitro and tumour growth in syngeneic rats (Sioud and Sørensen 1998). In addition to being involved in tumour growth

(Dean et al. 1996), PKC-α isoenzyme was found to play an important role in the neoplastic progression of tumours and drug resistant (MDR) phenotype (Ways et al. 1995; Caponigro et al. 1997). Furthermore, PKC-α was found to activate Raf-1 (Sözeri et al. 1992).

In order to identify the PKC-α target substrates, we have used a human gene array of 588 key genes involved in DNA synthesis, apoptosis, cell–cell communication, intracellular signal-transduction pathways, cell-surface molecules and transcription factors. Downregulation of PKC-α by a siRNA inhibited the expression of several target proteins, in particular the p21 WAF1/CIP1 protein (Leirdal and Sioud 2004). Thus, in glioma cells it appears that the activation of PKC-α induces the expression of the p21 WAF1/CIP1, which in turn activates, rather inhibits, the cyclin-dependent kinases. The positive effects of p21 on G1 phase progression are largely due to its function as an assembly factor for active cyclin D/cdk complexes. Our findings should validate PKC-α as an attractive therapeutic target.

5.1.3
Wnt Proteins and Their Receptors

The Wnt signalling pathway is crucial in regulating cell fate during embryogenesis, cell proliferation in adults tissues and carcinogenesis. Wnt proteins released from or presented on the surface of signalling cells act on target cells by binding to the Frizzled (Fz) low density lipoprotein (LDL) receptor-related protein (LRP) complex at the cell surface (Logan and Nusse 2004). Activation of the Wnt receptors transduces a signal to several intracellular proteins, including Dishevelled, which via its association with axin prevents glycogen synthase kinase (GSK) 3β from phosphorylating critical protein substrates. Cytoplasmic β-catenin levels are normally kept low via continuous proteasome-mediated degradation, which is controlled by a complex containing GSK/APC/Axin.

When cells receive Wnt signals, the degradation pathway is inhibited, and consequently β-catenin accumulates in the cytoplasm and nucleus. Nuclear β-catenin interacts with transcription factors such as T cell-specific transcription factor (TCF) to affect transcription. Mutations and/or signals that promote constitutive activation of the Wnt signalling pathway lead to cancer. The best-known example of a disease involving a Wnt pathway mutation that produces tumours is familial adenomatous polyposis (FAP), an autosomal, dominantly inherited disease in which patients display hundreds or thousands of polyps in the colon and rectum (Nishisho et al. 1991). Mutations were also found in several cancers, including hepatocellular carcinoma (HCC) and sporadic colon cancer, indicating the uncoupling of normal β-catenin regulation from Wnt signalling control is an important event in the genesis of many cancers (Logan and Nusse 2004).

So far RNAi has emerged as a versatile strategy to study gene function and validate therapeutic targets. To translate this technology to medical use, an immediate challenge is to determine the efficacy of siRNAs in vivo. As a first step, we have evaluated the ability of liposomal carriers such as DOTAP (N-[1-(2,3-dioleoyloxy)propyl]-N,N,N-trimethylammonium methylsulphate) to deliver active siRNA in vivo. We have assessed the intravenous delivery and found that a substantial majority of siRNA molecules were localized around the vessels 6 h after intravenous injection via the tail vein (Sioud and Sørensen 2003). Interestingly, intravenous co-administration of an anti-green fluorescent protein (anti-GFP) siRNA and a plasmid-encoding GFP-inhibited GFP expression in various organs, such as the liver and spleen (Sørensen et al. 2003). Similarly, intraperitoneal delivery of anti-TNF-α siRNA inhibited TNF-α expression in vivo and delayed the onset of septic shock following LPS injection (Sørensen et al. 2003). Of 12 recently investigated siRNAs, a siRNA targeting the mRNA site 5′-CCAACGGCAUGGAUCUCAA-3′ exhibited the greatest protective effect (Sioud and Sørensen 2004). These early findings represent a proof of principle to apply siRNA technology to disease associated with infections and/or other diseases such as cancer. Recent in vitro studies indicated that siRNA can activate immune response. In contrast to human cells, however, we have found that chemically synthesized siRNAs can be used in mice to silence gene expression without triggering an immune response. Following tail vein injection, larger dsRNAs and LPS induced inflammatory cytokine responses but not siRNA as compared to control mice (Sioud and Sørensen 2003).

Because the Wnt pathway plays an important role in tumour cell proliferation, we investigated its involvement in neuroblastoma survival. Neuroblastomas are the most frequently occurring solid tumours in children under 5 years of age. Unfortunately, the full clinical spectrum of neuroblastomas includes very aggressive tumours, unresponsive to multi-modality treatment. Thus, novel approaches targeting molecular defects may be beneficial to neuroblastoma patients. Targeting Wnt-1 or its receptor Fz-2 with siRNAs suppressed growth of human neuroblastoma Xenografts in mice (R. Schafer et al., submitted). Thus, interfering with Wnt signalling pathway might be a beneficial modality in neuroblastoma.

6
Tumour–Stroma Interaction: A Novel Target in Cancer

In addition to defects in signalling pathways, the microenvironment of the tumour host interface also plays a proactive role during malignant disease progression, including the transition from carcinoma in situ to invasive cancer and metastasis (Mueller and Fusenig. 2004). On the host side, the stroma cells together with extracellular matrix components provide the microenvironment that is important for cancer cell growth, invasion and metastasis. In general, the stroma compartment contains a variety of cell types, including immune cells, muscle and fibroblast cells and vascular cells (De Wever and Mareel 2003). These cells are able to modify the phenotype of the tumour cells by direct cell-to-cell contacts, via soluble factors or by modification of extracellular matrix components similar to wound healing (Werner and Grose 2003). Furthermore, cancer cells themselves express molecules, either secreted or presented on cell surface, to interact with the surrounding stromal cells. These include basic fibroblast growth factor (bFGF), members of the vascular endothelial growth factor (VEGF), platelet-derived growth factor (PDGF), epidermal growth factor receptor (EGFR) binding proteins, cytokines, chemokine, chemokine receptors, colony-stimulating factors (CSF-1) and transforming growth factors-β (TGF-β). In addition to altering tumour growth, these factors induce angiogenesis, a crucial process for tumour growth. Angiogenesis promotes not only tumour growth, but also progression from a pre-malignant to a malignant and invasive tumour phenotype. Both VEGF and bFGF are highly expressed in malignant gliomas. Thus, we have investigated whether their inhibition would inhibit tumour growth in vivo. Targeting these factors by nuclease resistance ribozymes reduced glioma growth in vivo (Sioud and Leirdal 2000).

In the case of breast cancer, stroma macrophages play a unique role because they are recruited into mammary gland carcinomas. The macrophages probably enhance tumour progression through the paracrine pathway, involving the production of CSF-1 by tumour cells (Pollard 2004). Recent studies have demonstrated the contribution of the haematopoietic growth factors granulocytes colony-stimulating factor (G-CSF) and GM-CSF to tumour progression in a broad spectrum of human tumours, including gliomas (Mueller et al. 1999). These growth factors are more likely to contribute to in vivo tumour progression via the recruitment of monocytes, macrophages and neutrophils into the tumour vicinity. Interestingly, targeting either the expression of CSF-1 or its receptor *c-fms* by siRNA suppressed growth of human mammary tumour xenografts in mice by inhibiting the recruitment of tumour-associated macrophages with concomitant reduction in local production of VEGF (Aharinejad et al. 2004). Thus, the development of a new class of therapy targeting, for example, intracellular mediators that act at the tumour–host communication interface could be of benefit. Taken together, these observations indicate that the tumour microenvironment is a potential therapeutic target. The advantages to targeting stroma include the fact that these cells are not as genetically unstable as cancer cells and are therefore less likely to develop drug resistance (Folkman 2003). However, when developing stroma cell-based therapies, we need to keep in mind that targeting just one aspect of the tumour stroma, and doing this in patients with late-stage cancer, is not likely to be successful. Therefore, we should combine strategies that target both stroma and cancer cells. Successful clinical trials may thus require inhibition of more than one step in the signalling cascade. In this regard, combination of nucleic acid enzyme therapy with existent pharmacological drugs may provide improvements in cancer therapy. Therefore, sensitization of tumour cell by nucleic acid enzymes to chemotherapy should be investigated. In this regard, the PKC-α ribozyme and DNA enzymes sensitized a breast cancer cell line, SKBR3, to taxol treatment (M. Leirdal and M. Sioud, unpublished results).

7
Conclusions

Despite significant success in achieving gene inhibition by ribozymes and siRNAs in vitro and in animals, it is still a challenge to translate these results into effective therapy in humans. Given the history of the antisense field, caution may be appropriate at this stage. The cellular heterogeneity of most cancer cells indicates that they use different signalling pathways for proliferation, migration and metastasis. For clinical applications, therefore, combination therapy targeting several target genes and/or signalling pathways might be required for effective therapy. Additionally, ribozymes and siRNAs might provide additive or synergistic treatment benefits if used in combination with conventional therapeutics. Finally, let us not forget that target identification and validation using either ribozymes or siRNAs in animals should facilitate the development of conventional drugs, which is the task of pharmaceutical companies.

References

Aharinejad S, Paulus P, Sioud M, Hofmann M, Zins K, Schafer R, Stanley ER, Abraham D (2004) Colony-stimulating factor-1 blockade by antisense oligonucleotides and small interfering RNAs suppresses growth of human mammary tumor xenografts in mice. Cancer Res 64:5378–5384
Arico M, Biondi A, Pui CH (1997) Juvenile myelomonocytic leukemia. Blood 90:479–488
Avgeropoulos NG, Batchelor TT (1999) New treatment strategies for malignant gliomas. Oncologist 4:209–224
Bassi GS, Mollegaard NE, Murchie AI, Lilley DM (1999) RNA folding and misfolding of the hammerhead ribozyme. Biochemistry 38:3345–3354
Beale G, Hollins AJ, Benboubetra M, Sohail M, Fox SP, Benter I, Akhtar S (2003) Gene silencing nucleic acids designed by scanning arrays: anti-EGFR activity of siRNA, ribozyme and DNA enzymes targeting a single hybridization-accessible region using the same delivery system. J Drug Target 11:449–456
Beigelman L, McSwiggen JA, Draper KG, Gonzalez C, Jensen K, Karpeisky AM, Modak AS, Matulic-Adamic J, DiRenzo AB, Haeberli P (1995) Chemical modification of hammerhead ribozymes. J Biol Chem 270:25702–25708
Bertrand E, Pictet R, Grange T (1994) Can hammerhead ribozymes be efficient tools to inactivate gene function? Nucleic Acids Res 22:293–300
Bode AM, Dong Z (2004) Targeting signal transduction pathways by chemopreventive agents. Mutat Res 555:33–51

Boyd MP, Ngok FK, Todd AV et al (2004) Critical steps in the implementation of haematopoietic progenitor-cell gene therapy using ribozyme vectors. In: Sioud M (ed) Methods in molecular biology, ribozymes and siRNA protocols, 2nd edn, vol. 252. Humana Press, New Jersey, pp 599–616

Caponigro F, French RC, Kaye SB (1997) Protein kinase C: a worthwhile target for anticancer drugs? Anticancer Drugs 8:26–33

Carr KM, Rosenblatt K, Petricoin EF, Liotta LA (2004) Genomic and proteomic approaches for studying human cancer: prospects for true patient-tailored therapy. Annu Rev Genomics Hum Genet 1:134–140

De Wever O, Mareel M (2003) Role of tissue stroma in cancer cell invasion. J Pathol 200:429–447

Dean N, McKay R, Miraglia L, Howard R, Cooper S, Giddings J, Nicklin P, Meister L, Ziel R, Geiger T, Muller M, Fabbro D (1996) Inhibition of growth of human tumor cell lines in nude mice by an antisense of oligonucleotide inhibitor of protein kinase C-alpha expression. Cancer Res 56:3499–3507

Dorsett Y, Tuschl T (2004) siRNAs: applications in functional genomics and potential as therapeutics. Nature 3:318–329

Eister G, Tuschl T (2004) Mechanisms of gene silencing by double-stranded RNA. Nature 431:343–349

Elbashir SM, Harborth J, Lendeckel W, Yalcin A, Weber K, Tuschl T (2001) Duplexes of 21-nucleotide RNAs mediate RNA interference in cultured mammalian cells. Nature 411:494–498

Fedor MJ, Uhlenbeck OC (1990) Substrate sequence effects on "hammerhead" RNA catalytic efficiency. Proc Natl Acad Sci USA 87:1668–1672

Fire A, Xu S, Montgomery MK, Kostas SA, Driver SE, Mello CC (1998) Potent and specific genetic interference by double-stranded RNA in Caenorhabditis elegans. Nature 391:806–811

Flory CM, Pavco PA, Jarvis TC, Lesch ME, Wincott FE, Beigelman L, Hunt SW 3rd, Schrier DJ (1996) Nuclease-resistant ribozymes decrease stromelysin mRNA levels in rabbit synovium following exogenous delivery to the knee joint. Proc Natl Acad Sci USA 93:745–748

Folkman J (2003) Fundamental concepts of the angiogenic process. Curr Mol Med 3:643–651

Grunweller A, Wyszko E, Bieber B, Jahnel R, Erdmann VA, Kurreck J (2003) Comparison of different antisense strategies in mammalian cells using locked nucleic acids, 2'-O-methyl RNA, phosphorothioates and small interfering RNA. Nucleic Acids Res 31:3185–3193

Guschlbauer W, Jankowski K (1980) Nucleoside conformation is determined by the electronegativity of the sugar substituent. Nucleic Acids Res 8:1421–1433

Hannon GJ, Rossi JJ (2004) Unlocking the potential of the human genome with RNA interference. Nature 431:371–378

Haseloff J, Gerlach WL (1988) Simple RNA enzymes with new and highly specific endoribonuclease activity. Nature 334:585–591

Heidenreich O, Benseler F, Fahrenholz A, Eckstein F (1994) High activity and stability of hammerhead ribozymes containing 2'-modified pyrimidine nucleosides and phosphoro-thioates. J Biol Chem 269:2131–2138

Hertel KJ, Pardi A, Uhlenbeck OC et al. (1992) Numbering system for the hammerhead. Nucleic Acids Res 20:3252

Iversen PO, Emanuel PD, Sioud M (2002) Targeting Raf-1 gene expression by a DNA enzyme inhibits juvenile myelomonocytic leukemia cell growth. Blood 99:4147–4153

Kashani-Sabet M, Scanlon KJ (1995) Application of ribozymes to cancer therapy. Cancer Gene Ther 2:213–223

Kijima H, Bouffard DY, Scanlon KJ (1996) Ribozyme-mediated reversal of human pancreatic carcinoma phenotype. In: Ikehara S, Takaku F, Good RA (eds) Bone marrow transplantation-basic and clinical studies. Springer-Verlag, Tokyo, pp 153–163

Koizumi M, Ohtsuka E (1991) Effects of phosphorothioate and 2′-amino goups in hammerhead ribozymes on cleavage rates and Mg^{2+} binding. Biochemistry 30:5145–5150

Layzer JM, McCaffrey AP, Tanner A, Huang Z, Kay MA, Sullenger BA (2004) In vivo activity of nuclease-resistant siRNAs. RNA 10:766–771

Leirdal M, Sioud M (1999) High cleavage activity and stability of hammerhead ribozymes with a uniform 2′-amino pyrimidine modification. Biochem Biophys Res Commun 250:171–174

Leirdal M, Sioud M (2004) Gene-array analysis of glioma cells after treatment with an anti-PKCα siRNA. In: Sioud M (ed) Methods in molecular biology, ribozymes and siRNA protocols, 2nd edn, vol 252. Humana Press, Totowa, pp 493–500

Lilley DMJ (2004) Analysis of global conformational transitions in ribozymes. In Sioud M (ed) Methods in molecular biology, ribozymes and siRNA protocols. 2nd edn, vol 252. Humana Press, Totowa, pp 77–108

Logan CY, Nusse R (2004) The Wnt signaling pathway in development and disease. Annu Rev Cell Dev Biol 20:781–810

Menger M, Tuschl T, Eckstein F, Porschke D (1996) Mg^{2+}-dependent conformational changes in the hammerhead ribozyme. Biochemistry 35:14710–14716

Mueller MM, Fusenig NE (2004) Friends or foes—bipolar effects of the tumor stroma in cancer. Nat Rev Cancer 4:839–849

Mueller MM, Herold-Mende CC, Riede D, Lange M, Steiner HH, Fusenig NE (1999) Autocrine growth regulation by granulocyte colony-stimulating factor and granulocyte macrophage colony-stimulating factor in human gliomas with tumor progression. Am J Pathol 155:1557–1567

Murray JB, Terwey DP, Maloney L, Karpeisky A, Usman N, Beigelman L, Scott W (1998) The structural basis of hammerhead ribozyme self-cleavage. Cell 92:665–673

Nishisho I, Nakamura Y, Miyoshi Y, Miki Y, Ando H, Horii A, Koyama K, Utsuno J, Baba S, Hedge P (1991) Mutations of chromosome 5q21 genes in FAP and colorectal cancer patients. Science 253:665–669

Olsen DB, Benseler F, Aurup H, Pieken WA, Eckstein F (1991) Study of a hammerhead ribozyme containing 2′-modified adenosine residues. Biochemistry 30:9735–9741

Parekh DB, Ziegler W, Parker PJ (2000) Multiple pathways control protein kinase C phosphorylation. EMBO J 19:496–503

Parry TJ, Cushman C, Gallegos AM, Agrawal AB, Richardson M, Andrews LE, Maloney L, Mokler VR, Wincott FE, Pavco PA (1999) Bioactivity of anti-angiogenic ribozymes targeting Flt-1 and KDR mRNA. Nucleic Acids Res 27:2569–2577

Patzke S, Hauge H, Sioud M, Finne EF, Sivertsen EA, Delabie J, Stokke T, Aasheim HC (2004) Identification of a novel centrosome/microtubule-associated coiled-coil protein involved in cell-cycle progression and spindle organization. Oncogene 24:1159–1173

Pieken WA, Olsen DB, Benseler F, Aurup H, Eckstein F (1991) Kinetic characterization of ribonuclease-resistant 2′-modified hammerhead ribozymes. Science 253:314–317

Pritchard C, McMahon M (1997) Raf revealed in life-or-death decisions. Nat Genet 16:214–215

Pruitt K, Der CJ (2001) Ras and Rho regulation of the cell cycle and oncogenesis. Cancer Lett 171:1–10

Rao JS (2003) Molecular mechanisms of glioma invasiveness: the role of proteases. Nat Rev Cancer 3:489–501

Raynolds A, Leake D, Boese Q et al. (2004) Rational siRNA design for RNA interference. Nat Biotechnol 22:326–330

Rebollo A, Martinez CA (1999) Ras proteins: recent advances and new functions. Blood 94:2971–2980

Scott WG, Finch JT, Klug A (1995) The crystal structure of an all-RNA hammerhead ribozyme: a proposed mechanism for RNA catalytic cleavage. Cell 81:991–1002

Sen GC (2001) Viruses and interferons. Annu Rev Microbiol 55:255–281

Sioud M (1994) Interaction between tumor necrosis factor α ribozyme and cellular proteins. J Mol Biol 242:619–629

Sioud M (1997) Effects of variations in length of hammerhead ribozyme antisense arms upon the cleavage of longer RNA substrates. Nucleic Acids Res 25:333–338

Sioud M (2004) Therapeutic siRNAs. Trends Pharmacol Sci 25:22–28

Sioud M, Jespersen L (1996) Enhancement of hammerhead ribozyme catalysis by glyceraldeyde-3-phosphate dehydrogenase. J Mol Biol 257:775–789

Sioud M, Leirdal M (1999) Substitution of the 2'-hydroxyl group at position 2.1 by an amino group interferes with Mg^{2+} binding and efficient cleavage by hammerhead ribozyme. Biochem Biophys Res Commun 262:461–466

Sioud M, Leirdal M (2000) Therapeutic RNA and DNA enzymes. Biochem Pharmacol 60:1023–1026

Sioud M, Sørensen DR (1998) A nuclease-resistant protein kinase Cα ribozyme blocks glioma cell growth. Nat Biotechnol 16:556–561

Sioud M, Sørensen DR (2003) Cationic liposome-mediated delivery of siRNAs in adult mice. Biochem Biophys Res Commun 312:1220–1225

Sioud M, (2004) Therapeutic potential of siRNAs. Drugs of the future 29:1–10

Sørensen DR, Leirdal M, Sioud M (2003) Gene silencing by systemic delivery of synthetic siRNAs in adult mice. J Mol Biol 327:761–766

Sözeri O, Vollmer K, Liyanage M, Frith D, Kour G, Mark GE 3rd, Stabel S (1992) Activation of the c-Raf protein kinase by protein kinase C phosphorylation. Oncogene 7:2259–2262

Ui-Tei K, Naito Y, Takahashi F, Haraguchi T, Ohki-Hamazaki H, Juni A, Ueda R, Saigo K (2004) Guidelines for the selection of highly effective siRNA sequences for mammalian and chick RNA interference. Nucleic Acids Res 32:936–948

Vickers TA, Koo S, Bennett CF, Crooke ST, Dean NM, Baker BF (2003) Efficient reduction of target RNAs by small interfering RNA and RNAse H-dependent antisense agents. A comparative analysis. J Biol Chem 278:7108–7118

Ways DK, Kukoly CA, deVente J, Hooker JL, Bryant WO, Posekany KJ, Fletcher DJ, Cook PP, Parker PJ (1995) MCF-7 breast cancer cells transfected with protein kinase C-α exhibit altered expression of other protein kinase C isoforms and display a more aggressive neoplastic phenotype. J Clin Invest 95:1906–1915

Werner S, Grose R (2003) Regulation of wound healing by growth factors and cytokines. Physiol Rev 83:835–870

Strategies to Identify Potential Therapeutic Target Sites in RNA

M. Lützelberger · J. Kjems (✉)

Department of Molecular Biology, University of Aarhus, C. F. Møllers Allé 130,
8000 Aarhus C, Denmark
jk@mb.au.dk

1	Introduction	244
2	Selecting Antisense Oligonucleotides	246
2.1	Random Antisense Oligonucleotide Libraries for Target Selection	246
2.2	Genomic Libraries for Target Selection	248
2.3	DNA Arrays as a Tool for Target Selection	249
3	Selecting Functional Ribozymes	250
4	Selecting Functional siRNA	252
4.1	Construction of Genomic siRNA Libraries	253
5	Application in Disease Treatment	254
6	Conclusions	255
	References	255

Abstract Antisense agents are powerful tools to inhibit gene expression in a sequence-specific manner. They are used for functional genomics, as diagnostic tools and for therapeutic purposes. Three classes of antisense agents can be distinguished by their mode of action: single-stranded antisense oligodeoxynucleotides; catalytic active RNA/DNA such as ribozymes, DNA- or locked nucleic acid (LNA)zymes; and small interfering RNA molecules known as siRNA. The selection of target sites in highly structured RNA molecules is crucial for their successful application. This is a difficult task, since RNA is assembled into nucleoprotein complexes and forms stable secondary structures in vivo, rendering most of the molecule inaccessible to intermolecular base pairing with complementary nucleic acids. In this review, we discuss several selection strategies to identify potential target sites in RNA molecules. In particular, we focus on combinatorial library approaches that allow high throughput screening of sequences for the design of antisense agents.

Keywords Antisense oligodeoxynucleotides · siRNA · Ribozymes · DNA/LNAzymes · SELEX

1
Introduction

Antisense oligodeoxynucleotides (As-ODNs), which belong to the simplest class of antisense agents, have been found useful for a wide range of applications in biochemical sciences, biotechnology and molecular medicine, both for research and diagnostic purposes. They were the first agents used to 'knock down' gene expression in a sequence-specific manner, providing a simple alternative to more difficult methods of creating a 'gene knockout' in cells and organisms. Because of their apparent sequence specificity, they were considered potent drugs for the treatment of diseases.

The inhibition of gene expression by As-ODNs is based on the simple principle of complementary base pairing between the oligonucleotide and its target RNA. Once bound, gene expression is repressed through mechanisms such as ribonuclease H (RNAse H)-mediated RNA cleavage (Fig. 1a), destabilization of the target mRNA and block of translation (Good 2003a,b). Moreover, As-ODNs are capable of redirecting pre-messenger RNA splicing (Sazani and Kole 2003; Vacek et al. 2003).

Despite the seemingly simple idea of reducing translation by complementary base pairing, antisense technology has never fulfilled its initially anticipated promise as a therapeutic tool. The major obstacles are generally believed to be poor intracellular delivery, the in vivo stability and toxicity of As-ODNs, and the lack of target accessibility. The latter problem is caused by the formation of stable secondary structures and assembly of the messenger (m)RNA into ribonucleoprotein complexes. More recently, the application of small interfering (si)RNAs has been developed as a highly potent approach to inhibit gene expression (Elbashir et al. 2001; McManus and Sharp 2002; Pickford and Cogoni 2003). Although it was initially suggested that siRNA applications are less sensitive to RNA structure in the target, it was recently demonstrated that the efficiency of RNA interference (RNAi)-mediated 'knock down' is also influenced by RNA structure (Kretschmer-Kazemi Far and Sczakiel 2003; Vickers et al. 2003; Westerhout et al. 2005).

Thus, it is well recognized that the identification of accessible sites in target RNA is crucial for the successful application of antisense agents. Therefore, a number of theoretical and practical approaches to determine accessible regions in RNA have been developed. Several computational models have been employed predicting secondary structures of the target RNA and estimating hybridization rates between As-ODNs and RNA (Wang and Drlica 2004). Although these approaches have been applied successfully in several studies, they were often oversimplified, disregarding tertiary structure and the fact that RNAs are incorporated into ribonucleoprotein complexes and may obtain different structures in vivo. Another approach, which has been used extensively, is the mapping of nucleotides that are accessible to chemical reagents or nucleases (Ehresmann et al. 1987), often in combination with a method called

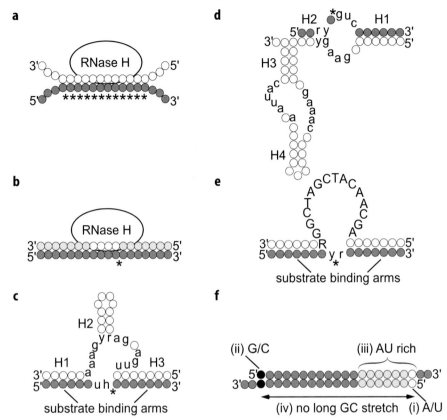

Fig. 1a–f Generalized structures of different antisense agents. *Circles* depict variable nucleotides. Cleavage sites in the substrates are indicated by *asterisks*. **a** As-ODN bound target RNA and RNase H. **b** Chimeric As-ODN (gap-mer) bound to its substrate and RNase H. *Light grey circles*, 2′-methoxyribonucleotides; *white circles*, DNA nucleotides. **c** Hammerhead ribozyme with target RNA. **d** Hairpin ribozyme in a *trans* configuration with target RNA. **e** '10–23' DNAzyme bound to its substrate RNA. **f** Small interfering RNA molecule (siRNA). For further details, see text. **a–e** *White* and *light-grey circles*, antisense agent; *dark-grey circles*, substrate RNA; *uppercase letters*, DNA; *lowercase letters*, RNA; Y, pyrimidine; R, purine nucleotides; H, 'A', 'C' or 'U'

'sequence-walking' (Monia et al. 1996; Peyman et al. 1995). In this approach, several oligonucleotides (usually around 10–100) targeted to various regions of an mRNA, are synthesized individually and their antisense activity or accessibility to the target site is measured. However, this method is costly and may not reveal a complete picture of all accessible sites on a target molecule since only 2%–5% of randomly chosen oligonucleotides usually have an effect on gene expression (Sohail and Southern 2000).

To approach this problem, more sophisticated methods that examine every sequence register of the target molecule have been developed by using libraries of random As-ODNs or genomic fragments. Here we compare these selection strategies and discuss their potential use in the context of different antisense agents.

2
Selecting Antisense Oligonucleotides

As-ODNs usually consist of 10–30 nt that are complementary to their target RNA. They can be used to inhibit gene expression by different mechanisms. Binding to the target can induce RNase H cleavage of the RNA and its subsequent degradation (Fig. 1a). Furthermore, As-ODNs that are targeted to the 5′ untranslated region (UTR) of an mRNA can prevent binding and assembly of the ribosome, thereby inhibiting translation (Good 2003b). Inhibition can be also achieved by steric hindrance of translational elongation, if the As-ODN is targeted to the coding region of an mRNA. However, the ribosome is highly processive, and only very tight-binding antisense agents, such as locked nucleic acids (LNA), can block progression of an assembled ribosome (Petersen and Wengel 2003).

2.1
Random Antisense Oligonucleotide Libraries for Target Selection

The involvement of ribonuclease H in the antisense-mediated degradation of mRNA in vivo led to the development of several procedures that make use of this enzyme to identify accessible target sites in mRNAs in vitro. Accessible sites determined by RNase H cleavage were found to be in good agreement with the antisense activity in cells (Matveeva et al. 1998). Furthermore, Vickers et al. (2003) found a significant degree of correlation between RNase H susceptible sites and active siRNAs.

RNase H is an endonuclease that specifically hydrolyses the phosphodiester bonds of RNA in DNA:RNA hybrids (Fig. 1a,b). In vitro, hybrids as short as 4 bp are sufficient for RNase H-mediated cleavage (Donis-Keller 1979). Thus, RNase H is often used in combination with random or semi-random oligonucleotide libraries (Birikh et al. 1997a). Random libraries usually consist of a pool of 10^6–10^{12} different oligonucleotides of a defined length, which can be between 10 and 20 nt. For selection, the target RNA is transcribed in vitro, end-labelled and mixed with the library. Oligonucleotides that are complementary to accessible sites form hybrids with the RNA and induce cleavage at places where base pairing occurs (Fig. 1a). To address the problem of protein binding to the target RNA, the mapping can be performed in the presence of cell-free extracts (Chen et al. 1996; Gee et al. 1998; Minshull and Hunt 1986; Scherr and Rossi 1998). Analysis of the cleaved products is usually carried out by gel electrophoresis.

Several problems are inherent to this methodology and the similar approach employing ribozyme libraries (see Sect. 3). For example, the presence of a highly accessible site close to the labelled terminus of the molecule may obscure other, potentially more accessible sites located more distal. Moreover, the precise cleavage site is often difficult to determine, since RNase H cleavage may occur at more than one position within the RNA:DNA hybrid (Fig. 1a). This can be important because shifting the register just a few nucleotides may result in significantly different antisense activity. An efficient way to address this problem is the use of 'gap-mers' (Fig. 1b), which are chimeric oligonucleotides consisting of a DNA core flanked by $2'$-methoxyribonucleotides (Ho et al. 1996). The DNA core, which may be only 4 nt long, restricts cleavage to a site-specific position (Inoue et al. 1988; Shibahara et al. 1987).

Another method, which is not dependent on RNase H cleavage of the target RNA, has been described by Allawi et al. (2001). It uses random oligonucleotides to prime the extension of complementary (c)DNA molecules using reverse transcriptase. The extension products are subsequently PCR amplified and their length is determined by gel electrophoresis with single nucleotide resolution to identify the sites where extension occurred. These priming sites were found in good correlation with sites susceptible to RNase H cleavage (Allawi et al. 2001).

A method solely based on hybridization of a random oligonucleotide library to the target RNA has been developed by Zhang et al. (2003). The target mRNA was biotinylated and conjugated to paramagnetic beads. After incubation with the library, the beads were washed, and the oligonucleotides specifically bound to the mRNA were eluted, cloned and sequenced. Similarly, gel-based methods to fractionate bound from non-bound oligonucleotides have been applied, but due to the limited resolving power of the acrylamide gels used, long RNAs must be mapped in fragments of roughly 0.5 kb in length.

Common to all these methods is the problem in finding suitable hybridization conditions, mild enough not to disrupt the authentic secondary structure of the mRNA, but sufficiently strong for selection of oligonucleotides that match their target completely.

In addition to the accessibility of the target molecule, binding affinity of an antisense oligonucleotide is also an important parameter. Stull et al. (1996) used a gel-shift binding assay to determine the dissociation constants for each of 37 different As-ODNs. An alternative approach to select As-ODNs with a high binding affinity is the iterative selection and amplification of sequences, termed SELEX (the acronym derives from systematic evolution of ligands by exponential enrichment) (Tuerk and Gold 1990). This method requires that the random nucleotides be flanked by fixed sequences to allow their amplification by PCR. At the end of the selection procedure, typically after 7–15 rounds, the winning candidates are cloned and sequenced. The advantage of this method is that the selection conditions can be strengthened after each round (Dausse et al. 2005; Pan et al. 2001). The frequency by which a particular site is selected may reveal, although indirectly, information about the strength of the interaction. A more direct approach for the measurement of binding affinities is the use of DNA arrays, which is discussed in Sect. 2.3.

2.2
Genomic Libraries for Target Selection

The high complexity of random oligonucleotide libraries and their high toxicity in cell-based assays make their use in many applications impractical. To address this problem, genomic libraries created from the target molecule have been used for site selection (Matveeva et al. 1997; Singer et al. 1997). They are constructed by random degradation or amplification of the target molecule and subsequent ligation of fixed linker sequences to both sides of the fragments. Libraries consisting of no more than 1 pmol DNA/RNA fragments are sufficiently large to contain the vast majority of all possible sequences of variable length occurring within the human genome. Plasmid DNA, genomic DNA and cDNA, as well as single-stranded DNA and RNA, may serve as input sequences. Their much lower complexity compared to completely random libraries can significantly simplify and accelerate the selection process. However, the difficulty in constructing such libraries is to restrict the fragment length to a fixed size that is suitable for the desired antisense application. This can be achieved by minimal size selection, as described in different studies, where a pool of fragments is generated by digestion with a cocktail of different restriction enzymes or by DNase I treatment (Brunel et al. 2001; Matveeva et al. 1997). A more elegant method developed by Jakobsen et al. (2004) solves this problem by ligating an upstream linker containing an *Mme*I restriction site to the random fragments, which were generated by sonication or DNase I treatment (Jakobsen et al. 2004; Luetzelberger et al. 2005). Cleavage with *Mme*I restricts the fragment size to 20 nt (Tucholski et al. 1995). After ligation to a downstream linker, the library can be transcribed and the RNA used as in the SELEX approach described above.

Another method called ROLL (random oligonucleotide ligated libraries), has been developed by Vlassov et al. (2004). It requires two oligonucleotides each consisting of a fixed primer sequence linked to a randomized 10-nt sequence at its 5'- or 3'-end. Annealing of a pair of these oligonucleotides to adjacent sites on the target DNA or RNA and their subsequent ligation by the addition of DNA ligase allows creating libraries of 20 nt in length. The length of the library fragments is controlled by the length of the random part of the oligonucleotides. However, sequencing of randomly selected clones of this library revealed that only a relatively small number of the analysed molecules had a near perfect 20-nt match, whereas about 50% had only 14–20 nt complementarity to the target. Furthermore, the authors observed that sequences derived from the ends of the target molecule were overrepresented in the library, which is a common problem when libraries are made from short linear target molecules. To avoid this problem, the molecules must be circularized prior to their random fragmentation or amplification (Jakobsen et al. 2004; Luetzelberger et al. 2005).

Another difficulty using libraries from short target molecules such as cDNA is that the complement of each fragment is also present in the library. During hybridization, both can form stable duplexes, thus reducing their availability for interaction with the target RNA. Due to their higher complexity, genomic libraries are less affected by dimer formation. However, it may still pose a problem after several rounds of SELEX selection, when the number of sequences in the library is reduced. The dimerization problem may be circumvented by constructing a directional library from single-stranded DNA, which can be obtained by asymmetric PCR (Gyllensten and Erlich 1988). In order to obtain single-stranded DNA molecules from such a PCR-generated library, a terminator primer can be used for its amplification (Williams and Bartel 1995). The 5'-end of this primer is made heavier by an extension with six C3 links, coupled to a stretch of 20 DNA nucleotides. *Taq* DNA polymerase cannot amplify the C3 region, thus producing a PCR product with two strands of unequal length. The strands can be subsequently purified by denaturing polyacrylamide gel electrophoresis.

Minor problems can be caused by the linker sequences that are flanking the genomic fragments, but in most cases they do not interfere with the selection process. For applications where additional sequences are not practical, Wen and Grey (2004) developed a method to construct primer-free libraries. However, such libraries cannot be transcribed into RNA, as they lack a promoter sequence.

In addition to their use for accessible target site selection in vitro, genomic libraries generated from random fragments can be applied to isolate genetic suppressor elements (GSEs) in vivo (Dunn et al. 1999; Gudkov et al. 1993; Holzmayer et al. 1992). For this purpose, the fragments are ligated into an expression vector and transferred into target cells. After the desired phenotype is selected, vectors are recovered and the fragments are sequenced. This technology has been used to isolate potent genetic inhibitors against human immunodeficiency virus (HIV)-1 (Dunn et al. 1999). In contrast to the antisense agents discussed above, GSEs may inhibit gene expression also by other mechanisms than the antisense effect, such as overexpression of regulatory gene products or a dominant-negative effect on wild-type protein function. Thus, after a GSE has been identified, its mode of action must be further characterized.

2.3
DNA Arrays as a Tool for Target Selection

The most elaborate technology to identify target sites both in terms of cost and equipment is the application of DNA arrays. With DNA arrays, accessible sites in the target mRNA can be detected by hybridization, but in contrast to random or genomic libraries, no cloning and sequencing of the selected oligonucleotides is needed. An advantage of scanning arrays is that they allow parallel measurement of the binding affinities for a large number of oligonucleotides to the target RNA. Conversely, they help in detecting unspecific binding.

In addition to these methods, oligonucleotide arrays can be designed comprising sets of oligonucleotides with variable length, which is important for the systematic analysis of RNA accessibility, as described by Milner et al. (1997). They used an array of 1,938 oligodeoxynucleotides ranging in length from monomers to 17-mers to measure the potential of each oligonucleotide to hybridize with rabbit beta-globin mRNA. For the selected oligonucleotides, the concentration required to inhibit translation by 50% was on average five times less than for randomly chosen oligonucleotides. Results obtained from other studies showed also good correlation between binding strength and antisense activity. For example, mapping of the 335-nt untranslated leader of HIV-1 revealed accessible sites in agreement with the RNA structure determined by chemical and enzymatic probing (Ooms et al. 2004).

In conclusion, microarrays that completely cover an mRNA sequence with oligonucleotides are an excellent tool for screening of accessible sites in a rapid way, but they have to be individually designed and impose very high costs, which in turn have prevented this method from being widely used.

3
Selecting Functional Ribozymes

The second type of antisense agents that have been used to map accessible sites in RNA belongs to the group of catalytic active nucleic acids such as ribozymes, DNAzymes or LNAzymes (Fig. 1c–e). Ribozymes have the advantage that they can be delivered to cells with plasmids or viral vectors and their expression can be controlled by fusion to a regulable promoter sequence. For example, Kawasaki et al. (2003a) made use of randomized hybrid-ribozyme libraries to target genes involved in the Fas-mediated pathway to apoptosis. A prerequisite of such an approach is the existence of a cell-based assay system, which allows identification of the desired phenotype. However, the use of ribozymes is often limited, not only by the accessibility of their target site, but also due to the sequence requirements of their catalytic core.

Hammerhead ribozymes are one of the smallest types of catalytic RNA (Fig. 1c). They are composed of approximately 40 nt and are able to induce site-specific cleavage of phosphodiester-bonds in RNA (Birikh et al. 1997b). Formally, they can be divided into two parts, a conserved stem-loop forming the catalytic core and flanking sequences that are reverse complementary to sequences surrounding their target site. The flanking sequences confer specificity and generally contain 14–16 nt. Using complementary base pairing, the hammerhead ribozyme can be designed to cleave any target RNA that contains a 'UH' sequence motif (where H is U, C or A; Fig. 1c; Ruffner et al. 1990).

As for As-ODNs, several strategies have been applied to select accessible target sites for ribozymes. Lieber and Strauss (1995) constructed a library of hammerhead ribozymes consisting of 13 random nucleotides on both sides of the stable stem-loop structure. These library ribozymes were targeted to a preselected 'UH' motif and allowed screening of accessible sites in the target-RNA molecule. The selected ribozymes were able to repress synthesis of human growth hormone RNA in a cellular assay by more than 99% (Lieber and Strauss 1995). An alternative approach finding optimal hammerhead ribozyme cleavage sites has been developed by Pan et al. (2001). They developed a SELEX-based strategy to select 'guide RNA' sequences that lack the stable stem-loop structure, but still contain a fixed sequence resembling the 'UH' motif. Then ribozymes were designed, based on the obtained sequences after several selection rounds.

A related strategy, based on the selection of 'external guide sequences' (EGS) has been applied by Kilani et al. (2000). To inhibit expression of herpes simplex virus (HSV)-1 and human cytomegalovirus (HCMV) mRNA in vivo, they used the M1 ribozyme, derived from the catalytic RNA subunit of RNase P (Trang et al. 2000). The M1 ribozyme cleaves an RNA helix that resembles its natural transfer (t)RNA substrate. When covalently linked to an EGS, M1 RNA can be engineered into a sequence-specific endonuclease, the M1GS RNA, which cleaves any target RNA that can form base pairs with the guide sequence. To date only selection strategies to find guide sequences for RNase P or M1GS RNA variants with high catalytic activity have been applied, but its potential utility for mapping accessible target sites was already proposed (Raj and Liu 2003, 2004).

Compared to the selection strategies described above, target site selection relying upon catalytic action of a self-cleaving ribozyme library seems to be more practical, since it allows optimizing ribozyme activity at the same time. In the approach invented by Barroso-DelJesus and Berzal-Herranz (2001), this was achieved by splitting a hairpin ribozyme (Fig. 1d) into two halves. The 5'-half of the ribozyme, containing the randomized substrate-binding region, was attached to the 3'-end of the target gene and fused to a T7 promoter sequence to allow in vitro transcription. Incubation of RNA transcribed from this template, together with the in vitro transcribed 3'-half of the ribozyme, yielded a library of self-cleaving RNA molecules. The cleavage products of this library were reverse transcribed and reused for another round of selection.

Another type of antisense agent, similar to ribozymes, is catalytic active DNA or LNA, known as DNAzymes and LNAzymes (Fig. 1e). In contrast to ribozymes, they can only be delivered exogenously. They are often chemically modified to increase their stability and activity in vivo, which makes them a useful substitute for ribozymes that are often unstable in the cellular environment. However, they have significant slower catalytic properties than ribozymes. Since DNAzymes do not exist in nature, they have been developed by in vitro evolution (Santoro and Joyce 1997). The best-studied enzyme is the '10–23' DNAzyme, which preferably cleaves in between pyrimidine-purine residues engulfed by the two binding arms (Fig. 1e). Similar techniques as for ribozymes have been applied to find accessible sites for DNAzymes in target mRNA (Cairns et al. 1999).

Taken together, ribozyme and DNAzyme libraries can be a useful tool to identify accessible sites in a RNA molecule both in vitro and in vivo. They are a good alternative to As-ODNs, but their application can be limited due to the sequence requirement and sensitivity to modification in the catalytic core.

4
Selecting Functional siRNA

RNAi induced by siRNA or short hairpin (sh)RNA is an important research tool in mammalian genetics. Together with sequence information gained by the human genome project, it has opened up the opportunity to silence practically any gene in the human genome. Since RNAi applications make use of a complex intracellular pathway that involves many different proteins (Matzke and Birchler 2005), their sequence requirements are much more stringent than those for the application of As-ODNs or ribozymes (Fig. 1f). For the design of siRNA in mammalian cells, four conditions must be met at the same time: (1) A/U at the $5'$-end of the antisense strand, (2) G/C at the $5'$-end of the sense strand, (3) A/U-richness in the $5'$-terminal third of the antisense strand and (4) the absence of any GC stretch over 9 bp in length (Ui-Tei et al. 2004). These rules appear to reflect the mechanism that determines which of the strands will be incorporated into the RNA-induced silencing complex (RISC) (Schwarz et al. 2003). Thus, in combination with the problem finding accessible target sites in mRNA, rational siRNA design can be a complicated task.

Although several computational methods have been developed to aid in the design of siRNA (Chalk et al. 2004; Naito et al. 2004; Yamada and Morishita 2004), it is often difficult to find siRNA that is both functional and free from 'off-target' effects, presumably because of similarity to other mRNA sequences. Recent studies using gene expression profiling have shown that silencing of non-targeted genes can occur even if they contain as few as 11 contiguous nucleotides of identity to the siRNA (Jackson et al. 2003; Jackson and Linsley 2004; Saxena et al. 2003; Scacheri et al. 2004). Conversely, single nucleotide polymorphisms (SNPs), occurring on average every 300–500 bases in the human genome, may result in variable RNAi efficiency when applied to different cell types. An attempt to circumvent this problem was the production of multiple siRNAs from large double-stranded RNA by recombinant human Dicer (Kawasaki et al. 2003b; Myers et al. 2003). However, this approach is not prac-

Strategies to Identify Potential Therapeutic Target Sites in RNA 253

tical for establishing the most effective siRNA molecule against a particular gene.

Recently, the first RNAi screens in mammalian cells have been accomplished (Berns et al. 2004; Paddison et al. 2004). The production of shRNA libraries for these screens is labour-intense and very expensive, since each shRNA was individually constructed from oligonucleotides, cloned into a vector and verified by sequencing. The library that has been constructed by Berns et al. (2004) contained 23,742 different shRNAs targeting 7,914 human genes. For each gene, no more than 3 to 9 siRNAs have been designed, imposing the risk that some genes are not sufficiently knocked down. Nonetheless, nearly 50% of the genes that were expected to target proteasomal proteins have been recovered when this library was evaluated in a screen for shRNAs that compromise proteasome function (Berns et al. 2004). Constructing libraries that target more sites in each gene could result in a higher sensitivity of these screening techniques and help to select more effective siRNAs. One attempt to achieve this goal is the construction of shRNA libraries from cDNA or genomic DNA, which at the same time avoids synthesis of individual shRNA clones.

4.1
Construction of Genomic siRNA Libraries

Similarly to the production of genomic libraries for As-ODN selection, genomic siRNA libraries can be made from a target cDNA or a pool of cDNAs. To date, two similar methods constructing such libraries have been published (Sen et al. 2004; Shirane et al. 2004). They are called REGS (restriction enzyme-generated siRNAs) and EPRIL (enzymatic production of RNAi libraries), and differ mainly in the way random fragments are generated and how the libraries are amplified. After random fragmentation of the cDNA using DNase I (EPRIL) or a mixture of frequently cutting restriction enzymes (REGS), the 100- to 200-bp fragments are ligated to a hairpin-shaped adapter oligonucleotide, which contains a recognition site for *Mme*I (Tucholski et al. 1995). After *Mme*I digestion, a second adapter is ligated to the 20-nt fragments, the libraries are amplified and ligated into a retroviral vector. Although both techniques are similar, REGS produced only an average of 34 shRNAs per kilobase of sequence, whereas EPRIL generated more than 200 unique siRNA constructs (Sen et al. 2004; Shirane et al. 2004). Thus, random fragmentation using DNase I resulted in a better coverage across the target gene than digestion with a mixture of different restriction enzymes.

Both methods have been evaluated by producing a library of siRNA constructs from a cDNA-encoding green fluorescent protein (GFP). Analysis of 262 nonredundant constructs revealed that about 56% of them had low RNAi efficiency, reducing GFP expression by only a factor of 1.5 (Shirane et al. 2004). Of the constructs, 30% reduced GFP expression by a factor 2 or more, and only a small percentage reduced expression by a factor of 8 or more. Comparable results silencing GFP expression were obtained by Sen et al. (2004), although they tested a significantly lower number of different shRNA clones. In both studies, siRNA constructs were also generated against endogenous genes, revealing similar results as for the GFP silencing experiments (Sen et al. 2004; Shirane et al. 2004).

Shirane et al. (2004) also developed a selection scheme that allows positive selection of efficient siRNA constructs in cells. For this purpose, a marker gene, which causes cell death by intracellular accumulation of a toxic derivative of ganciclovir, was fused to the target mRNA. Only cells transduced with an shRNA construct that is functional and effective were able to escape cell death. The shRNA was subsequently identified by analysing the expression constructs from surviving cells.

These authors also propose that siRNA libraries covering the entire transcriptome could be generated if cDNA libraries are used as the input sequence (Sen et al. 2004; Shirane et al. 2004). However, a major disadvantage of such an approach is that siRNA constructs with two different orientations are obtained. Taking into account that no more than 30%–40% of the constructs are effective, only 10%–15% of the siRNAs in such a library can be expected to mediate gene silencing. Nevertheless, genomic shRNA libraries seem to be more practical for RNAi-based screening techniques than rationally designed libraries due to their ease of construction and denser siRNA coverage per gene.

5
Application in Disease Treatment

Although the first publication about using As-ODNs in cells appeared more than two decades ago (Zamecnik and Stephenson 1978), antisense technology has not achieved the initially anticipated breakthrough as a therapeutic tool. To date only one antisense drug, Vitravene (Isis Pharmaceuticals, Carlsbad, CA, USA), which targets the cytomegalovirus *IE2* mRNA, has been approved by the Food and Drug Administration (FDA), and approximately 20 are in clinical trials (Crooke 2004). In addition to difficulties accessing the target RNA, problems with delivery, stability and toxicity have been hampering the development of antisense drugs. Recent progress to overcome these problems has been made by chemical modification of antisense agents. Chemical modifications were often used as part of a strategy to increase the stability of the antisense agent, thereby lowering its effective concentration and reducing its toxicity for in vivo applications (Kurreck 2003). For instance, Jakobsen et al. (M.R. Jakobsen et al., submitted) demonstrated that an LNA antisense oligonucleotide targeted to a pre-selected site efficiently blocked HIV-1 replication when applied in nanomolar scale, which is similar to the effective concentration generally reported for many siRNAs.

Furthermore, chemical modifications can significantly improve target accessibility, for instance by introduction of 2'-O-methyl RNA or LNA. Schubert et al. (2004) reported that incorporation of LNA monomers into the substrate recognition arms of a DNAzyme enabled degradation of previously inaccessible virus RNA at a high catalytic rate. In agreement with these studies, Jakobsen et al. (M.R. Jakobsen et al., submitted) reported a higher inhibitory effect on HIV-1 replication for LNAzymes than DNAzymes when targeted to the same accessible sites of the 5'-UTR. Thus, LNA nucleotides with increased target affinity are able to compete successfully with internal RNA structures.

6
Conclusions

Antisense technology has not yet reached its promised potential. The selection of target sites, which are accessible to antisense reagents, is a prerequisite for its application. Rational design based on structural information about the target RNA has generally proved to be an inefficient strategy. A more promising approach has been to use empirical methods based on random or genomic oligonucleotide libraries. It has been demonstrated that sites mapped in this way are more prone to antisense inhibition in vivo. These library-based methods are relatively simple to apply and are suitable for identifying experimental targets as well as for the development of therapeutic agents. Studies that compare these methods differ in their judgment, and it is difficult to decide impartially which methods are superior for a particular purpose. However, it seems obvious that the current techniques have to be refined and standardized in future to allow identification of more efficient antisense agents in a more effective way.

References

Allawi HT, Dong F, Ip HS, Neri BP, Lyamichev VI (2001) Mapping of RNA accessible sites by extension of random oligonucleotide libraries with reverse transcriptase. RNA 7:314–327

Barroso-DelJesus A, Berzal-Herranz A (2001) Selection of targets and the most efficient hairpin ribozymes for inactivation of mRNAs using a self-cleaving RNA library. EMBO Rep 2:1112–1118

Berns K, Hijmans EM, Mullenders J, Brummelkamp TR, Velds A, Heimerikx M, Kerkhoven RM, Madiredjo M, Nijkamp W, Weigelt B, Agami R, Ge W, Cavet G, Linsley PS, Beijersbergen RL, Bernards R (2004) A large-scale RNAi screen in human cells identifies new components of the p53 pathway. Nature 428:431–437

Birikh KR, Berlin YA, Soreq H, Eckstein F (1997a) Probing accessible sites for ribozymes on human acetylcholinesterase RNA. RNA 3:429–437

Birikh KR, Heaton PA, Eckstein F (1997b) The structure, function and application of the hammerhead ribozyme. Eur J Biochem 245:1–16

Brunel C, Ehresmann B, Ehresmann C, McKeown M (2001) Selection of genomic target RNAs by iterative screening. Bioorg Med Chem 9:2533–2541

Cairns MJ, Hopkins TM, Witherington C, Wang L, Sun LQ (1999) Target site selection for an RNA-cleaving catalytic DNA. Nat Biotechnol 17:480–486

Chalk AM, Wahlestedt C, Sonnhammer EL (2004) Improved and automated prediction of effective siRNA. Biochem Biophys Res Commun 319:264–274

Chen TZ, Lin SB, Wu JC, Choo KB, Au LC (1996) A method for screening antisense oligodeoxyribonucleotides effective for mRNA translation-arrest. J Biochem (Tokyo) 119:252–255

Crooke ST (2004) Progress in antisense technology. Annu Rev Med 55:61–95

Dausse E, Cazenave C, Rayner B, Toulme JJ (2005) In vitro selection procedures for identifying DNA and RNA aptamers targeted to nucleic acids and proteins. Methods Mol Biol 288:391–410

Donis-Keller H (1979) Site specific enzymatic cleavage of RNA. Nucleic Acids Res 7:179–192

Dunn SJ, Park SW, Sharma V, Raghu G, Simone JM, Tavassoli R, Young LM, Ortega MA, Pan CH, Alegre GJ, Roninson IB, Lipkina G, Dayn A, Holzmayer TA (1999) Isolation of efficient antivirals: genetic suppressor elements against HIV-1. Gene Ther 6:130–137

Ehresmann C, Baudin F, Mougel M, Romby P, Ebel JP, Ehresmann B (1987) Probing the structure of RNAs in solution. Nucleic Acids Res 15:9109–9128

Elbashir SM, Harborth J, Lendeckel W, Yalcin A, Weber K, Tuschl T (2001) Duplexes of 21-nucleotide RNAs mediate RNA interference in cultured mammalian cells. Nature 411:494–498

Gee JE, Robbins I, van der Laan AC, van Boom JH, Colombier C, Leng M, Raible AM, Nelson JS, Lebleu B (1998) Assessment of high-affinity hybridization, RNase H cleavage, and covalent linkage in translation arrest by antisense oligonucleotides. Antisense Nucleic Acid Drug Dev 8:103–111

Good L (2003a) Diverse antisense mechanisms and applications. Cell Mol Life Sci 60:823–824

Good L (2003b) Translation repression by antisense sequences. Cell Mol Life Sci 60:854–861

Gudkov AV, Zelnick CR, Kazarov AR, Thimmapaya R, Suttle DP, Beck WT, Roninson IB (1993) Isolation of genetic suppressor elements, inducing resistance to topoisomerase II-interactive cytotoxic drugs, from human topoisomerase II cDNA. Proc Natl Acad Sci U S A 90:3231–3235

Gyllensten UB, Erlich HA (1988) Generation of single-stranded DNA by the polymerase chain reaction and its application to direct sequencing of the HLA-DQA locus. Proc Natl Acad Sci U S A 85:7652–7656

Ho SP, Britton DH, Stone BA, Behrens DL, Leffet LM, Hobbs FW, Miller JA, Trainor GL (1996) Potent antisense oligonucleotides to the human multidrug resistance-1 mRNA are rationally selected by mapping RNA-accessible sites with oligonucleotide libraries. Nucleic Acids Res 24:1901–1907

Holzmayer TA, Pestov DG, Roninson IB (1992) Isolation of dominant negative mutants and inhibitory antisense RNA sequences by expression selection of random DNA fragments. Nucleic Acids Res 20:711–717

Inoue H, Hayase Y, Iwai S, Ohtsuka E (1988) Sequence-specific cleavage of RNA using chimeric DNA splints and RNase H. Nucleic Acids Symp Ser 135–138

Jackson AL, Linsley PS (2004) Noise amidst the silence: off-target effects of siRNAs? Trends Genet 20:521–524

Jackson AL, Bartz SR, Schelter J, Kobayashi SV, Burchard J, Mao M, Li B, Cavet G, Linsley PS (2003) Expression profiling reveals off-target gene regulation by RNAi. Nat Biotechnol 21:635–637

Jakobsen MR, Damgaard CK, Andersen ES, Podhajska A, Kjems J (2004) A genomic selection strategy to identify accessible and dimerization blocking targets in the 5′-UTR of HIV-1 RNA. Nucleic Acids Res 32:e67

Kawasaki H, Kuwabara T, Miyagishi M, Taira K (2003a) Identification of functional genes by libraries of ribozymes and siRNAs. Nucleic Acids Res Suppl 331–332

Kawasaki H, Suyama E, Iyo M, Taira K (2003b) siRNAs generated by recombinant human Dicer induce specific and significant but target site-independent gene silencing in human cells. Nucleic Acids Res 31:981–987

Kilani AF, Trang P, Jo S, Hsu A, Kim J, Nepomuceno E, Liou K, Liu F (2000) RNase P ribozymes selected in vitro to cleave a viral mRNA effectively inhibit its expression in cell culture. J Biol Chem 275:10611–10622

Kretschmer-Kazemi Far R, Sczakiel G (2003) The activity of siRNA in mammalian cells is related to structural target accessibility: a comparison with antisense oligonucleotides. Nucleic Acids Res 31:4417–4424

Kurreck J (2003) Antisense technologies. Improvement through novel chemical modifications. Eur J Biochem 270:1628–1644

Lieber A, Strauss M (1995) Selection of efficient cleavage sites in target RNAs by using a ribozyme expression library. Mol Cell Biol 15:540–551

Luetzelberger M, Jakobsen MR, Kjems J (2005) SELEX strategies to identify antisense and protein target sites in RNA or heterogeneous nuclear ribonucleoprotein complexes. In: Hartmann RK, Bindereif A, Schoen A, Westhof E (eds) Handbook of RNA biochemistry, vol 2. Wiley-VCH Verlag GmbH and Co., Weinheim, pp 878–894

Matveeva O, Felden B, Audlin S, Gesteland RF, Atkins JF (1997) A rapid in vitro method for obtaining RNA accessibility patterns for complementary DNA probes: correlation with an intracellular pattern and known RNA structures. Nucleic Acids Res 25:5010–5016

Matveeva O, Felden B, Tsodikov A, Johnston J, Monia BP, Atkins JF, Gesteland RF, Freier SM (1998) Prediction of antisense oligonucleotide efficacy by in vitro methods. Nat Biotechnol 16:1374–1375

Matzke MA, Birchler JA (2005) RNAi-mediated pathways in the nucleus. Nat Rev Genet 6:24–35

McManus MT, Sharp PA (2002) Gene silencing in mammals by small interfering RNAs. Nat Rev Genet 3:737–747

Milner N, Mir KU, Southern EM (1997) Selecting effective antisense reagents on combinatorial oligonucleotide arrays. Nat Biotechnol 15:537–541

Minshull J, Hunt T (1986) The use of single-stranded DNA and RNase H to promote quantitative 'hybrid arrest of translation' of mRNA/DNA hybrids in reticulocyte lysate cell-free translations. Nucleic Acids Res 14:6433–6451

Monia BP, Johnston JF, Geiger T, Muller M, Fabbro D (1996) Antitumor activity of a phosphorothioate antisense oligodeoxynucleotide targeted against C-raf kinase. Nat Med 2:668–675

Myers JW, Jones JT, Meyer T, Ferrell JE Jr (2003) Recombinant Dicer efficiently converts large dsRNAs into siRNAs suitable for gene silencing. Nat Biotechnol 21:324–328

Naito Y, Yamada T, Ui-Tei K, Morishita S, Saigo K (2004) siDirect: highly effective, target-specific siRNA design software for mammalian RNA interference. Nucleic Acids Res 32:W124–129

Ooms M, Verhoef K, Southern E, Huthoff H, Berkhout B (2004) Probing alternative foldings of the HIV-1 leader RNA by antisense oligonucleotide scanning arrays. Nucleic Acids Res 32:819–827

Paddison PJ, Silva JM, Conklin DS, Schlabach M, Li M, Aruleba S, Balija V, O'Shaughnessy A, Gnoj L, Scobie K, Chang K, Westbrook T, Cleary M, Sachidanandam R, McCombie WR, Elledge SJ, Hannon GJ (2004) A resource for large-scale RNA-interference-based screens in mammals. Nature 428:427–431

Pan WH, Devlin HF, Kelley C, Isom HC, Clawson GA (2001) A selection system for identifying accessible sites in target RNAs. RNA 7:610–621

Petersen M, Wengel J (2003) LNA: a versatile tool for therapeutics and genomics. Trends Biotechnol 21:74–81

Peyman A, Helsberg M, Kretzschmar G, Mag M, Grabley S, Uhlmann E (1995) Inhibition of viral growth by antisense oligonucleotides directed against the IE110 and the UL30 mRNA of herpes simplex virus type-1. Biol Chem Hoppe Seyler 376:195–198

Pickford AS, Cogoni C (2003) RNA-mediated gene silencing. Cell Mol Life Sci 60:871–882

Raj S, Liu F (2004) In vitro selection of external guide sequences for directing human RNase P to cleave a target mRNA. Methods Mol Biol 252:413–424

Raj SM, Liu F (2003) Engineering of RNase P ribozyme for gene-targeting applications. Gene 313:59–69

Ruffner DE, Stormo GD, Uhlenbeck OC (1990) Sequence requirements of the hammerhead RNA self-cleavage reaction. Biochemistry 29:10695–10702

Santoro SW, Joyce GF (1997) A general purpose RNA-cleaving DNA enzyme. Proc Natl Acad Sci U S A 94:4262–4266

Saxena S, Jonsson ZO, Dutta A (2003) Small RNAs with imperfect match to endogenous mRNA repress translation. Implications for off-target activity of small inhibitory RNA in mammalian cells. J Biol Chem 278:44312–44319

Sazani P, Kole R (2003) Modulation of alternative splicing by antisense oligonucleotides. Prog Mol Subcell Biol 31:217–239

Scacheri PC, Rozenblatt-Rosen O, Caplen NJ, Wolfsberg TG, Umayam L, Lee JC, Hughes CM, Shanmugam KS, Bhattacharjee A, Meyerson M, Collins FS (2004) Short interfering RNAs can induce unexpected and divergent changes in the levels of untargeted proteins in mammalian cells. Proc Natl Acad Sci U S A 101:1892–1897

Scherr M, Rossi JJ (1998) Rapid determination and quantitation of the accessibility to native RNAs by antisense oligodeoxynucleotides in murine cell extracts. Nucleic Acids Res 26:5079–5085

Schubert S, Furste JP, Werk D, Grunert HP, Zeichhardt H, Erdmann VA, Kurreck J (2004) Gaining target access for deoxyribozymes. J Mol Biol 339:355–363

Schwarz DS, Hutvagner G, Du T, Xu Z, Aronin N, Zamore PD (2003) Asymmetry in the assembly of the RNAi enzyme complex. Cell 115:199–208

Sen G, Wehrman TS, Myers JW, Blau HM (2004) Restriction enzyme-generated siRNA (REGS) vectors and libraries. Nat Genet 36:183–189

Shibahara S, Mukai S, Nishihara T, Inoue H, Ohtsuka E, Morisawa H (1987) Site-directed cleavage of RNA. Nucleic Acids Res 15:4403–4415

Shirane D, Sugao K, Namiki S, Tanabe M, Iino M, Hirose K (2004) Enzymatic production of RNAi libraries from cDNAs. Nat Genet 36:190–196

Singer BS, Shtatland T, Brown D, Gold L (1997) Libraries for genomic SELEX. Nucleic Acids Res 25:781–786

Sohail M, Southern EM (2000) Selecting optimal antisense reagents. Adv Drug Deliv Rev 44:23–34

Stull RA, Zon G, Szoka FC Jr (1996) An in vitro messenger RNA binding assay as a tool for identifying hybridization-competent antisense oligonucleotides. Antisense Nucleic Acid Drug Dev 6:221–228

Trang P, Kilani A, Kim J, Liu F (2000) A ribozyme derived from the catalytic subunit of RNase P from Escherichia coli is highly effective in inhibiting replication of herpes simplex virus 1. J Mol Biol 301:817–826

Tucholski J, Skowron PM, Podhajska AJ (1995) MmeI, a class-IIS restriction endonuclease: purification and characterization. Gene 157:87–92

Tuerk C, Gold L (1990) Systematic evolution of ligands by exponential enrichment: RNA ligands to bacteriophage T4 DNA polymerase. Science 249:505–510

Ui-Tei K, Naito Y, Takahashi F, Haraguchi T, Ohki-Hamazaki H, Juni A, Ueda R, Saigo K (2004) Guidelines for the selection of highly effective siRNA sequences for mammalian and chick RNA interference. Nucleic Acids Res 32:936–948

Vacek M, Sazani P, Kole R (2003) Antisense-mediated redirection of mRNA splicing. Cell Mol Life Sci 60:825–833

Vickers TA, Koo S, Bennett CF, Crooke ST, Dean NM, Baker BF (2003) Efficient reduction of target RNAs by small interfering RNA and RNase H-dependent antisense agents. A comparative analysis. J Biol Chem 278:7108–7118

Vlassov AV, Koval OA, Johnston BH, Kazakov SA (2004) ROLL: a method of preparation of gene-specific oligonucleotide libraries. Oligonucleotides 14:210–220

Wang JY, Drlica K (2004) Computational identification of antisense oligonucleotides that rapidly hybridize to RNA. Oligonucleotides 14:167–175

Wen JD, Gray DM (2004) Selection of genomic sequences that bind tightly to Ff gene 5 protein: primer-free genomic SELEX. Nucleic Acids Res 32:e182

Westerhout EM, Ooms M, Vink M, Das AT, Berkhout B (2005) HIV-1 can escape from RNA interference by evolving an alternative structure in its RNA genome. Nucleic Acids Res 33:796–804

Williams KP, Bartel DP (1995) PCR product with strands of unequal length. Nucleic Acids Res 23:4220–4221

Yamada T, Morishita S (2004) Accelerated off-target search algorithm for siRNA. Bioinformatics 21:1316–1324

Zamecnik PC, Stephenson ML (1978) Inhibition of Rous sarcoma virus replication and cell transformation by a specific oligodeoxynucleotide. Proc Natl Acad Sci U S A 75:280–284

Zhang HY, Mao J, Zhou D, Xu Y, Thonberg H, Liang Z, Wahlestedt C (2003) mRNA accessible site tagging (MAST): a novel high throughput method for selecting effective antisense oligonucleotides. Nucleic Acids Res 31:e72

Oligonucleotide-Based Antiviral Strategies

S. Schubert · J. Kurreck (✉)

Institute for Chemistry (Biochemistry), Free University Berlin, Thielallee 63,
14195, Berlin, Germany
jkurreck@chemie.fu-berlin.de

1	Introduction	262
2	Antisense Oligonucleotides	262
2.1	Development of Efficient Antisense Oligonucleotides	263
2.2	Antisense Oligonucleotides as Antiviral Agents	265
2.3	Antisense Oligonucleotides in Clinical Trials	266
3	Ribozymes	268
3.1	Development of Active Ribozymes	269
3.2	Cleavage of Viral RNA by Ribozymes	270
3.3	Ribozymes in Clinical Trials	271
4	RNA Interference	272
4.1	Designing Efficient siRNA Approaches	272
4.1.1	Selection of Efficient siRNAs and Suitable Targets	272
4.1.2	Delivery of siRNAs	273
4.1.3	Specificity of siRNAs	274
4.1.4	Viral Escape	275
4.2	Antiviral RNAi Approaches	276
4.2.1	Respiratory Diseases	276
4.2.2	Viral Hepatitis	278
4.2.3	HIV	279
4.2.4	Coxsackievirus B3	281
4.3	From Bench to Bedside	281
5	Conclusions	282
	References	282

Abstract In the age of extensive global traffic systems, the close neighborhood of man and livestock in some regions of the world, as well as inadequate prevention measures and medical care in poorer countries, greatly facilitates the emergence and dissemination of new virus strains. The appearance of avian influenza viruses that can infect humans, the spread of the severe acute respiratory syndrome (SARS) virus, and the unprecedented raging of human immunodeficiency virus (HIV) illustrate the threat of a global virus pandemic. In addition, viruses like hepatitis B and C claim more than one million lives every year for want of efficient therapy. Thus, new approaches to prevent virus propagation are urgently needed. Antisense strategies are considered a very attractive means of inhibiting viral replication, as oligonucleotides can be designed to interact with any viral RNA, provided its sequence is known. The ensuing targeted destruction of viral RNA should interfere with viral replication without entailing negative effects on ongoing cellular processes. In this review, we will give some examples of the employment of antisense oligonucleotides, ribozymes, and RNA interference strategies for antiviral purposes. Currently, in spite of encouraging results in preclinical studies, only a few antisense oligonucleotides and ribozymes have turned out to be efficient antiviral compounds in clinical trials. The advent of RNA interference now seems to be refueling hopes for decisive progress in the field of therapeutic employment of antisense strategies.

Keywords Antisense oligonucleotides · Antiviral agents · Ribozymes · RNA interference · RNAi

1
Introduction

In recent years, the prevalence of chronic infections with viruses such as human immunodeficiency virus (HIV) and the hepatitis B and C viruses (HBV and HCV, respectively) has been steadily increasing and new viruses like the severe acute respiratory syndrome (SARS) coronavirus have emerged. Thus, the demand for efficient antiviral treatments is obvious. Currently, approximately 40 small molecular compounds have been approved to treat viral infections, at least half of which are intended for patients with HIV infections (De Clercq 2004). The most prominent class of drugs used to inhibit viral propagation is the group of inhibitors of DNA or RNA synthesis, many of which are nucleoside analogs. These substances, however, are not fully specific for viral polymerases and cause severe side effects upon long-term treatment. For numerous viral infections, effective therapies are lacking altogether.

A strategy for the fast development of specific antiviral agents is therefore desirable. Antisense (AS) strategies employ oligonucleotides (ONs) complementary to a given target RNA. They offer the opportunity to fulfill demand for the development of an antiviral compound as soon as the sequence of a virus is known. In fact, the first AS study, published in 1978, describes the use of an AS ON to inhibit replication of Rous-Sarcoma virus (Zamecnik and Stephenson 1978). The following sections will deal with the use of AS ONs, ribozymes, and small interfering (si)RNAs as antiviral agents. Aptamers will not be addressed here, although there is no doubt about their usefulness to diagnose and treat viral infections (e.g., Darfeuille et al. 2004; De Beuckelaer et al. 1999; overview by McKnight and Heinz 2003). This type of ON is dealt with in the chapters by H.U. Göringer et al., M. Menger et al., M. Sprinzl et al., H. Ulrich, and A.K. Deisingh of the present volume.

2
Antisense Oligonucleotides

AS ONs are typically 15–20 nucleotides in length and bind to their cognate RNA via Watson-Crick base pairing. Since a DNA sequence of this length will statisti-

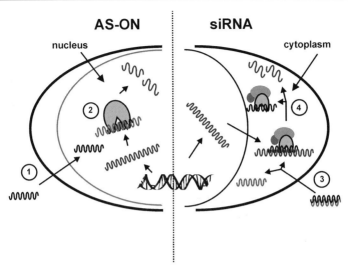

Fig. 1 Comparison of mechanisms of gene silencing by AS ONs and siRNAs. AS ONs (*left*) exert their effect predominantly in the nucleus, whereas siRNAs work mainly in the cytoplasm (*right*). *Left*: An AS ON is transported into the nucleus (*1*) and base-pairs with the complementary sequence of an mRNA. RNase H is recruited to the hybrid helix (*2*) and cleaves the RNA moiety. *Right*: An siRNA double helix reaches the cytoplasm (*3*). One of the strands is incorporated into a protein complex called RISC, while the other strand is discarded. The siRNA guides RISC to a complementary sequence on the target RNA. Upon binding, cleavage of the target molecule is induced (*4*). RISC can go on through multiple rounds of cleavage

cally occur only once in the human genome, the targeted RNA can be considered to be a highly specific receptor for the AS agent. AS ONs are known to act by two distinct mechanisms. (1) In the cell nucleus, the heteroduplex of a DNA ON bound to an RNA is recognized by RNase H, which cleaves the RNA moiety of the hybrid (Fig. 1). The RNA fragments are further degraded by exonucleases, whereas the ON is set free and can bind to new RNA molecules in a multiple turnover manner. (2) In the cytoplasm, AS ONs can disable messenger (m)RNAs and prevent protein synthesis by a steric blockade of the ribosome.

2.1
Development of Efficient Antisense Oligonucleotides

Theoretically, AS ONs can be directed against any region of the targeted RNA. In practice, however, long RNA molecules are known to form complex secondary and tertiary structures. Furthermore, various proteins bind to RNA molecules, precluding ONs from hybridizing. It is therefore necessary to select regions of the targeted RNA that are accessible to the AS ON. Various methods have been developed for this purpose and are summarized in the review by Sohail and Southern (2000).

Another major decision in the development of an AS ON is the choice of a suitable chemistry, i.e., nucleotide modification. Unprotected DNA ONs are degraded in blood serum within a few hours. Therefore, chemically modified building blocks are often used for AS ONs. The most widely employed DNA analogs are phosphorothioates in which one of the nonbridging oxygen atoms of the phosphodiester bond is replaced by a sulfur atom (Fig. 2). ONs consisting of phosphorothioates combine several advantages, including enhanced nuclease resistance and activation of RNase H cleavage (Eckstein 2000). Major disadvantages, however, are their decreased binding affinity to a complementary sequence as compared to an isosequential DNA molecule and their unintended propensity to interact with various proteins. Although binding to plasma proteins improves the pharmacokinetic profile by increasing serum half-life, interactions with other proteins may be disadvantageous and result in toxic side effects that have been observed when higher doses of phosphorothioates were applied (Levin 1999).

Due to the problems associated with phosphorothioates, other types of modifications have been developed. Nucleotides of the second generation carry a methyl or methoxy-ethyl group at the 2′ oxygen of the ribose (Fig. 2). In recent years, numerous nucleic acid analogs have been developed for applications in AS technology (Kurreck 2003). The types of alterations range from substitutions of functional groups of the ribose by fluoro- or amino-groups over bi- or tricycle nucleotides, to a complete replacement of the ribose-phosphate backbone by peptide bonds. Most of these nucleotides are less toxic and have a higher target affinity than phosphorothioates, but they lack the ability to induce RNase H cleavage of the complementary mRNA. Therefore, so-called gapmers have been developed, which consist of modified nucleotides at both ends to protect the ON from exonucleases and a central stretch of DNA or phosphorothioate monomers that is sufficient to activate RNase H.

Due to their favorable properties, locked nucleic acids (LNAs, Fig. 2) have increasingly been used for AS applications in recent years (Jepsen and Wengel 2004; and the chapter by S. Kauppinen et al., this volume). Gapmers consisting of LNA and DNA monomers in the center were found to exhibit desirable properties like improved nuclease stability and enhanced target affinity (Kurreck

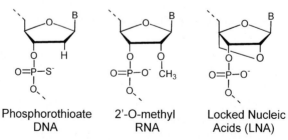

Fig. 2 Modified nucleotides that have been widely used for antisense approaches

et al. 2002) resulting in an almost 200-fold higher potency compared to an isosequential DNA ON (Grunweller et al. 2003).

An important hurdle that has to be overcome for successful AS applications is the cellular uptake of the ONs. DNAs are highly charged molecules that cannot cross hydrophobic membranes efficiently. Therefore, transfection agents are employed to facilitate entry of AS ONs into cells. The most widely used reagents are lipids with positively charged headgroups that neutralize the negative charge of the ONs. The ON–liposome complexes are thought to be taken up by endocytosis. Intracellular release of the ONs can be facilitated by the addition of helper lipids that interfere with the endosomal membrane. Further details about the use of cationic lipids as well as other types of transfection agents are described in a recent overview by Seksek and Bolard (2004).

2.2
Antisense Oligonucleotides as Antiviral Agents

A major advantage of AS ONs is their general applicability, because they can be directed against virtually any RNA of interest. Viruses with RNA genomes are particularly well suited to be targeted by AS ONs, since not only the mRNA but also the genomic RNA can be attacked and, at least theoretically, complete virus clearance can be achieved. In contrast, for DNA viruses or retroviruses with their proviral DNA stably integrated into the host genome, only mRNA (or newly synthesized genomic RNA in the latter case) can be targeted by AS ONs. Thus, only inhibition of virus spreading can be expected, and continuous treatment is required. Nevertheless, AS ONs have successfully been applied to inhibit many viruses of high medical relevance (McKnight and Heinz 2003). Due to space restraints, only a few recent examples can be discussed below to demonstrate the potential of AS ONs in treating virus replication.

Infections with HCV are a major health problem worldwide. Chronic infection with this plus-stranded RNA virus causes liver cirrhosis, liver failure, and hepatocellular carcinoma, often leading to the requirement of liver transplantation. Since current treatment of HCV is unsatisfactory, a need for new, specific anti-HCV drugs stands to reason. AS ONs have therefore widely been used with the intention of inhibiting HCV replication. The 5'-untranslated region (UTR) is one of the most highly conserved regions of the HCV genome and has most frequently been targeted with AS ONs. Since this region is strongly structured, intensive efforts were made to identify accessible target sites. The entire viral cycle of HCV is cytoplasmic and thus AS ONs do not necessarily need to be designed to activate RNase H, which is mainly located in the nucleus. Rather, it has been shown that AS ONs interfering with the assembly of a translation initiation complex on the internal ribosome entry site (IRES) inhibit translation of the viral polyprotein in cell-free translation assays and transfected hepatoma cell lines. Further details of the application of ON-based strategies to inhibit HCV are given in a review by Martinand-Mari et al. (2003).

With more than 40 million infected individuals worldwide, HIV is one of the most severe causes of infectious diseases. Although half of the substances that have been approved for the treatment of viral infections are intended to treat HIV infections, there is still an urgent need for new therapeutic approaches. Current drugs are not only too expensive for patients in poor countries; they also exert severe side effects upon long-term treatment and become ineffective due to the emergence of resistant mutants. Numerous efforts to prevent HIV replication with inhibitory ONs have been published (for a review, see Jing and Xu 2001). Only a few recent examples can be given here. As described for HCV, anti-HIV ONs are not necessarily required to induce virus RNA degradation by RNase H, as they may as well be employed to inhibit essential processes of the viral lifecycle by steric blocking. For example, Arzumanov et al. (2001) developed AS ONs against the HIV-1 *trans*-activating response region (TAR), a 59-residue stem-loop that interacts with the *trans*-activator protein Tat. Steric blockade of this interaction by a chimeric $2'$-O-methyl RNA/LNA ON prevented full-length HIV transcription. In another study, the HIV-1 dimerization initiation site was chosen to be targeted by AS ONs (Elmén et al. 2004): An LNA/DNA mix-mer directed against this region prevented the dimerization of the genome and inhibited replication of a clinical HIV-1 isolate in a human T cell line.

Working on a different class of viruses, Yuan et al. (2004) have recently described the inhibition of coxsackievirus B3 (CBV-3) replication in cardiomyocytes and in mouse hearts. CBV-3 is a member of the plus-stranded picornavirus family. It can infect multiple organs of humans and is considered to be one of the major causes of viral myocarditis, which may develop into dilated cardiomyopathy and eventually lead to heart failure. No specific antiviral treatments exist for this important pathogen to date. The authors found AS ONs targeting the proximal terminus of the $3'$-UTR to effectively inhibit CBV-3 replication in a cardiomyocyte cell line. The antiviral activity of the AS ON was further evaluated in a CBV-3 myocarditis mouse model, and a significant decrease of viral replication and virus titers was observed.

The examples given above as well as numerous further studies demonstrate the potential of AS ONs to act as specific antiviral agents. The AS approach is particularly appealing for diseases for which no satisfactory specific treatment is available. It is therefore not surprising that a rather high percentage of the AS ONs currently being tested in clinical trials are intended to treat patients with viral infections.

2.3
Antisense Oligonucleotides in Clinical Trials

Approximately 20 AS ONs have reached the stage of clinical testing, and one AS drug is currently on the market (a comprehensive overview is given in Crooke 2004). Here, we will focus only on those AS ONs that are directed against viral targets (Table 1): Isis Pharmaceuticals is working on a phosphorothioate AS

Table 1 Antisense oligonucleotides in clinical development to treat viral infections according to Crooke (2004) and companies' Web pages

Drug	Company	Virus	Status
Vitravene	Isis	CMV	Approved
ISIS 14803	Isis	HCV	Phase II
CpG 7909	Coley	HBV	Phase I/II
CpG 10101	Coley	HCV	Phase I/II
GPI-2A	Novopharm	HIV-1	Phase I
GEM92	Hybridon	HIV-1	Phase I
MBI 1121	Hybridon	HPV	Phase I

CMV, human cytomegalovirus; HBV, hepatitis B virus; HCV, hepatitis C virus; HPV, human papillomavirus; HIV, human immunodeficiency virus

ON targeting the IRES of HCV. In a phase II trial, patients with chronic HCV infection were treated with the AS ON, and several individuals experienced significant viral titer reductions. Further AS ONs were investigated with respect to their ability to treat HIV infections: GPI-2A, developed by Novopharm Biotech, exerted strange adverse effects in a phase I trial, most likely due to the cationic liposomal formulation used as delivery system. Hybridon is developing second-generation AS ONs targeting the gag gene of HIV-1 (GEM92) and the mRNA of the E1 protein of human papillomavirus (MBI 1121). Phase I trials with these ONs showed promising safety results and confirmed the possibility of second-generation AS ONs to be delivered orally. Further ONs are being tested in patients with HBV and HCV infections (CpG 7909 and CpG 10101 by Coley Pharmaceutical and 1018-ISS by Dynavax Technologies). The mechanism of action of the latter ONs is thought to be activation of the immune response via Toll-like receptors that recognize CpG motives rather than a classical AS mechanism (Agrawal and Kandimalla 2004).

The first AS drug approved by the U.S. Food and Drug Administration (FDA)—in 1998—is a phosphorothioate ON named Vitravene (Fomivirsen). This antiviral ON targets the RNA of the immediate-early mRNA of the human cytomegalovirus (CMV) mRNA and is intended to treat CMV-induced retinitis in immunodeficient patients with acquired immunodeficiency syndrome (AIDS). A major drawback of this drug is its mode of application, since it must be injected intravitreally. Although Vitravene meets an important need for affected patients, it is only of minor commercial significance. In late 2004 a second ON-base drug was approved by the FDA: Macugen is an aptamer that targets the vascular endothelial growth factor (VEGF) and provides an antiangiogenic treatment for patients with the wet form of age-related macular degeneration, an eye disease that leads to loss of central vision.

3
Ribozymes

In the early 1980s, the research groups of Thomas Cech and Sidney Altman discovered RNA molecules that possess catalytic activity in the absence of any protein moiety. These *ribo*nucleic acids with en*zym*atic properties were named ribozymes. Meanwhile, several classes of ribozymes have been identified that can roughly be divided into large ribozymes consisting of several hundreds to thousands of nucleotides and small ribozymes that range from 30 to about 150 nucleotides in length (Doudna and Cech 2002). In addition to being fascinating objects of basic research, some ribozymes have been employed for medical purposes (for reviews, see Sullenger and Gilboa 2002; Schubert and Kurreck 2004). Here, we will focus on the use of ribozymes as antiviral agents.

Hammerhead and hairpin ribozymes are the most intensively studied and widely used ribozymes to date. Both classes were originally isolated from plant pathogens. Their application as molecular tools became possible only after the development of variants capable of cleaving target RNAs in a multiple turnover manner. Secondary structures of a hammerhead ribozyme targeting the HIV-1 tat gene and a hairpin ribozyme directed against a site in the 5′ long terminal repeat of HIV-1 are depicted in Fig. 3a and b, respectively. As described for AS ONs, ribozymes bind to their specific target RNA by Watson-Crick base pairing. In addition, they possess the capability of cleaving a complementary RNA molecule without the aid of cellular proteins.

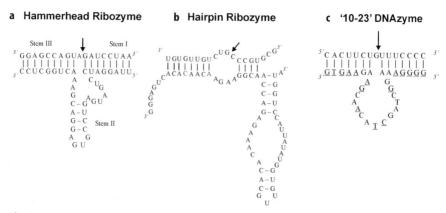

Fig. 3a–c Secondary structure of (**a**) a hammerhead ribozyme targeting HIV-1 *tat* gene (Ngok et al. 2004); (**b**) a hairpin ribozyme engineered to cleave a site in the 5′ long terminal repeat of HIV-1 (Ojwang et al. 1992); and (**c**) a 10-23 DNAzyme against a consensus sequence in the 5′-UTR of picornaviruses (Schubert et al. 2003). The optimized DNAzyme containes 2′-O-methyl RNA monomers at the underlined positions. *Arrows* indicate cleavage sites

3.1
Development of Active Ribozymes

The challenges that have to be faced when developing therapeutic ribozymes are similar to those for the optimization of efficient AS ONs, but some additional difficulties have to be met. At first, a suitable target site has to be identified. This region has to be accessible to the ribozyme employed and, in addition, it has to fulfill certain sequence requirements for efficient cleavage by ribozymes. For example, the hammerhead ribozyme cleaves preferably after AUG or GUC triplets, whereas the 10-23 DNAzyme described below processes any junction of a purine and pyrimidine.

The second important step is to stabilize the ribozyme, since RNA molecules are even more susceptible to nucleolytic degradation than DNA ONs. This task is complicated by the fact that the introduction of modified nucleotides very often leads to a severe loss of catalytic activity. Beigelman and colleagues (1995) performed a systematic optimization study for hammerhead ribozymes and developed a design with modified nucleotides in all positions except for five essential ribonucleotides. This molecule retained high catalytic activity and displayed a serum half-life of more than 10 days. It has also been used in clinical trials, as will be described in Sect. 3.3.

To date, only catalytically active ONs composed of RNA have been discovered in nature. Enzymatic DNA ONs, referred to as DNAzymes, DNA enzymes, or deoxyribozymes, however, have been obtained by in vitro selection, since they can be expected to be more stable against nucleases. The most prominent representative of this class of molecules is the so-called '10-23' DNAzyme (Santoro and Joyce 1997), the secondary structure of which is shown in Fig. 3c.

Although catalytically active ONs consisting of DNA have a higher intrinsic resistance against nucleolytic degradation, they still need further protection for applications in biological systems. We have systematically improved the properties of a 10-23 DNAzyme targeting the 5'-UTR of the human rhinovirus (Schubert et al. 2003). The resulting DNAzyme, containing modified nucleotides (underlined) at both termini and in the catalytic core, is depicted in Fig. 3c. It displays a tenfold increased catalytic activity under multiple turnover conditions and a substantially improved resistance against endo- and exonucleases.

The DNAzyme was directed against a consensus sequence of picornaviruses, which occurs not only in the rhinovirus but also in various polio-, echo-, entero-, and coxsackieviruses. Surprisingly, the unmodified DNAzyme that cleaves the rhinovirus RNA with reasonable rates was incapable of cleaving coxsackievirus A21 RNA, despite full sequence homology and a similar overall structure. After the introduction of nucleotides with high target affinity (2'-O-methyl RNA and LNA), however, the DNAzyme regained the capability of cleaving the coxsackievirus RNA with high catalytic turnover rates by successfully competing with local structures of the target RNA (Schubert et al.

2004). The modified DNAzyme can now be used to degrade a broad range of picornavirus RNAs. Furthermore, we demonstrated that the strategy of introducing nucleotides with high target affinity may generally be applicable to cleave seemingly unsuitable target sites.

As described for AS ONs, cellular uptake of ribozymes and deoxyribozymes has to be enhanced by the use of transfection reagents. Ribozymes, however, cannot only be exogenously applied as chemically presynthesized molecules, but they can also be encoded on plasmids and expressed endogenously. This opens the road to the use of viral vectors that have been employed for gene therapeutic purposes, e.g., retroviruses, adenoviruses, and adeno-associated viruses.

3.2
Cleavage of Viral RNA by Ribozymes

Ribozymes and deoxyribozymes have widely been used to inhibit virus replication in cell culture and in vivo. Here, only a few examples will be discussed; a comprehensive overview has recently been given by Peracchi (2004). Early on, various regions of HIV-1 have been targeted by hammerhead and hairpin ribozymes (e.g., Zhou et al. 1994; Ojwang et al. 1992). Meanwhile, the protective effect of ribozymes against HIV-1 infection has also been demonstrated in vivo: In a SCID-hu mouse model, $CD34^+$ hematopoietic progenitor cells transduced with anti-HIV-1 *tat-rev* or *env* hammerhead ribozymes as well as Rev aptamers were significantly protected against HIV-1 infection upon challenge (Bai et al. 2002). The 10–23 DNAzyme has also been successfully employed to inhibit HIV-1 by either directly targeting the viral RNA (Zhang et al. 1999) or by preventing virus entry by downregulation of the CCR5 coreceptor (Goila and Banerjea 1998).

In addition to HIV-1, hepatitis B and C viruses have been major targets of antiviral ribozymes. For example, a chemically stabilized hammerhead ribozyme has been designed that targets the highly conserved 5′-UTR of HCV. In cell culture, it inhibited replication of an HCV/poliovirus chimera by up to 90% (Macejak et al. 2000). This ribozyme, dubbed Heptazyme, has subsequently been employed in clinical trials as will be outlined below. A different approach has been developed to generate a ribozyme targeting HBV by expression of a triple-ribozyme cassette that undergoes intracellular self-processing (Pan et al. 2004). This sophisticated construct was suitable to reduce viral DNA in the liver of transgenic mice by greater than 80%, as measured 2 weeks after infection. Furthermore, a DNAzyme containing phosphoroamidate nucleotides was capable of inhibiting the replication of influenza A viruses by more than 99% (Takahashi et al. 2004).

3.3
Ribozymes in Clinical Trials

Several ribozymes have already been tested in clinical trials (summarized in Sullenger and Gilboa 2002; Schubert and Kurreck 2004). Most of these studies were intended to treat viral infections, while some ribozymes were used in attempts to inhibit cancer growth. For clinical trials, ribozymes were either expressed intracellularly from vectors or delivered as chemically presynthesized ONs.

The first phase I clinical trials in the mid 1990s were conducted with retroviral vectors that delivered hammerhead and hairpin ribozymes targeting HIV-1 RNA. Lymphocytes or hematopoietic precursor cells that had been isolated from infected individuals or their healthy twins were treated with the ribozyme vectors ex vivo. After selection and expansion, the transduced cells were infused into the bloodstream of the infected patient. The treatment was supposed to increase the resistance of hematopoietic cells against spreading of the virus. Although the procedure was found to be safe and ribozyme expression in transduced cells could be detected, the duration of the protective effect was unsatisfactory. It will therefore be necessary to develop and improve methods to transduce pluripotent hematopoietic stem cells in order to achieve long-term resistance against HIV (Michienzi et al. 2003).

In addition to these studies to treat patients with HIV infections, one clinical trial was initiated with a hammerhead ribozyme targeting HCV. In contrast to the trials described in the preceding paragraph, in which viral vectors were employed to deliver ribozymes, the aforementioned chemically presynthesized ribozyme Heptazyme was used for these tests (Usman and Blatt 2000). Encouraging results of initial clinical studies led to the initiation of a phase II trial with Heptazyme alone and in combination with interferon (IFN) in 2001. The finding that the HCV RNA level in the serum of patients treated with the ribozyme was reduced by only 10%, along with results from toxological studies, led to the decision to stop clinical experimentation of Heptazyme (Peracchi 2004).

Taken together, ribozymes to treat viral infections or cancer were well tolerated in clinical settings, but their therapeutic efficiency was rather low. Recent findings indicate that the minimized ribozyme variants used in these trials lacked sequence elements that are essential for high catalytic activity under physiological conditions (Khvorova et al. 2003a). Since a new technique believed to be much more efficient has emerged with the discovery of RNA interference (RNAi), most approaches to develop RNA-based therapeutics have now been switched to the use of siRNA rather than ribozymes.

4
RNA Interference

RNAi denotes the disruption of RNA molecules induced by double-stranded (ds)RNA of the same sequence. The process was first identified in the nematode *Caenorhabditis elegans* (Fire et al. 1998). Subsequent work elucidated the biochemical basis underlying this mechanism (for a recent review, see Meister and Tuschl 2004): Long dsRNA is cut up by an RNase III-type endonuclease termed Dicer, resulting in 19mer RNA duplexes with symmetrical two-nucleotide overhangs at both 3'-ends, known as "small interfering RNAs" (siRNAs). One of the strands, preferably the AS strand, is subsequently incorporated into a proteinaceous complex called "RNA-induced silencing complex," RISC, whose exact composition is still under investigation. The RNA strand programs RISC to act specifically on RNAs of the complementary sequence. The targeted molecule is bound and cleaved in the center of the target sequence. Afterwards, the damaged RNA molecule is quickly degraded by cellular nucleases, while RISC can go on through several rounds of cleavage (Haley and Zamore 2004). In contrast to RNase H-mediated cleavage induced by AS ONs, this process takes place predominantly in the cytoplasm (Fig. 1).

The overall mechanism of gene silencing mediated by dsRNA is conserved in virtually all eukaryotes. In mammals, however, the presence of dsRNA exceeding 30 bp in the cytoplasm can initiate the innate IFN immune response, resulting in extensive cell death. This unspecific reaction, which has hampered the use of RNAi in mammalian cells, can be avoided if 21mer siRNA molecules are employed (Elbashir et al. 2001). siRNAs were shown to be extremely potent inhibitors of target gene expression (e.g., Grunweller et al. 2003). They have meanwhile been used to knock down a plethora of individual genes in cell culture experiments in order to investigate gene function or to trace molecular pathways (Silva et al. 2004). RNAi has also been applied in mouse models of human disease (summarized in Sioud 2004; and the chapter by M. Sioud, this volume). Comprehensive overviews on the current status and application of RNAi methods are given in numerous recent reviews (e.g., Mittal 2004; Dorsett and Tuschl 2004).

4.1
Designing Efficient siRNA Approaches

4.1.1
Selection of Efficient siRNAs and Suitable Targets

Targeted degradation of an RNA molecule triggered by siRNAs is a multistep process. At least two factors are critical to its efficacy:

1. siRNA structure: The assembly of the RNAi enzyme complex RISC has been shown to be dependent on thermodynamic characteristics of the siRNA

with respect to the relative stability of both ends of the duplex (Khvorova et al. 2003b; Schwarz et al. 2003). Based on these findings, several algorithms have been developed to predict efficient siRNAs (e.g., Reynolds et al. 2004).

2. Accessibility of the target sequence: We and others have observed that even a well-designed siRNA with excellent thermodynamic features may show no silencing activity when the target region is sequestered in stable internal structures (Kretschmer-Kazemi Far and Sczakiel 2003; Schubert et al. 2005b). For example, targeting of the 5′-UTR of picornaviruses has turned out to be particularly inefficient (Schubert et al. 2005a; Phipps et al. 2004; Saleh et al. 2004). This region is known to be highly structured, and it associates with cellular and viral proteins, possibly precluding siRNAs from hybridizing. Similarly, some viral RNA species have been found to be resistant to degradation by RNAi: siRNAs directed against HIV-1, rotavirus, and respiratory syncytial virus each afforded degradation only of the mRNA species of the viruses, but did not interfere with viral genomic RNA of the same sequence (Bitko and Barik 2001; Arias et al. 2004; Hu et al. 2002). This observation is most likely due to extensive association of the viral genomic RNAs with proteins or its localization to compartments inaccessible to RISC.

4.1.2
Delivery of siRNAs

Looking back on 20 years of effort with the delivery of ribozymes and AS ONs, researchers today can benefit from a rich body of experience regarding ON transfer into cells (see above and Seksek and Bollard 2004 for a review). Similar to ribozymes, siRNAs can be delivered as presynthesized dsRNA molecules, or they can be expressed intracellularly from plasmids. Transfection efficiency for presynthesized siRNAs is usually quite high in standard cell lines, but the silencing effect wears off in a matter of days as the intracellular concentration of siRNA decreases due to cell division and degradation by cellular nucleases. In addition, siRNAs have to be modified chemically to increase their nuclease resistance in body fluids.

To achieve longer lasting knockdown, plasmid vectors have been designed that afford continuous expression of short hairpin (sh)RNAs, which are intracellularly processed to give siRNAs (Fig. 4; Mittal 2004). The use of plasmid vectors allows for the selection of stably transfected cells that have been shown to exert gene-silencing activity, even 2 months after vector delivery (Brummelkamp et al. 2002). Delivery of plasmid DNA using standard transfection protocols, however, is often unsatisfactory, particularly when primary cells or stem cells are concerned. More efficient uptake can be achieved by exploiting the natural ability of viruses to infect host cells. Oncoviruses, lentiviruses, adenoviruses, and recombinant adeno-associated viruses are the viral classes most often employed as shuttles for delivery of transgenes into cells. An excel-

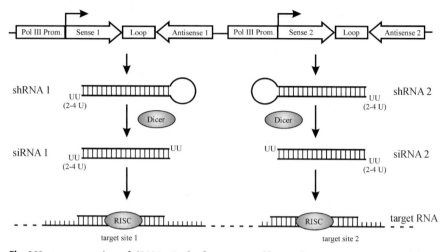

Fig. 4 Vector expression of siRNAs. In the first step, a self-complementary RNA, named short hairpin RNA (shRNA), is expressed under control of a polymerase III promoter. The shRNA is subsequently processed to give an siRNA by the endonuclease Dicer. An siRNA double expression (SiDEx) vector can be employed to express two shRNAs directed against different sites on the same target RNA simultaneously (Schubert et al. 2005a). Thus, mutations in one of the target sites leading to abrogation of silencing can be compensated by the unaltered efficiency of the second siRNA (for further details, see text)

lent comprehensive overview on viral delivery systems is given in the paper by Thomas et al. (2003).

Systemic delivery into mice is usually performed by injecting the siRNA into the tail vein in a large volume of liquid using high pressure, a method termed hydrodynamic tail vein injection (Fig. 5). siRNAs are subsequently found to be enriched in the liver. High-pressure injection, however, is not applicable to humans, and more subtle ways of systemic administration need to be sought. One of the most promising reports yet has come from Alnylam Pharmaceuticals. In their study, an siRNA was conjugated with a cholesterol molecule. This construct had substantially improved pharmacokinetic properties and efficiently reduced plasma levels of endogenous apolipoprotein B in mice after intravenous administration (Soutschek et al. 2004).

4.1.3
Specificity of siRNAs

siRNAs are considered highly specific molecular tools, but their capacity to discriminate against mismatches seems to be dependent on the siRNA sequence and position of the mismatched nucleotides. Unspecific effects of siRNAs have been described in several reports. Semizarov et al. (2003) found induction of genes related to stress and apoptosis at higher siRNA concentrations. Further-

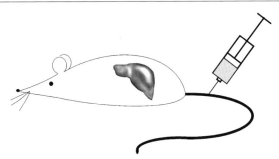

Fig. 5 Hydrodynamic tail vein injection: siRNAs are rapidly injected into the tail vein of mice in a large volume of fluid. The siRNAs accumulate preferentially in the liver

more, off-target effects were detected by expression profiling, and induction of the IFN response by shRNAs or siRNAs has also been reported (Jackson et al. 2003; Bridge et al. 2003; Sledz et al. 2003). Moreover, siRNAs were shown to inhibit translation of partially complementary mRNA molecules by imperfect hybridization in a microRNA-like manner (Doench and Sharp 2004; Scacheri et al. 2004). Most siRNAs will have partial complementarity with one or more human genes and are thus prone to unintended gene silencing. It is therefore advisable to employ only highly efficient siRNAs, so that low doses yield sufficient results and side effects are expected to be minimized. For a further discussion of relevant issues concerning efficiency and specificity of RNAi approaches, see Hannon and Rossi (2004).

4.1.4
Viral Escape

Viruses often elude long-term inhibition by acquiring mutations that render the antiviral compound ineffective. This problem has been observed in RNAi approaches for sustained inhibition of both poliovirus and HIV-1 (Gitlin et al. 2002; Boden et al. 2003; Das et al. 2004). Although the siRNAs employed in these studies inhibited viral propagation initially, virus titers increased again upon prolonged incubation. Analysis of the resistant mutants revealed either a point mutation in the target region or extensive rearrangements of the viral RNA. Several strategies may be useful to prevent the enrichment of escape mutants. One is targeting of conserved sequences that are less likely to yield viable virus after mutation. Another is reducing levels of host cellular genes necessary for viral entry or propagation, which has led to substantial reduction of virus titers (Zhang et al. 2004; Novina et al. 2002). Cellular genes are less prone to mutation and are therefore less likely to allow viral escape of silencing. When pursuing such an approach, however, the risk of side-effects must obviously be addressed thoroughly. In analogy to conventional combination therapy, a third strategy to prevent escape-mutant enrichment involves two or more siRNAs being used simultaneously. We have recently described an siRNA double expression vector (SiDEx) containing genes for two different siRNAs both targeted against the RNA-dependent RNA polymerase of coxsackievirus B3 (Schubert et al. 2005a; Fig. 4). Even when an artificial point mutation was introduced into the target gene so that the respective mono-expression vector was no longer capable of silencing, the double expression vector maintained high silencing activity. Although we did not find any deleterious effect of the simultaneous use of two siRNAs, it has been observed that different siRNAs present in one cell may compete for access to the RNAi machinery (Bitko et al. 2005). Thus, the efficiency of one siRNA may be compromised due to the presence of a second species. Another difficulty in preventing viral escape may be the amount of viral particles present in the blood of patients in the state of viremia. In HCV infection, up to 10^7 virions per milliliter may be present in the bloodstream, so that even 90% reduction of the titer may still leave enough virus particles intact to allow escape. The highest efficiency knockdown is therefore even more desirable.

4.2
Antiviral RNAi Approaches

A vast body of studies reporting the employment of RNAi for antiviral purposes in mammalian cells has already been published. The extent of inhibition of viral propagation measured in these studies is dependent on several factors, some of which are choice of cell line, virulence and titer of the virus strain, siRNA delivery method, time between infection and siRNA delivery, and the time and method of readout and detection. It is thus difficult to make quantitative comparisons between the studies or to draw conclusions as to a possible clinical utility of individual compounds from cell-culture experiments. In the following paragraphs, some studies will be discussed that are likely to lead to clinical applications in the nearer future. We are aware of the fact that our selection is biased. Comprehensive overviews of the classes of viruses that have been targeted by RNAi are provided in the papers by Haasnoot et al. (2003) and Ryther et al. (2005).

4.2.1
Respiratory Diseases

Influenza is a major threat for societies worldwide. At least 20 million people died in the catastrophic outbreak of the virus in 1918. Influenza viruses are enveloped, single-stranded RNA viruses whose genome is segmented into eight parts of negative polarity. They are notorious for changing their outward shape due to small mutations in the principal antigens. Thus, the virus may escape protective immunity induced by a previously prevalent viral strain. Vaccination can provide effective protection against influenza virus infection.

In older or immunocompromised people, however, vaccination's efficacy is rather limited. Recently, warnings about the threat of an influenza pandemic have been discussed in a news focus by Enserink (2004).

Ge and colleagues (2003) tested the antiviral effect of 20 siRNAs directed against conserved regions of influenza A virus in cell culture and embryonated chicken eggs. Three of these siRNAs were potent inhibitors of viral propagation. While one siRNA reduced only the level of its specific target mRNA, two others resulted in reduced levels of all viral RNA species. The authors conclude that the targeted proteins have critical roles in viral transcription and replication.

About 1 year later, successful application of the previously identified siRNAs in mice was reported in two studies. Tompkins and coworkers (2004) administered siRNAs by hydrodynamic tail vein injection followed by viral infection 16–24 h later. After 2 days, virus titers in the lungs were 10- to 56-fold lower than in mice receiving an unrelated control siRNA. By day 18 post challenge, 60% of the animals treated with control siRNA had died, whereas mortality was reduced to 10%–20% in siRNA-treated animals. A combination of both siRNAs yielded 100% survival.

In the second study, siRNAs were administered complexed with polyethyleneimine (PEI) by intravenous injection, a method compatible with use in humans (Ge et al. 2004). PEI-complexed nucleic acids were found predominantly in the lungs. Reduction of virus titers 24 h after infection was approximately tenfold, even when siRNAs were administered 5 h after infection. Also, plasmid DNA coding for shRNAs was found to inhibit viral propagation in mice when administered intravenously in complex with PEI or instilled in the lungs.

Recently, intranasal administration of siRNAs to the lungs has been reported by two groups (Bitko et al. 2005; Zhang et al. 2005). The siRNAs were directed against different regions of respiratory syncytial virus (RSV), a major respiratory pathogen of the family Paramyxoviridae that contains nonsegmented (−) strand RNA genomes (Barik 2004). When 5 nmol of an efficient siRNA was administered intranasally with or without transfection agents, viral titers were reduced in the lungs of mice by up to three orders of magnitude. In addition, siRNA-treated mice showed no signs of respiratory distress during up to 6 weeks of observation. When used against an ongoing infection, the antiviral effect decreased, but remained clearly visible even when siRNAs were given 4 days after viral challenge.

The siRNA applied by Zhang and coworkers (2005) was directed against the viral NS1 gene that antagonizes IFN-β signaling in the host. shRNA-expressing plasmids complexed with nanochitosan, a natural, biodegradable polymer, were delivered 2 days before viral inoculation. NS1 expression in the lungs was diminished 18 h later, whereas other viral genes remained unaffected. This observation lends further proof to the finding discussed above, namely that the viral genomic RNA may be inaccessible to RNAi. IFN-β levels in the lung tissue of mice were significantly increased in animals that received anti-NS1 shRNA before infection. Treatment with siRNA up to 3 days after infection still reduced virus titer in the lungs. Moreover, mice treated with NS1 shRNA 2 days after infection showed a substantial decrease in lung inflammation compared to control mice.

4.2.2
Viral Hepatitis

HBV is a partially double-stranded DNA virus of the Hepadnaviridae family that primarily infects the liver. Although efficient vaccines are available, an estimated one million people die every year of HBV-related diseases, and the number of carriers lays around 350 million. Chronic infection can give rise to liver cirrhosis and hepatocellular carcinoma. After having entered the cell, the viral genome is transported to the nucleus, where four RNA species are transcribed and exported to the cytoplasm. One 3.5-kb transcript, referred to as pregenomic RNA, also serves as a template for reverse transcription. RNAi cannot target the viral circular dsDNA that resides in cell nuclei of chronically infected patients, but interference with one or more of the mRNA species has been reported repeatedly. Replication of HBV is not supported by most animals. Therefore, replication-competent plasmids have been developed that mimic most of the steps of the viral life cycle in cell culture and small animals.

In what was one of the first reports on functional antiviral RNAi in mammals, McCaffrey and colleagues (2003) co-injected shRNA expression vectors and a plasmid containing an HBV replicon into the tail vein of mice, using hydrodynamic transfection. After 7 days, the livers of the animals were probed for replicated genomic viral DNA and RNA transcripts by Northern and Southern blotting. The shRNAs significantly reduced the level of HBV RNA and DNA. Also, the amount of HBV surface antigen HBsAg in serum was decreased by up to 85% using the most effective of the shRNAs under investigation. In addition, however, two unexplained observations were reported: (1) An shRNA directed against a site present only in the longest of the four mRNAs was found to reduce not only the level of the targeted mRNA, but also levels of the other viral transcripts. (2) Treatment with any shRNA including an unrelated control shifted the balance of single-stranded and double-stranded genomic DNA towards the single-stranded form. These data raise the possibility that shRNAs may have sequence-independent antiviral effects whose mechanisms are unknown as yet.

Chemically synthesized siRNAs were employed by Klein and coworkers (2003) for tail vein injection together with the replication-competent HBV plasmid. Both siRNAs inhibited expression of their target proteins in the liver of mice. While the effect of one of the siRNAs was sustained for at least 11 days, inhibition by the second siRNA ceased around day 5.

Uprichard and coworkers (2005) succeeded in clearing a preexisting HBV gene expression from the liver of mice. Antigen levels in the serum, RNA expression, and the presence of replicative DNA intermediates were all significantly reduced in transgenic mice containing an HBV-plasmid. shRNA-expressing vectors were administered by hydrodynamic tail vein injection of recombinant adenoviruses. The antiviral effect was sustained for at least 26 days. Interestingly, in this study the level of 3.5-kb RNA was somewhat less reduced than the level of the 2.1-kb transcripts, although both species contained the target sites. The reason for this observation lends itself to some speculation.

Another major hepatic pathogen is HCV (see above), which also cannot easily be grown in cell culture. For this reason, plasmid replicons containing subgenomic fragments or the whole genome of HCV are used. To show the general applicability of antiviral RNAi approaches in mice, McCaffrey and colleagues (2002) used high-pressure tail vein injection to deliver a plasmid expressing a fusion construct of luciferase and the viral protein NS5B to the livers of mice. When siRNAs or shRNA-expressing plasmids were co-transfected with the reporter plasmid, 80% and 90% reduction of luciferase expression over the whole body of the animals were measured, respectively. Several groups have reported on siRNAs directed against different regions on the mRNA of HCV that reduced viral RNA and protein levels (for a listing, see Radhakrishnan et al. 2004; Randall and Rice 2004). Randall and colleagues (2003) found a greater than 98% inhibition of RNA levels on transfection of an efficient siRNA. This decrease may have its cause either in equal reduction of HCV RNA in all cells, or in complete destruction of viral RNA in most of the cells. Using immunofluorescence techniques, the authors found that indeed almost all cells were completely clear of the HCV NS5A protein. Few cells, however, fluoresced brightly, indicating that no detectable reduction in HCV levels had taken place. To achieve sustained inhibition, stable transfection of shRNA expression vectors has been performed. Wilson and colleagues (2004) found a 70% reduction of HCV-gene expression when cells were challenged with the replicon that had been stably transfected with shRNA 3 weeks earlier. Zhang and coworkers (2004) used adenoviral delivery to knock down cellular targets indispensable for HCV replication and achieved significant reductions in HCV-expressing cells.

4.2.3
HIV

The greatest body of work regarding antiviral RNAi approaches has been performed on HIV. As outlined above, efforts have already been taken to inhibit the spread of this devastating virus by means of gene therapy employing ribozymes. RNAi has proved significantly more potent than ribozymes in the reduction of virus titers. Therefore, employing shRNA-expressing vectors for gene therapy in settings that have been validated with ribozymes may boost the efficiency of the approach.

Numerous siRNAs and shRNAs have been directed against different regions on HIV RNA that successfully suppressed viral gene expression and replication (see, e.g., the comprehensive list given in Haasnoot et al. 2003; Hannon and

Rossi 2004). Early on, the question was raised which step of the viral life cycle siRNAs would interfere with. Jacque and colleagues (2002) tested a number of presynthesized siRNAs and plasmid-expressed shRNAs against the viral vif and nef genes as well as the long terminal repeat (LTR). These siRNAs suppressed the production of viral particles from an infectious HIV clone 20- to 50-fold. The authors also found a dramatic inhibition of viral genomic RNA, complementary (c)DNA, and integrated DNA. From these studies, it can be concluded that the siRNAs employed in this case inhibited the viral life cycle at an early stage. Somewhat different results, however, were presented by Hu and coworkers (2002) who found no reduction of viral cDNA in cells that had been transfected with potent siRNAs against viral gag and pol RNA. Coburn and Cullen (2002) used siRNAs against tat and rev and noticed a slight decrease in proviral DNA production that does not seem to be sufficient to explain the massive drop in viral gene expression brought about by the siRNAs. If and to what extent incoming viral RNA is accessible to RNAi is thus still under discussion.

Viral escape has been shown to be a major problem when trying to inhibit HIV-1 replication (see above). Several groups have therefore turned to silencing cellular genes necessary for viral propagation (Arteaga et al. 2003). Qin and colleagues (2003) transduced T lymphocytes from uninfected individuals with a lentiviral vector coding for an siRNA against CCR5, the viral coreceptor. A tenfold downregulation of the expression of the receptor was observed that led to a 3- to 7-fold protection against viral challenge. CCR5 may be a suitable target for interference with HIV-1 entry because it is not essential for normal immune function.

An elegant method of targeted delivery of a transgene coding for an siRNA is to isolate hematopoietic progenitor cells that give rise to the hematopoietic cells capable of being infected by the virus. These cells are then transduced with the transgene of choice and reinfused into the patient. Banerjea and coworkers (2003) have used a lentiviral vector to introduce an anti-rev siRNA into primary $CD34^+$ hematopoietic progenitor cells. The cells were subsequently expanded in SCID-hu thy/liv grafts. After 60 days, the mice were sacrificed and thymocytes were collected. In one animal, 3% of the cells were positive for the transgene, whereas another animal showed 53% positivity. The in vivo-derived thymocytes that were positive for the transgene were subsequently enriched and challenged with HIV-1. siRNA-expressing thymocytes showed significant protection against the viral challenge as long as 10 days post infection.

Because of the significant experience obtained with anti-HIV ribozymes and the high relevance and urgent need for new and cheap HIV-therapeutics, human clinical trials employing the RNAi approach against viruses are expected to begin soon (Hannon and Rossi 2004).

4.2.4
Coxsackievirus B3

In our own research, we focus on the siRNA-induced inhibition of CBV-3, which is a major myocardial pathogen that causes severe heart muscle infections (see Sect. 2.2). In our opinion, CBV-3 is particularly well suited to be targeted by siRNAs, since the virus has a cytoplasmic lifecycle. Complete virus clearance by RNAi can be expected only for this type of virus, because viral DNA genomes or DNA intermediates of retroviruses are not susceptible to RISC-mediated cleavage (Fig. 1). Initial experiments with siRNAs directed against the 5′-UTR of CBV-3 were unsuccessful, most likely due to the stable secondary structure of this region of the viral genome. Subsequent efforts to target the 3D RNA-dependent RNA polymerase with siRNAs, however, led to a concentration-dependent inhibition of the virus. A plaque assay revealed a reduction of virus propagation by one $\log_{10}$ step (Schubert et al. 2005a). Another aspect supports the idea to use RNAi approaches against coxsackieviruses. In a clinical phase II trial, Kühl et al. (2003) demonstrated that IFN-β treatment eliminates cardiotropic viruses and improves left ventricular function. The effect of siRNAs is more specific than that of IFN-β, but the observed induction of the IFN response by siRNAs (Sledz et al. 2003), an undesired side effect in most cases, may prove beneficial in this application.

4.3
From Bench to Bedside

In evaluating the state of development that RNAi techniques have achieved, we must ask: How long will it take until the first drug is approved that acts by means of RNAi? Enthusiasm about this novel and highly efficient technique and skepticism about its applicability are in balance for the time being. Expectations may be somewhat limited, as 20 years of research on the medical application of AS molecules has led to no more than a single approved AS drug of minor commercial significance and one approved aptamer.

A number of companies like Alnylam Pharmaceuticals, Benitec, International Therapeutics, CytRx Corporation, Sirna Therapeutics, and Nucleonics are developing siRNAs for the treatment of viral infections such as HIV, HCV, and HBV, according to the companies' respective Web sites. Two companies, Sirna Therapeutics and Acuity Pharmaceuticals, have recently initiated phase I clinical studies of siRNAs interfering with vascularization for treatment of age-related macular degeneration. Results from these studies would be the first clinical data available for siRNAs and may allow better-founded predictions on the future of RNAi in therapeutic approaches.

5
Conclusions

ONs are a valuable alternative to low molecular weight compounds for the treatment of viral infections. AS ONs and ribozymes have been used for many years to inhibit virus replication. Some of the most advanced molecules have even been evaluated in clinical trials, and the only approved antisense drug to date is an antiviral AS ON. Despite being well tolerated in the doses employed, AS ONs and ribozymes frequently failed to provide convincing evidence for their therapeutic benefit. The more recently developed RNAi technology, however, is likely to overcome most of the problems of the aforementioned approaches due to its higher efficiency. Only three and a half years after the initial description of gene silencing by siRNAs in mammalian cells, the first clinical trials based on the principle of RNAi have commenced. Furthermore, siRNAs against cancer and viral targets can be expected to enter the clinic soon. One of the major problems of long-term inhibition of viruses is their ability to escape any kind of treatment due to their high mutation rate. It will therefore be advantageous to select sequences of the virus that are highly conserved. In addition, several distinct regions of the virus should be targeted simultaneously with either siRNA double expression (SiDEx) vectors (Schubert et al. 2005a) or by combining different types of antiviral agents, e.g., ribozymes and siRNAs (Michienzi et al. 2003). After the principle establishment of ONs as antiviral agents, these techniques will offer the opportunity to rapidly develop inhibitors of newly discovered viruses or variants of well-known viruses.

Acknowledgements The authors wish to thank Volker A. Erdmann for his continuous support and inspiration and all lab members for their help and engagement for the scientific work. We are furthermore thankful to our collaborators Heinz Zeichhardt and Hans-Peter Grunert for introducing us into the field of virology. Financial support by the Deutsche Forschungsgemeinschaft (DFG Ku-1436, SFB/TR19 TP C1) and the Fonds der Chemischen Industrie is gratefully acknowledged.

References

Agrawal S, Kandimalla ER (2004) Antisense and siRNA as agonists of toll-like receptors. Nat Biotechnol 22:1533–1537

Arias F, Dector MA, Segovia L, Lopez T, Camacho M, Isa P, Espinosa R, Lopez S (2004) RNA silencing of rotavirus gene expression. Virus Res 102:43–51

Arteaga HJ, Hinkula J, van Dijk-Härd I, Dilber MS, Wahren B, Christensson B, Mohamed AJ, Smith CIE (2003) Choosing CCR5 or Rev siRNA in HIV-1. Nat Biotechnol 21:230–231

Arzumanov A, Walsh AP, Rajwanshi VK, Kumar R, Wengel J, Gait MJ (2001) Inhibition of HIV-1 Tat-dependent trans activation by steric block chimeric 2′-O-methyl/LNA oligoribonucleotides. Biochemistry 40:14645–14654

Bai J, Banda N, Lee NS, Rossi J, Akkina R (2002) RNA-based anti-HIV gene therapeutic constructs in SCID-hu mouse model. Mol Ther 6:770–782

Banerjea A, Li M-J, Bauer G, Remling L, Lee N-S, Rossi J, Akkina R (2003) Inhibition of HIV-1 by lentiviral vector-transduced siRNAs in T lymphocytes differentiated in SCID-hu mice and CD34$^+$ progenitor cell-derived macrophages. Mol Ther 8:62–71

Barik S (2004) Control of nonsegmented negative-stand RNA virus replication by siRNA. Virus Res 102:27–35

Beigelman L, McSwiggen JA, Draper KG, Gonzalez C, Jensen K, Karpeisky AM, Modak AS, Matulic-Adamic J, DiRenzo AB, Haeberli P, Sweedler D, Tracz D, Grimm S, Wincott FE, Thackaray VG, Usman N (1995) Chemical modification of hammerhead ribozymes. J Biol Chem 270:25702–25708

Bitko V, Barik S (2001) Phenotypic silencing of cytoplasmic genes using sequence-specific double-stranded short interfering RNA and its application in the reverse genetics of wild type negative-strand RNA viruses. BMC Microbiol 1:34

Bitko V, Musiyenko A, Shulyayeva O, Barik S (2005) Inhibition of respiratory viruses by nasally administered siRNA. Nat Med 11:50–55

Boden D, Pusch O, Lee F, Tucker L, Ramratnam B (2003) Human immunodeficiency virus type 1 escapes from RNA interference. J Virol 77:11531–11535

Bridge AJ, Pebemard S, Ducraux A, Nicoulaz L, Iggo R (2003) Induction of an interferon response by RNAi vectors in mammalian cells. Nat Genet 34:263–264

Brummelkamp TR, Bernards R, Agami R (2002) A system for stable expression of short interfering RNAs in mammalian cells. Science 296:550–553

Coburn GA, Cullen BR (2002) Potent and specific inhibition of human immunodeficiency virus type 1 replication by RNA interference. J Virol 76:9225–9231

Crooke ST (2004) Progress in antisense technology. Annu Rev Med 55:61–95

Darfeuille F, Hansen JB, Orum H, Di Primo C, Toulme J-J (2004) LNA/DNA chimeric oligomers mimic RNA aptamers targeted to the TAR RNA element of HIV-1. Nucleic Acids Res 32:3101–3107

Das AT, Brummelkamp TR, Westerhout EM, Vink M, Madiredjo M, Bernards R, Berkhout B (2004) Human immunodeficiency virus type 1 escapes from RNA interference-mediated inhibition. J Virol 78:2601–2605

De Beuckelaer A, Fürste J-P, Gruszecka M, Wittmann-Liebold B, Erdmann VA (1999) Selection of RNA aptamers that bind to a peptide of the canyon region of human rhinovirus 14. In: Wagner E et al. (eds) Proceedings of the International Congress on Endocytobiology, Symbiosis and Biomedicine, Freiburg im Breisgau, Germany, 5–9 April 1998, Endocytobiology, VII. "From Symbiosis to Eukaryotism". Freiburg, pp 565–577

De Clercq E (2004) Antivirals and antiviral strategies. Nat Rev Microbiol 2:704–720

Doench JG, Sharp PA (2004) Specificity of microRNA target selection in translational repression. Genes Dev 18:504–511

Dorsett Y, Tuschl T (2004) siRNAs: applications in functional genomics and potential as therapeutics. Nat Rev Drug Discov 3:318–329

Doudna JA, Cech TR (2002) The chemical repertoire of natural ribozymes. Nature 418:222–228

Eckstein F (2000) Phosphorothioate oligonucleotides: what is their origin and what is unique about them? Antisense Nucleic Acid Drug Dev 10:117–121

Elbashir SM, Harborth J, Lendeckel W, Yalcin A, Weber K, Tuschl T (2001) Duplexes of 21-nucleotide RNAs mediate RNA interference in cultured mammalian cells. Nature 411:494–498

Elmén J, Zhang H-Y, Zuber B, Ljungberg K, Wahren B, Wahlestedt C, Liang Z (2004) Locked nucleic acids containing antisense oligonucleotides enhance inhibition of HIV-1 genome dimerization and inhibit virus replication. FEBS Lett 578:285–290

Enserink M (2004) Looking the pandemic in the eye. Science 306:392–394

Fire A, Xu S, Montgomery MK, Kostas SA, Driver SE, Mello CC (1998) Potent and specific genetic interference by double-stranded RNA in Caenorhabditis elegans. Nature 391:806–811

Ge Q, McManus MT, Nguyen T, Shen C-H, Sharp PA, Eisen HN, Chen J (2003) RNA interference of influenza virus production by directly targeting mRNA for degradation and indirectly inhibiting all viral transcription. Proc Natl Acad Sci USA 100:2718–2723

Ge Q, Filip L, Bai A, Nguyen T, Eisen HN, Chen J (2004) Inhibition of influenza virus production in virus-infected mice by RNA interference. Proc Natl Acad Sci USA 101:8676–8681

Gitlin L, Karelsky S, Andino R (2002) Short interfering RNA confers intracellular antiviral immunity in human cells. Nature 418:430–434

Goila R, Banerjea A (1998) Sequence specific cleavage of the HIV-1 coreceptor CCR5 gene by a hammer-head ribozyme and a DNA-enzyme: inhibition of the coreceptor function by a DNA-enzyme. FEBS Lett 435:233–238

Grunweller A, Wyszko E, Bieber B, Jahnel R, Erdmann VA, Kurreck J (2003) Comparison of different antisense strategies in mammalian cells using locked nucleic acids, 2′-O-methyl RNA, phosphorothioates and siRNA. Nucleic Acids Res 31:3185–3193

Haasnoot PCJ, Cupac D, Berkhout B (2003) Inhibition of virus replication by RNA interference. J Biomed Sci 10:607–616

Haley B, Zamore PD (2004) Kinetic analysis of the RNAi enzyme complex. Nat Struct Mol Biol 11:599–606

Hannon GJ, Rossi JJ (2004) Unlocking the potential of the human genome with RNA interference. Nature 431:371–377

Hu W-Y, Myers CP, Kilzer JM, Pfaff SL, Bushman FD (2002) Inhibition of retroviral pathogenesis by RNA interference. Curr Biol 12:1301–1311

Jackson AL, Bartz SR, Schelter J, Kobayashi SV, Burchard J, Mao M, Li B, Cavet G, Linsey PS (2003) Expression profiling reveals off-target gene regulation by RNAi. Nat Biotechnol 21:635–637

Jacque J-M, Triques K, Stevenson M (2002) Modulation of HIV-1 replication by RNA interference. Nature 418:435–438

Jepsen JS, Wengel J (2004) LNA-antisense rivals siRNA for gene silencing. Curr Opin Drug Discov Devel 7:188–194

Jing N, Xu X (2001) Rational drug design of DNA oligonucleotides as HIV inhibitors. Curr Drug Targets 1:79–90

Khvorova A, Lescoute A, Westhof E, Jayasena SD (2003a) Sequence elements outside the hammerhead ribozyme catalytic core enable intracellular activity. Nat Struct Biol 10:708–712

Khvorova A, Reynolds A, Jayasena SD (2003b) Functional siRNAs and miRNAs exhibit strand bias. Cell 115:209–216

Klein C, Bock CT, Wedemeyer H, Wustefeld T, Locarnini S, Dienes HP, Kubicka S, Manns MP, Trautwein C (2003) Inhibition of hepatitis B virus replication in vivo by nucleoside analogues and siRNA. Gastroenterology 125:9–18

Kretschmer-Kazemi Far R, Sczakiel G (2003) The activity of siRNA in mammalian cells is related to structural target accessibility: a comparison with antisense oligonucleotides. Nucleic Acids Res 31:4417–4424

Kühl U, Pauschinger M, Schwimmbeck PL, Seeberg B, Lober C, Noutsias M, Poller W, Schultheiss H-P (2003) Interferon-beta treatment eliminates cardiotropic viruses and improves left ventricular function in patients with myocardial persistence of viral genomes and left ventricular dysfunction. Circulation 107:2793–2798

Kurreck J (2003) Antisense technologies: improvement through novel chemical modifications. Eur J Biochem 270:1628–1644

Kurreck J, Wyszko E, Gillen C, Erdmann VA (2002) Design of antisense oligonucleotides stabilized by locked nucleic acids. Nucleic Acids Res 30:1911–1918

Levin AA (1999) A review of issues in the pharmacokinetics and toxicology of phosphorothioate antisense oligonucleotides. Biochim Biophys Acta 1489:69–84

Macejak D, Jensen KL, Jamison S, Domenico K, Roberts EC, Chaudhary N, von Carlowitz I, Bellon L, Tong MJ, Conrad A, Pavco PA, Blatt LM (2000) Inhibition of Hepatitis C Virus (HCV)-RNA-dependent translation and replication of a chimeric HCV Poliovirus using synthetic stabilized ribozymes. Hepatology 31:769–776

Martinand-Mari, C, Lebleu B, Robbins I (2003) Oligonucleotide-based strategies to inhibit human hepatitis C virus. Oligonucleotides 13:539–548

McCaffrey AP, Meuse L, Pham TT, Conklin DS, Hannon GJ, Kay MA (2002) RNA interference in adult mice. Nature 418:38–39

McCaffrey AP, Nakai H, Pandey K, Huang Z, Salazar FH, Xu H, Wieland SF, Marion PL, Kay MA (2003) Inhibition of hepatitis B virus in mice by RNA interference. Nat Biotechnol 21:639–644

McKnight KL, Heinz BA (2003) RNA as a target for developing antivirals. Antivir Chem Chemother 14:61–73

Meister G, Tuschl T (2004) Mechanisms of gene silencing by double-stranded RNA. Nature 431:343–349

Michienzi A, Castonotto D, Lee N, Li S, Zaia J, Rossi JJ (2003) RNA-mediated inhibition of HIV in a gene therapy setting. Ann N Y Acad Sci 1002:63–71

Mittal V (2004) Improving the efficiency of RNA interference in mammals. Nat Rev Genet 5:355–365

Ngok FK, Mitsuyasu RT, Macpherson JL, Boyd MP, Symonds GP, Amado RG (2004) Clinical gene therapy research utilizing ribozymes. Methods Mol Biol 252:581–598

Novina CD, Murray MF, Dykxhoorn DM, Beresford PJ, Riess J, Lee S-K, Collman RG, Lieberman J, Shankar P, Sharp PA (2002) siRNA-directed inhibition of HIV-1 infection. Nat Med 8:681–686

Ojwang JO, Hampel A, Looney DJ, Wong-Staal F, Rappaport J (1992) Inhibition of the human immunodeficiency virus type 1 expression by a hairpin ribozyme. Proc Natl Acad Sci USA 89:10802–10806

Pan W-H, Xin P, Morrey JD, Clawson GA (2004) A self-processing ribozyme cassette: utility against human papillomavirus 11 E6/E7 mRNA and hepatitis B virus. Mol Ther 9:596–606

Peracchi A (2004) Prospects for antiviral ribozymes and deoxyribozymes. Rev Med Virol 14:47–64

Phipps KM, Martinez A, Lu J, Heinz BA, Zhao G (2004) Small interfering RNA molecules as potential anti-human rhinovirus agents: in vitro potency, specificity and mechanism. Antiviral Res 61:49–55

Qin X-F, An DS, Chen ISY, Baltimore D (2003) Inhibiting HIV-1 infection in human T cells by lentiviral-mediated delivery of small interfering RNA against CCR5. Proc Natl Acad Sci USA 100:183–188

Radhakrishnan SK, Layden TJ, Gartel AL (2004) RNA interference as a new strategy against viral hepatitis. Virology 323:173–181

Randall G, Rice CM (2004) Interfering with hepatitis C virus RNA replication. Virus Res 102:19–25

Randall G, Grakoui A, Rice CM (2003) Clearance of replicating hepatitis C virus replicon RNAs in cell culture by small interfering RNAs. Proc Natl Acad Sci USA 100:235–240

Reynolds A, Leake D, Boese Q, Scaringe S, Marshall WS, Khvorova A (2004) Rational siRNA design for RNA interference. Nat Biotechnol 22:326–330

Ryther RCC, Flynt AS, Phillips III JA, Patton JG (2005) siRNA therapeutics: big potential from small RNAs. Gene Ther 12:5–11

Saleh M-C, Van Rij RP, Andino R (2004) RNA silencing in viral infections: insights from poliovirus. Virus Res 102:11–17

Santoro SW, Joyce GF (1997) A general purpose RNA-cleaving DNA enzyme. Proc Natl Acad Sci USA 94:4262–4266

Scacheri PC, Rozenblatt-Rosen O, Caplen NJ, Wolfsberg TG, Umayam L, Lee JC, Hughes CM, Shanmugam KS, Bhattacharjee A, Meyerson M, Collins FS (2004) Short interfering RNAs can induce unexpected and divergent changes in the levels of untargeted proteins in mammalian cells. Proc Natl Acad Sci USA 101:1892–1897

Schubert S, Kurreck J (2004) Ribozyme- and deoxyribozyme-strategies for medical applications. Curr Drug Targets 5:667–681

Schubert S, Gül DC, Grunert H-P, Zeichhardt H, Erdmann VA, Kurreck J (2003) RNA cleaving "10–23" DNAzymes with enhanced stability and activity. Nucleic Acids Res 31:5982–5992

Schubert S, Fürste JP, Werk D, Grunert H-P, Zeichhardt H, Erdmann VA, Kurreck J (2004) Gaining target access for deoxyribozymes. J Mol Biol 339:355–363

Schubert S, Grunert H-P, Zeichhardt H, Werk D, Erdmann VA, Kurreck J (2005a) Maintaining Inhibition: siRNA double expression vectors against coxsackieviral RNAs. J Mol Biol 346:457–465

Schubert S, Grünweller A, Erdmann VA, Kurreck J (2005b) Local RNA target structure influences siRNA efficacy: systematic analysis of intentionally designed binding regions. J Mol Biol 348:883–893

Schwarz DS, Hutvágner G, Du T, Xu Z, Aronin N, Zamore PD (2003) Asymmetry in the assembly of the RNAi enzyme complex. Cell 115:199–208

Seksek O, Bolard J (2004) Delivery agents for oligonucleotides. Methods Mol Biol 252:545–568

Semizarov D, Frost L, Sarthy A, Kroeger P, Halbert DN, Fesik SW (2003) Specificity of short interfering RNA determined through gene expression signatures. Proc Natl Acad Sci USA 100:6347–6352

Silva J, Chang K, Hannon GJ, Rivas FV (2004) RNA-interference-based functional genomics in mammalian cells: reverse genetics coming of age. Oncogene 23:8401–8409

Sioud M (2004) Therapeutic siRNAs. Trends Pharmacol Sci 25:22–28

Sledz CA, Holko M, deVeer MJ, Silverman RH, Williams BR (2003) Activation of the interferon system by short-interfering RNAs. Nat Cell Biol 5:834–839

Sohail M, Southern EM (2000) Selecting optimal antisense reagents. Adv Drug Deliv Rev 44:23–34

Soutschek J, Akinc A, Bramlage B, Charisse K, Constien R, Donoghue M, Elbashir S, Geick A, Hadwiger P, Harborth J, John M, Kesavan V, Lavine G, Pandey RK, Racie T, Rajeev KG, Röhl I, Toudjarska I, Wang G, Wuschko S, Bumcrot D, Koteliansky V, Limmer S, Manoharan M, Vornlocher H-P (2004) Therapeutic silencing of an endogenous gene by systemic administration of modified siRNAs. Nature 432:173–178

Sullenger BA, Gilboa E (2002) Emerging clinical applications of RNA. Nature 418:252–258

Takahashi H, Hamazaki H, Habu Y, Hayashi M, Abe T, Miyano-Kurosaki N, Takaku H (2004) A new modified DNA enzyme that targets influenza virus A mRNA inhibits viral infection in cultured cells. FEBS Lett 560:69–74

Thomas CE, Ehrhardt A, Kay MA (2003) Progress and problems with the use of viral vectors for gene therapy. Nat Rev Genet 4:346–358

Tompkins SM, Lo C-Y, Tumpey TM, Epstein SL (2004) Protection against lethal influenza virus challenge by RNA interference in vivo. Proc Natl Acad Sci USA 101:8682–8686

Uprichard SL, Boyd B, Althage A, Chisari FV (2005) Clearance of hepatitis B virus from the liver of transgenic mice by short hairpin RNAs. Proc Natl Acad Sci USA 102:773–778

Usman N, Blatt LM (2000) Nuclease-resistant synthetic ribozymes: developing a new class of therapeutics. J Clin Invest 106:1197–1202

Wilson JA, Jayasena S, Khvorova A, Sabatinos S, Rodrigue-Gervais IG, Arya S, Sarangi F, Harris-Brandts M, Beaulieu S, Richardson CD (2003) RNA interference blocks gene expression and RNA synthesis from hepatitis C replicons propagated in human liver cells. Proc Natl Acad Sci USA 100:2783–2788

Yuan J, Cheung PKM, Zhang H, Chau D, Yanagawa B, Cheung C, Luo H, Wang Y, Suarez A, McManus BM, Yang D (2004) A phosphorothioate antisense oligodeoxynucleotide specifically inhibits coxsackievirus B3 replication in cardiomyocytes and mouse hearts. Lab Invest 84:703–714

Zamecnik PC, Stephenson ML (1978) Inhibition of Rous sarcoma virus replication and cell transformation by a specific oligodeoxynucleotide. Proc Natl Acad Sci USA 75:280–284

Zhang J, Yamada O, Sakamoto T, Yoshida H, Iwai T, Matsushita Y, Shimamura H, Araki H, Shimotohno K (2004) Down-regulation of viral replication by adenoviral-mediated expression of siRNA against cellular cofactors for hepatitis C virus. Virology 320:135–143

Zhang W, Yang H, Kong X, Mohapatra S, San Juan-Vergara H, Hellermann G, Behera S, Singam R, Lockey RF, Mohapatra SS (2005) Inhibition of respiratory syncytial virus infection with intranasal siRNA nanoparticles targeting the viral NS1 gene. Nat Med 11:56–62

Zhang X, Xu Y, Ling H, Hattori T (1999) Inhibition of infection of incoming HIV-1 virus by RNA cleaving DNA enzyme. FEBS Lett 458:151–156

Zhou C, Bahner IC, Larson GP, Zaia JA, Rossi JJ, Kohn EB (1994) Inhibition of HIV-1 in human T-lymphocytes by retrovirally transduced anti-tat and rev hammerhead ribozymes. Gene 149:33–39

Gene-Expressed RNA as a Therapeutic: Issues to Consider, Using Ribozymes and Small Hairpin RNA as Specific Examples

G. C. Fanning · G. Symonds (✉)

Johnson Johnson Research, The Australian Technology Park, Strawberry Hills, Locked Bag 4555, 2012 Sydney NSW, Australia
gsymonds@medau.jnj.com

1	Gene Regulation by Ribozymes and RNAi	289
2	Delivery and In Vivo Toxicity	292
3	Expression	296
4	Potency, Dose and Toxicity	297
5	Commercial Development	298
6	Summary	298
	References	299

Abstract In recent years there has been a greater appreciation of both the role of RNA in intracellular gene regulation and the potential to use RNA in therapeutic modalities. In the latter case, RNA can be used as a therapeutic target or a drug. The chapters in this volume cover the varied and potent actions of RNA as antisense, ribozymes, aptamers, microRNA and small hairpin RNA in gene regulation, as well as their use as potential therapeutics for metabolic and infectious diseases. Our group has been involved in the development of anti-HIV gene expression constructs to treat HIV. In this chapter, we address the relevant scientific and some of the commercial issues in the use of RNA as a therapeutic. Specifically, the chapter discusses delivery, expression, potency, toxicity and commercial development using, as examples, hammerhead ribozymes and small hairpin RNA.

Keywords Ribozyme · RNAi · Gene therapy · HIV

1
Gene Regulation by Ribozymes and RNAi

Nucleic acid-based agents such as ribozymes have been used to regulate gene expression and thereby dissect the function of biological pathways (Takagi et al. 2002; Kamath et al. 2003; Waninger et al. 2004). In most recent times, RNA interference (RNAi) has been highlighted due to its ease of use and efficacy of gene suppression.

Hammerhead ribozymes are so named for the secondary structure of the minimized sequence shown by Haseloff and Gerlach (1988) to be required for the *trans* cleavage of target RNA species. The hammerhead ribozyme is composed of three helices: the first and third provide target specificity as binding arms upstream and downstream of the target NUX triplet; the second forms the characteristic catalytic core that binds magnesium and cleaves immediately after the target NUX (Fig. 1a). The length and composition of each of the hammerhead ribozyme helices has been subjected to in vitro analysis and selection, thereby providing a number of catalytic variations on the original minimized ribozyme (Eckstein et al. 2001; McCall et al. 1992; Persson et al. 2002). The intracellular efficiency of ribozyme cleavage is dependent on (1) the degree of ribozyme expression (a major determinant of intracellular concentration) and (2) co-localisation of the target and ribozyme. In terms of expression, a recent

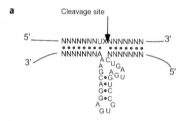

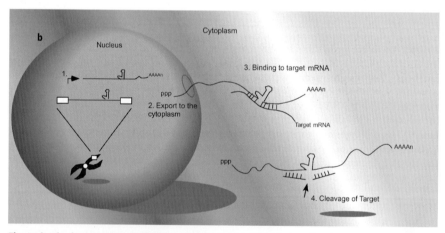

Fig. 1a,b The hammerhead ribozyme (**a**) is shown, illustrating the characteristic hammerhead catalytic motif, binding arms and the target mRNA indicating the NUX recognition and cleavage site (**b**). An expressed hammerhead ribozyme needs to be integrated within the host genome via gene transfer as part of, for example, a retroviral recombinant virus (see Fig. 3 for procedure). Once integrated, the ribozyme is transcribed (*1*) and, if under the control of a Pol II promoter, encodes a poly A tail signalling export from the nucleus into the cytoplasm (*2*). In the cytoplasm, the ribozyme-containing RNA transcript binds to the target RNA by Watson-Crick base pairing (*3*), resulting in the cleavage of the target RNA (*4*)

study has demonstrated an additional loop sequence in helix I that appears to enhance *cis* cleavage under intracellular conditions (Khvorova et al. 2003).

RNAi has proved very effective at gene suppression. For the most part, this has been due to the high success rate of designing and applying RNAi to a wide variety of cell culture and animal experiments. The RNAi pathway exists in both plants and animals; in plants it is an integral part of providing protection from viruses (Waterhouse et al. 2001). In contrast, mammals have higher order pathways to deal with viral infection [interferon, double-stranded (ds)RNA-activated protein kinase (PKR), $2'$-$5'$ oligo adenylate synthetase (OAS)] and in this case the RNAi pathway appears to have an important role in the processing of microRNAs (miRNA)—shown to regulate host gene expression during development and cellular differentiation (Lau et al. 2001). The mode of action is shown in Fig. 2 and summarised briefly here. RNA that is expressed by a polymerase III promoter as a hairpin or as separate sense and antisense

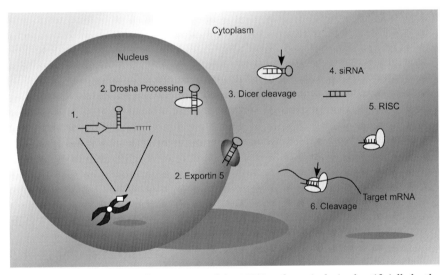

Fig. 2 The active and targeted component of the RNAi pathway is derived artificially by the integration of an expression cassette aimed at producing a hairpin RNA transcript that has homology to the RNA marked for down regulation. Most applications of this technology utilize Pol III promoters where transcription (*1*) results in a small hairpin RNA retained in the nucleus. The hairpin is recognised by an enzyme called Drosha (and other unknown proteins) and undergoes modification before being exported from the nucleus by Exportin 5 (*2*). In the cytoplasm, the dsRNA portion of the hairpin is recognised by an enzyme known as Dicer that cleaves the RNA 21–23 bases from the $5'$-end of the RNA transcript (*3*), forming the characteristic siRNA molecule (*4*). Some investigators deliver siRNA directly to the cell to intervene in the RNAi pathway at this point although the processing of small hairpin (sh)RNA is thought to be a more efficient way of using this pathway for gene suppression. The siRNA is included into an RNA protein complex known as the RNA-induced silencing complex (*RISC*) that uses the antisense portion of the siRNA as a guide for binding target RNA (*5*). Once bound the target RNA is cleaved and degraded (*6*)

RNA is recognised in the nucleus by an RNAse III enzyme, Drosha. The partially processed hairpin is then exported from the nucleus to the cytoplasm by Exportin 5 (Yi et al. 2003) and binds with a protein complex that contains Dicer (Bernstein et al. 2001). Dicer cleaves the hairpin into the effector RNA fragments termed small interfering (si)RNA. These are 19-base dsRNA species with two $3'$ overhangs (Elbashir et al. 2001). Direct delivery of siRNA to a cell bypasses the nuclear processing steps and, like the processed siRNAs, enters the RNA-induced silencing complex (RISC). RISC mediates cleavage of the target RNA sequence using the specificity provided by the bound antisense RNA strand and the catalytic activity of the RISC protein component Ago2 (Meister et al. 2004; Liu et al. 2004). Research continues towards the discovery of other cellular proteins involved in this pathway (Denli et al. 2004).

2
Delivery and In Vivo Toxicity

Delivery has long been the major issue for gene therapy. A search of one of the major gene therapy clinical trial databases (http://www.wiley.co.uk/genetherapy/clinical/) shows that most gene transfer studies aim to treat cancer using direct injection of viral or naked DNA preparations (Edelstein et al. 2004). The goal of these trials is to either increase the sensitivity of tumour cells to apoptosis or deliver tumour antigens to promote an anti-cancer immune response. Another major mode of gene transfer investigated to date is the ex vivo transduction and subsequent reinfusion of autologous cells into patients. This delivery approach using haematopoietic stem/progenitor cells [with the caveat of observed toxicity in certain X-linked severe combined immune deficiency syndrome (SCID-X1) patients] was successful in treating SCID-X1 (Hacein-Bey-Abina et al. 2002; Cavazzana-Calvo et al. 2004; Gaspar et al. 2004) and adenosine deaminase deficient (ADA)-SCID (Bordignon et al. 1995; Aiuti et al. 2002a,b) patients. It has also been applied clinically to potentially generate human immunodeficiency virus (HIV)-protected cells (Kohn et al. 1999; Kang et al. 2002; Amado et al. 2004). Tempering gene therapy successes to date, a patient death after direct injection of an adenoviral preparation and the development of lymphoproliferative disease/leukaemia in three of the patients treated in the SCID-X1 clinical trial have highlighted some of the potential risks of this new form of treatment (discussed later in this section).

The ex vivo transduction approach is illustrated in Fig. 3. The patient receives granulocyte colony stimulating factor (G-CSF) to mobilise haematopoietic stem cells from the bone marrow into the peripheral blood (Hubel and Engert 2003). The protocols for this procedure were pioneered in the bone marrow transplantation setting, and a basic protocol involves the systemic administration of 10 µg/kg of G-CSF followed by blood collection (apheresis). Apheresis is performed as an outpatient procedure and nucleated (predomi-

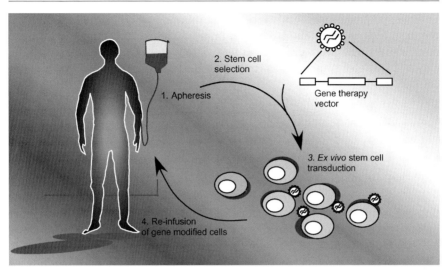

Fig. 3 Flow diagram of the out-patient procedure for the ex vivo transduction and re-infusion of genetically modified cells. Prior to apheresis the patient receives granulocyte colony stimulating factor to mobilise stem cells from the bone marrow into the peripheral blood system. After mobilization, the cells in peripheral blood are apheresed (*1*) and the stem cells separated by the CD34$^+$ surface marker (*2*). The stem cells are then cultured with recombinant virus that encodes the gene therapy. The recombinant virus illustrated contains a simple transfer vector composed of a long terminal repeat (LTR) (*hatched*) driving expression of a therapeutic gene (*cross-hatched*). In the final step, gene-modified cells are re-infused into the patient

nantly mononuclear) cells from the equivalent of one to several blood volumes can be collected. Following this, the haematopoietic stem/progenitor cells are collected, generally using a CD34 antibody/bead complex and the cells cultured with cytokines to drive them into cycle. They are then transduced, harvested and reinfused to the patient. Reinfusing 1×10^6 CD34$^+$ progenitor cells per kilogram is enough cells to rescue a person after bone marrow ablation (Chao and Blume 1990). In gene-therapy settings, larger doses of cells ranging from 2×10^6 to 5×10^6 cells/kg can be achieved to improve engraftment with gene-modified cells (Cavazzana-Calvo et al. 2000; Bordignon et al. 1995; Amado et al. 2004; Gaspar et al. 2004). For most applications, ex vivo transduction has utilised retroviral vectors encoding the therapeutic gene; this requires the cells be induced into cycle prior to exposure to the recombinant virus (Nolta and Kohn 1990). The induction of cell cycling does not appear to compromise stem/progenitor cell pluripotency—gene-modified cells of myeloid and lymphoid lineages have been reported in the peripheral blood system using this protocol (Kobari et al. 2000; Amado et al. 2004). In our own studies using this method, we have observed cell marking and gene expression greater than 3 years post infusion in T lymphocytes and monocytes (Amado et al. 2004).

Important aspects of future research in this field will include the search for the most appropriate stem/progenitor cells for a given application and the use of other retroviral vectors, such as lentiviral vectors that do not require the cycling of cells for transduction (Uchida et al. 1988; Naldini 1998; Douglas et al. 2001).

The occurrence of two cases of leukaemia (and a third with lymphoproliferative disease at the time of writing) in SCID-X1 children receiving retroviral gene therapy has been discussed within the field since early 2003. The theoretical occurrence of a mutagenic event caused by the integration of a transgene into or near to an oncogene was known; however, it was thought to be negligible using retroviral-incompetent vectors, given the data showing no leukaemia developing in mice, large animals and approximately 1,800 patients that had received retrovirus-modified cells. There was a previous single report of leukaemia development in mice. A retroviral vector containing the nerve growth factor receptor (NGFr) gene was shown to integrate within the ecotropic viral integration site-1 (Evi-1) in haematopoietic progenitor cells, giving rise to a single leukaemic clone within one of the five primary transplanted mice. Bone marrow was collected from primary transplanted mice, pooled and then re-transplanted into secondary mice. All 10 secondary transplant mice developed haematopoietic disorders within 22 weeks. In these mice the development of leukaemia appeared to be due to the combination of NGFr and Evi-1 (Li et al. 2002). In the SCID-X1 trial, the two children who presented with leukaemia were the youngest, while the recent third child was 8 months of age at the time of infusion. SCID-X1 is an immune deficiency caused by the absence of a gene located on the X chromosome encoding interleukin (IL)-2γc, an essential component of the IL-2, -4, -7, -9, -15 and -21 receptors and therefore essential to the maturation of T cells during thymopoiesis (Buckley 2004). This clinical work is considered the first major success for gene therapy with all patients developing T cells and 9 out of 11 a large number of T cells with diverse T cell repertoire that was curative (Cavazzana-Calvo et al. 2000; Hacein-Bey-Abina et al. 2002). For the first two children affected by the leukaemic serious adverse event, the gene therapy procedure was directly implicated in the development of leukaemia. The malignant cells from both patients were characterised by an insertion of the retrovirus bearing the γ common chain complementary (c)DNA in or near to the LMO2 gene locus, a gene known to be involved in T cell leukaemia in mice and humans (Hacein-Bey-Abina et al. 2003). However, this finding alone could not fully explain the development of leukaemia, as the other patients in retroviral-based clinical trials had not developed leukaemia after retroviral gene transfer, and the probability of each patient bearing at least one *LMO2* insertion was high. Additional contributing factors have been summarised as (1) the transgene itself (Dave et al. 2004), (2) the age of the children (relatively immature haematopoietic system), (3) the massive proliferation of T cells in response to the introduction of the γ common gene in these cells and (4) the disease itself. Up to the recent report on the third child, these events did not preclude the continuation of SCID-X1 trials, although there was a requirement for additional patient screening and the FDA evaluated it on a case-by-case basis (UK clinical trials were also successful and did not actually stop) (Berns et al. 2004; Gaspar et al. 2004).

Direct administration of vector to the patient is conceptually and logistically simpler; the vector can be introduced to the site (e.g. cancer) or systemically introduced. However, targeting gene delivery vectors to appropriate body sites is not a trivial undertaking and there has been very limited success to date. This is reflected in the majority of in vivo procedures being dependent on localising the delivery vehicle to the site of action (e.g. a tumour) by direct injection.

Adenoviral vectors have been used extensively for this purpose and their use was subject to considerable debate following the death of an adolescent in Philadelphia who was enrolled in a gene therapy trial for the treatment of partial ornithine transcarboxylase deficiency. This condition, due to the inborn error of urea synthesis (Lindgren et al. 1984; McCullough et al. 2000), leads to the build-up of ammonia causing death if not controlled. The patient received 6×10^{11} particles/kg, (the highest dose in the clinical trial) by direct injection into the right hepatic artery (Raper et al. 2001). Within 18 h, clinical signs of an adverse reaction were apparent and this led to systemic inflammatory response syndrome and ultimate failure of the lungs, resulting in death (Raper et al. 2003). In addition to clinical documentation that was found to be deficient, relevant issues included the relationship between dose of vector and toxicity—this did not appear to be linear and had substantial subject-to-subject variability. The NIH report (2002) *inter alia*, stated that deciphering the reasons why one patient reacted so violently has yet to be determined but will be important to the future of this form of gene therapy.

In 2003, a drug license was granted in China for a recombinant Ad-53 gene therapy for squamous cell carcinoma of the head and neck (SCCHN: Pearson and Jia 2004). SCCHN is a common cancer and, despite recent advances in treatment, still presents a major clinical challenge. This is the world's first commercially available gene therapy and we review here the important points associated with this product as an example of a directly injected gene therapy. p53 is a critical modulator of the cellular response to exogenous and endogenous stress and has a central role as a tumour suppressor. Inactivation of one or more components of the p53 network is a common event in human tumours (for review see Gasco and Crook 2003). There are several biotechnology firms in phase II or later clinical development based on a similar principle of delivering a gene directly to the tumour to restore or initiate apoptotic processes. In data directly related to the approved gene therapy, adenovirus encoding the p53 gene was administered weekly by direct injection of 1×10^{12} viral particles into the SCCHN. In combination with radiotherapy, 64% of late stage tumours completely regressed and 32% showed partial regression. (These data are from the SiBiono Web site www.Sibio.com, and to date they have not appeared in the peer-reviewed literature.)

As for classical small molecules, toxicity is also an issue for gene therapy; it is being addressed both in pre-clinical models and by the close monitoring of patients throughout the clinical trial period and in follow-up (Nyberg et al. 2004).

3
Expression

The expression of non-host RNA in a cell must also be considered, given that cells have evolved elaborate mechanisms for detecting and eliminating foreign RNA and DNA, for example, from viruses. The measures evolved by viruses to ensure the uptake and expression of their own genetic material perhaps highlight the need to be sophisticated in attempts to transfer genetic material to human cells. There has been a relatively high degree of success using constitutive and inducible expression systems in human cells. A wide variety of promoters have been used to constitutively express genes including ribozymes and small hairpin (sh)RNAs that are Pol II- and Pol III-dependent (Brummelkamp et al. 2002). It is not just expression per se; a key issue appears to be the intracellular location, in that investigators have demonstrated the need to co-localise the target and therapeutic RNA (Kuwabara et al. 2001; Bertrand et al. 2001). This is particularly true for ribozymes that do not utilise a host protein system for activity. Examples of promoters used for ribozymes include the murine leukaemia virus (MLV) LTR (Sun et al. 1994), transfer (t)RNA (Thompson et al. 1995), viral-associated RNA I (VAI) (Cagnon and Rossi 2000), small nuclear RNA U1 (De Young et al. 1994), small nuclear RNA U6 (U6) (Ilves et al 1996), cytomegalovirus (CMV) (Mahieu et al. 1994) and simian virus 40 (SV40) (Chen et al. 1992). For shRNA constructs, Pol III promoters (H1 and U6) have been the promoters of choice; they act to release the RNA transcript in the nucleus, after which the therapeutic is transported to the site of action by the RNAi cellular machinery (Brummelkamp et al. 2002; Miyagishi and Taira 2002). The nuclear localisation of shRNA, particularly those longer than 30 bases, is important to avoid the anti-viral response proteins, such as PKR, located in the cytoplasm (Wang et al 2004).

Tight control of transgene expression represents a major goal for this field. This has become increasingly apparent with recent publications focusing on the off-target effects that can occur with any RNA-based therapy where partial homology with host sequence results in activity not predetermined by design (Jackson et al. 2003).

In addition, the development of inducible promoter systems is an area of active research likely to be applied to clinical applications where expression needs to be modulated rather than constitutive. In terms of induction, these expression systems may respond to a disease stimulus—an example being the HIV-LTR that would express in response to an actively replicating (tat producing) HIV (Weerasinghe et al. 1991; Unwalla et al. 2004). In another form, expression responds to an administered pharmacological agent such as an antibiotic or a small molecule.

There are numerous examples of inducible systems in the literature (for review see Toniatti et al. 2004). By way of example, the development of the Tet Off/On system is worthy of summary here. The Tet system is based on a DNA-binding protein that binds to a sequence in the prokaryotic Tet-resistance operon. In its native form, the Tet repressor (TetR) binds to the DNA sequence preventing transcription and in the presence of tetracycline (or derivatives such as doxycycline) undergoes a conformation change, removing its DNA-binding capacity. When bound to the gene activation domain VP16, the TetR–VP16 protein induced expression and in the presence of the antibiotic transcription was shut down. This so-called Tet-Off system was not ideal, as the shutting down of the system required the presence of the exogenous agent (tetracycline) (Baron and Bujard 2000). Mutagenesis of the TetR–VP16 fusion protein reversed the binding properties such that in the presence of the exogenous agent, TetR–VP16 bound to the DNA and induced expression (Tet-On system) (Urlinger et al. 2000). While this system was shown to be specific in its gene induction, there was considerable basal gene expression in the absence of bound TetR–VP16. The isolation of variants of the TetR–VP16 greatly enhanced the induction of the expression system and eliminated the basal gene expression to a point where it could be considered for clinical application (Urlinger et al. 2000; Lamartina et al. 2003). Importantly, the fusion protein TetR–VP16 has been shown to be non-immunogenic in mice, a feature that needs to be demonstrated in humans for it to be widely applied. Antibiotics have been used extensively in humans, and tetracycline and doxycycline have both been shown to have good safety profiles, making their use as activators of gene therapeutics feasible (Klein and Cunha 2001). The relevant features of the Tet-On system are more broadly applicable and can be summarised as: a switched on rather than a switched off system; specificity for transgene expression cassette; bioavailability and reversibility of induction for safety; low basal activity; high inducibility; dose dependence; and low immunogenicity.

4
Potency, Dose and Toxicity

Potency and toxicity is the double-edged sword of dosing that is negotiated for all drugs. For gene therapy, dose can refer to the number of viral particles directly injected into a tumour, the number of stem/progenitor cells re-infused into a patient, the number of cells derived from gene-modified stem cells over time or the amount of intracellular expression of a gene therapeutic. The balance of potency and toxicity is well documented in the small-molecule literature, and the relatively transient nature of these types of drugs, and the ability to determine the duration of the effect in animal models, make at least some of the issues related to dose quantifiable. The effects that are not always readily quantifiable are those that include long-term biological effects, for example on-going liver damage. Gene therapy using expression constructs aims to give long-lasting effects from a single or infrequent dose; this makes measuring the relevant dose and potential toxicities in a way more complex and less well defined. The NIH has recommended guidelines for the long-term monitoring of patients as a way to address the presently unknown aspects of gene therapy.

5
Commercial Development

As in any new field, there has been considerable commentary on the potential of gene therapy. Such commentary has been punctuated by the hype of this new technology and the lows of unrealised expectations and adverse events. Throughout, the themes of safety, efficacy, cost and availability have come to the fore. However, the commercial model that would apply is not obvious. Is it a 'bottle' of drug that is administered, or should one consider the delivery and logistics? In particular, what is the product/service and how is the treatment going to be given to patients? The overall cost of a product can be divided into the following areas: (1) composition, (2) development, including clinical trials and (3) administration/monitoring. The first cost (composition) needs to take into account that any one gene therapeutic will be composed of 'bits and pieces' derived from different researchers each with its own intellectual property; for example, a relatively 'simple' gene therapeutic would still be composed of a delivery vector, a promoter and a therapeutic gene. In terms of the second cost, the pre-clinical and clinical work will have determined the means by which patients are treated, and in a way this will lead to the commercial model. The logistics of treating patients is a major issue, particularly in procedures that require the ex vivo manipulation of cells (refer to Fig. 3).

The economic reality of this process has not yet been trialed, and there are obvious advantages for gene therapies that are directly injected into the patient. Another issue common to all forms of gene therapy is the requirement for long-term monitoring of patients that have received therapy (Nyberg et al. 2004). This is a significant commitment.

6
Summary

Hammerhead ribozymes and expressed dsRNA have shown activity for a wide range of individual mRNA targets in tissue culture and as randomised libraries in in vitro selection systems. As such, they have demonstrated their potential as regulators of gene expression in the context of disease. The intracellular activity of a gene therapeutic is the first significant step in the design and development of a product based on genetic manipulation. In this chapter we have reviewed the additional importance of delivery and expression, the potential toxicities of expressing therapeutic RNA in a cell and some of the commercial issues associated with the development of RNA as a drug.

References

Aiuti A, Vai S, Mortellaro A, Casorati G, Ficara F, Andolfi G, Ferrari G, Tabucchi A, Carlucci F, Ochs HD, Notarangelo LD, Roncarolo MG, Bordignon C (2002a) Immune reconstitution in ADA-SCID after PBL gene therapy and discontinuation of enzyme replacement. Nat Med 5:423–425

Aiuti A, Slavin S, Aker M, Ficara F, Deola S, Mortellaro A, Morecki S, Andolfi G, Tabucchi A, Carlucci F, Marinello E, Cattaneo F, Vai S, Servida P, Miniero R, Roncarolo MG, Bordignon C (2002b) Correction of ADA-SCID by stem cell gene therapy combined with nonmyeloablative conditioning. Science 296:2410–2413

Amado RG, Mitsuyasu RT, Rosenblatt JD, Ngok FK, Bakker A, Cole S, Chorn N, Lin LS, Bristol G, Boyd MP, MacPherson JL, Fanning GC, Todd AV, Ely JA, Zack JA, Symonds GP (2004) Anti-human immunodeficiency virus hematopoietic progenitor cell-delivered ribozyme in a phase I study: myeloid and lymphoid reconstitution in human immunodeficiency virus type-1-infected patients. Hum Gene Ther 3:251–262

Baron U, Bujard H (2000) Tet repressor-based system for regulated gene expression in eukaryotic cells: principles and advances. Methods Enzymol 327:401–421

Baum C, von Kalle C, Staal FJ, Li Z, Fehse B, Schmidt M, Weerkamp F, Karlsson S, Wagemaker G, Williams DA (2004) Chance or necessity? Insertional mutagenesis in gene therapy and its consequences. Mol Ther 1:5–13

Bernstein E, Caudy AA, Hammond SM, Hannon GJ (2001) Role for a bidentate ribonuclease in the initiation step of RNA interference. Nature 409:363–366

Bertrand E, Castanotto D, Zhou C, Carbonnelle C, Lee NS, Good P, Chatterjee S, Grange T, Pictet R, Kohn D, Engelke D, Rossi JJ (1997) The expression cassette determines the functional activity of ribozymes in mammalian cells by controlling their intracellular localization. RNA 1:75–88

Bordignon C, Notarangelo LD, Nobili N, Ferrari G, Casorati G, Panina P, Mazzolari E, Maggioni D, Rossi C, Servida P (1995) Gene therapy in peripheral blood lymphocytes and bone marrow for ADA-immunodeficient patients. Science 270:470–475

Brummelkamp TR, Bernards R, Agami R (2002) A system for stable expression of short interfering RNAs in mammalian cells. Science 296:550–553

Buckley RH (2004) Molecular defects in human severe combined immunodeficiency and approaches to immune reconstitution. Annu Rev Immunol 22:625–655

Cagnon L, Rossi JJ (2000) Downregulation of the CCR5 beta-chemokine receptor and inhibition of HIV-1 infection by stable VA1-ribozyme chimeric transcripts. Antisense Nucleic Acid Drug Dev 4:251–261

Cavazzana-Calvo M, Fischer A (2004) Efficacy of gene therapy for SCID is being confirmed. Lancet 364:2155–2156

Cavazzana-Calvo M, Hacein-Bey S, de Saint Basile G, Gross F, Yvon E, Nusbaum P, Selz F, Hue C, Certain S, Casanova JL, Bousso P, Deist FL, Fischer A (2000) Gene therapy of human severe combined immunodeficiency (SCID)-X1 disease. Science 288:669–672

Chao NJ, Blume KG (1990) Bone marrow transplantation. Part II-autologous. West J Med 152:46–51

Chen CJ, Banerjea AC, Harmison GG, Haglund K, Schubert M (1992) Multitarget-ribozyme directed to cleave at up to nine highly conserved HIV-1 env RNA regions inhibits HIV-1 replication-potential effectiveness against most presently sequenced HIV-1 isolates. Nucleic Acids Res 20:4581–4589

Dave UP, Jenkins NA, Copeland NG (2004) Gene therapy insertional mutagenesis insights. Science 2303:333

De Young MB, Kincade-Denker J, Boehm CA, Riek RP, Mamone JA, McSwiggen JA, Graham RM (1994) Functional characterization of ribozymes expressed using U1 and T7 vectors for the intracellular cleavage of ANF mRNA. Biochemistry 33:12127–12138

Denli AM, Tops BB, Plasterk RH, Ketting RF, Hannon GJ (2004) Processing of primary microRNAs by the Microprocessor complex. Nature 432:231–235

Douglas JL, Lin WY, Panis ML, Veres G (2001) Efficient human immunodeficiency virus-based vector transduction of unstimulated human mobilized peripheral blood CD34+ cells in the SCID-hu Thy/Liv model of human T cell lymphopoiesis. Hum Gene Ther 12:401–413

Eckstein F, Kore AR, Nakamaye KL (2001) In vitro selection of hammerhead ribozyme sequence variants. Chembiochem 2:629–635

Edelstein ML, Abedi MR, Wixon J, Edelstein RM (2004) Gene therapy clinical trials worldwide 1989-2004—an overview. J Gene Med 6:597–602

Elbashir SM, Lendeckel W, Tuschl T (2001) RNA interference is mediated by 21- and 22-nucleotide RNAs. Genes Dev 15:188–200

Gasco M, Crook T (2003) The p53 network in head and neck cancer. Oral Oncol 39:222–231

Gaspar HB, Parsley KL, Howe S, King D, Gilmour KC, Sinclair J, Brouns G, Schmidt M, Von Kalle C, Barington T, Jakobsen MA, Christensen HO, Al Ghonaium A, White HN, Smith JL, Levinsky RJ, Ali RR, Kinnon C, Thrasher AJ (2004) Gene therapy of X-linked severe combined immunodeficiency by use of a pseudotyped gammaretroviral vector. Lancet 364:2181–2187

Hacein-Bey-Abina S, Le Deist F, Carlier F, Bouneaud C, Hue C, De Villartay JP, Thrasher AJ, Wulffraat N, Sorensen R, Dupuis-Girod S, Fischer A, Davies EG, Kuis W, Leiva L, Cavazzana-Calvo M (2002) Sustained correction of X-linked severe combined immunodeficiency by ex vivo gene therapy. N Engl J Med 346:1185–1193

Hacein-Bey-Abina S, Von Kalle C, Schmidt M, McCormack MP, Wulffraat N, Leboulch P, Lim A, Osborne CS, Pawliuk R, Morillon E, Sorensen R, Forster A, Fraser P, Cohen JI, de Saint Basile G, Alexander I, Wintergerst U, Frebourg T, Aurias A, Stoa-Lyonnet D, Romana S, Radford-Weiss I, Gross F, Valensi F, Delabesse E, Macintyre E, Sigaux F, Soulier J, Leiva LE, Wissler M, Prinz C, Rabbitts TH, Le Deist F, Fischer A, Cavazzana-Calvo M (2003) LMO2-associated clonal T cell proliferation in two patients after gene therapy for SCID-X1. Science 302:415–419

Haseloff J, Gerlach WL (1988) Simple RNA enzymes with new and highly specific endoribonuclease activities. Nature 334:585–591

Hubel K, Engert A (2003) Clinical allocations of granulocyte colony-stimulating factor: an update and summary. Ann Hematol 82:207–213

Ilves H, Barske C, Junker U, Bohnlein E, Veres G (1996) Retroviral vectors designed for targeted expression of RNA polymerase III-driven transcripts: a comparative study. Gene 171:203–208

Jackson AL, Bartz SR, Schelter J, Kobayashi SV, Burchard J, Mao M, Li B, Cavet G, Linsley PS (2003) Expression profiling reveals off-target gene regulation by RNAi. Nat Biotechnol 21:635–637

Kamath RS, Fraser AG, Dong Y, Poulin G, Durbin R, Gotta M, Kanapin A, Le Bot N, Moreno S, Sohrmann M, Welchman DP, Zierlen P, Ahringer J (2003) Systematic functional analysis of the Caenorhabditis elegans genome using RNAi. Nature 421:231–237

Kang EM, de Witte M, Malech H, Morgan RA, Phang S, Carter C, Leitman SF, Childs R, Barrett AJ, Little R, Tisdale JF (2002) Nonmyeloablative conditioning followed by transplantation of genetically modified HLA-matched peripheral blood progenitor cells for hematologic malignancies in patients with acquired immunodeficiency syndrome. Blood 99:698–701

Khovorova A, Lescoute A, Westhof E, Jayasena SD (2003) Sequence elements outside the hammerhead ribozyme catalytic core enable intracellular activity. Nat Struct Mol Biol 10:708–712

Klein NC, Cunha BA (2001) New uses of older antibiotics. Med Clin North Am 85:125–132

Kobari L, Pflumio F, Giarratana M, Li X, Titeux M, Izac B, Leteurtre F, Coulombel L, Douay L (2000) In vitro and in vivo evidence for the long-term multilineage (myeloid, B, NK, and T) reconstitution capacity of ex vivo expanded human CD34(+) cord blood cells. Exp Hematol 28:1470–1480

Kohn DB, Bauer G, Rice CR, Rothschild JC, Carbonaro DA, Valdez P, Hao Q, Zhou C, Bahner I, Kearns K, Brody K, Fox S, Haden E, Wilson K, Salata C, Dolan C, Wetter C, Aguilar-Cordova E, Church J (1999) A clinical trial of retroviral-mediated transfer of a rev-responsive element decoy gene into CD34(+) cells from the bone marrow of human immunodeficiency virus-1-infected children. Blood 94:368–671

Kuwabara T, Warashina M, Koseki S, Sano M, Ohkawa J, Nakayama K, Taira K (2001) Significantly higher activity of a cytoplasmic hammerhead ribozyme than a corresponding nuclear counterpart: engineered tRNAs with an extended 3′ end can be exported efficiently and specifically to the cytoplasm in mammalian cells. Nucleic Acids Res 29:2780–2788

Lamartina S, Silvi L, Roscilli G, Casimiro D, Simon AJ, Davies ME, Shiver JW, Rinaudo CD, Zampaglione I, Fattori E, Colloca S, Gonzalez Paz O, Laufer R, Bujard H, Cortese R, Ciliberto G, Toniatti C (2003) Construction of an rtTA2(s)-m2/tts(kid)-based transcription regulatory switch that displays no basal activity, good inducibility, and high responsiveness to doxycycline in mice and non-human primates. Mol Ther 7:271–280

Lau NC, Lim LP, Weinstein EG, Bartel DP (2001) An abundant class of tiny RNAs with probable regulatory roles in Caenorhabditis elegans. Science 294:858–862

Li Z, Dullmann J, Schiedlmeier B, Schmidt M, von Kalle C, Meyer J, Forster M, Stocking C, Wahlers A, Frank O, Ostertag W, Kuhlcke K, Eckert HG, Fehse B, Baum C (2002) Murine leukemia induced by retroviral gene marking. Science 296:497

Lindgren V, DeMartinville B, Horwich AL, Rosenberg LE, Francke U (1984) Human ornithine transcarbamylase locus mapped to band Xp21.1 near Duchenne muscular dystrophy locus. Science 226:698–700

Mahieu M, Deschuyteneer R, Forget D, Vandenbussche P, Content J (1994) Construction of a ribozyme directed against human interleukin-6 mRNA: evaluation of its catalytic activity in vitro and in vivo. Blood 84:3758–3765

McCall MJ, Hendry P, Jennings PA (1992) Minimal sequence requirements for ribozyme activity. Proc Natl Acad Sci U S A 89:5710–5714

McCall MJ, Hendry P, Mir AA, Conaty J, Brown G, Lockett TJ (2000) Small, efficient hammerhead ribozymes. Mol Biotechnol 14:5–17

McCullough B, Yuddkoff M, Batshaw ML, Wilson JM, Raper SE, Tuchman M (2000) Genotype spectrum of ornithine transcarbamylase deficiency: correlation with clinical and biochemical phenotype. Am J Med Genet 93:313–319

Miyagishi M, Taira K (2002) U6 promoter-driven siRNAs with four uridine 3′ overhangs efficiently suppress targeted gene expression in mammalian cells. Nat Biotechnol 20:497–500

Naldini L (1998) Lentiviruses as gene transfer agents for delivery to non-dividing cells. Curr Opin Biotechnol 9:457–463

NIH Report (2002) Assessment of adenoviral vector safety and toxicity: report of the national institutes of health recombinant DNA advisory committee. Hum Gene Ther 13:3–13

Nolta JA, Kohn DB (1990) Comparison of the effects of growth factors on retroviral vector-mediated gene transfer and the proliferative status of human hematopoietic progenitor cells. Hum Gene Ther 1:257–268

Nyberg K, Carter BJ, Chen T, Dunbar C, Flotte TR, Rose S, Rosenblum D, Simek SL, Wilson C (2004) Workshop on long-term follow-up of participants in human gene transfer research. Mol Ther 10:976–980

Pearson and Jia (2004) China approves first gene therapy. Nature 22:3–4

Persson T, Hartmann RK, Eckstein F (2002) Selection of hammerhead ribozyme variants with low Mg^{2+} requirement: importance of stem-loop II. Chembiochem 3:1066–1071

Raper SE, Yudkoff M, Chirmule N, Gao GP, Nunes F, Haskal ZJ, Furth EE, Propert KJ, Robinson MB, Magosin S, Simoes H, Speicher L, Hughes J, Tazelaar J, Wivel NA, Wilson JM, Batshaw ML (2002) A pilot study of in vivo liver-directed gene transfer with an adenoviral vector in partial ornithine transcarbamylase deficiency. Hum Gene Ther 13:163–175

Raper SE, Chirmule N, Lee FS, Wivel NA, Bagg A, Gao GP, Wilson JM, Batshaw ML (2003) Fatal systemic inflammatory response syndrome in a ornithine transcarbamylase deficient patient following adenoviral gene transfer. Mol Genet Metab 80:148–158

Sun LQ, Warrilow D, Wang L, Witherington C, Macpherson J, Symonds G (1994) Ribozyme-mediated suppression of Moloney murine leukemia virus and human immunodeficiency virus type I replication in permissive cell lines. Proc Natl Acad Sci U S A 91:9715–9719

Takagi Y, Suyama E, Kawasaki H, Miyagishi M, Taira K (2002) Mechanism of action of hammerhead ribozymes and their allocations in vivo: rapid identification of functional genes in the post-genome era by novel hybrid ribozyme libraries. Biochem Soc Trans 30:1145–1149

Thompson JD, Ayers DF, Malmstrom TA, McKenzie TL, Ganousis L, Chowrira BM, Couture L, Stinchcomb DT (1995) Improved accumulation and activity of ribozymes expressed from a tRNA-based RNA polymerase III promoter. Nucleic Acids Res 23:2259–2268

Toniatti C, Bujard H, Cortese R, Ciliberto G (2004) Gene therapy progress and prospects: transcription regulatory systems. Gene Ther 11:649–657

Uchida N, Sutton RE, Friera AM, He D, Reitsma MJ, Chang WC, Veres G, Scollay R, Weissman IL (1998) HIV, but not murine leukemia virus, vectors mediate high efficiency gene transfer into freshly isolated G0/G1 human hematopoietic stem cells. Proc Natl Acad Sci U S A 95:11939–11944

Unwalla HJ, Li MJ, Kim JD, Li HT, Ehsani A, Alluin J, Rossi JJ (2004) Negative feedback inhibition of HIV-1 by TAT-inducible expression of siRNA. Nat Biotechnol 22:1573–1578

Urlinger S, Baron U, Thellmann M, Hasan MT, Bujard H, Hillen W (2000) Exploring the sequence space for tetracycline-dependent transcriptional activators: novel mutations yield expanded range and sensitivity. Proc Natl Acad Sci U S A 97:7963–7968

Wang Q, Carmichael GG (2004) Effects of length and location on the cellular response to double-stranded RNA. Microbiol Mol Biol Rev 68:432–452

Waninger S, Kuhen K, Hu X, Chatterton JE, Wong-Staal F, Tang H (2004) Identification of cellular cofactors for human immunodeficiency virus replication via a ribozyme-based genomics approach. J Virol 78:12829–12837

Waterhouse PM, Wang MB, Lough T (2001) Gene silencing as an adaptive defense against viruses. Nature 411:834–842

Weerasinghe M, Liem SE, Asad S, Read SE, Joshi S (1991) Resistance to human immunodeficiency virus type 1 (HIV-1) infection in human CD4+ lymphocyte-derived cell lines conferred by using retroviral vectors expressing an HIV-1 RNA-specific ribozyme. J Virol 65:5531–5534

Yi R, Qin Y, Macara IG, Cullen BR (2003) Exportin-5 mediates the nuclear export of pre-microRNAs and short hairpin RNAs. Genes Dev 17:3011–3016

RNA Aptamers: From Basic Science Towards Therapy

H. Ulrich

Department of Biochemistry, Instituto de Química, Universidade de São Paulo,
Caixa Postal 26077, São Paulo 05513-970, Brazil
henning@iq.usp.br

1	Introduction	306
2	In Vitro Selection of Functional Nucleic Acids	307
3	Applications of SELEX	309
3.1	Development of Aptazymes	309
3.2	Development of Novel Binding Species	310
3.2.1	Aptamers as Modulators of Protein Function In Vitro	310
3.2.2	Intracellularly Expressed Aptamers for Dissection of Signal Transduction Pathways	313
3.2.3	Fluorescence-Tagged Aptamers for Target Quantification and Cytometry Applications	314
4	Optimization of RNA Aptamers Towards Therapeutic Approaches	315
5	Conclusions	321
	References	322

Abstract The SELEX technique (systematic evolution of ligands by exponential enrichment) provides a powerful tool for the in vitro selection of nucleic acid ligands (aptamers) from combinatorial oligonucleotide libraries against a target molecule. In the beginning of the technique's use, RNA molecules were identified that bind to proteins that naturally interact with nucleic acids or to small organic molecules. In the following years, the use of the SELEX technique was extended to isolate oligonucleotide ligands (aptamers) for a wide range of proteins of importance for therapy and diagnostics, such as growth factors and cell surface antigens. These oligonucleotides bind their targets with similar affinities and specificities as antibodies do. The in vitro selection of oligonucleotides with enzymatic activity, denominated aptazymes, allows the direct transduction of molecular recognition to catalysis. Recently, the use of in vitro selection methods to isolate protein inhibitors has been extended to complex targets, such as membrane-bound receptors, and even entire cells. RNA aptamers have also been expressed in living cells. These aptamers, also called intramers, can be used to dissect intracellular signal transduction pathways. The utility of RNA aptamers for in vivo experiments, as well as for diagnostic and therapeutic purposes, is considerably enhanced by chemical modifications, such as substitutions of the 2′-OH groups of the ribose backbone in order to provide resistance against enzymatic degradation in biological fluids. In an alternative approach, Spiegelmers are identified through in vitro selection of an unmodified D-RNA molecule against a mirror-image (i.e. a D-peptide) of a selection target, followed by synthesis of the unnatural nuclease-resistant L-configuration of the RNA aptamer that recognizes the natural configuration of its selection target (i.e. a L-peptide). Recently, nuclease-resistant inhibitory RNA aptamers have been developed against a great variety of targets implicated in disease. Some results have already been obtained in animal models and in clinical trials.

Keywords SELEX (systematic evolution of ligands by exponential enrichment) · In vitro selection · RNA aptamer · Aptazyme · Intramer · Spiegelmer · Therapeutic applications of aptamers

1
Introduction

All RNA molecules and proteins existing in living organisms are the product of a natural selection process. This process has been reproduced in in vitro experiments, also denominated as directed evolution or evolutionary engineering, where heterogeneous populations of RNAs in a defined environment are selected for desired properties. Molecules possessing these properties have a selection advantage and are therefore enriched in the heterogeneous population, whereas non-functional molecules are being gradually eliminated. Directed evolution methods can be based on site-directed mutagenesis or recombination of already existing molecules or the identification of functional RNAs from combinatorial synthesized libraries. Directed evolution methods such as the SELEX (systematic evolution of ligands by exponential enrichment) technique have been employed in the last decade to develop RNA and DNA molecules with desired binding properties. In this chapter, we shall only refer to the in vitro selection of high-affinity RNA molecules or of RNA molecules with enzymatic activities.

The SELEX technique goes back to the original work of Gold and co-workers (Tuerk and Gold 1990) who used a partial random RNA pool to select high-affinity binders against the RNA binding site of T4-DNA polymerase. One of the selected RNA species was identical to the wild-type sequence found in the bacteriophage messenger (m)RNA, whereas the second selected sequence differed from the wild-type sequence. Based on these findings, the authors suggested that affinity RNA ligands could conceivably be developed for any target molecule. In fact, over the following years, RNAs were identified by in vitro selection, binding to target molecules that are not known to naturally bind RNA. As these high-affinity RNA-target binders possess a unique binding specificity and are capable of differentiating between very similar target molecules, they are also referred to as aptamers (from the Latin root aptus, meaning fit) (Ellington and Szostak 1990; Sassanfar and Szostak 1990). Targets against those RNA aptamers have been developed include small organic molecules such as ATP and organic dyes (Ellington and Szostak 1990; Sassanfar and Szostak 1990), growth factors such as nerve growth factor (NGF) (Binkley et al. 1995), basic fibroblast growth factor (b-FGF) (Jellinek et al. 1995), vascular endothelial growth factor (VEGF) (Ruckman et al. 1998), hormones such as substance P (Nieuwlandt et al. 1995) and neuropeptide Y (Proske et al. 2002), antibodies (Hamm 1996; Lee and Sullenger 1997), enzymes (Zhang et al. 2003), disease-related prion and amyloid proteins (Ylera et al. 2002; Weiss et al. 1997), and cell-surface antigens (Davis et al. 1998). In vitro selection procedures for RNA aptamers have been developed for complex targets including receptor proteins that are only functional in their membrane environment (Ulrich et al. 1998, 2004; Homann and Göringer 1999), and for even entire cells (Homann and Göringer 1999; Ulrich et al. 2002).

RNA aptamers with catalytic activity (aptazymes) have also been identified by in vitro selection processes. This approach is based on the observation that in addition to the already binding properties, RNAs may possess catalytic activity. Natural-occurring RNA enzymes (ribozymes) include various self-cleaving RNAs, i.e. hammerhead, hairpin and hepatitis-δ virus ribozymes (reviewed by Doherty and Doudna 2000). SELEX procedures aim at changing the function of already-known ribozymes, at understanding the mechanism by which ribozymes catalyse chemical reactions (Michel et al. 1990) and at the development of new aptazymes (Robertson and Joyce 1990; Wilson and Szostak 1995; Hager et al. 1996; Hager and Szostak 1997; Tuschl et al. 1998; Hesselberth et al. 2003; Robertson et al. 2004).

2
In Vitro Selection of Functional Nucleic Acids

The SELEX method (Tuerk and Gold 1990; Ellington and Szostak 1990) is a powerful tool for the in vitro selection of nucleic acids capable of a desired function from combinatorial oligonucleotide libraries (DNA or RNA) with diversities up to 10^{15} different molecules and possible secondary and tertiary structures. This technique involves reiterative selections with increasing stringency and amplification of only those sequences that display a target function within a huge population of random RNA molecules (see Fig. 1 for a scheme of the SELEX procedure).

For SELEX, a partial random oligonucleotide is created by chemical synthesis containing an inner region, commonly containing 15–75 random position, that is flanked on both sides by constant sequences, one of them containing a T7 promoter site. Primers, complementary to the constant site, are designed for second-strand synthesis and PCR amplification of the chemically synthesized DNA pool. In vitro transcription of the double-stranded DNA template using T7-RNA polymerase yields the original RNA pool containing at least 10^{12} different sequences and possible secondary and tertiary structures. $2'$-F- or $2'$-NH_2-modified pyrimidines instead of $2'$-OH-pyrimidines may be used for the in vitro transcription reaction in order to produce nuclease-resistant RNA molecules. RNA molecules are purified from protein contaminations, heat-

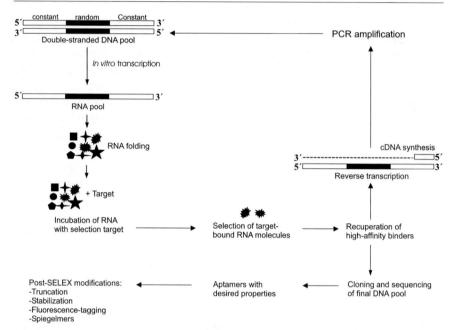

Fig. 1 In vitro selection of RNA aptamers by systematic evolution of ligands by exponential enrichment. High-affinity binders (aptamers) are identified from an RNA library containing up to 10^{15} random sequences and different shapes by reiterative selection against a target (see text for further explanation). The RNA can be stabilized against nuclease attack before the selection process by incorporating modified pyrimidines into the transcripts. Post-SELEX stabilization of already-selected RNA aptamers includes O-methylation of purines and chemical synthesis of L-RNA enantiomers (Spiegelmers) (see also Fig. 4)

denatured and renatured at room temperature to induce proper folding. This RNA pool is now presented to its target molecules in an excess of RNA over possible target-binding sites to ensure binding and subsequent selection of only high-affinity binders. High-affinity binders are eluted from the target using an antagonist competing for the same binding site on the target. These selected ligands are amplified using RT-PCR procedures to give the second-generation DNA pool. In vitro transcription of the obtained DNA provides the RNA pool for the next SELEX cycle. During initial SELEX rounds, a target-binding population is established in the heterogeneous RNA pool. The following reiterative cycles are carried out with increasing stringency until non-functional RNA molecules are entirely removed and only the highest-affinity RNA molecules have survived the selection process. At this stage, no further increase in binding affinity of the RNA population towards the target can be achieved. This final RNA pool that has been purified to a homogenous population of target binders is reverse-transcribed to DNA, and individual clones are identified from isolated clones by DNA sequencing. One expects that selected functional RNA aptamers share common structural motifs, i.e. short conserved nucleotide sequences that are found in most of the previous random sequences of the cloned aptamers. Aptamers falling into structural classes are screened for binding and desired biological activity. As conserved motifs are often located in stem-loop regions, which are responsible for aptamer-target interactions, final efforts lie in truncation of the full-length aptamer to a minimal size that is still active. Truncated aptamers can be easily modified at precise positions and reporter groups, i.e. fluorescence tags can be attached (reviewed by Ulrich et al. 2004).

3
Applications of SELEX

3.1
Development of Aptazymes

A variation of the aptamer selection strategies can be used to identify novel RNA molecules with catalytic activity (aptazymes), thereby transducing molecular recognition of aptamers to catalytic activity. In this case, catalytically active RNA molecules within in a pool of largely inactive RNA molecules need to be selected. As aptazymes bind to their substrates with high affinity, it is possible to combine SELEX procedures based on affinity selection and identification of catalytic properties.

The majority of in vitro selections for aptazymes have targeted phosphoester transfer reactions. Therefore, most RNA molecules with enzymatic activity developed by an in vitro selection possess either RNA cleavage activity, such as hammerhead ribozymes (Koizumi et al. 1999) and group I self-splicing introns (Beaudry and Joyce 1992; Thompson et al. 2002), or RNA ligase activity (Ekland et al. 1995; Robertson and Ellington 1999, 2001; Robertson et al. 2004).

Wilson and Szostak (1995) selected a self-alkylating biotin-utilizing ribozyme. The process for enzymatic self-alkylation occurs by a S_N2 mechanism that is in principle the same as the mechanism of hydrolysis, phosphoryl and acyl transfers of RNA molecules already mentioned.

As a first step, an RNA species was enriched in the pool, binding to the substrate (biotin). The conserved region of the selected aptamer was again randomized and a second process of SELEX cycles was initiated in order to identify self-alkylating ribozymes using N'-biotinoyl-N'-iodoacetyl-ethylenediamine (BIE) as an internal substrate. In vitro transcription reactions in the presence of 8-mercaptoguanosine yielded RNA transcripts with a single free thiol position at its 5'-end. Aptazymes catalysing self-alkylation (covalently linking biotin to their 5'-end) were subsequently identified by retaining biotinylated RNA molecules on streptavidin agarose.

As a further achievement, allosteric-activated ribozymes (RNA molecular switches), were identified by in vitro selection from random RNA libraries (Koizumi et al. 1999; Robertson and Ellington 1999, 2001; Robertson et al. 2004).

Robertson and Ellington (1999, 2001) developed an in vitro selection procedure for ribozyme ligases, which are activated by binding of a second molecule to an allosteric site. After randomizing an already selected ribozyme ligase, variants of this enzyme were selected in the presence of a protein cofactor such as tyrosyl-t RNA synthetase or hen egg-white lysozyme. Ribozyme ligases were identified whose enzymatic activities were enhanced several thousand fold in the presence of their respective protein cofactor. The identification of RNA enzymes containing molecular switches may be interesting for therapeutic approaches. For instance, hammerhead ribozymes are promising for cleaving oncogene-coding RNA and blocking replication of RNA viruses and retroviruses.

3.2
Development of Novel Binding Species

RNA molecules binding to their targets with high affinity were first developed for proteins that are known to naturally bind RNA (Tuerk and Gold 1990). In the following years, the technique was extended for isolating high-affinity binders against growth factors and other molecules that can be presented as purified protein for aptamer selection (reviewed by Jayasena 1999). The notion that aptamers recognize particular epitopes in a similar way as antibodies do, and bind with dissociation constants (K_d) of up to a picomolar range to their targets led to the suggestion that aptamers could substitute antibodies in certain applications (Xu and Ellington 1996; Tasset et al. 1997; reviewed by Ulrich et al. 2004). Recently, aptamers against complex targets such as membrane proteins or cell membranes have been developed by subtractive SELEX procedures in which unspecific binding oligonucleotides are discarded and only target binders are enriched in the selection pools (Ulrich et al. 2002; Daniels et al. 2002). The introduction of modified pyrimidines and the synthesis of L-forms of aptamers have largely increased nuclease-resistance of RNA aptamers and subsequently their applications for in vitro and in vivo studies (reviewed by Kusser 2000; Vater and Klussmann 2003; details in Sect. 4).

3.2.1
Aptamers as Modulators of Protein Function In Vitro

Aptamers bind to a particular epitope on their target proteins and have been developed into effective tools for studying mechanisms of receptors and enzymes, as well as for dissecting intracellular signal transduction pathways. Depending on the competitor used for displacement of target-bound RNA molecules during in vitro selection, aptamers inhibit target-protein function by either binding to the catalytic site of an enzyme or the agonist-binding site of a receptor or binding to an allosteric site on the protein. In most cases, aptamers selected against regulatory sites act as inhibitors by affecting target-protein conformations and stabilizing inactive protein forms. However, aptamers that displace inhibitors from a regulatory site of the receptors and are biologically inactive by themselves protect normal protein function as shown for RNA aptamers alleviating cocaine- and MK-801-induced inhibition of the nicotinic acetylcholine receptor (nAChR). Both the abused drug cocaine and the anticonvulsant MK-801 compete for binding to the same site on the nAChR (Ulrich et al. 1998; Hess et al. 2000; Ulrich and Gameiro 2001; Krivoshein and Hess 2004; Cui et al. 2004).

Following selection and identification of a pool of cocaine-displaceable RNA aptamers, two classes of RNA aptamers with different conserved sequences in the previous random region, denominated as class I and class II aptamers, were identified (Ulrich et al. 1998). The most active aptamers of each class, aptamer I-14 and II-3, were characterized for their action on nAChR function by whole-cell recording electrophysiology in cell cultures expressing the muscle-type nAChR. Aptamer I-14 competed with cocaine for binding on the receptor and was a potent inhibitor of agonist (carbamoylcholine)-induced nAChR activity. Aptamer II-3, however, did not inhibit nAChR function, although it competed with cocaine for receptor binding. When aptamer II-3 was co-applied with MK-801 or cocaine, it alleviated the inhibitory effect of MK-801 and cocaine on carbamoylcholine-induced ion flow (Hess et al. 2000). The results of this research are summarized in Fig. 2. In addition to the identification of a compound that alleviates cocaine-induced inhibition of the receptor, the results of this study were taken as proof that cocaine is not a sterical channel blocker but binds to an allosteric site different from the channel lumen. Recently, nuclease-resistant RNA aptamers have been developed that protect the nAChR against inhibition by cocaine (Cui et al. 2004).

Transmissible spongiform encephalopathies result from modification of the prion precursor protein (PrP^c) to the pathological form (PrP^{sc}) and involve spongiform degeneration and astrocyte gliosis leading to neurodegeneration. Membrane-proximal domains of the precursor PrP^c contain mainly α-helical/random coil structures, whereas the pathological form is dominated by β-sheet structures. As a mechanism for propagation of PrP^{sc}, it has been proposed that PrP^{sc} forms a heterodimer with PrP^c, followed by PrP^{sc}-dependent conversion of PrP^c to PrP^{sc} and formation of a PrP^{sc} homodimer (Cohen et al. 1994).

Rhie et al. (2003) selected a nuclease-resistant RNA aptamer that bound with tenfold higher affinity to disease-related conformations of PrP^{sc} than to the non-pathogenic form of the protein (PrP^c). The aptamer was used to map binding sites on the prion protein. The difference in binding affin-

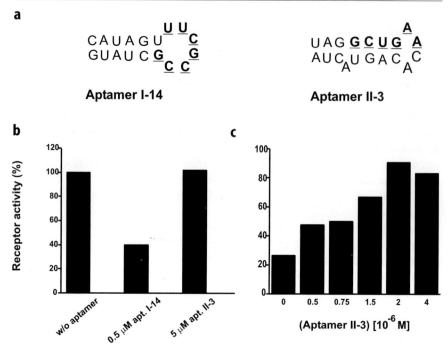

Fig. 2a–c Structure and action of class I and II cocaine-displaceable RNA aptamers. **a** Prediction of secondary structures of both aptamers revealed the localization of conserved sequences in the previous random regions (in **bold** and *underlined*) in stem-loop regions that are believed to be essential for target recognition. **b** The effects of RNA aptamers I-14 and II-3 on agonist (100 µM carbamoylcholine)-induced stimulation of the acetylcholine receptor were determined by whole-cell recording electrophysiology. **c** Aptamer II-3 alleviates inhibition of the acetylcholine receptor by MK-801 (500 µM). The effect of the aptamer is dose-dependent (Ulrich et al. 1998; Hess et al. 2000)

ity of the aptamer to PrP^c and PrP^{sc} was explained by the existence of two binding sites, one site accessible on both conformations of the protein, and another site that is only exposed in the disease-associated protein conformation. The work of Proske et al. (2002) resulted in the development of an RNA aptamer that reduced PrP^{sc} formation in persistently prion-infected neuroblastoma cells. As a mechanism of aptamer action, it has been suggested that aptamer binding alters β-sheet structure, resulting in less tight folding of PrP^{sc} aggregation and increased susceptibility of the aggregates for protease degradation. In addition to having developed an important tool for studying the molecular mechanism of prion conversion and propagation, anti-prion aptamers may gain importance as diagnostics and therapeutics for prion disease.

3.2.2
Intracellularly Expressed Aptamers for Dissection of Signal Transduction Pathways

RNA aptamers directed against extracellular targets need to be chemically modified in order to resist degradation by nucleases present in biological fluids. However, various RNA aptamers have been developed against various intracellular targets, such as, i.e. transcription factors and signalling proteins including nuclear factor (NF)-κB, Erk 1/2 mitogen-activated protein kinases and guanine nucleotide exchange factors (Lebruska and Maher 1999; Seiwert et al. 2000; Mayer et al. 2001; Theis et al. 2004). In addition to studying aptamer-target interactions in cell-free systems, RNA aptamers can be constitutively expressed in cells in order to knock down the gene expression of a protein of interest. Cell cultures can be directly transfected with RNA aptamers without any need of prior protection against nuclease activity (Theis et al. 2004). Intracellularly expressed aptamers (intramers) have been used for dissecting intracellular signal transduction pathways and for inhibiting viral activity. Intramers provide an alternative to the RNA interference (RNA_i) technology that is often used to study the biological function of proteins. The difference between both techniques is that RNA_i destroys target protein-encoding mRNA, whereas intramers aim at inhibition of protein function. As the knockdown of target-protein gene expression by RNA_i may be incomplete in some cases (Elbashir et al. 2001), an intramer may be used to achieve complete inhibition of target-protein function.

Blind and co-workers (1999) selected RNA aptamers binding to the cytoplasmatic domain of the human β-2 integrin lymphocyte function-associated antigen (LFA)-1 and expressed them in leukocytes using a vaccinia virus-based expression system. These aptamers, also denominated as intramers, blocked signal transduction between the inside of the cell and the extracellular domain of the integrin.

Aptamers have been developed against several proteins that are essential for the human immunodeficiency virus (HIV) life cycle, such as the HIV-nucleocapsid protein, reverse transcriptase, integrase and the Rev protein (Berglund et al. 1997; Joshi and Prasad 2002; Allen et al. 1995; Chaloin et al. 2002). Intramers have been used to inhibit Rev protein activity in vitro. For instance, when put under the control of RNA polymerase III-dependent expression vectors, in vivo-transcribed anti-Rev intramers inhibited HIV replication and production in cell cultures (Good et al. 1997). Chaloin et al. (2002) selected intramers against the HIV Rev protein that possessed powerful antiviral activity in human T lymphoid cell lines.

A transgenic intramer-expressing Drosophila model was developed to knock down the activity of the nuclear B52 protein in vivo (Shi et al. 1999). B52 has been shown to participate in splicing, and its expression level is critical for development. Intramer expression was driven under the control of a heat-shock inducible RNA polymerase II promoter. A pentameric aptamer was designed and co-localized with the B-52 protein at its insertion site in the polytene chromosome. A hammerhead ribozyme sequence was positioned at the 3'-end of each pentavalent aptamer to induce self-cleavage and release of functional intramers. Intramer-expressing animals showed a reduced survival rate during development compared to unaffected wild-type animals, confirming the importance of B52 for Drosophila development.

The importance of applications of intramers as tools to study intracellular regulatory circuits has been increased by the development of aptamers whose activities can be switched on or off at any time. Ligand-inactivated intramers with controllable activity can be created by selecting aptamers that bind with high affinity to their target protein, but also possess an allosteric binding site for regulation of aptamer activity. In the absence of its regulator, the intramer exerts its function by inhibiting target protein function. As a result of the binding of the regulator, aptamer conformation is changed and the aptamer-target protein complex dissociates. As a proof of principle, RNA aptamers against the nuclear enzyme formamidopyrimidine glycosylase were isolated whose activity was regulated by binding of neomycin to an allosteric site (Vuyisich and Beal 2002).

Werstuck and Green (1998) developed a method for controlling gene expression by ligand-activated aptamers. RNA aptamers were selected against antibiotics such as kanamycin and tobramycin. Insertion of the aptamer sequence into the 5'-untranslated region of an mRNA allowed mRNA translation by addition of the ligand (kanamycin and tobramycin) to be repressible in vitro as well as in mammalian cells. Intramers with controllable activity are particularly useful for the study of cell functions when timing is critical.

3.2.3
Fluorescence-Tagged Aptamers for Target Quantification and Cytometry Applications

The breakthrough in genomics and proteomic research has led to demand for identification of fluorescence-labelled ligands for a vast number of target proteins. Fluorescent ligands are used for target protein quantification and studying protein function in a cellular context. Mostly mono- and polyclonal antibodies have been developed to fluorescent ligands. Although being the method of choice for detecting antigen concentration in biological fluids, tissue and cells, antibodies have some limitations, which have led to the search for alternative strategies for the identification of high-affinity ligands for research and diagnostics.

As already mentioned, the capability of aptamers to recognize their target is similar to that of antibodies (Xu and Ellington 1996; Tasset et al. 1997). For instance, aptamers were able to recognize a specific isoform of protein kinase C (Conrad et al. 1994), to differentiate between a phosphorylated and unphosphorylated protein (Seiwert et al. 2000) and to discriminate ATP from other nucleotides (Sassanfar and Szostak 1990). Aptamers are developed by an in vitro selection method and therefore can be evolved against every target, including toxins and compounds, that does not elicit any immune response. The use of RNA aptamers for diagnostic and in vivo applications was limited in the past due to their poor nuclease resistance. These initial limitations have been overcome by the development of modified nucleotides that largely enhance half-times of RNA aptamers in biological fluids (reviewed by Kusser 2000). Post-SELEX modifications of selected aptamers, such as attaching fluorescent reporters, are done at the researcher's will (reviewed by Ulrich et al. 2004).

As an alternative approach, fluorescence-labelled RNA aptamers against ATP were selected from an RNA pool in which UTP had already been substituted for fluorescein-UTP. The initial pool had been synthesized in such a way that uridine residues would be poorly represented. These fluorescein-UTP-labelled aptamers detected ATP at a concentration of 25 µM in complex mixtures (Jhaveri et al. 2000).

An example for the development of fluorescent-tagged RNA aptamers for flow cytometry comes via the work of Davis et al. (1998). Aptamers selected against CD4 were linked to a biotin moiety, coupled to streptavidin-fluorescein or streptavidin-phycoerythrin and then used for separation of CD4-expressing T cells from those lacking CD4. Homann and Göringer (1999) have used fluorescent-labelled RNA ligands to identify a 42-kDa protein within the flagellar pocket of the blood-stream form of *Trypanosoma brucei*. In addition to being used for target imaging in cells or for target separation, fluorescent-tagged aptamers can substitute antibodies in dot blot, enzyme-linked immunosorbent (ELISA), and Western-blot assays and aptamer-chip based biosensors (Drolet et al. 1996; Ulrich et al. 2004; McCauley et al. 2003). In the future, aptamer-coupling to quantum dots that allow the use of various intensity levels and colours may become important for high-throughput screening in chip-based technologies (reviewed by Ulrich et al. 2004).

4
Optimization of RNA Aptamers Towards Therapeutic Approaches

The possible therapeutic potential of RNA aptamers depends on many aspects that apply to conventional pharmaceuticals.

First, the large surface of aptamers permits more interactions with their target molecules than small molecules share with their receptors. Reiterative in vitro selection cycles against the target, using a huge excess of possible ligands over target-binding sites, results in amplification of the highest-affinity binders. Binding affinities of aptamers to their targets are in many cases higher than the binding affinity of natural ligands used as competitors in the selection process (Ulrich et al. 1998; O'Connell et al. 1998). For instance, Ulrich et al. (1998) used cocaine as a displacement agent for SELEX for RNA aptamers

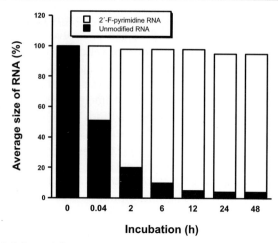

Fig. 3 Stability of 2′-fluoro (2′-F)-pyrimidine modified RNA pool in biological fluids. Both, an unmodified and a 2′-F-modified RNA pool (length of 90 nucleotides) were internally radiolabelled. Five microcuries of radiolabelled RNA were incubated for several time periods at 37°C with 95% bovine calf serum. Samples were separated at indicated time points and analysed for the average size by denaturing polyacrylamide gel electrophoresis and autoradiography. The average size of the remaining RNA fragments was plotted as a percentage of the 90-nucleotide full-length RNA at time 0, which was considered to be 100% (Ulrich et al. 2004)

binding to the nicotinic acetylcholine receptor. The selected aptamers were 200-fold more active in inhibiting the acetylcholine receptor than cocaine was (Ulrich et al. 1998; Ulrich and Gameiro 2001).

Second, RNA aptamers need to be stabilized against degradation in biological fluids. Particularly, modifications at the 2′-OH-function of ribose enhance the stability of RNA (Pieken et al. 1991). Most commonly, in vitro transcription reactions using T7-RNA polymerase are set up in the presence of 2′-fluoro (2′-F)- or 2′-amino (2′-NH$_2$)-pyrimidines instead of 2′-OH-pyrimidines which are prone to nuclease attack. The resulting modified RNA transcripts, which are much more stable against nuclease activity compared to unprotected ones, are used for SELEX against the target molecule (Fig. 3). Eluted 2′-F- or 2′-NH$_2$-pyrimidine RNAs are accepted by avian myeloblastosis virus (AMV) reverse transcriptase as templates for enzymatic complementary (c)DNA synthesis.

Further stabilization of already-selected RNA aptamers for in vivo applications includes O-methyl-substitutions in purine nucleotides, requiring chemical synthesis of modified RNA molecules. The in vitro evolution of a T7-RNA polymerase variant that transcribes 2′-O-methyl RNA is an important step towards enzymatic synthesis of a stable RNA with modified pyrimidines and purines (Chelliserrykattil and Ellington 2004; see Fig. 4a for structures of modified purines and pyrimidines).

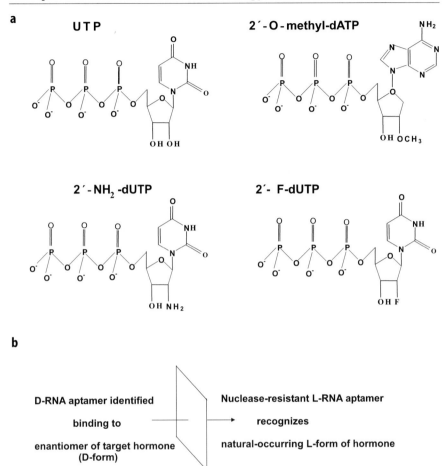

Fig. 4a,b Chemical modifications of RNA aptamers lending resistance against nuclease attacks. **a** In vivo and possible therapeutic applications of selected RNA aptamers have been improved by using SELEX libraries containing RNA molecules with back-bone modifications, such as substitution of the 2′-OH position of the ribose of pyrimidines by 2′-fluoro- or 2′-amino-functions. These modified nucleotides that are accepted as substrates for in vitro transcription reactions can be used to protect RNA molecules against degradation along reiterative SELEX cycles. O′-methyl-substituted purines granting additional nuclease resistance, however, have to be inserted into RNA molecules by chemical synthesis, and are therefore only used for post-SELEX modifications of already-selected RNA aptamers. **b** Spiegelmers are L-enantiomers (mirror images) of identified D-RNA aptamers. They are assembled by chemical synthesis and recognize the enantiomer (mirror image) of the selection target. Spiegelmers are nuclease resistant

An alternative to chemically modifying purines and pyrimidines of D-RNA molecules during and after in vitro selection is the application of L-nucleic acids (Klussmann et al. 1996; Nolte et al. 1996). During these studies, D-RNA ligands were selected against L-adenosine and D-arginine, in turn. Following chiral inversion, the mirror image of the selected RNA aptamer, termed Spiegelmer (in German *Spiegel* means mirror), binds with high affinity to the natural configuration of the target (D-adenosine or L-arginine). Spiegelmers are extremely resistant against degradation by nucleases (Fig. 4b).

Examples for L-RNA Spiegelmers that may gain therapeutic importance are anti-nociceptin/orphanin (N/OFQ) and anti-ghrelin L-aptamers (Faulhammer et al. 2004; Heimling et al. 2004).

The aim of the study of Faulhammer et al. (2004) was to identify a Spiegelmer that prevents binding of the endogenous ligand neuropeptide N/OFQ to the opioid receptor-like 1 receptor that is implicated in the regulation of pain and stress. Spiegelmers selected against N/OFQ inhibited receptor-induced G protein activation in cell culture and antagonized receptor-induced potassium currents in *Xenopus* oocytes.

The Spiegelmers selected against ghrelin had a nanomolar binding affinity to its target in vitro and inhibited ghrelin-binding to its growth hormone secretagogue receptor 1a (GHS-R1a), thereby abolishing receptor-induced calcium mobilization in cell culture. Systematic administration of ghrelin and subsequent GHS-R1a receptor activation is known to trigger the release of growth hormone in rats and humans. When a polyethylene glycol (PEG)-linked anti-ghrelin L-RNA Spiegelmer was injected at 5–10 times the dose of ghrelin administration into the blood stream of rats, no increase of growth hormone in the plasma was detected, implicating the inhibition of ghrelin-induced GHS-R1a receptor activation (Heimling et al. 2004). Spiegelmers blocking ghrelin binding to its receptor may become important for treatment of diseases that are associated with high concentrations of circulating ghrelin, such as Prader-Willi syndrome (Cummings et al. 2002).

Third, towards therapeutic application and optimization aptamer effects, the time of residence of the aptamer in the plasma needs to be increased, as small molecules are eliminated immediately. Formulations are being developed that liberate aptamers over an adequate period of time and in a controllable manner.

The aptamer showing the farthest development towards therapy is the anti-VEGF D-RNA aptamer that has been made nuclease-resistant by chemical modification of the 2′-OH-groups of purines and pyrimidines (Ruckman et al. 1998). VEGF is an angiogenic promoter present in a wide variety of human tumours, inducing the development of tumour-associated vasculature. The blocking of VEGF actions limits both the proliferation of primary tumours as well as the development of metastasis in immunosuppressed mice (Warren et al. 1995; Huang et al. 2001; Kim et al. 2002). One of the four isotypes of VEGF, $VEGF_{165}$ (a dimeric protein composed of 165 amino acid subunits) appears to predominate in adult tumours and was therefore used as a target for the production of a VEGF-specific aptamer. RNase-resistant 2′-F-RNA aptamers were selected that bound with high affinity (K_d of 200 pM) to the human

VEGF$_{165}$-isoform (Ruckman et al. 1998; Tucker et al. 1999). Representative aptamers were truncated to the minimal sequence necessary for high-affinity binding and then were further stabilized by replacement of 2'-OH by 2'-O-methyl groups at all riboses of purine positions. These aptamers significantly reduced intradermal VEGF-induced vascular permeability in adult guinea-pigs in vivo. However, compared to its good stability in biological liquids, the half time of the residence of anti-VEGF aptamers in plasma in vivo was found to be very short, mainly as a result of renal clearance (Willis et al. 1998). The half-life of aptamers in the plasma is increased by coupling them to high molecular linkers, such as PEG or liposomes (Willis et al. 1998; Tucker et al. 1999; Farokhzad et al. 2004; Healy et al. 2004; Heimling et al. 2004).

Liposome- and PEG-anchored anti-VEGF aptamers, denominated DAG-NX213 and NX-1838 (Gilead Sciences; www.gilead.com), were tested for improved pharmacokinetics and biological activity. These improved aptamers remained in the plasma for about 9 h, whereas aptamers that had not been coupled to high molecular weight linkers were eliminated within minutes (Willis et al. 1998; Tucker et al. 1999). A controlled-drug delivery system, using poly-(co-glycolic) acid microspheres containing the anti-VEGF aptamer, was developed for long-term inhibition of VEGF-mediated responses in vivo in order not to depend on daily repeated injections of the aptamer drug (Carrasquillo et al. 2003). In vitro release studies revealed a controlled liberation of a pharmacologically active dose of 2 µg/day aptamer over a period of more than 20 days. The activity of the release aptamer was confirmed by inhibition of VEGF-induced proliferation in human umbilical vein endothelial cell (HUVEC) cultures.

The VEGF antagonist NX-1838 has also been shown to be an anticancer agent, as it blocks tumour growth in experimental animal models (Huang et al. 2001; Kim et al. 2002). In a mouse model of neuroblastoma, administration of the anti-VEGF aptamer resulted in a partial inhibition of tumour growth after 6 weeks compared to untreated control animals (Kim et al. 2002).

Huang et al. (2001) injected nude mice with cultured Wilms tumour cells and maintained the mice for 1 week before administration of anti-VEGF aptamers. Daily injections of aptamers for 5 weeks resulted in 80% loss of tumour weight compared to control animals. Future perspectives will lie in the clinical testing of the effectiveness of NX-1838 in decreasing tumour size and the frequency of metastasis.

In addition to its role in tumour growth, VEGF is correlated with other disease-related processes of neovascularization, such as pathological proliferative retinopathy (Ozaki et al. 2000). The anti-VEGF aptamer is a promising candidate for drug development, as it blocks only pathological, not physiological, retinal vascularisation (Oshida et al. 2001).

The 27 nucleotide-containing NX-1838 anti-VEGF RNA aptamer, 5'CGGA-AUCAGUGAAUGCUUAUACAUCCG 3', containing a 40-kDa PEG-moiety (Tucker et al. 1999; Carrasquillo et al. 2003) is an injectable angiogenesis

Table 1 In vivo applications of RNA aptamers

Aptamer effect	Experimental system	Accomplishments	Reference
Inhibition of VEGF-mediated angiogenesis	Inhibition of VEGF-induced vascular permeability in rats and guinea-pigs	Identification of a minimal necessary for biological activity in vivo	Ruckman et al. 1998
	Pharmacokinetics in monkeys	Improved pharmacokinetics by attachment of PEG-linker	Tucker et al. 1999
	Mice infected with cultured Wilms' tumour cells	Reduction of tumour	Huang et al. 2001
	Inhibition of VEGF-induced proliferation in HUVEC cells	Development of controlled aptamer-delivery system	Carrasquillo et al. 2003
	Rats with proliferative retinopathy	Pathological, but not physiological neovascularization inhibited	Ishida et al. 2003
	Clinical studies (phases I and II)	Blocking pathologic neovascularization in human vascular eye disease	Eyetech Group 2002
	Phase II study: early treatment of diabetic retinopathy	Visible effect in nearly 90% of treated patients	Eyetech Group 2003
Inhibition of angiopoietin 2-induced potentiation of angiogenesis	Rat corneal pocket assay	Inhibition of b-FGF-mediated neovascularization in vivo	White et al. 2003
Inhibition of GHS-R1a Activation by ghrelin	Receptor activation in vitro	Inhibition of GHS-R1a receptor activation	Heimling et al. 2004
	Rats injected with ghrelin	Inhibition of GHS-R1a-induced secretion of growth hormone in vivo by a PEG 1-linked Spiegelmer	
Inhibition of factor IXa-induced coagulation	Porcine model of coagulation	Anti-coagulative effect in vivo	Rusconi et al. 2004
		Aptamer inactivation in vivo by addition of antidote-oligonucleotide	
Prevention of autoimmune myasthenia gravis	Rat model of EAMG	Reversal of clinical symptoms of EAMG by a PEG-coupled aptamer	Hwang et al. 2004
		Prevention of loss of nicotinic acetylcholine receptors in vivo	

EAMG, experimental myasthenia gravis; HUVEC, human umbilical vein endothelial cells inhibitor that is undergoing clinical tests in the United States for treatment of various forms of pathologic ocular neovascularization, including age-related macular degeneration. In phase II clinical studies of early treatment of diabetic retinopathy, nearly 90% of patients who had received the aptamer formulation showed stabilized or improved vision 3 months after treatment (Eyetech Study Group 2002, 2003; www.agingeye.net/mainnews/may2003eyetechstudy.php).

Further RNA aptamers that have been active in vivo models represent promising candidates for clinical testing. An aptamer binding to angiopoietin-2 was developed in order to block angiopoietin-induced potentiation of actions of proangiogenic growth factors (White et al. 2003). In vivo experiments confirmed the activity of the aptamer in a rat corneal micropocket angiogenesis assay, where the aptamer inhibited basic fibroblast growth factor-mediated neovascularization.

RNA aptamers with controllable activity were developed, binding to the coagulation factor IXa. The anti-IXa RNA aptamers showed anti-coagulant activity in vivo, and the addition of an antidote oligonucleotide inactivated the aptamer-induced effect (Rusconi et al. 2002, 2004). In Table 1, the effects of RNA aptamers that have been tested in animals and humans are summarized.

5
Conclusions

The targeting of proteins by the formation of stable complexes between them and selected oligonucleotides is the key feature of the SELEX technology. The selection of an RNA aptamer obeys the same rules that apply for in vivo evolution of a bioactive molecule. Reiterative rounds of in vitro selection ensure that only those RNA molecules with high-affinity binding to their targets or desired enzymatic activity are amplified.

Due to their binding specificities and affinities of up to the picomolar range, RNA aptamers rival antibodies in certain applications. Advantages of aptamers over antibodies lie mostly in their non-protein nature and their resistance against degradation in biological fluids. Denaturing of aptamers is reversible. Aptamers can be transported at room temperature, and denatured aptamers can be regenerated within minutes.

Although the large-scale synthesis of non-natural RNA aptamer sequences remains expensive, aptamers have been turned into promising therapeutic agents. The in vitro selection of aptamers has been automated, permitting the large-scale production of aptamers for pharmaceutical and therapeutic applications (Cox et al. 1998; Brody et al. 1999).

Clinical applications have now been planned or initiated using RNA against targets implicated in pathological ocular neovascularization, cancer, allergic disease, thrombosis, HIV and hepatitis C (reviewed by Sun 2000; Rimmele 2003; Thiel 2004).

Acknowledgements The author would like to acknowledge Dr. George P. Hess (Cornell University, Ithaca, N.Y.), who has guided the development and characterization of cocaine-displaceable RNA aptamers, done by this author and others in his laboratory. H.U.'s current research on aptamer development is financed by FAPESP (Fundação de Amparo à Pesquisa do Estado de São Paulo) and CNPQ (Conselho Nacional de Desenvolvimento Cientifico e Tecnológico), Brazil.

References

Allen P, Worland S, Gold L (1995) Isolation of high-affinity RNA ligands to HIV-integrase from a random pool. Virology 209:327–336

Beaudry AA, Joyce GF (1992) Directed evolution of an RNA enzyme. Science 257:635–641

Berglund JA, Charpentier B, Rosbash M (1997) A high affinity site for the HIV-1 nucleocapsid protein. Nucleic Acids Res 25:1042–1049

Binkley J, Allen P, Brown DM, Green L, Tuerk C, Gold L (1995) RNA ligands to human nerve growth factor. Nucleic Acids Res 23:3198–3205

Blind M, Kolanus W, Famulok M (1999) Cytoplasmatic RNA modulators of an inside-out signal transduction cascade. Proc Natl Acad Sci USA 96:3606–3610

Brody EN, Willis MC, Smith JD, Jayasena SD, Zichi D, Gold L (1999) The use of aptamers in large arrays for molecular diagnostics. Mol Diagn 4:381–388

Carrasquillo KG, Ricker JA, Rigas JK, Miller JW, Gragoudas ES, Adamis AP (2003) Controlled delivery of the anti-VEGF aptamer EYE001 with poly(lactic-co-glycolic) acid microspheres. Invest Ophthalmol Vis Sci 44:290–299

Chaloin L, Lehmann MJ, Scakiel G, Restle T (2002) Endogenous expression of a high-affinity pseudoknot RNA aptamer suppresses replication of HIV-1. Nucleic Acids Res 30:4001–4008

Chelliserrykattil J, Ellington AD (2004) Evolution of a T7 RNA polymerase variant transcribes 2′-O-methyl RNA. Nat Biotechnol 22:1155–1160

Cohen FE, Pan KM, Huang Z, Baldwin M, Fletterick RJ, Prusiner SB (1994) Structural clues to prion replication. Science 264:530–531

Conrad R, Keranen LM, Ellington AD, Newton AC (1994) Isozyme-specific inhibition of protein kinase C by RNA aptamers. J Biol Chem 269:32051–32054

Cox JC, Rudolph P, Ellington AD (1998) Automated RNA selection. Biotechnol Prog 14:845–850

Cui Y, Ulrich H, Hess GP (2004) Selection of 2′-fluoro-modified RNA aptamers for alleviation of cocaine and MK-801 inhibition of the nicotinic acetylcholine receptor. J Membr Biol 202:137–149

Cummings DE, Clement K, Purnell JQ, Vaisse C, Foster KE, Frayo RS, Schwartz MW, Basdevant A, Weigle DS (2002) Elevated plasma ghrelin levels in Prader Willi syndrome. Nat Med 8:643–644

Daniels DA, Chen H, Hicke BJ, Swiderek KM, Gold L (2003) A tenascin-C aptamer identified by tumor cell SELEX: systematic evolution of ligands by exponential enrichment. Proc Natl Acad Sci USA 100:15416–15421

Davis KA, Lin Y, Abrams B, Jayasena SD (1998) Staining of cell surface CD4 with 2′-F-pyrimidine-containing RNA aptamers for flow cytometry. Nucleic Acids Res 26:3915–3924

Doherty EA, Doudna JA (2000) Ribozyme structures and mechanisms. Annu Rev Biochem 69:597–615

Drolet DW, Moon-McDermott L, Romig TS (1996) An enzyme-linked oligonucleotide assay. Nat Biotechnol 14:1021–1025

Ekland EH, Szostak JW, Bartel DP (1995) Structurally complex and high active RNA ligases derived from random RNA sequences. Science 269:364–370

Elbashir SM, Harborth J, Lendeckel W, Yalcin A, Weber K, Tuschl T (2001) Duplexes of 21-nucleotide RNA mediate RNA interference in cultured mammalian cells. Nature 411:494–498

Ellington AD, Szostak JW (1990) In vitro selection of RNA molecules that bind specific ligands. Nature 346:818–822

Eyetech Study Group (2002) Preclinical and phase 1A clinical evaluation of an anti-VEGF pegylated aptamer (EYE001) for the treatment of exudative age-related macular degeneration. Retina 22:143–152

Eyetech Study Group (2003) Anti-vascular endothelial growth factor therapy for subfoveal choroidal neovascularization secondary to age-related macular degeneration: phase II study results. Ophthalmology 110:979–986

Farokhzad OC, Jon S, Khademhosseini A, Tran TN, LaVan DA, Langer R (2004) Nanoparticle-aptamer bioconjugates: a new approach for targeting prostate cancer cells. Cancer Res 64:7668–7672

Faulhammer D, Eschgfäller B, Stark S, Burgstaller P, Englberger W, Erfurth J, Kleinjung F, Rupp J, Dan Vulcu S, Schröder W, Vonhoff S, Nawrath H, Gillen C, Klussmann S (2004) Biostable aptamers with antagonistic properties to the neuropeptide nociceptin/orphanin FQ. RNA 10:516–527

Good PD, Krikos AJ, Li SXL, Bertrand E, Lee NS, Giver L, Ellington AD, Zaia JA, Rossi JJ, Engelke DR (1997) Expression of small, therapeutic RNAs in human cell nuclei. Gene Ther 4:45–54

Hager AJ, Szostak JW (1997) Isolation of novel ribozymes that ligate AMP-activated RNA substrates. Chem Biol 4:607–617

Hager AJ, Pollard JD, Szostak JW (1996) Ribozymes: aiming at RNA replication and protein synthesis. Chem Biol 3:717–725

Hamm J (1996) Characterisation of antibody-binding RNAs selected from structurally constrained libraries. Nucleic Acids Res 24:2220–2227

Healy JM, Lewis SD, Kurz M, Boomer RM, Thompson KM, Wilson C, McCauley TG (2004) Pharmacokinetics and biodistribution of novel aptamer compositions. Pharm Res 21:2234–2246

Heimling S, Maasch C, Eulberg D, Buchner K, Schröder W, Lange C, Vonhoff S, Wlotzka B, Tschöp MH, Rosewicz S, Klussmann S (2004) Inhibition of ghrelin action in vitro and in vivo by an RNA-Spiegelmer. Proc Natl Acad Sci USA 101:13174–13179

Hess GP, Ulrich H, Breitinger H-G, Niu L, Gameiro AM, Grewer C, Srivastava S, Ippolito JE, Lee SM, Jayaraman V, Coombs SE (2000) Mechanism-based discovery of ligands that prevent inhibition of the nicotinic acetylcholine receptor. Proc Natl Acad Sci USA 97:13895–13900

Hesselberth JR, Robertson MP, Knudsen SM, Ellington AD (2003) Simultaneous detection of ligase analytes with an aptazyme ligase array. Anal Biochem 312:106–112

Homann M, Göringer HU (1999) Combinatorial selection of high affinity RNA ligands to live African trypanosomes. Nucleic Acids Res 27:2006–2014

Huang J, Moore J, Soffer S, Kim E, Rowe D, Manley CA, O'Toole K, Middlesworth W, Stolar C, Yamashiro D, Kandel J (2001) Highly specific antiangiogenic therapy is effective in suppressing growth of experimental Wilms tumors. J Pediatr Surg 36:357–361

Hwang B, Han K, Lee S-W (2003) Prevention of passively transferred experimental autoimmune myasthenia gravis by an in vitro selected RNA aptamer. FEBS Lett 548:85–89

Ishida S, Usui T, Yamashiro K, Kaji Y, Amano S, Ogura Y, Hida T, Oguchi Y, Ambati J, Miller JW, Gragoudas ES, Ng YS, D'Amore PA, Shima DT, Adamis AP (2003) VEGF164-mediated inflammation is required for pathological, ischemia-induced retinal neovascularization. J Exp Med 198:483–489

Jayasena SD (1999) Aptamers: an emerging class of molecules that rival antibodies in diagnostics. Clin Chem 45:1628–1650

Jellinek D, Lynott CK, Rifkin DB, Janjic N (1993) High-affinity RNA ligands to basic fibroblast growth factor inhibit receptor binding. Proc Natl Acad Sci USA 90:11227–11231

Jhaveri S, Rajendran M, Ellington AD (2000) In vitro selection of signaling aptamers. Nat Biotechnol 18:1293–1297

Joshi P, Prasad VR (2002) Potent inhibition of human immunodeficiency virus type 1 replication by template analog reverse transcriptase inhibitors derived by SELEX (systematic evolution of ligands by exponential enrichment). J Virol 76:6545–6557

Kim ES, Serur A, Huang J, Manley CA, McCrudden KW, Frischer JS, Soffer SZ, Ring L, New T, Zabski S, Rudge JS, Holash J, Yancopoulos GD, Kandel JJ, Yamashiro DJ (2002) Potent VEGF blockade causes regression of coopted vessels in a model of neuroblastoma. Proc Natl Acad Sci USA 99:11399–11404

Klussmann S, Nolte A, Bald R, Erdmann VA, Fürste JP (1996) Mirror-image RNA that binds adenosine. Nat Biotechnol 14:1112–1115

Koizumi M, Kerr JN, Soukup GA, Breaker RR (1999) Allosteric ribozymes sensitive to the second messengers cAMP and cGMP. Nucleic Acids Symp Ser 42:275–276

Krivoshein AV, Hess GP (2004) Mechanism-based approach to the successful prevention of cocaine inhibition of the neuronal (alpha 3 beta 4) nicotinic acetylcholine receptor. Biochemistry 43:481–489

Kusser W (2000) Chemically modified nucleic acid aptamers for in vitro selections: evolving evolution. J Biotechnol 74:27–38

Lee SW, Sullenger BA (1997) Isolation of a nuclease-resistant decoy RNA that can protect human acetylcholine receptors from myasthenic antibodies. Nat Biotechnol 15:41–45

Lebruska LL, Maher LJ (1999) Seclection and characterization of RNA decoy for transcription factor NF-kappa B. Biochemistry 38:3168–3174

Mayer G, Blind M, Nagel W, Bohm T, Knorr T, Jackson CL, Kolanus W, Famulok M (2001) Controlling small guanine-nucleotide-exchange factor function through cytoplasmic RNA intramers. Proc Natl Acad Sci USA 98:4961–4965

McCauley TG, Hamaguchi N, Stanton M (2003) Aptamer-based biosensor arrays for detection and quantification of biological macromolecules. Anal Biochem 319:244–250

Michel F, Netter P, Xu MQ, Shub DA (1990) Mechanism of 3' splice site selection by the catalytic core of bacteriophage T4: the role of a novel base-pairing interaction in group I introns. Genes Dev 4:777–788

Nieuwlandt D, Wecker M, Gold L (1995) In vitro selection of ligands to substance P. Biochemistry 34:5651–5659

Nolte A, Klussmann S, Bald R, Erdmann VA, Fürste JP (1996) Mirror-design of L-oligonucleotide ligands binding to L-arginine. Nat Biotechnol 14:1116–1119

O'Connell D, Koenig A, Jennings S, Hicke B, Han H-L, Fitzwater T, Chang Y-F, Varki N, Parma D, Varki A (1996) Calcium-dependent oligonucleotide antagonists against L-selectin. Proc Natl Acad Sci USA 93:5883–5887

Ozaki H, Seo MS, Ozaki K, Yamada H, Yamada E, Okamoto N, Hofmann F, Wood JM, Campochiaro PA (2000) Blockade of vascular endothelial cell growth factor receptor signaling is sufficient to completely prevent retinal neovascularization. Am J Pathol 156:697–707

Pieken WA, Olsen DB, Benseler F, Aurup H, Eckstein F (1991) Kinetic characterization of ribonuclease-resistant 2′-modified hammer head ribozymes. Science 253:314–317

Proske D, Höfliger M, Söll RM, Beck-Sickinger AG, Farmulok M (2002) A Y2 receptor-mimetic aptamer directed against neuropeptide Y. J Biol Chem 277:11416–11422

Rhie A, Kirby L, Sayer N, Wellesley R, Disterer P, Sylvester I, Gill A, Hope J, James W, Tahiri-Alaoui A (2003) Characterization of 2′-fluoro-RNA aptamers that bind preferentially to disease-related conformations of prion protein and inhibit conversion. J Biol Chem 278:39697–39705

Rimmele M (2003) Nucleic acid aptamers as tools and drugs: Recent developments. Chembiochem 4:963–971

Robertson DL, Joyce GF (1990) Selection in vitro of an RNA enzyme that specifically cleaves single-stranded DNA. Nature 344:467–468

Robertson MP, Ellington AD (1999) In vitro selection of an allosteric ribozyme that transduces analytes to amplicons. Nat Biotechnol 17:62–66

Robertson MP, Ellington AD (2001) In vitro selection of nucleoprotein enzymes. Nat Biotechnol 19:650–655

Robertson MP, Knudsen SM, Ellington AD (2004) In vitro selection of ribozymes dependent on peptides for activity. RNA 10:114–127

Ruckman J, Green LS, Beeson J, Waugh S, Gillette WL, Henninger DD, Claesson-Welsh L, Janjic N (1998) 2′-Fluoropyrimidine RNA-based aptamers to the 165-amino acid form of vascular endothelial growth factor (VEGF165). J Biol Chem 273:20556–20567

Rusconi CP, Scardino E, Layzer J, Pitoc GA, Ortel PA, Monroe D, Sullenger BA (2002) RNA aptamers as reversible antagonists of coagulation factor IXa. Nature 419:90–94

Rusconi CR, Roberts JD, Pitoc AG, Nimjee SM, White RR, Quick G, Scardino E, Fay WP, Sullenger BA (2004) Antidote-mediated control of an anticoagulant aptamer in vivo. Nat Biotechnol 22:1423–1428

Sassanfar M, Szostak JW (1990) An RNA motif that binds ATP. Nature 364:550–553

Seiwert SD, Stines Nahreini T, Aigner S, Ahn NG, Uhlenbeck OC (2000) RNA aptamers as pathway-specific MAP kinase inhibitors. Chem Biol 7:833–843

Shi H, Hoffman BE, Lis JT (1999) RNA aptamers as effective protein antagonists in a multicellular organism. Proc Natl Acad Sci USA 96:10033–10038

Sun S (2000) Technology evaluation: SELEX, Gilead Sciences Inc. Curr Opin Mol Ther 2:100–105

Tasset DM, Kubik MF, Steiner W (1997) Oligonucleotide inhibitors of human thrombin that bind distinct epitopes. J Mol Biol 272:688–698

Theis MG, Knorre A, Kellersch B, Moelleken J, Wieland F, Kolanud W, Famulok M (2004) Discriminatory aptamer reveals serum response element transcription regulated by cytohesin-2. Proc Natl Acad Sci USA 101:11221–11226

Thiel K (2004) Oligo oligarchy—the surprisingly small word of aptamers. Nat Biotechnol 22:649–651

Thompson KM, Syrett HA, Knudsen SM, Ellington AD (2002) Group I aptazymes as genetic regulatory switches. BMC Biotechnol 2:21

Tucker CE, Chen LS, Judkins MB, Farmer JA, Gill SC, Drolet DW (1999) Detection and pharmacokinetics of an anti-vascular endothelial growth factor oligonucleotide-aptamer (NX1838) in rhesus monkeys. J Chromatogr B Biomed Sci Appl 732:203–212

Tuerk C, Gold L (1990) Systematic evolution of ligands by exponential enrichment: RNA ligands to bacteriophage T4 DNA polymerase. Science 249:505–510

Tuschl T, Sharp PA, Bartel DP (1998) Selection in vitro of ribozymes from a partially randomized U2 and U6 snRNA library. EMBO J 17:2637–2650

Ulrich H, Gameiro AM (2001) Aptamers as tools to study dysfunction in the neuronal system. Curr Med Chem Cent Nerv Sys Agents 1:125–132

Ulrich H, Ippolito JE, Pagan OR, Eterovic VE, Hann RM, Shi H, Lis JT, Eldefrawi ME, Hess GP (1998) In vitro selection of RNA molecules that displace cocaine from the nicotinic acetylcholine receptor. Proc Natl Acad Sci USA 95:14051–14056

Ulrich H, Magdesian MH, Alves MJM, Colli W (2002) In vitro selection of RNA aptamers that bind to cell adhesion receptors of Trypanosoma cruzi and inhibit cell invasion. J Biol Chem 277:20756–20762

Ulrich H, Martins AHB, Pesquero JB (2004) RNA and DNA aptamers in cytomics analysis. Cytometry 59A:220–231

Vater A, Klussmann S (2003) Toward third-generation aptamers: Spiegelmers and their therapeutic prospects. Curr Opin Drug Discov Devel 6:253–261

Vuyisich M, Beal PA (2002) Controlling protein activity with ligand-regulated RNA aptamers. Chem Biol 9:907–913

Warren RS, Yuan H, Matli MR, Gillett NA, Ferrara N (1995) Regulation by vascular endothelial growth factor of human colon cancer tumorigenesis in a mouse model of experimental liver metastasis. J Clin Invest 95:1789–1797

Weiss S, Proske D, Neumann M, Groschup MH, Kretzschmar HA, Famulok M, Winnacker EL (1997) RNA aptamers specifically interact with the prion protein PrP. J Virol 71:8790–8797

Werstuck G, Green MR (1998) Controlling gene expression in living cells through small molecule-RNA interactions. Science 282:296–298

White RB, Shan S, Rusconi CP, Shetty G, Dewhirst MW, Kontos CD, Sullenger BA (2003) Inhibition of rat corneal angiogenesis by a nuclease-resistant RNA aptamer specific for angiopoietin-2. Proc Natl Acad Sci U S A 100:5028–5033

Willis MC, Collins BD, Zhang T, Green LS, Sebesta DP, Bell C, Kellogg E, Gill SC, Magallanez A, Knauer S, Bendele RA, Janjic N (1998) Liposome-anchored vascular endothelial growth factor aptamers. Bioconjug Chem 9:573–582

Wilson C, Szostak JW (1995) In vitro evolution of a self-alkylating ribozyme. Nature 374:777–782

Xu W, Ellington AD (1996) Anti-peptide aptamers recognize amino acid sequence and bind a protein epitope. Biochemistry 93:7475–7480

Ylera F, Lurz R, Erdmann VA, Fürste JP (2002) Selection of RNA aptamers to the Alzheimer's disease amyloid peptide. Biochem Biophys Res Commun 290:1583–1588

Zhang XM, Shao NS, Chi MG, Sun MJ (2003) Screening of RNA molecules inhibiting human acetylcholinesterase by virtue of systematic evolution of ligands by exponential enrichment. Acta Pharmacol Sin 24:711–714

RNA Aptamers Directed Against Oligosaccharides

M. Sprinzl (✉) · M. Milovnikova · C. S. Voertler

Laboratorium für Biochemie, Universität Bayreuth, 95440 Bayreuth, Germany
mathias.sprinzl@uni-bayreuth.de

1	Introduction	327
2	RNA Aptamers	328
2.1	In Vitro Selection of Nucleic Acids	328
2.2	Nucleic Acid Aptamers Directed Against Carbohydrates	329
2.3	RNA Aptamers Directed Against Synthetic Oligosaccharides	333
3	Summary and Outlook	335
	References	339

Abstract Nucleic acid molecules are designed to interact predominantly with proteins or complementary nucleic acids. Interaction of nucleic acids with carbohydrates, abundant constituents of glycoproteins and glycolipids, are not common in cells. Biomedical applications of nucleic acids targeted against oligosaccharides, which are involved in the function of receptors, immune answer, host interaction with invading infectious agents, and cancer metastasis, are feasible. In vitro selection of nucleic acids interacting with oligo- and polysaccharides is a promising strategy to identify potential inhibitors of biochemical recognition processes in which carbohydrates are involved. Several RNA and DNA aptamers directed against carbohydrates have already been isolated and characterized. The results are summarized in this article, and an attempt is made to draw initial conclusions concerning the perspectives of the outlined approach.

Keywords RNA and DNA aptamer · Saccharide · Carbohydrate nucleic acid interaction · Glycoprotein · Glycolipid · Synthesis of oligosaccharides

1
Introduction

Carbohydrates conjugated to lipids and proteins are abundant constituents of living cells. They play an important role in numerous biochemical and cellular processes including the glycoprotein/protein-interaction, recognition of various cell receptors, cells adhesion, and cell/pathogen interaction. Based on our knowledge of the molecular mechanisms of these processes, lipopolysaccharides and glycoproteins are considered a promising target for biomedical intervention (Weintraub 2003).

There are several strategies available to interfere with the function of glycoproteins. Binding of small molecules to the carbohydrate moiety of a glycoprotein is one possibility. It is, however, usually hampered by the low specificity of such inhibitors and their inability to distinguish between the great varieties of oligosaccharides that possess mostly uncharged functional groups in their structure. The opposite approach relies on small molecules acting as carbohydrate mimetics that block the receptor macromolecules and interfere with glycoprotein binding. This strategy usually fails to produce promising solutions due to low specificity and low affinity of such mimetics to their targets caused, most probably, by the limited amount of intramolecular interactions possible for small molecules.

Carbohydrates are, more efficiently than small molecules, recognized by macromolecules. Among the different proteins that recognize carbohydrates are particularly lectins, receptor proteins, enzymes, and antibodies. All of them are of high biochemical importance. Antibodies fulfill the demand to be directed against a variety of glycoprotein structures, depending on the chosen antigen. However, the affinity of antibodies to glycoproteins is not sufficiently high and the specificity is surprisingly low (Weis and Drickamer 1996). This may hinder their use as specific and tight-binding molecules.

2
RNA Aptamers

2.1
In Vitro Selection of Nucleic Acids

About one and a half decades ago the method of in vitro selection, or SELEX (systemic evolution of ligands by exponential enrichment), of oligonucleotides capable of fulfilling a given biochemical task (enzymatic activity, binding to a ligand, introducing a conformational change, RNA switches) from libraries of random sequences was developed (Tuerk and Gold 1990; Ellington and Szostak 1990). This easy, practicable laboratory protocol allowed for the isolating of thousands of RNA molecules that bind low molecular mass ligands, proteins, nucleic acids, specific cellular structures, and even whole cells. Some of these binding nucleic acids, named aptamers, served as leads to develop stable variants of an active structural principle for use as therapeutics or analytical agents. Some potential drugs and promising concepts had already been developed by this in vitro selection-based technology (Nimjee et al. 2005; Rusconi et al. 2004; Dougan et al. 2003; White et al. 2003). Although both DNA and RNA have a potential to bind ligands, the structural diversity and intrinsic ability of RNA to form a variety of tertiary structures constitutes a greater potential for RNA compared to DNA molecules to recognize and bind to a variety of ligands. However, the high price for this favorable structural variability is the low stability of RNA, hampering the direct use of RNA aptamers in in vivo applications. This drawback can be overcome by stabilizing the RNA, usually by replacing the 2′-OH group by another functional group unable to participate on a phosphate transfer from the vicinal 3′-position (Osborne et al. 1997).

A more sophisticated but also much more complicated and expensive approach is the identification of the RNA sequence that binds to the enantiomer opposite to the naturally occurring target by SELEX and chemical synthesis of RNA build-up from L-ribose-phosphate backbone. These so-called spiegelmers are able to bind to the naturally occurring enantiomer of the target and at the same time are resistant to nuclease attack (Klussmann et al. 1996; Nolte et al. 1996).

The in vitro selection of RNA aptamers has several obvious advantages compared to in vivo selection of antibodies. In contrast to the production of antibodies in vivo, there is a the possibility of using a wide variety of ligands to which the RNA aptamers may be directed, including labile and toxic compounds, endogenous cellular components, and ligands with limited availability, since relatively small amounts are required to perform the in vitro RNA selection. Furthermore, after identification of the RNA lead structures isolated by in vitro selection, a stable and active chemical variant of the aptamer can be prepared by chemical synthesis, a possibility that is not available for protein antibodies.

2.2
Nucleic Acid Aptamers Directed Against Carbohydrates

Nucleic acid interactions with polymeric cellular components, particularly with proteins, are dominated by hydrogen bonds, hydrophobic and ionic interactions between phosphate residues and positively charged amino acids. This is defined by the sequence and tertiary structure of the nucleic acid. Carbohydrate residues are void of basic functional groups that are capable of undergoing ionic interaction with nucleic acids. In addition, they possess a very flexible structure. Such molecules are not expected to be a good choice for specific interactions with nucleic acids (Hermann and Westhof 1998; Herrmann 2005).

Despite the high conformational freedom of their carbohydrate residues, very efficient and specific interactions occur between aminoglycoside antibiotics (Fig. 1) and ribonucleic acids. Several aminoglycoside antibiotics capable of preventing conformational changes in ribosomal RNAs were functionally characterized. They bind to specific regions of ribosomal RNA, where they block important ribosomal functions, like peptidyl transferase reaction or decoding of mRNA. Aminoglycoside antibiotics are excellent ligands for in vitro selection. Examples are RNA aptamers directed against kanamycin B (Kwon et al. 2001), tobramycin (Wang et al. 1996), streptomycin (Tereshko et al. 2003), and neomycin B (Cowan et al. 2000). The interactions of aminogly-

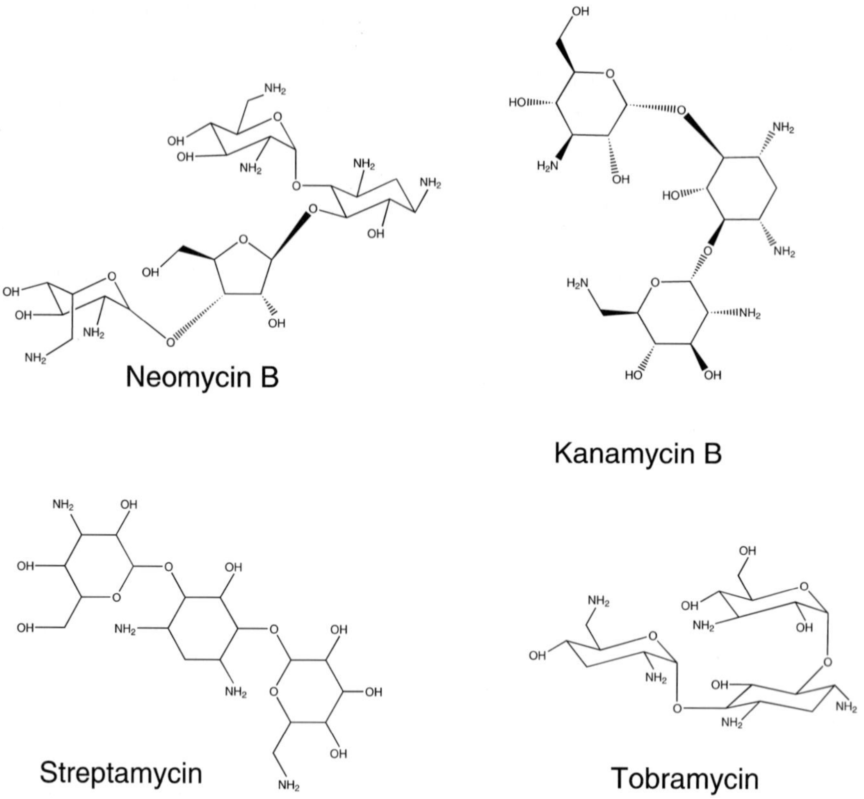

Fig. 1 Structures of aminoglycoside antibiotics used as ligands for in vitro RNA selection

coside antibiotics with RNA are usually specific and provide stable complexes with dissociation constants in the range of 10^{-4} to 10^{-10} M. The reason for this high "aptamerogenic" activity lies in their structure that contains, besides carbohydrate residues with hydroxyl groups, basic functional groups that may potentially interact as protonated cations with the phosphate anion of RNA backbone, particularly replacing the structurally crucial Mg^{2+} ions (Hermann and Westhof 1998; Hermann and Westhof 1999, 2000). In the complex with RNA, the carbohydrate moieties of the aminoglycoside antibiotics are usually placed into a pocket formed by the tertiary or secondary structure elements of RNA and stabilized by hydrogen bonds. This is documented by structures of several antibiotic RNA complexes that were solved by nuclear magnetic resonance (NMR) and crystallographic analyses (Hermann and Patel 2000; Patel et al. 1997; Patel and Suri 2000).

Aminoglycosides with their positively charged amino groups are, however, a special case. In the last decade several attempts have also been made to isolate RNA aptamers against uncharged oligosaccharides and polysaccharides.

In vitro selection of RNA directed against galactose, glucose, and mannose provided aptamers with relatively low affinities in the range between 10^{-4} and 10^{-5} M (Kawakami et al. 1998). Oligosaccharides and polysaccharides are significantly more aptamerogenic compared to monosaccharides. RNA aptamers possessing high affinity for chitin were isolated (Fukusaki et al. 2000). Engelke and coworkers (Yang et al. 1998) reported the isolation of single-stranded DNA specifically interacting with the disaccharide cellobiose. In vitro selection of these DNA aptamers was performed by selection of DNA that binds to the cellulose surface and was followed by elution of retarded oligonucleotides with cellobiose. The isolated DNA aptamers were specific for cellobiose [two (1→4)β-glucose units] and cellotetraose [four (1→4)β-glucose units] but did not interact with disaccharides lactose, maltose, or gentobiose (Fig. 2). These experiments clearly demonstrated the capability of single stranded DNA to interact with carbohydrates containing only carbon, hydrogen, and oxygen atoms. Later, this study was extended for isolation of RNA aptamers. They were identified by selection for binding to commercial Sephadex G100 chromatography matrix (Srisawat et al. 2001). Sephadex is a polysaccharide composed of α-(1→6)-D-glucopyranose units (95%) and α-(1→3)D-glucopyranose branch points (5%). Competition experiments against the basic components of dextrans isomaltose, isomaltotriose, and isomaltotetraose indicated that the optimal binding site for the isolated aptamer should contain more then four glucose units. Remarkably, one of the isolated RNA aptamers was highly selective toward Sephadex, whereas other common supporting polysaccharides like Sepharose, Sephacryl, cellulose, and pustulan did not bind to this aptamer. Given this high specificity in regard to polysaccharide structure, the authors suggested the application of the Sepharose-binding aptamer as an RNA-tag for use in affinity chromatog-

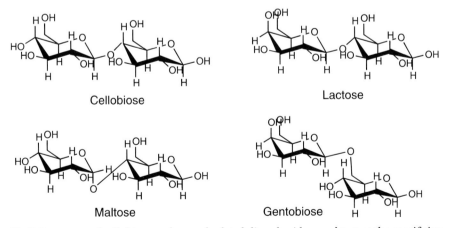

Fig. 2 Structures of cellobiose and several related disaccharides used to test the specificity of cellulose directed DNA aptamers

raphy to purify RNAs or ribonucleoprotein complexes on Sephadex affinity matrices.

Attempts to isolate nucleic acid aptamers applicable for biomedical purposes were also undertaken. Sialyl Lewis X (sLe^x) is a tetrasaccharide glycoconjugate of many membrane proteins (Fig. 3). It acts as a ligand for the selectin proteins in cell adhesion during inflammatory processes (Lasky 1995). Overexpression of sLe^x is also characteristic for various cancer cells (Hanski et al. 1995). Yu and coworkers isolated RNA aptamers against this tetrasaccharide immobilized to bovine serum albumin (Jeong et al. 2001). After 17 cycles of selection, six clones were isolated that produced RNA with specific affinity to sLe^x. These aptamers did not bind to unmodified agarose but efficiently interacted with agarose to which sLe^x was conjugated. The binding constants for the interaction of the RNA aptamers with sLe^x were determined by surface plasmon resonance using bovine sLe^x immobilized on serum albumin. All selected RNA aptamers had a similar or even better binding affinities to sLe^x ($K_d = \sim 10^{-9}$–10^{-10} M) than the commercially available monoclonal anti-sLe^x antibody. This suggested that the in vitro RNA selection might generate high-affinity aptamers that could substitute for antibodies directed against small carbohydrates with poor antigenic properties. However, the specificity of anti-sLe^x aptamers is not very high. Even the best RNA-aptamer from this study had only 10- and 100-fold higher affinity to sLe^x compared to the structurally related sLe^A and lactose, respectively.

A promising new approach to overcome the low "aptamerogenic" properties of nucleic acids against carbohydrates was recently published by Sawai and coworkers (Masud et al. 2004). These authors applied the classical DNA-SELEX method to isolate aptamers that bind to sialyllactose (Fig. 4), a main constituent of many glycoprotein conjugates. To achieve this, they replaced thymidine-5′-triphosphate by 5-[N-(6-aminohexyl)carbamoylmethyl]-2′-deoxyuridine-5′-

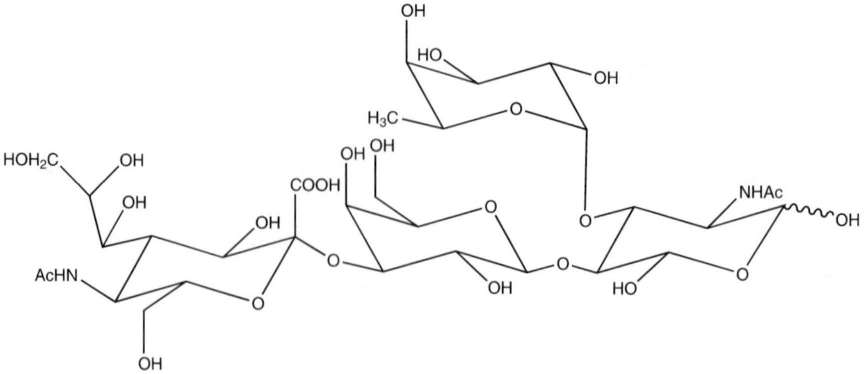

Fig. 3 Structure of sialyl Lewis X, a glycoconjugate of many membrane proteins

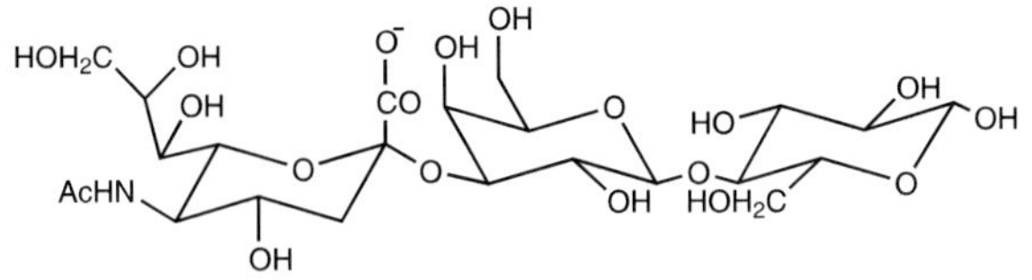

Fig. 4 Sialyllactose used for isolation of DNA aptamers containing 5-[N-(6-aminohexyl)carbamoylmethyl]-2′-deoxyuridine (*small insert*)

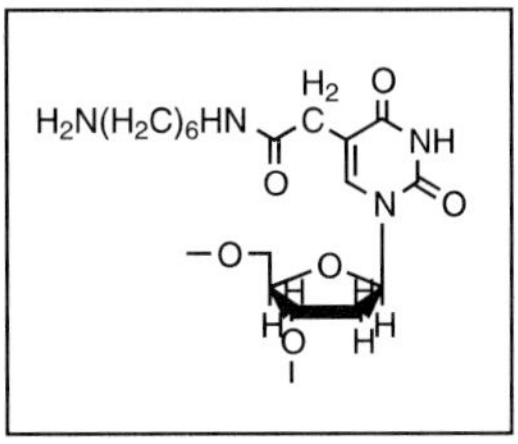

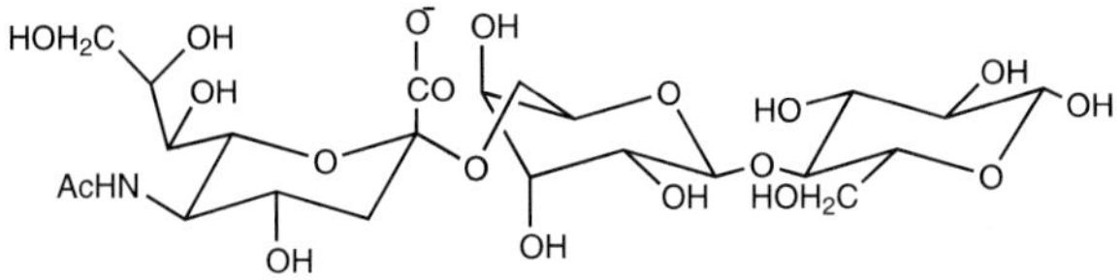

triphosphate (Fig. 4) in the PCR protocol. A DNA polymerase incorporated this nucleotide with high efficiency and produced DNA with an additional amino group on all thymidine residues. The potential of such a modified DNA to interact with negatively charged sialyl residue increased as expected. Several single-stranded, about 100 nucleotide-long DNA aptamers that bind sialyllactose with dissociation constant in the micromolar range could indeed be isolated. Based on a nucleic acid folding algorithm, these deoxyribo-oligonucleotides form secondary structures with three-way junctions. Several "amino" thymidine derivatives are located in the vicinity of these junctions. The work demonstrates the ability of nucleic DNA for selective recognition of oligosaccharides, provided the interaction is additionally supported by complementary ionic interactions between the carbohydrate and nucleic acid.

2.3
RNA Aptamers Directed Against Synthetic Oligosaccharides

Recently, in vitro selection of RNA aptamers against chemically synthesized oligosaccharides derived from naturally occurring glycosyl residues were systematically investigated in our laboratory (M. Milovnikova, unpublished). A pool of RNA oligonucleotides obtained from synthetic DNA containing a 50-nucleotide-long random sequence was used for this study. Different tri-, tetra-, penta-, and heptasaccharides (Fig. 5) were immobilized via a carbohydrate linker containing a cleavable, S-S-linkage on Sepharose matrix and alternatively on glass surface. After interaction with a pool of about 10^{13} RNA molecules with random sequences, the oligosaccharide RNA complexes were eluted from the matrix by cleavage of the disulfide bond. This procedure was

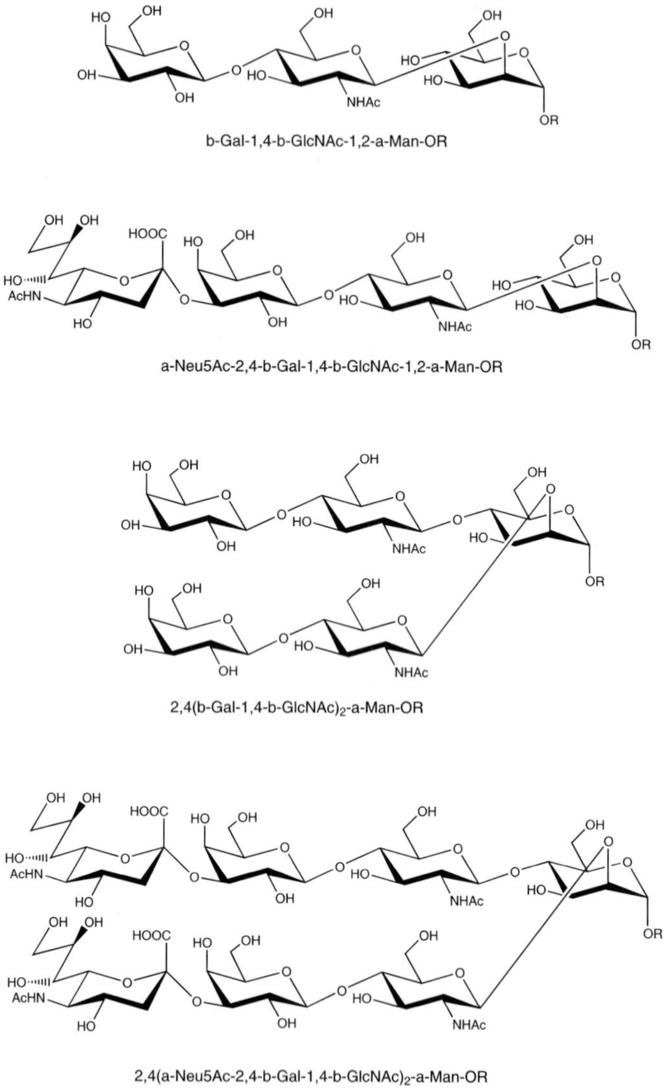

Fig. 5 Structure of synthetic oligosaccharides used for isolation of RNA aptamers

chosen to prevent isolation of RNAs that bind nonspecifically to the polysaccharide matrix. For even better control of selectivity, a counter selection step on unsubstituted matrices without oligosaccharide ligands was performed.

Several RNAs with defined sequences that interacted with oligosaccharide ligands were isolated. Their interaction with oligosaccharides ligands was quantified by surface plasmon resonance and filter binding assays.

The results of this study demonstrate that RNA can interact with oligosaccharides with binding constants reaching the nanomolar range. The specificity of the interactions is, however, surprisingly low. Isolated RNA aptamers usually possessed a significant cross-reactivity between the different immobilized oligosaccharide ligands and the polysaccharide matrix to which these ligands were immobilized. This was not surprising since a sequence homology between isolated aptamers, regardless of the oligosaccharide ligand against which they were selected, could be observed. Interesting in this regard was the comparison of RNA aptamer sequences obtained by in vitro selection against oligosaccharide immobilized on Sepharose and glass, respectively. Besides sequences that appeared only in selections on glass-immobilized oligosaccharides, one sequence was isolated that bound to all used substrates, unmodified Sepharose, Sepharose-immobilized heptasaccharide, and glass-immobilized heptasaccharide. Thus, this carbohydrate binder recognizes the heptasaccharide with high affinity, as was confirmed by surface plasmon resonance experiments, but is not able to differentiate between particular carbohydrate structures.

3
Summary and Outlook

Aminoglycoside antibiotics with carbohydrate residues that contain positively charged functional groups are ideal ligands for in vitro selection of RNA aptamers.

DNA and RNA aptamers can, however, also recognize carbohydrates possessing exclusively hydroxyl functions and form with them stable complexes with K_ds in the nanomolar range.

Table 1 summarizes the previous work on in vitro selection of nucleic acids aptamers towards carbohydrate targets. The following conclusions can be drawn from these results:

- It is possible to isolate RNA and DNA aptamers directed against polysaccharides.

- Selection leads to more specific aptamers when additional substituents with a potential to participate in ionic interactions are integrated into carbohydrate and/or nucleic acid structures. In addition, functional groups capable of undergoing hydrogen bond interactions common to carbohydrates are important.

- The specificity of RNA (DNA) aptamers increases with the number of carbohydrate units of the oligosaccharide ligand used in the in vitro selection protocol.

- Strategies for protection of RNA aptamers against enzymatic degradation during in vivo applications and delivery system have to be considered.

Table 1 Aptamer selection against carbohydrate targets

Target/ligand	Chemical nature of carbohydrate binder	Affinity to target	Specificity[a]	Binding motif of the aptamer, minimal size	Reference
Sialyllactose	DNA	4.9 μM[b]	n.d.	Three-way junction with amino-modified thymidines 42 nt	Masud et al. 2004
Sialyl Lewis X	RNA	3.3 nM[c]	lactose[c]: 290 nM Le^X: 24 nM Le^A: 14 nM sLe^A: 17 nM sLe^X-BSA: 0.057 nM BSA: 100 nM	n.d.	Jeong et al. 2001
Sephadex G100 (1–3 branched α-1-6 glucose, crosslinked with epichlorohydrin)	RNA	n.d.	Sepharose CL-4B[d]: – Sephacryl S-500[d]: – Cellulose[d]: – Pustulan[d]: – Isomaltose[d]: – Isomaltotriose[d]: – Isomaltotetraose[d]: – Dextran B512[d]: +++	Internal bulge, 33 nt	Srisawat et al. 2001
Cellulose	DNA	Low μM	Cellulose[d]: +++ Cellotetraose[e]: 0.6 μM Cellobiose[e]: ≤0.3 μM Lactose[e]: – Maltose[e]: – Gentiobiose[e]: –	G-rich sequence stretches, n.d.	Yang et al. 1998

Table 1 (continued)

Target/ligand	Chemical nature of carbohydrate binder	Affinity to target	Specificity[a]	Binding motif of the aptamer, minimal size	Reference
Trisaccharide β-Gal-1,4-β-GlcNAc-1,2-Man	RNA	≤1 µM[c]	Sephadex G25[d]: – Sephacryl S200 HR[d]: – Biogel P6[d]: – Superdex 200[d]: – Sepharose 6B[d]: + Sepharose 6B-trisaccharide[d]: ++	A+G rich stem-loop structures	M. Milovnikova, unpublished
Tetrasaccharide α-Neu5Ac-2,3-β-Gal-1,4-β-GlcNAc-1,2-Man	RNA	n.d.	Sepharose 6B[d]: – Sepharose 6B-tetrasaccharide[d]: ++	A+G rich stem-loop structures	M. Milovnikova, unpublished
Pentasaccharide 2,4-(β-Gal-1,4-β-GlcNAc)2-α-Man	RNA	n.d.	Sepharose 6B[d]: – Sepharose 6B-pentasaccharide[d]: ++	A+G rich stem-loop structures	M. Milovnikova, unpublished
Heptasaccharide 2,4-(α-Neu5Ac-2,3-β-Gal-1,4-β-GlcNAc)2-α-Man	RNA	n.d.	Sepharose 6B[d]: – Sepharose 6B-heptasaccharide[d]: ++	A+G rich stem-loop structures	M. Milovnikova, unpublished
Chitin (β-1,4-N-acetylglucosamine)	DNA	n.d.	Chitin[d]: ++ Cellulose[d]: +	G-rich stem loops	Fukusaki et al. 2000
Kanamycin B[i]	RNA	180 nM[c,f]	Tobramycine[c,f]: 12 nM Kanamycin A[c,f]: 4.4 µM Neomycine[c,f]: 1.1 µM Paramomycine[c,f]: 1.5 µM	Hairpin stem loop motif with bulged nucleotides	Kwon et al. 2001

Table 1 (continued)

Target/ligand	Chemical nature of carbohydrate binder	Affinity to target	Specificity[a]	Binding motif of the aptamer, minimal size	Reference
Neomycin B[i]	RNA	115 nM[h]	Paramomycin[h]: ≥ 10 µM	Distorted hairpin stem loop, 23 nt	Cowan et al. 2000
Tobramycin[i]	RNA	7 nM[f]	Neomycine B[f]: 1 µM Gentamycine[f]: 7.81 µM Erythromycin[f]: 923 µM D-Glucosamine[f]: –	Stem loop structures with bulged-out nucleotides forming a deep binding pocket, ≈ 30 nt	Jiang and Patel 1998
Streptomycin[i]	RNA	≤1 µM[h]	Bluensomycin[h]: –	Binding pocket of two intertwined asymmetrical internal loops, ≈ 46 nt	Tereshko et al. 2003

[a] Specificity was tested by determination of K_d or qualitative assessment of the affinity to potential competitors; –, no binding; +, ++, +++: weak, strong, very strong binding, respectively; n.d., not determined [b] K_d estimated by equilibrium filtration [c] K_d estimated by surface plasmon resonance (Biacore) [d] Qualitative comparison by determining the binding of radiolabeled aptamer to the solid-phase material and/or in presence of additional competitor [e] Change of the absorption at 260 nM due to structural rearrangements upon target or competitor binding [f] K_d estimated by fluorescence anisotropy changes of dye-labeled target and/or aptamer [g] K_d estimated by competitive affinity elution from a target-immobilized affinity column; qualitative binding of competitor tested by elution from target-matrices: +++, ≥75%; ++, 33%–75%; +, 11%–33%; –, ≤11% of initial bound aptamer were eluted [h] K_d or qualitative test of specificity estimated from intensity of protection of adenosine in the aptamer bindings site against DMS modification [i] Representative aminoglycoside target; more aptamers binding to other aminoglycosides exist, but are beyond the scope of this article

Isolation of RNA aptamers that specifically interact with oligosaccharides, the main prerequisite for biomedical applications, has been difficult to achieve. Promising in the future are the selection strategies with oligonucleotides that contain 2'-amino groups on ribose residues, or ω-aminoalkyl substituents on nucleobases. Such aptamers, constituted of modified nucleotides, may contribute to the stability of aptamers and at the same time provide the necessary structural element to interact with negatively charged carbohydrate residues.

Despite several unsolved problems, obvious at the present time, the identification of specific RNA aptamers targeted against carbohydrates remains a challenge for bioorganic chemists, biochemists, and molecular biologists working in the field of biomedical research.

References

Cowan JA, Ohyama T, Wang D, Natarajan K (2000) Recognition of a cognate RNA aptamer by neomycin B: quantitative evaluation of hydrogen bonding and electrostatic interactions. Nucleic Acids Res 28:2935–2942

Dougan H, Weitz JI, Stafford AR, Gillespie KD, Klement P, Hobbs JB, Lyster DM (2003) Evaluation of DNA aptamers directed to thrombin as potential thrombus imaging agents. Nucl Med Biol 30:61–72

Ellington AD, Szostak JW (1990) In vitro selection of RNA molecules that bind specific ligands. Nature 346:818–822

Fukusaki E, Kato T, Maeda H, Kawazoe N, Ito Y, Okazawa A, Kajiyama S, Kobayashi A (2000) DNA aptamers that bind to chitin. Bioorg Med Chem Lett 10:423–425

Hanski C, Hanski ML, Zimmer T, Ogorek D, Devine P, Riecken EO (1995) Characterization of the major sialyl-Lex-positive mucins present in colon, colon carcinoma, and sera of patients with colorectal cancer. Cancer Res 55:928–933

Hermann T (2005) Drugs targeting the ribosome. Curr Opin Struct Biol 15:355–366

Hermann T, Patel DJ (2000) Adaptive recognition by nucleic acid aptamers. Science 287:820–825

Hermann T, Westhof E (1998) Saccharide-RNA recognition. Biopolymers 48:155–165

Hermann T, Westhof E (1999) Docking of cationic antibiotics to negatively charged pockets in RNA folds. J Med Chem 42:1250–1261

Hermann T, Westhof E (2000) Rational drug design and high-throughput techniques for RNA targets. Comb Chem High Throughput Screen 3:219–234

Jeong S, Eom T, Kim S, Lee S, Yu J (2001) In vitro selection of the RNA aptamer against the Sialyl Lewis X and its inhibition of the cell adhesion. Biochem Biophys Res Commun 281:237–243

Jiang L, Patel DJ (1998) Solution structure of the tobramycin-RNA aptamer complex. Nat Struct Biol 5:769–774

Kawakami J, Kawase Y, Sugimoto N (1998) In vitro selection of aptamers that recognize a monosaccharide. Anal Chim Acta 365:95–100

Klussmann S, Nolte A, Bald R, Erdmann VA, Furste JP (1996) Mirror-image RNA that binds D-adenosine. Nat Biotechnol 14:1112–1115

Kwon M, Chun SM, Jeong S, Yu J (2001) In vitro selection of RNA against kanamycin B. Mol Cells 11:303–311

Lasky LA (1995) Selectin-carbohydrate interactions and the initiation of the inflammatory response. Annu Rev Biochem 64:113–139

Masud MM, Kuwahara M, Ozaki H, Sawai H (2004) Sialyllactose-binding modified DNA aptamer bearing additional functionality by SELEX. Bioorg Med Chem 12:1111–1120

Nimjee SM, Rusconi CP, Sullenger BA (2005) Aptamers: an emerging class of therapeutics. Annu Rev Med 56:555–583

Nolte A, Klussmann S, Bald R, Erdmann VA, Furste JP (1996) Mirror-design of L-oligonucleotide ligands binding to L-arginine. Nat Biotechnol 14:1116–1119

Osborne SE, Matsumura I, Ellington AD (1997) Aptamers as therapeutic and diagnostic reagents: problems and prospects. Curr Opin Chem Biol 1:5–9

Patel DJ, Suri AK (2000) Structure, recognition and discrimination in RNA aptamer complexes with cofactors, amino acids, drugs and aminoglycoside antibiotics. J Biotechnol 74:39–60

Patel DJ, Suri AK, Jiang F, Jiang L, Fan P, Kumar RA, Nonin S (1997) Structure, recognition and adaptive binding in RNA aptamer complexes. J Mol Biol 272:645–664

Rusconi CP, Roberts JD, Pitoc GA, Nimjee SM, White RR, Quick G Jr, Scardino E, Fay WP, Sullenger BA (2004) Antidote-mediated control of an anticoagulant aptamer in vivo. Nat Biotechnol 22:1423–1428

Srisawat C, Goldstein IJ, Engelke DR (2001) Sephadex-binding RNA ligands: rapid affinity purification of RNA from complex RNA mixtures. Nucleic Acids Res 29:E4

Tereshko V, Skripkin E, Patel DJ (2003) Encapsulating streptomycin within a small 40-mer RNA. Chem Biol 10:175–187

Tuerk C, Gold L (1990) Systematic evolution of ligands by exponential enrichment: RNA ligands to bacteriophage T4 DNA polymerase. Science 249:505–510

Wang Y, Killian J, Hamasaki K, Rando RR (1996) RNA molecules that specifically and stoichiometrically bind aminoglycoside antibiotics with high affinities. Biochemistry 35:12338–12346

Weintraub A (2003) Immunology of bacterial polysaccharide antigens. Carbohydr Res 338:2539–2547

Weis WI, Drickamer K (1996) Structural basis of lectin-carbohydrate recognition. Annu Rev Biochem 65:441–473

White RR, Shan S, Rusconi CP, Shetty G, Dewhirst MW, Kontos CD, Sullenger BA (2003) Inhibition of rat corneal angiogenesis by a nuclease-resistant RNA aptamer specific for angiopoietin-2. Proc Natl Acad Sci U S A 100:5028–5033

Yang Q, Goldstein IJ, Mei HY, Engelke DR (1998) DNA ligands that bind tightly and selectively to cellobiose. Proc Natl Acad Sci U S A 95:5462–5467

Aptamer-Based Biosensors: Biomedical Applications

A. K. Deisingh

Caribbean Industrial Research Institute, The University of the West Indies, St. Augustine, Trinidad and Tobago
anildeisingh@aol.com

1	Aptamer Selection and Properties	342
1.1	Properties of Aptamers	343
2	Biosensor Theory and Design	344
3	Aptasensors	345
4	Biomedical Applications Based on Transduction Methods	347
4.1	Optical	347
4.1.1	Surface Plasmon Resonance	347
4.1.2	Colourimetry	348
4.1.3	Fibre Optics	349
4.2	Fluorescence and Luminescence	349
4.2.1	Fluorescence	349
4.2.2	Luminescence	351
4.2.3	Chip-Based Assays	351
4.3	Acoustic Wave Devices	352
4.4	Other Methods	354
4.4.1	Electrochemical Aptasensors	354
4.4.2	Micromachined Sensors	354
5	Concluding Remarks	355
	References	355

Abstract This chapter considers the use of aptamer-based biosensors (generally termed 'aptasensors') in various biomedical applications. A comparison of antibodies and aptamers is made with respect to their use in the development of biosensors. A brief introduction to biosensor design and theory is provided to illustrate the principles of the field. Various transduction approaches, viz. optical, fluorescence, acoustic wave and electrochemical, are discussed. Specific biomedical applications described include RNA folding, high-throughput screening of drugs, use as receptors for measuring biological concentrations, detection of platelet-derived growth factor, protein binding and detection of HIV-1 Tat protein.

Keywords Biosensor · Aptasensor · Molecular beacon · Fluorescence · Acoustic wave

1
Aptamer Selection and Properties

Aptamers are macromolecules composed of nucleic acids that bind tightly to a specific molecular target (Archemix Corporation 2003). The normal approach for the generation of aptamers is the SELEX (systematic evolution of ligands by exponential enrichment) method, independently developed by G. Joyce, F. Szostak and L. Gold in 1990 (Famulok 2005). This technique, also called in vitro selection, allows the simultaneous screening of diverse pools of RNA or DNA molecules for a particular feature, e.g. binding to small organic molecules, large proteins or generation of ribozyme catalysis (Famulok 2005). For this last application, it is possible to utilize ribozymes with ligation activity and isomerases and ribozymes that catalyse the ATP-dependent phosphorylation of RNA oligonucleotides (Famulok 2005). Although aptamers have many advantages over antibodies for similar applications, their adoption has been slow because: (1) many researchers believe SELEX is the only way to select aptamers, (2) SELEX is considered to be specialized, cumbersome, labour-intensive and time-consuming and (3) SELEX is protected by patents that are rigorously enforced (Guthold 2005).

Recently, however, AptaRes (2005) has introduced a new approach to aptamer production that they claim is much faster than the SELEX approach. This Monolex technology is a one-round procedure with the steps involving the synthesis of an oligonucleotide library with regions of random sequence, affinity adsorption of the oligonucleotides to a target, affinity sorting of the oligonucleotides along an affinity resin and separation of oligonucleotides with different levels of affinity into different pools, amplification of the separated oilgonucleotide pools (polyclonal aptamers) and identification of individual oligonucleotides by cloning and sequencing (monoclonal aptamer). Particular advantages of the Monolex aptamers include (AptaRes 2005):

a. Low target quantity required for isolation

b. In vitro selection process requires only one step

c. Each pool contains many target-specific aptamers

d. Narrow range of affinity level in each aptamer pool

e. Production and modification by enzymatic amplification for immediate use

f. Fast availability of aptamers for drug quantification after high-throughput drug identification

This new technology can prove useful in biosensor application especially with respect to (AptaRes 2005):

a. Site-directed coupling of aptamers to the biosensor surface.

b. Strong denaturing conditions can be applied to dissociate the target complex.

c. Renaturation capabilities of the oligonucleotide aptamers.

d. Thermostability of aptamers.

e. Quick application of polyclonal aptamers.

f. Confirmation of identity by structural analysis.

Insufficient stability of nucleic acids as therapeutic agents is often a major potential disadvantage but could, to some extent, be overcome by using libraries of chemically modified nucleic acids, e.g. $2'$-fluoro-or $2'$-amino-$2'$-deoxypyrimidine nucleic acids (Famulok and Mayer 1999). Another strategy that may overcome the stability issue makes use of the mirror-image or 'Spiegelmer' approach that exploits nuclease resistance of the enantiomer of naturally-occurring nucleic acids (Famulok and Mayer 2004). These topics are described in earlier chapters.

1.1
Properties of Aptamers

Aptamers are chemically stable and can be boiled or frozen without loss of activity. This feature makes them suitable for use in biosensor applications where stability is a prime requirement. Additionally, they can undergo a variety of modifications to optimize their properties for specific applications. They can be circularized, linked in pairs or clustered on to the surfaces of molecules (Archemix Corporation 2003). Aptamers can also be modified to reduce their degradation by enzymes in in vivo situations. This plasticity provides an advantage for aptamers over antibodies where chemical modification is difficult to control.

The binding properties of aptamers allow them to distinguish between closely related compounds such as families of proteins or between different conformational states of the same protein. This feature arises since the surface area of interaction between an aptamer and its molecular target is large, with the result that small changes in the target molecule can disrupt association (Archemix Corporation 2003). This confers a high degree of specificity to aptamer association. Moreover, aptamers have high affinities, typically in the picomolar to low nanomolar range, to their target molecules.

These important chemical and binding properties allow aptamers to be used for various biosensor applications, several of which will be discussed in this chapter.

2
Biosensor Theory and Design

Over the past two decades, many attempts have been made to use biosensor technology as a sensitive and reliable detection protocol. Advantages to this approach include (Turner and Newman 1998; Cunningham 1998):

a. The ability to provide continuous data with respect to a specific analyte.

b. Targeted specificity.

c. Fast response times.

d. The possibility of mass production.

e. The elimination of extensive sample preparation.

f. Measurements are obtained without disturbance of the sample.

All sensors are composed of two main regions: the site for incorporation of an agent for selective chemical recognition and a physico-chemical transducer. The chemical reaction produces a signal such as a colour change, fluorescence, change in the oscillator frequency of a crystal or changes in conductivity. In turn, the transducer responds to this signal and provides some indication of the amount of analyte present (Cattrall 1997). Biosensors incorporate a biological sensing element (e.g. enzyme, antibody, DNA, aptamer) close to the signal transducer to give a reagent-free sensing system for the target analyte (Hall 1990).

For biosensors to be useful, there should be: a high degree of specificity; good stability to operating conditions such as pH, temperature and ionic strength; retention of biological activity in the immobilized state; and no undesirable sample contamination (Hall 1990). A typical biosensor is represented in Fig. 1.

Many types of transducers are available for use in biosensors, and the most important of these are summarized in Table 1. These will be considered during the discussion of aptasensor development.

Table 1 Transducers used in biosensor devices

Transducer	Examples
Electrochemical	Ion-selective electrodes, electronic noses, ion-selective field effect transistors
Optical	Optical fibres, surface plasmon resonance, fluorescence, luminescence
High frequency	Piezoelectric devices, surface acoustic wave sensors
Heat sensitive	Calorimetric sensors
Miscellaneous	Whole cells, single molecules

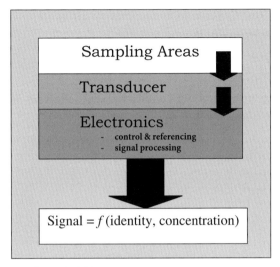

Fig. 1 The main parts of a typical biosensor

3
Aptasensors

As mentioned previously, aptamers possess several advantages over antibodies, and it is expected that biosensors using aptamers as the molecular recognition elements (aptasensors) will become more important in biomedical applications. Some of these advantages are listed in Table 2.

Table 2 Advantages of aptamers over antibodies in biosensor applications (based on O'Sullivan 2002)

Antibodies	Aptamers
Limited shelf-life and sensitive to temperature, which may lead to denaturation	Denatured aptamers can be regenerated in a few minutes, they are stable to long-term storage and can be transported at ambient temperature
May have batch-to-batch variation	Generally, no variation among batches
Labelling can lead to loss in affinity	Reporter molecules can be attached without loss in affinity
Require the use of strict physiological conditions	Non-physiological conditions, e.g. buffer, temperature, can be readily used
Requires the use of animals	Generated by an in vitro process that does not involve animal use
Kinetic parameters cannot be modified	Kinetic parameters can be readily changed as required

The application of aptamers as recognition elements in biosensors offers several advantages such as (O'Sullivan 2002):

a. Small molecules can be detected using sandwich formats that eliminate the need for laborious competitive assays.
b. Aptamers can be selected in conditions similar to that of the real matrix making them ideal for food applications.
c. Aptamers can be modified during immobilization or labelling without effects on affinity.
d. Aptamers can be subjected to repeated cycles of denaturation and regeneration.

Aptamers can also be made to possess a high binding specificity that, combined with target versatility and relative simplicity of in vitro selection, make these molecules suitable for bioanalytical applications. A further advantage is obtained if real-time fluorescence signalling is utilized, which eliminates the need for complex separation steps. Standard RNA and DNA do not contain fluorescent groups, and so it is necessary to modify aptamers with extrinsic fluorophores to make them display fluorescence characteristics (Nutui and Li 2003). This can be achieved in several ways including covalent attachment of a fluorophore to a region of an aptamer that will undergo target-induced conformational change (Jhaveri et al. 2000).

Recently, molecular beacons have become more important in the production of aptasensors. These are single-stranded oligonucleotides that fluoresce when they hybridize. They possess a stem-loop structure in which the stem is composed of complementary sequences close to the probe sequence in the loop. A fluorophore is bound to one end of the stem and a quencher is attached to the other. The unhybridized molecule does not fluoresce, but when the molecular beacon encounters a target molecule, the base-pairing is disrupted such that the quenching group is too far away to quench the fluorophore. The hybrid thereby fluoresces (Anderson 1999).

Molecular beacon-based signalling aptamers (sometimes termed 'aptamer beacons') have been generated by designing the aptamer in such a fashion that the addition of analyte results in a large conformational change with a resultant increase in fluorescent signal (Nutui and Li 2003; Rajendran and Ellington 2003). These methods, however, require knowledge of the detailed structure and sequence of the aptamer. To ensure that molecular beacons could be made more suitable for compounds other than nucleic acids, Rajendran and Ellington (2003) developed a method that directly couples selection for ligand binding to a nucleic acid conformational change. This in turn leads to a fluorescent signal. These designed molecular beacons are capable of mismatch discrimination and can be used in real-time PCR. These features can be useful in biosensor development, but more research is needed to ascertain if molecular beacon signalling can be adapted to such applications (Rajendran and Ellington 2003).

According to O'Sullivan (2002), molecular beacons can be utilized for aptasensor development with several modes of assay being possible. The most common of these is having the aptamer incorporated in an immobilized hairpin structure. Such a system was described where an immobilized molecular beacon aptamer showed fluorescence in the presence of Tat-1 protein but not in the presence of RNA-binding proteins (Yamamoto and Kumar 2000).

The development of aptasensors is being enhanced by three main methods: (1) optical, (2) acoustic and (3) fluorescent to analyse biological phenomena either in real time or by immobilizing the aptamer onto a solid support (Luzi et al. 2003).

4
Biomedical Applications Based on Transduction Methods

4.1
Optical

Of the optical methods, surface plasmon resonance (SPR), colourimetry and fluorescence have been the most widely used to date. Because fluorescence approaches have been quite extensive, they will be discussed in a separate sub-section.

4.1.1
Surface Plasmon Resonance

In SPR, a selective surface is formed by immobilizing the aptamer on to the surface of a sensor-chip. The analyte is then injected at a constant flow rate while the instrument measures changes in the resonance angle that occur at the sensor-chip surface. The angle varies when the aptamer binds to the analyte (Luzi et al. 2003). It has been found that the signal is proportional to the bound molecules.

According to Luzi and co-workers (2003), SPR technology has been used to study the selection of aptamers against specific targets that may be utilized as therapeutics. This involves the use of a commercially available SPR system (BIACore, Biacore AB, Sweden) that allows the calculation of kinetic parameters of the interaction between the aptamers and the target analyte, i.e. association (K_a) and dissociation (K_d) rate constants and affinity constants (K_A). In one example, the aptamers were selected from a library of 10^4 variants through their interaction with S-adenosylhomocysteine (SAH) and the K_a, K_d and K_A values were calculated (Gebhart et al. 2000). Although the complexes were stable, association rates were slow, indicating micro- to sub-micromolar affinities. In such a case, biosensors can prove advantageous, as the interactions are studied in real time and data are easily obtained (Luzi et al. 2003).

Biosensors based on the use of SPR technology have also been used to study molecules with activity dependent on certain conformations, e.g. ribozymes and deoxyribozymes. Japanese researchers (Okumoto et al. 2002) have reported that SPR can be important for studying RNA folding. A Ca^{2+}-dependent deoxyribozyme that was immobilized on an SPR-sensor chip was used to distinguish higher ordered structures of RNA.

4.1.2
Colourimetry

Recently, Ho and Leclerc (2004) have described the development of optical sensors based on hybrid aptamer-conjugated polymer complexes. A water-soluble cationic polythiophene was used as a 'polymeric stain' to specifically transduce the binding of an aptamer to its target into a clear optical signal. This approach does not require chemical modification on the probes or analytes and is based on different electrostatic interactions and conformational structures between the polythiophene and anionic single-stranded oligonucleotides. The polythiophene exhibits colour changes due to conformational changes of the flexible conjugated backbone. Additionally, it displays optical changes when complexed to single-stranded DNA or double-stranded DNA, allowing it to be a good molecule for transduction of an aptamer to a given target. This methodology has allowed the detection of human thrombin in the femtomole range (2×10^{-15} mol) within a few minutes. The authors claimed that it could be adapted for other chemical or biochemical targets (such as pathogenic proteins) and that it could be useful in the high-throughput screening of new drugs (Ho and Leclerc 2004).

A colourimetric adenosine biosensor based on the aptazyme-directed assembly of gold nanoparticles has been reported (Liu and Lu 2004). The aptazyme is based on the 8–17 DNAzyme with an adenosine aptamer motif to modulate the DNAzyme activity as a result of allosteric interactions. In the absence of adenosine, the aptazyme is inactive and the substrate strands serve as linkers to assemble DNA-functionalized 13 nm-diameter gold nanoparticles leading to a blue colour. The presence of adenosine activates the aptazyme, which cleaves the substrate strand leading to disruption of the formation of nanoparticle aggregates. A red colour is thereby observed. Up to 1 mM concentrations of adenosine could be detected semi-quantitatively by the degree of blue to red colour changes or quantitatively by the extinction ratio at 520 and 700 nm (Liu and Lu 2004). The authors reported that the sensor displayed good selectivity as, under the same conditions, 5 mM guanosine, cytidine or uridine resulted in a blue colour only. Colour differences can be observed by spotting the sensor solution onto alumina thin layer chromatography (TLC) plates. This approach allows the design of colourimetric biosensors for many analytes of biochemical importance.

With respect to sensor development, colourimetric detection provides several advantages (Liu and Lu 2004). These include the avoidance of difficulties associated with handling and disposal of radioisotopes and the elimination of many costs associated with instrumentation and operation in fluorescence, and real-time, on-site detection and quantification are easy.

4.1.3
Fibre Optics

Recently a fibre optic biosensor based on DNA aptamers has been used for the measurement of thrombin concentration (Spiridonova and Kopylov 2002). Anti-thrombin DNA aptamers were immobilized on silica microspheres, placed inside 3 µm-diameter microwells in the distal tip of an imaging optical fibre and coupled to a modified epifluorescence microscope at the distal tip. The thrombin concentration was determined by a competitive binding assay using a fluorescein-labelled competitor. The biosensor was found to be selective and, moreover, could be reused without any change in sensitivity. The limit of detection for thrombin was reported to be 1 nM, with assay time being 15 min per sample. Additionally, it was indicated that each microbead was identified with 90% precision and that, although signals from individual microspheres showed significant dispersion, the averaged signals from 100 microspheres had a reliable detection (relative standard deviation of 3%).

Previously, Kleinjung and co-workers (1998) immobilized a biotinylated RNA aptamer selected to L-adenosine on an optical fibre surface containing streptavidin. This was a similar approach to that described above in that measurements were obtained using total internal reflectance fluorescence (TIRF) in a fibre optic format with the sensor being based on a competitive system using a fluorescein isothiocyanate (FITC)-labelled L-adenosine as the receptor.

4.2
Fluorescence and Luminescence

4.2.1
Fluorescence

Most aptasensors reported to date have been based on fluorescence transduction, although it will be appreciated that considerable research is being conducted with other transduction methods.

In 1998, Hieftje's group reported on the development of a flexible biosensor that utilizes immobilized nucleic acid aptamers to detect free non-labelled non-nucleic acid targets such as proteins (Potyrailo et al. 1998). This was the first aptamer-based biosensor that was used to detect nucleic acid targets. An anti-thrombin DNA aptamer was fluorescently labelled with FITC, covalently immobilized on a microscope cover slip, and protein binding was detected by monitoring the evanescent wave-induced fluorescence anisotropy of the immobilized aptamer. The biosensor was able to detect 0.7 aM of thrombin in a 140 pl volume of solution with a precision of better than 4% relative standard deviation over the range 0 to 200 nM in less than 10 min.

Fluorescence anisotropy offers several advantages over SPR and TIRF spectroscopy. The detection of anisotropy discriminates between different surface-bound targets and is insensitive to variations in the refractive index of the sample solution (Potyrailo et al. 1998). Both SPR and TIRF are affected by these factors.

Tan and colleagues recently described the first application of a protein aptamer for the detection of an oncoprotein, platelet-derived growth factor (PDGF), by fluorescence anisotropy (Fang et al. 2001). The aptamer was labelled with fluorescein to bind to the PDGF protein. When the labelled aptamer is bound with its target protein, the rotational motion of the fluorophore attached to the complex becomes much slower due to an increase in molecular weight, resulting in a significant fluorescence anisotropy change. This allows the detection of the binding events between the aptamer and the protein in real time and in homogeneous solutions. The PDGF was detected in the sub-nanomolar range. The authors predict that efficient oncoprotein detection by fluorescence anisotropy will find applications in protein monitoring and cancer diagnosis.

Yamana and co-workers (2003) have reported on the use of a bis-pyrene-labelled DNA aptamer as an intelligent fluorescent biosensor. They have indicated that there is high signal intensity and specificity for detecting the target ligand in a homogeneous system. The bis-pyrene fluorophore is easily incorporated as a fluorescent non-nucleosidic linker at internal or terminal positions of aptamers. The labelled aptamers display large fluorescence changes in intensity when they bind to ligands (Yamana et al. 2003). This feature allows them to be useful in the development of an intelligent biosensor. This is brought about by the excimer (480 nm) and monomer (380 nm) fluorescence emissions of the bis-pyrene fluorophore being sensitive to local structural changes as a result of base-pairing and/or nucleotide sequence variations near the bis-pyrene label attached to oligonucleotide duplexes.

In an interesting development, Merino and Weeks (2003) have used the differential activity of $2'$-amine-substituted nucleotides in flexible versus constrained nucleic acid structures to resolve bound and ligand-free states of an ATP aptamer in biologically complex solutions. Fluorogenic compounds such as fluorescamine (FCM) are ideal for converting aptamers into small-molecule sensors via reaction at the $2'$-amine. FCM reacts with aliphatic amines in milliseconds to form a stable fluorescent product. In addition, it undergoes hydrolysis to form non-fluorescent products that will quench the reagent rapidly (Merino and Weeks 2003). The ligand-FCM reaction is readily detected by fluorescence resonance energy transfer, which is a main advantage of this approach. Moreover, the development of sensors by this technique requires little information about the aptamer in that only identification of a ligand-induced conformational change is necessary. The authors suggest that this method will be useful for developing sensors from diverse aptamers that bind small-molecule ligands via an induced fit.

Stanton and colleagues (Hamaguchi et al. 2001) have described the adaptation of molecular beacon techniques to aptamers that specifically bind protein targets. These aptamer beacons can adopt two or more conformations, one of which will allow ligand binding. An anti-thrombin aptamer was developed into an aptamer beacon by adding nucleotides to the 5′-end that are complementary to nucleotides at the 3′-end of the aptamer. In the absence of thrombin, the added nucleotides will form a stem-loop structure, while in the presence of thrombin the beacon forms the ligand-binding structure. The resultant conformational change can be measured by fluorescence, thus allowing the aptamer beacon to be a sensitive device for detecting proteins. The authors speculate that since aptamers can be immobilized without significant loss of activity, then it should be possible to do the same with aptamer beacons to generate chip arrays to detect individual proteins. However, there are limitations to the use of aptamer beacons. The major factor is that several proteins (e.g. lactose dehydrogenase) bind non-specifically to single-stranded DNA, leading to non-specific increases in fluorescent intensity. Another limitation is that metal ions can affect nucleic acid conformation (Hamaguchi et al. 2001).

4.2.2
Luminescence

Bruno and Kiel (1999) have used electrochemiluminescence to detect anthrax spores. SELEX was used to select and amplify aptamers capable of binding to and detecting non-pathogenic Sterne strain *Bacillus anthracis* spores. The method involved the use of an aptamer-magnetic bead-electrochemiluminescence (AM-ECL) sandwich assay scheme that was able to detect at least three distinct populations of single-stranded DNA aptamers. This, in effect, indicates a detection limit of 10^6 anthrax spores. Consequently, aptamer biosensors can be useful in biological warfare and bioterrorism detection.

4.2.3
Chip-Based Assays

A chip-based biosensor for multiplex analysis of protein analytes has been developed by researchers at Archemix Corporation (2003). The biosensor utilizes immobilized DNA and RNA aptamers, selected against different protein targets, to simultaneously detect and quantify levels of individual proteins in complex biological mixtures (McCauley et al. 2003). Aptamers were fluorescently labelled and immobilized on a glass substrate and fluorescence polarization anisotropy was used for solid and solution-phase measurements of target-protein binding. The system was used to detect and quantify cancer-related proteins such as vascular endothelial factor and basic fibroblast growth factor. The biosensor was multiplexed to detect these factors even in the presence of complex biological matrices such as serum.

In a related development, researchers at the University of Texas have reported on the production of a chip-based sensor array composed of individually addressable agarose microbeads to detect DNA oligonucleotides (Ali et al. 2003). Biotinylated DNA capture probes were incorporated into the bead microreactors, which were derivatized with avidin docking sites. The probes are then arranged in micromachined cavities on silicon wafers that allow for fluid flow through the microreactors. This allows the detection of target DNA by fluorescence changes that occur upon binding to the target. It has been reported that the limit of detection was approximately 10^{-13} M. Although aptamers were not studied here, this approach could be suitable for these targets.

Finally, Ellington and colleagues (Kirby et al. 2004) have immobilized aptamers in chip arrays for the detection of multiple analytes for biothreat agents. This involved specific enzymatic incorporation of linkers into the aptamers and then using DNA arrays to generate and analyse aptamer beacon chips. This approach allows the aptamers to be generated with differing affinities and specificities for a particular target. Furthermore, by immobilizing a series of aptamers with different affinities, it will be possible to recognize altered or modified analytes.

4.3
Acoustic Wave Devices

Acoustic wave devices are usually based on resonant oscillating quartz sensors that can detect small changes in series resonant frequency due to changes in mass and other variables of the oscillating system as a result of a binding or dissociation event (caesar Foundation 2005). Piezoelectric materials are widely used in the fabrication of biosensors, as these respond sensitively to the perturbation of the acoustic properties at the sensor-liquid interface. Quartz is the most frequently used piezoelectric material because of its ready availability and relatively low temperature dependence (Deisingh and Thompson 2002). A typical acoustic wave device is shown in Fig. 2.

The binding of an analyte and its receptor (aptamer) by the quartz sensor has also been studied. The aim was to develop an aptamer-based microbalance and to use this as a biosensor to transform a binding event directly into an electronic signal. This would allow sensing in solution along with electronic detection and control. However, this is an on-going process that requires much further research (caesar Foundation 2005).

Fig. 2 An acoustic wave sensor. (Reprinted by permission of Dr. Shakour Ghafouri, University of Toronto)

Mascini's group has developed biosensors to detect the human immunodeficiency virus (HIV)-1 Tat protein (Tombelli and Mascini 2005; Tombelli et al. 2004). An RNA aptamer has been used as the biorecognition element to detect the *trans*-activator of transcription (Tat) protein. Tat is a protein of 86 to 101 amino acids that controls the early phase of the HIV-1 replication cycle. A specific RNA aptamer was immobilized on the gold surface of quartz crystals and surface plasmon resonance chips, and the interaction with Tat protein in solution was investigated. Parameters such as sensitivity, selectivity and reproducibility were studied and a comparison with immunosensors was performed.

A quartz crystal microbalance (QCM) was used to evaluate the suitability of immobilized aptamers as ligands in a biodetector. The coated crystal is placed in a flowthrough system that allows the injection of a probe and, also, the monitoring of the association and dissociation of the analyte (Liss et al. 2004). Binding of the analyte to the quartz surface leads to a decrease in the resonance frequency that can be readily monitored. The quartz crystals were activated with 3,3-dithiodipropionic acid that crosslinks the free amino groups of antibodies or streptavidin. Aptamer oligonucleotides were either 5′- or 3′-biotinylated and immobilized on streptavidin-coated surfaces (Liss et al. 2004). The specificity of antibody and aptamer-coated biosensors were compared by coating the quartz crystals with either a monoclonal antibody against human IgE or a 37-nt single-stranded DNA aptamer selected against the F_c domain of IgE respectively. The authors reported that the aptamer is equivalent to the antibody with respect to sensitivity and specificity. This system can prove useful to simultaneously analyse the expression levels of tumour and apoptosis proteins.

4.4
Other Methods

4.4.1
Electrochemical Aptasensors

Electrochemical detection has the advantages of high sensitivity, fast response, robustness, low cost and the potential for miniaturization (Mir and Katakis 2004). An electrochemical aptasensor was reported for the detection of thrombin where the thrombin was immobilized on the surface and detected by horseradish peroxidase (HRP)-labelled aptamer or biotin-labelled aptamer after reaction with streptavidin-HRP. Other configurations investigated include the immobilization of the aptamer on the surface by biotin or by a thiol group. The thrombin, when bound to the aptamer, was detected by a chromogenic substrate or by labelling with HRP. Thrombin was also detected by use of a sandwich format with a second labelled aptamer and, additionally, a reagent-less electrochemical aptabeacon was studied. In each case, the conformational change induced when the aptamer changes from a duplex structure to a 3-D quadruplex structure upon binding to thrombin, was detected electrochemically.

4.4.2
Micromachined Sensors

Micromachined sensors have low noise and scalability due to their small size. The low noise will result in higher resolution, while scalability allows many sensors to be used in parallel while also using small volumes for point-of-care systems (Savran 2005). Savran's group at Purdue University (Indiana, USA) is interested in using aptamers with micromachined sensors that do not require the use of labelling. A micromechanical silicon-nitride device that allows differential detection of biomolecules has been reported (Savran et al. 2002, 2003). The sensor has two adjacent flexible cantilevers that bend in response to the surface stress resulting from the adsorption of biomolecules. Interdigitated fingers between the tip areas of the two cantilevers allow optical detection of the relative bending between the two cantilevers. One cantilever contains the receptor molecules that bind to target molecules and is used as the main sensor, while the other is blocked with non-specific molecules (Savran 2005). When receptor-target binding occurs, the sensor cantilever bends but the reference does not. An advantage of this methodology is that both cantilevers bend by the same amount for disturbances such as temperature changes and non-specific binding. Thus, only the response to biomolecular adsorption is detected.

This approach was used to detect proteins using aptamer-based receptor molecules. The ligand was Taq DNA polymerase while the receptor of the sensor was an anti-Taq aptamer modified with a thiol group at one end to allow for covalent binding onto a gold surface (Savran et al. 2004). One cantilever was functionalized with aptamers selected for Taq DNA polymerase while the other was blocked with single-stranded DNA. The polymerase-aptamer binding induces a change in surface stress that causes a differential cantilever bending that ranges from 3 to 32 nm depending on ligand concentration. The authors concluded that these sensors are sensitive, specific and can be repeatable for protein detection.

5
Concluding Remarks

This chapter considered the potential role of aptasensors in biomedical applications. Since the field is quite new, not many applications have been described in the literature, but the outlook is becoming more positive. The introduction of automated platforms for aptamer selection along with developments in increasing the stability of aptamers to nuclease aptamers will become more accessible to workers in the biosensor area. This is clearly seen from the chip-based and micromachined sensors that are currently being analysed.

However, before it is accepted that aptamers can be the solution to bioanalytical problems, it has to be noted that these molecules have their limitations. Enzymes in cells can break down natural RNA molecules, and therefore substitutes for natural RNA components need to be found. Thus, Spiegelmers, made of novel RNA-type building blocks but which are mirror-reversed, can be a useful possibility. These molecules offer high specificity and are resistant to enzyme breakdown.

Regardless of the problems being experienced and the slow pace of development, it appears that aptasensors, especially with the increase of transduction approaches, will provide much scope for biomedical applications in the next decade. This will also be enhanced by the use of aptamer beacons that allow reagentless, one-step analyses.

Acknowledgements I would like to express my gratitude to Ms. Satie Siewah of the University of the West Indies for her assistance in the preparation of this manuscript.

References

Ali MF, Kirby R, Goodey AP, Rodriguez MD, Ellington AD, Neikirk DP, McDevitt JT (2003) DNA hybridization and discrimination of single-nucleotide mismatches using chip-based microbead arrays. Anal Chem 75:4732–4739
Anderson MLM (1999) Nucleic acid hybridization. Bios Publishers, Oxford, pp 209–210
AptaRes (2005) MonoLex information Web page: http://www.aptares.de/html/technology.html. Cited 2 July 2005

Archemix Corporation (2003) Company's home Web page: http://www.archemix.com. Cited 2 July 2005
Bruno JG, Kiel JL (1999) In vitro selection of DNA aptamers to anthrax spores with electrochemiluminescent detection. Biosens Bioelectron 14:457–464
caesar Foundation (2005) Aptamer Biosensors Web page: http://www.caesar.de/568.0.html. Cited 2 July 2005
Cattrall RW (1997) Chemical sensors. Oxford University Press, Oxford, pp 1–8
Cunningham AJ (1998) Introduction to bioanalytical sensors. Wiley, New York, pp 1–18
Deisingh AK, Thompson M (2002) Detection of toxigenic and infectious bacteria. Analyst 127:567–581
Famulok M (2005) Rheinische Friedrich-Wilhelms-Universität Bonn, Kekulé-Institut für Organische Chemie und Biochemie, Michael Famulok workgroup homepage, http://famulok.chemie.uni-bonn.de/. Cited 13 July 2005
Famulok M, Mayer G (1999) Aptamers as tools in molecular biology and immunology. In: Famulok M, Wong C-H, Winnacker E-L (eds) Combinatorial chemistry in biology. Current topics in microbiology and immunology. Springer-Verlag, Berlin, Heidelberg, New York, pp 123–136
Fang X, Cao Z, Beck T, Tan W (2001) Molecular aptamer for real-time oncoprotein platelet-derived growth factor monitoring by fluorescence anisotropy. Anal Chem 73:5752–5757
Gebhart K, Shokraei A, Babaie E, Linquist B (2000) RNA aptamers to S-adenosylhomocysteine: kinetic properties, divalent cation dependency and comparison with anti-S-adenosylhomocysteine antibody. Biochemistry 39:7255–7265
Guthold M (2005) Novel, single-molecule aptamer selection method [abstract]. NCI Office of Technology and Industrial Relations: http://otir.nci.nih.gov/abstracts/imat_guthold1102.html. Cited 2 July 2005
Hall EAH (1990) Biosensors. Open University Press, Milton Keynes, pp 1–45
Hamaguchi N, Ellington A, Stanton M (2001) Aptamer beacons for the direct detection of proteins. Anal Biochem 294:126–131
Ho H-A, Leclerc M (2004) Optical sensors based on hybrid aptamer/conjugated polymer complexes. J Am Chem Soc 126:1384–1387
Jhaveri S, Rajendran M, Ellington AD (2000) In vitro selection of signaling aptamers. Nat Biotechnol 18:1293–1297
Kirby R, Cho E J, Gehrke B, Bayer T, Park YS, Neikirk DP, McDevitt JT, Ellington AD (2004) Aptamer-based sensor arrays for the detection and quantitation of proteins. Anal Chem 76:4066–4075
Kleinjung F, Klussmann S, Erdmann VA, Scheller FW, Furste JP, Bier FF (1998) High-affinity RNA as a recognition element in a biosensor. Anal Chem 70:328–331
Liu J, Lu Y (2004) Adenosine-dependent assembly of aptazyme-functionalized gold nanoparticles and its application as a colorimetric biosensor. Anal Chem 76:1627–1632
Liss M, Petersen B, Prohaska E, Wolf H (2004) Preparation of optimized high affinity receptors for a biochip designed to simultaneously analyze expression products. German Human Genome Project: http://www.dhgp.de/research/projects/abstracts/pdf/9949.pdf. Cited 2 July 2005
Luzi E, Minunni M, Tombelli S, Mascini M (2003) New trends in affinity sensing: aptamers for ligand binding. Trends Anal Chem 22:810–818
McCauley TG, Hamaguchi N, Stanton M (2003) Aptamer-based biosensor arrays for detection and quantification of biological macromolecules. Anal Biochem 319:244–250
Merino EJ, Weeks KM (2003) Fluorogenic resolution of ligand binding by a nucleic acid aptamer. J Am Chem Soc 125:12370–12371

Mir M, Katakis I (2004) Electrochemical aptasensors [abstract]. Second International Workshop on Multianalyte Biosensing Devices, 18–20 February 2004. http://www.etseq.urv.es/dinamic/congres/novel/poster8.htm. Cited 2 July 2005

Nutui R, Li Y (2003) Structure-switching signaling aptamers. J Am Chem Soc 125:4771–4778

O'Sullivan CK (2002) Aptasensors—the future of biosensing? Anal Bioanal Chem 372:44–48

Okumoto Y, Ohmichi T, Sugimoto N (2002) Immobilized small deoxyribozyme to distinguish RNA secondary structures. Biochemistry 41:2769–2773

Potyrailo RA, Conrad RC, Ellington AD, Hieftje GM (1998) Adapting selected nucleic acid ligands (aptamers) to biosensors. Anal Chem 70:3419–3425

Rajendran M, Ellington AD (2003) In vitro selection of molecular beacons. Nucleic Acids Res 31:5700–5713

Savran CA, Sparks AW, Sihler J, Li J, Wu W-C, Berlin DE, Burg TP, Fritz J, Schmidt MA, Manalis SR (2002) Fabrication and characterization of a micromechanical sensor for differential detection of nanoscale motions. J Microelectromech Syst 11:703–708

Savran CA, Burg TP, Fritz J, Manalis SR (2003) Microfabricated mechanical biosensor with inherently differential readout. Appl Phys Lett 83:1659–1661

Savran CA, Knudsen SM, Ellington AD, Manalis SR (2004) Micromechanical detection of proteins using aptamer-based receptor molecules. Anal Chem 76:3194–3198

Savran CA (2005) Personal Web site, Purdue University, Mechanical Engineering Department: http://widget.ecn.purdue.edu/~savran. Cited 2 July 2005

Spiridonova VA, Kopylov AM (2002) DNA aptamers as radically new recognition elements for biosensors. Biochemistry (Mosc) 67:706–709

Tombelli S, Mascini M (2005) Aptamer biosensors as DNA/RNA-based tools for proteomics. Web page: http://www.montefiore.ulg.ac.be/services/microelec/materials_dna2004/tombelli.pdf. Cited 13 July 2005

Tombelli S, Minunni M, Gullotto A, Luzi E, Mascini M (2004) Aptamers for HIV-1 Tat protein. Eighth World Congress on Biosensors, 24–26 May 2004 [abstract]. http://www.sparksdesigns.co.uk/biopapers04/papers/bs420.pdf. Cited 2 July 2005

Turner APF, Newman JD (1998) An introduction to biosensors. In: Scott AO (ed) Biosensors for food analysis. Royal Society of Chemistry, Cambridge, pp 1–20

Yamamoto R, Kumar PKR (2000) Molecular beacon aptamer fluoresces in the presence of Tat protein of HIV-1. Genes Cells 5:389–396

Yamana K, Ohtani Y, Nakano H, Saito I (2003) Bis-pyrene labeled DNA aptamer as an intelligent fluorescent biosensor. Bioorg Med Chem Lett 13:3429–3431

Application of Aptamers in Therapeutics and for Small-Molecule Detection

M. Menger · J. Glökler · M. Rimmele (✉)

RiNA Netzwerk RNA-Technologien GmbH, Takustr. 3, 14195 Berlin, Germany
information@rna-network.com

1	Properties of Aptamers	359
2	Aptamers in Therapeutics	362
3	Detection Methods for Small Molecules and Advantages of Aptamers	363
4	Aptamer Applications in Biotechnology and Medicine	366
	References	369

Abstract Nucleic acids that can bind with high affinity and specificity to target molecules are called "aptamers". Aptamers recognise a large variety of different molecule classes. The main focus of this chapter is small molecules as targets. Aptamers are applied complementarily to antibody technologies and can substitute antibodies or small molecules wherever their different properties, such as biochemical nature or highly discriminating capacities, are advantageous. Examples of promising applications of these versatile molecules are discussed in the field of therapeutics and biotechnology with a special view to small-molecule detection.

Keywords Aptamers · Small molecules · Detection · Low molecular weight compounds · Biosensor

1
Properties of Aptamers

The Greek word *haptein* meaning "to attach to" is the origin of the word "aptamer". Aptamers can be peptides or nucleic acids. Nucleic acid aptamers are single-stranded (ss)DNA or ssRNA molecules that bind to a target with high affinity and specificity, having dissociation constants down to the picomolar range (Jayasena 1999). They typically contain between 25 and 100 nucleotides. DNA or RNA nucleic acid aptamers recognise a large variety of low molecular weight compounds (Gold et al. 1995; Hermann and Patel 2000; Patel 1997; Wilson and Szostak 1999). The range and size of aptamer target molecules, however, spread from mere ions such as Zn^{2+} over nucleotides, antibiotics, small molecules, RNA structures, proteins, receptors, viruses and cells up to whole organisms (Famulok and Mayer 1999; Famulok et al. 2001; Homann and

Goringer 2001; Jayasena 1999; Kawakami et al. 2000; Rimmele 2003; Ulrich et al. 2002; Toulme et al. 2004).

In contrast to antisense, ribozyme and small interfering (si)RNA, aptamers can bind not only on the protein coding level, but can also bind to the protein itself. Thereby they are able to distinguish between highly homologous epitopes or isotypes and are able to influence certain activities of a multifunctional target molecule.

Aptamers can greatly enhance the efficiency of modern drug development because their binding characteristics, especially to small molecules, are comparable to or even better than monoclonal antibodies. Just like antibodies, aptamers have the capacity to form elaborate three-dimensional structures and shapes. They are often used complementarily to antibody technologies and can substitute antibodies or small molecules wherever their distinct properties—due to their different biochemical nature and to the different techniques of their development—are advantageous. So far, the widest-spread method to find high affinity nucleic acids with high specificity for a certain target is a technology called SELEX (systematic evolution of ligands by exponential enrichment) (Tuerk and Gold 1990; Ellington and Szostak 1990). SELEX comprises an iterative process of in vitro selections using nucleic acid to target binding and partitioning events of unbound nucleic acids in order to isolate high-affinity nucleic acid ligands from large pools of randomised sequence libraries. Starting libraries for SELEX usually contain more than 10^{15} different sequences. The technology has been further developed and improved, even automated (Cox and Ellington 2001; Cox et al. 2002), and has been administered to find aptamers for various application needs (Brody and Gold 2002; Bruno and Kiel 2002; Famulok and Mayer 2001; Jayasena 1999; Martell et al. 2002; Nimjee et al. 2005; Rajendran and Ellington 2002; Sun 2000; Vuyisich and Beal 2002; Wang et al. 2003; White et al. 2000, 2001; Zhang et al. 2004). Additionally, new aptamer-development processes and applications are constantly being explored and tested in many laboratories.

The multiple advantages of aptamers include their high specificity and affinity, their easy and highly reliable production by enzymatic or chemical synthesis, their regenerability by simple means and their storability. In addition, advantages over monoclonal antibodies are the higher inhibitory potential and the wider range of chemical modification possibilities because they can be synthesised enzymatically and chemically. Aptamers are also stable, with no loss of activity under a wide range of buffer conditions, and resistant to harsh treatments such as physical or chemical denaturation. Different chemical modification possibilities offer diverse immobilisation properties using numerous additional molecules. Because it is possible to develop aptamers entirely in vitro without the need of cells or animal immune systems, aptamer generation offers the choice of a great diversity of binding conditions. Aptamers can be developed against all different kinds of targets, including cell-toxic molecules, compounds that can only be solubilised in solvents other than water, and targets against which an immune response cannot be elicited (Famulok et al. 2001; Jayasena 1999). Interestingly, to date there has been no evidence for immunogenicity of aptamers when applied in vivo (Rimmele 2003). In most cases, aptamers need, however, metal ions in order to attain their functional shape and activity, much like ribozymes (Fedor 2002; Hanna and Doudna 2000). Intriguingly, aptamers can be functionalised in connection with ribozymes as "allosteric aptamers" for many powerful applications (Breaker 2002).

One of the three main disadvantages of aptamer technologies is the unfortunate finding that aptamers (much like antibodies) cannot be developed for every target molecule. Target properties such as an isoelectric low point (pI) can be a challenge. In many cases the successful development of an aptamer with the right properties seems to greatly depend on the use of the exact conditions (i.e. assay temperature) during the development process in which the aptamer will later be used (Daniels et al. 2002).

A caveat of nucleic acid aptamers also used to involve their nucleases sensitivity. Fortunately, many ways have been found to stabilise RNA aptamers, such as chemical modifications at the $2'$ position of the ribonucleotide moieties (for instance $2'$-amino, $2'$-deoxy or $2'$-methoxy), circularisation of the molecules or capping to avoid exonuclease attack (Kim et al. 2002; Jellinek et al. 1995; Menger et al. 1996; Pieken et al. 1991; White et al. 2000). Some of these modifications can be introduced during the selection itself by using tolerant T7 RNA polymerase mutants; others have to be inserted by chemical synthesis after the final sequence and structure has been analysed (Aurup et al. 1992; Chelliserrykattil and Ellington 2004; Meis and Chen 2002; Sousa and Padilla 1995). An alternative method is "Spiegelmer" technology (Eulberg and Klussmann 2003; Klussmann et al. 1996). Spiegelmers are mirror-image aptamers composed of L-ribose or L-$2'$-deoxyribose units and are not degradable by nucleases.

Still, one of the biggest challenges for aptamer (as well as antibody) application in medicine is their efficient delivery into cells. This type of research is finding increasing interest. Among incorporating them into liposome vesicles and applying those vesicles (White et al. 2000), gene therapeutic approaches seem to be the most successful techniques at present to import aptamer function in vivo for intracellular protein targets (Famulok and Verma 2002; Good et al. 1997; Iyo et al. 2002).

Whereas most aptamers exert their function not on a protein coding level, aptamers are being explored to also regulate eukaryotic gene expression directly in response to the binding of their target molecule (Toulme et al. 2004). Intriguingly, mounting evidence is now being presented of naturally occurring aptamers as part of "riboswitches" in prokaryotes as well as eukaryotes, and the importance of their regulatory functions in gene expression is being elucidated (Breaker 2004; Hentze and Kuhn 1996; Sudarsan et al. 2003; Winkler et al. 2002).

In this chapter, we want to present some examples of aptamers in therapeutics and some important biotechnological detection platforms for the specific detection of analytes, comparing previous technologies and aptamer technologies. We will focus on the detection of low molecular weight compounds (including metal ions) up to small molecules (including peptides and enzymatic cofactors). In addition, we will present, as an example, one of our own new aptamers against a low molecular weight compound and toxin in a biosensor system (E. Ehrentreich-Förster, D. Orgel, A. Krause-Griep, B. Cech, V.A. Erdmann, F. Bier, F.W. Scheller and M. Rimmele, submitted for publication).

2
Aptamers in Therapeutics

One of the first aptamers selected against proteinaceous targets was the anti-thrombin aptamer (Bock et al. 1992). At the same time it has been the first ssDNA aptamer described in the literature. Because of its naturally existing heparin binding site, thrombin is an ideal target for the development of an aptamer ligand. DNA aptamers directed against thrombin can at the same time inhibit the enzymatic function and thereby become a valuable anticoagulant (Bock et al. 1992; Griffin et al. 1993).

New approaches have been taken to make use of a so-called "antidote" mechanism in order to specifically control anticoagulant aptamer activity in vivo. This has been demonstrated using another anticoagulant 2'-fluoro-pyrimidine-stabilised RNA aptamer directed against factor IXa. The antidote effect could be demonstrated to work also if administered to patients (Rusconi et al. 2002). The antidote works by hybridising the aptamer to a complementary strand, thus reducing the active structure of the aptamer to an inactive double helix. Pharmacokinetics of the aptamer were improved by attachment of a cholesterol moiety to the 5' end instead of the previously used polyethylene glycol (PEG) and by further stabilisation with an inverted deoxythymidine additional to the inverted deoxythymidine at the 3' end (Rusconi al. 2004). The antidote itself is composed of a 2'-methoxy derivative of RNA.

The most advanced aptamer drug currently under investigation is Macugen (pegaptanib) by Eyetech Pharmaceuticals/Pfizer, an anti-vascular endothelial growth factor (VEGF) aptamer used as an angiogenesis inhibitor for the potential treatment of age-related macular degeneration (AMD) and diabetic macular edema (Vinores 2003). The VEGF aptamer was initially selected from a 2'-fluoro-pyrimidine RNA library to yield clones with K_d values of 5–50 pM (Ruckman et al. 1998). A high specificity was confirmed and further modifications like the introduction of 2'-methoxy-purine nucleotides along with a conjugation to PEG improved the pharmacokinetics of the aptamer (Tucker et al. 1999). Especially the treatment of exudative AMD and diabetic macular edema via intravitreal injection was successful (Eyetech Study Group 2002). Finally, Macugen was approved as the first aptamer by the Food and Drug Administration (FDA) in December 2004 for the treatment of neovascular AMD.

Furthermore, Eyetech Pharmaceuticals/Pfizer made Macugen available to treat AMD in January 2005.

It is conceivable from these recent data that other aptamers in current trials will also meet all the demands to become available as therapeutics. The unusual properties of aptamers that allow the reversible action by a specific antidote is a further asset in comparison to conventional drugs. Especially the recent FDA approval of Macugen will pave the way for further acceptance of oligonucleotide-based therapeutics.

3
Detection Methods for Small Molecules and Advantages of Aptamers

Small molecules are involved in innumerable biochemical processes in life and can operate as substrate, catalyst or inhibitor in biochemical reactions. Because of these properties, small molecules are valuable as drugs in medicine or in biotechnological applications. In addition, in all cases the feasibility of quickly and reliably detecting them is essential.

Different and versatile techniques have been developed for the detection of small molecules. Classic detection methods are based on optical spectroscopic techniques. In these, electromagnetic radiation with a particular wavelength and intensity is applied to an object by which it is adsorbed, dispersed or emitted. Techniques like X-ray, ultraviolet-visible (UV/VIS), infrared (IR), electron paramagnetic resonance (EPR) and nuclear magnetic resonance (NMR) spectroscopy are used for small-molecule detection as well as for the structural analysis of molecules (Lottspeich and Zorbos 1998).

Other methods involve radioactive and non-radioactive-labelling. In recent years fluorescence, chemifluorescence and chemiluminescence have emerged as alternative technologies to traditional radioisotope-based systems. Convenience, speed and safety are strong arguments for non-radioactive labelling techniques, but use of radioisotopes may still offer significant advantages, i.e. because the insertion of a radioisotope does not change the structure of a molecule. In both application forms, the radioisotopes or the chromophores are directly embedded in the analyte or in the ligand, which exerts high affinity and high specificity for the analyte. The scintillation proximity assay (SPA) and fluorescence resonance energy transfer (FRET)-based assay exemplify the application of both techniques in high-throughput drug screening strategies (Clegg 1995; Woodbury and Venton 1999).

A further method is the surface plasmon resonance (SPR)-based interaction analysis technique. SPR is independent of any labelling. However, low molecular weight compounds can only be detected to a certain limit because they do not create strong enough signals (Szabo et al. 1995; Woodbury and Venton 1999).

Other techniques include chromatographic and electrophoretic methods, capillary electrophoresis (CE) and mass and microscope spectroscopy (Altria and Elder 2004; Lottspeich and Zorbos 1998; Woodbury and Venton 1999).

Most recently, the microarray technology has become a crucial tool for large-scale and high-throughput analysis (Glokler and Angenendt 2003; Walter et al. 2002; Zhu and Snyder 2003), its greatest advantage being the considerable reduction of sample consummation and the possibility to screen in parallel.

Using antibodies, rapid and simple immunoassays have been widely used for the detection of macromolecules like proteins, the most common format being the enzyme-linked immunosorbent assay (ELISA), performed as a sandwich assay. However, as the analyte becomes smaller, it is sterically impossible to be bound by two antibodies simultaneously. Also, small haptens often escape the immune system and specific antibodies are difficult to obtain via immunisation strategies. Technologies like phage display and ribosome display have been developed to overcome these difficulties, but there still is a constant search for a proper scaffold to recognise small molecules in a specific manner (Cicortas Gunnarsson et al. 2004; Vogt and Skerra 2004).

A competitive assay as described by us (see below and E. Ehrentreich-Förster, D. Orgel, A. Krause-Griep et al., submitted for publication) using aptamers can, however, detect a small molecule or a low molecular weight compound with high sensitivity and specificity. In recent years many aptamers as well as "aptazymes" (a combination of aptamer and ribozyme) have been developed as an alternative for small molecule detection (Burgstaller et al. 2002; Famulok 1999). Among those, aptamers against small molecules have been identified ranging from metabolic cofactors like Flavine adenine dinucleotide (FAD) (Clark and Remcho 2003; Roychowdhury-Saha et al. 2002), biotin (Nix et al. 2000; Wilson et al. 1998), vitamin B12 (Lorsch and Szostak 1994; Sussman et al. 1999) and elicitors like $3'$-$5'$-cyclic adenosine monophosphate (cAMP) (Nonin-Lecomte et al. 2001; Koizumi and Breaker 2000) to toxins and drugs like antibiotics (Gold et al. 1995; Famulok 1999; Wilson and Szostak 1999).

We experienced that aptamers developed against a low molecular weight compound alone will hardly recognise the conjugated compound on a protein surface. We assume that the conjugated compound is probably buried or not fully accessible in its three-dimensional structure for aptamer binding, due to protein side chains or its being masked by the protein's immanent charges. To render aptamers capable of recognising a low molecular weight compound in a variety of milieus, it is necessary to develop the desired aptamers through varying steps and surroundings according to the anticipated chemical conditions and platforms.

An example of an aptamer developed against a low molecular weight compound is an aptamer that has been generated by us in co-operation with the Fraunhofer Institute for Biomedical Engineering (IBMT). We developed aptamers against the toxic agent trinitrotoluene (TNT) (E. Ehrentreich-Förster, D. Orgel, A. Krause-Griep et al., submitted for publication; Rimmele 2003; Rimmele and Ehrentreich-Förster 2004). This work opens a new field of aptamer applications for environmental analytics and chemical-process controlling using a biosensor approach. TNT and especially its degradation products are very toxic and can be recognised with antibodies only with low affinity (K_d value estimated in the millimolar range). Aptamers developed against TNT have a much higher affinity and specificity than antibodies, and they could be administered in a portable biosensor system (Fig. 1). A reason for the significantly higher affinity of aptamers to TNT in the sensor could be the highly flexible three-dimensional structure of nucleic acids (Hermann and Patel 2000; Rimmele 2003). The presented system also clearly demonstrates a further advantage of aptamers compared to proteinaceous antibody molecules. Aptamers are compatible with organic solvents that are needed for the solubilisation and detection of organic molecules like TNT (Baldrich et al. 2004; O'Sullivan 2001; Rimmele 2003). Our aptamers against TNT could be developed in buffers containing considerable amounts of methanol.

In the established biosensor assay, aptamers with high affinity have a high specificity for TNT, since the structurally similar explosive N-methyl-N-2,4,6-tetranitroaniline (Tetryl) displays no affinity to the TNT aptamers (data not shown).

The principle of measurement of the fibre-optic field biosensor is an indirect competitive assay. A fluorescence signal is detected with a photomultiplier tube (PMT). A scheme of the experimental set-up of the biosensor is shown in Fig. 2. The analyte TNT is covalently bound to the previously activated surface (glass fibre) of the measuring cell. The aptamers coupled with fluorescence beads and

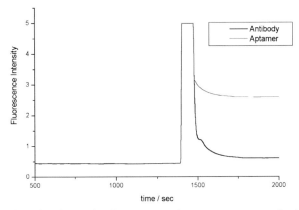

Fig. 1 Overlay plot of real-time binding curves of aptamer versus antibody in the fibre-optic field sensor. The toxin (trinitrotoluene) was immobilised on a sensory glass fibre. Fluorescence-labelled aptamer or fluorescence-labelled antibody is pumped into the measuring cell and binds to the toxin on the fibre. Binding is shown as fluorescence intensity. The different curve progression mirrors a significant difference in toxin affinity of aptamer versus antibody. Binding measurements were performed by Eva Ehrentreich-Förster at the IBMT

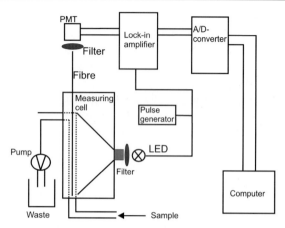

Fig. 2 Instrumental set-up of the fibre-optic field biosensor. The sample is pumped through the measuring cell with an embedded sensing glass fibre. The excitation light passes through an interference filter and is guided by a fibre bundle to the measuring cell. The sensory glass fibre leads the fluorescence through a filter to a photomultiplier tube (*PMT*). The PMT signal is collected, converted and connected to a computer for data sampling

the analysed probe are injected together into the measuring cell. In the absence of TNT in the analysed probe the aptamers bind to the TNT molecules coupled on the glass fibre. The result is a strong fluorescence signal on the PMT. In the presence of TNT in the analysed probe, a number of aptamer molecules binds to the free TNT molecules of the probe. This leaves fewer aptamer molecules, coupled with fluorescence beads, that can bind to the TNT coupled on the glass fibre. The result is a reduced fluorescence signal on the PMT (Fig. 3).

4
Aptamer Applications in Biotechnology and Medicine

Small-molecule recognition by aptamers is gaining increasing importance in biotechnology and medicine. In the field of biotechnology, many processes like fermentation in bioreactors have to constantly be surveyed to yield products that are well defined and safe for humans. Of interest is the metabolic state of the fermenting organism and the appearance of unwanted by-products, as well as the accumulation of the product itself. Monitoring results often need to be available in real time to allow decisions to be taken on a process (Hewitt and Nebe-Von-Caron 2004). Biosensors recognising small molecules on an enzymatic basis are already widely applied; however, not all molecules have a specific and simple enzyme assay available (Inaba et al. 2003; Zaydan et al. 2004). Biosensors based on aptamers as described above can be designed to employ the same principle for a multitude of different molecules, thus reducing the complexity for establishing such assays.

Application of Aptamers in Therapeutics and for Small-Molecule Detection

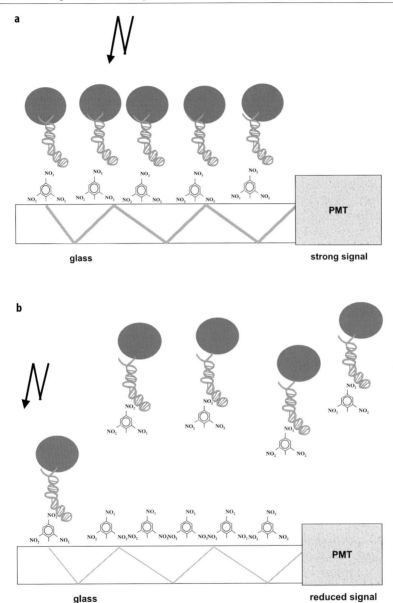

Fig. 3a,b The principle of measurement in a fibre-optic field biosensor. **a** Sensor fibre with immobilised toxin trinitrotoluene (TNT). Fluorescence-labelled aptamer (depicted as *coiled structure with a spheric bead attached*) is added to the flow cell and binds to its target molecules on the surface. The fluorescence of the bead is generated by laser light (shown as an *arrow*). The PMT transduces the light at the fibre. **b** Sensor fibre with immobilised toxin TNT. Fluorescence-labelled aptamer and probe are mixed and then added to the flow cell. If the sample contains TNT, most aptamers will bind to the free TNT instead of the immobilised TNT on the surface, which results in a reduced light signal at the fibre

In medicine, aptamers could be employed for the rapid detection of metabolic disease indicators and to study the pharmacokinetics of drugs and their degradation products. Examples of molecules in typical metabolic diseases that are currently monitored include advanced glycosylation end products in diabetes and citrullination in rheumatoid arthritis (Ahmed and Thornalley 2003; Rubin and Sonderstrup 2004). However, many metabolic disease markers, like phenylalanine and tyrosine in phenylketonuria, may not be identified as conjugates to proteins alone but would have to be identified in solution (Wibrand 2004).

Not only in phase trials of patients are drug pharmacokinetics important to study. In so-called personalised medicine, the genetic backgrounds of patients are screened, since there are differences between individuals in reaction to certain compounds (Nicholls 2003). But rather than rely on genetic information only, it may be even more important to monitor the fate and derivatives of the administered drug itself.

In addition to the simple detection of small molecules, aptamers in the form of riboswitches can be applied to control transcriptional or translational activity in vivo. In biotechnology this may be used in cases where a well-defined and controlled expression of several genes simultaneously is important. So far, microbial recombinant gene expression is mostly regulated by the classical operon model using repressor molecules inducible by substances like isopropyl-β-D-thiogalactopyranoside (IPTG), arabinose and tetracycline (Lutz and Bujard 1997). Engineered riboswitches offer an additional supply of regulating elements controlled by otherwise inert substances like theophylline and malachite green (Grate and Wilson 2001; Suess et al. 2004). The discovery of natural riboswitches in eukaryotic organisms points to a similar direction of engineering gene regulation in higher cells (Kubodera et al. 2003; Ooms et al. 2003).

These developments suggest that the range of applications using aptamers is yet to increase. Future discoveries in the field of RNA molecules in vivo, as in RNA interference, will add to the scope of things to be developed on the basis of engineered RNA, including aptamer elements.

Due to the superior performance of aptamers in small-molecule assays in comparison to conventional immunoassays, we envision that biosensors based on aptamers will become increasingly applicable not only in toxin screening of the environment and food, but also in many biotechnological and medicinal applications.

Acknowledgements Aptamer development against environmental toxin was funded by the German Federal Ministry of Education and Technology (BMBF) and the City of Berlin. The development of aptamers was accomplished by D. Orgel and B. Cech in the laboratory "Functional RNAs" at RiNA GmbH. The biosensor development and binding measurements were realised by E. Ehrentreich-Förster at the Fraunhofer Institute for Biomedical Engineering. The authors would like to thank J. Klein and A. Wochner for critically reading the manuscript.

References

Ahmed N, Thornalley PJ (2003) Quantitative screening of protein biomarkers of early glycation, advanced glycation, oxidation and nitrosation in cellular and extracellular proteins by tandem mass spectrometry multiple reaction monitoring. Biochem Soc Trans 31:1417–1422

Altria KD, Elder D (2004) Overview of the status and applications of capillary electrophoresis to the analysis of small molecules. J Chromatogr A 1023:1–14

Aurup H, Williams DM, Eckstein F (1992) 2′-Fluoro- and 2′-amino-2′-deoxynucleoside 5′-triphosphates as substrates for T7 RNA polymerase. Biochemistry 31:9636–9641

Baldrich E, Restrepo A, O'Sullivan CK (2004) Aptasensor development: elucidation of critical parameters for optimal aptamer performance. Anal Chem 76:7053–7063

Bock LC, Griffin LC, Latham JA, Vermaas EH, Toole JJ (1992) Selection of single-stranded DNA molecules that bind and inhibit human thrombin. Nature 355:564–566

Breaker RR (2002) Engineered allosteric ribozymes as biosensor components. Curr Opin Biotechnol 13:31–39

Breaker RR (2004) Natural and engineered nucleic acids as tools to explore biology. Nature 432:838–845

Brody EN, Gold L (2000) Aptamers as therapeutic and diagnostic agents. J Biotechnol 74:5–13

Bruno JG, Kiel JL (2002) Use of magnetic beads in selection and detection of biotoxin aptamers by electrochemiluminescence and enzymatic methods. Biotechniques 32:178–80, 182–183

Burgstaller P, Jenne A, Blind M (2002) Aptamers and aptazymes: accelerating small molecule drug discovery. Curr Opin Drug Discov Devel 5:690–700

Chelliserrykattil J, Ellington AD (2004) Evolution of a T7 RNA polymerase variant that transcribes 2′-O-methyl RNA. Nat Biotechnol 22:1155–1160

Cicortas Gunnarsson L, Nordberg Karlsson E, Albrekt AS, Andersson M, Holst O, Ohlin M (2004) A carbohydrate binding module as a diversity-carrying scaffold. Protein Eng Des Sel 17:213–221

Clark SL, Remcho VT (2003) Electrochromatographic retention studies on a flavin-binding RNA aptamer sorbent. Anal Chem 75:5692–5696

Clegg RM (1995) Fluorescence resonance energy transfer. Curr Opin Biotechnol 6:103–110

Cox JC, Ellington AD (2001) Automated selection of anti-protein aptamers. Bioorg Med Chem 9:2525–2531

Cox JC, Hayhurst A, Hesselberth J, Bayer TS, Georgiou G, Ellington AD (2002) Automated selection of aptamers against protein targets translated in vitro: from gene to aptamer. Nucleic Acids Res 30:e108

Daniels DA, Sohal AK, Rees S, Grisshammer R (2002) Generation of RNA aptamers to the G-protein-coupled receptor for neurotensin, NTS-1. Anal Biochem 305:214–226

Ellington AD, Szostak JW (1990) In vitro selection of RNA molecules that bind specific ligands. Nature 346:818–822

Eulberg D, Klussmann S (2003) Spiegelmers: biostable aptamers. Chembiochem 4:979–983

Eyetech Study Group (2002) Preclinical and phase 1A clinical evaluation of an anti-VEGF pegylated aptamer (EYE001) for the treatment of exudative age-related macular degeneration. Retina 22:143–152

Famulok M (1999) Oligonucleotide aptamers that recognize small molecules. Curr Opin Struct Biol 9:324–329

Famulok M, Mayer G (1999) Aptamers as tools in molecular biology and immunology. Curr Top Microbiol Immunol 243:123–136

Famulok M, Verma S (2002) In vivo-applied functional RNAs as tools in proteomics and genomics research. Trends Biotechnol 20:462–466

Famulok M, Blind M, Mayer G (2001) Intramers as promising new tools in functional proteomics. Chem Biol 8:931–939

Fedor MJ (2002) The role of metal ions in RNA catalysis. Curr Opin Struct Biol 12:289–295

Glokler J, Angenendt P (2003) Protein and antibody microarray technology. J Chromatogr B Analyt Technol Biomed Life Sci 797:229–240

Gold L, Polisky B, Uhlenbeck O, Yarus M (1995) Diversity of oligonucleotide functions. Annu Rev Biochem 64:763–797

Good PD, Krikos AJ, Li SX, Bertrand E, Lee NS, Giver L, Ellington A, Zaia JA, Rossi JJ, Engelke DR (1997) Expression of small, therapeutic RNAs in human cell nuclei. Gene Ther 4:45–54

Grate D, Wilson C (2001) Inducible regulation of the S. cerevisiae cell cycle mediated by an RNA aptamer-ligand complex. Bioorg Med Chem 9:2565–2570

Griffin LC, Tidmarsh GF, Bock LC, Toole JJ, Leung LL (1993) In vivo anticoagulant properties of a novel nucleotide-based thrombin inhibitor and demonstration of regional anticoagulation in extracorporeal circuits. Blood 81:3271–3276

Hanna R, Doudna JA (2000) Metal ions in ribozyme folding and catalysis. Curr Opin Chem Biol 4:166–170

Hentze MW, Kuhn LC (1996) Molecular control of vertebrate iron metabolism: mRNA-based regulatory circuits operated by iron, nitric oxide, and oxidative stress. Proc Natl Acad Sci USA 93:8175–8182

Hermann T, Patel DJ (2000) Adaptive recognition by nucleic acid aptamers. Science 287:820–825

Hewitt CJ, Nebe-Von-Caron G (2004) The application of multi-parameter flow cytometry to monitor individual microbial cell physiological state. Adv Biochem Eng Biotechnol 89:197–223

Homann M, Goringer HU (2001) Uptake and intracellular transport of RNA aptamers in African trypanosomes suggest therapeutic "piggy-back" approach. Bioorg Med Chem 9:2571–2580

Inaba Y, Hamada-Sato N, Kobayashi T, Imada C, Watanabe E (2003) Determination of D- and L-alanine concentrations using a pyruvic acid sensor. Biosens Bioelectron 18:963–971

Iyo M, Kawasaki H, Taira K (2002) Allosterically controllable maxizymes for molecular gene therapy. Curr Opin Mol Ther 4:154–165

Jayasena SD (1999) Aptamers: an emerging class of molecules that rival antibodies in diagnostics. Clin Chem 45:1628–1650

Jellinek D, Green LS, Bell C, Lynott CK, Gill N, Vargeese C, Kirschenheuter G, McGee DP, Abesinghe P, Pieken WA, et al (1995) Potent 2′-amino-2′-deoxypyrimidine RNA inhibitors of basic fibroblast growth factor. Biochemistry 34:11363–11372

Kawakami J, Imanaka H, Yokota Y, Sugimoto N (2000) In vitro selection of aptamers that act with Zn2+. J Inorg Biochem 82:197–206

Kim SJ, Kim MY, Lee JH, You JC, Jeong S (2002) Selection and stabilization of the RNA aptamers against the human immunodeficiency virus type-1 nucleocapsid protein. Biochem Biophys Res Commun 291:925–931

Klussmann S, Nolte A, Bald R, Erdmann VA, Furste JP (1996) Mirror-image RNA that binds D-adenosine. Nat Biotechnol 14:1112–1115

Koizumi M, Breaker RR (2000) Molecular recognition of cAMP by an RNA aptamer. Biochemistry 39:8983–8992

Kubodera T, Watanabe M, Yoshiuchi K, Yamashita N, Nishimura A, Nakai S, Gomi K, Hanamoto H (2003) Thiamine-regulated gene expression of Aspergillus oryzae thiA requires splicing of the intron containing a riboswitch-like domain in the 5′-UTR. FEBS Lett 555:516–520

Lorsch JR, Szostak JW (1994) In vitro selection of RNA aptamers specific for cyanocobalamin. Biochemistry 33:973–982

Lottspeich F, Zorbos H (1998) Bioanalytik. Spektrum Akad. Verlag, Heidelberg, Berlin, New York, pp 9–463

Lutz R, Bujard H (1997) Independent and tight regulation of transcriptional units in Escherichia coli via the LacR/O, the TetR/O and AraC/I1-I2 regulatory elements. Nucleic Acids Res 25:1203–1210

Meis JE, Chen F (2002) In vitro synthesis of 2′-fluoro-modified RNA transcripts that are completely resistant to RNase A digestion using the DuraScribe ^{TM}T7 transcription kit. Epicentre Forum 9:10–11

Menger M, Tuschl T, Eckstein F, Porschke D (1996) Mg(2+)-dependent conformational changes in the hammerhead ribozyme. Biochemistry 35:14710–14716

Nicholls H (2003) Improving drug response with pharmacogenomics. Drug Discov Today 8:281–282

Nimjee SM, Rusconi CP, Sullenger BA (2005) APTAMERS: An Emerging Class of Therapeutics. Annu Rev Med 56:555–583

Nix J, Sussman D, Wilson C (2000) The 1.3 A crystal structure of a biotin-binding pseudoknot and the basis for RNA molecular recognition. J Mol Biol 296:1235–1244

Nonin-Lecomte S, Lin CH, Patel DJ (2001) Additional hydrogen bonds and base-pair kinetics in the symmetrical AMP-DNA aptamer complex. Biophys J 81:3422–3431

O'Sullivan CK (2001) Aptasensors—the future of biosensing? Anal Bioanal Chem 372:44–48

Ooms M, Huthoff H, Russell R, Liang C, Berkhout B (2004) A riboswitch regulates RNA dimerization and packaging in human immunodeficiency virus type 1 virions. J Virol 78:10814–10819

Patel DJ (1997) Structural analysis of nucleic acid aptamers. Curr Opin Chem Biol 1:32–46

Pieken WA, Olsen DB, Benseler F, Aurup H, Eckstein F (1991) Kinetic characterization of ribonuclease-resistant 2′-modified hammerhead ribozymes. Science 253:314–317

Rajendran M, Ellington AD (2002) Selecting nucleic acids for biosensor applications. Comb Chem High Throughput Screen 5:263–270

Rimmele M (2003) Nucleic acid aptamers as tools and drugs: recent developments. Chembiochem 4:963–971

Rimmele M, Ehrentreich-Förster E (2004) Nukleinsäuren als biochemische Fängermoleküle. BIOforum 04:68–69

Roychowdhury-Saha M, Lato SM, Shank ED, Burke DH (2002) Flavin recognition by an RNA aptamer targeted toward FAD. Biochemistry 41:2492–2499

Rubin B, Sonderstrup G (2004) Citrullination of self-proteins and autoimmunity. Scand J Immunol 60:112–120

Ruckman J, Green LS, Beeson J, Waugh S, Gillette WL, Henninger DD, Claesson-Welsh L, Janjic N (1998) 2′-Fluoropyrimidine RNA-based aptamers to the 165-amino acid form of vascular endothelial growth factor (VEGF165). Inhibition of receptor binding and VEGF-induced vascular permeability through interactions requiring the exon 7-encoded domain. J Biol Chem 273:20556–20567

Rusconi CP, Scardino E, Layzer J, Pitoc GA, Ortel TL, Monroe D, Sullenger BA (2002) RNA aptamers as reversible antagonists of coagulation factor IXa. Nature 419:90–94

Rusconi CP, Roberts JD, Pitoc GA, Nimjee SM, White RR, Quick G Jr, Scardino E, Fay WP, Sullenger BA (2004) Antidote-mediated control of an anticoagulant aptamer in vivo. Nat Biotechnol 22:1423–1428

Sousa R, Padilla R (1995) A mutant T7 RNA polymerase as a DNA polymerase. EMBO J 14:4609–4621

Sudarsan N, Barrick JE, Breaker RR (2003) Metabolite-binding RNA domains are present in the genes of eukaryotes. RNA 9:644–647

Suess B, Fink B, Berens C, Stentz R, Hillen W (2004) A theophylline responsive riboswitch based on helix slipping controls gene expression in vivo. Nucleic Acids Res 32:1610–1614

Sun S (2000) Technology evaluation: SELEX, Gilead Sciences Inc. Curr Opin Mol Ther 2:100–105

Sussman D, Greensides D, Reilly K, Wilson C (1999) Preliminary characterization of crystals of an in vitro evolved cyanocobalamin (vitamin B12) binding RNA. Acta Crystallogr D Biol Crystallogr 55:326–328

Szabo A, Stolz L, Granzow R (1995) Surface plasmon resonance and its use in biomolecular interaction analysis (BIA). Curr Opin Struct Biol 5:699–705

Toulme JJ, Di Primo C, Boucard D (2004) Regulating eukaryotic gene expression with aptamers. FEBS Lett 567:55–62

Tucker CE, Chen LS, Judkins MB, Farmer JA, Gill SC, Drolet DW (1999) Detection and plasma pharmacokinetics of an anti-vascular endothelial growth factor oligonucleotide-aptamer (NX1838) in rhesus monkeys. J Chromatogr B Biomed Sci Appl 732:203–212

Tuerk C, Gold L (1990) Systematic evolution of ligands by exponential enrichment: RNA ligands to bacteriophage T4 DNA polymerase. Science 249:505–510

Ulrich H, Magdesian MH, Alves MJ, Colli W (2002) In vitro selection of RNA aptamers that bind to cell adhesion receptors of Trypanosoma cruzi and inhibit cell invasion. J Biol Chem 277:20756–20762

Vinores SA (2003) Technology evaluation: pegaptanib, Eyetech/Pfizer. Curr Opin Mol Ther 5:673–679

Vogt M, Skerra A (2004) Construction of an artificial receptor protein ("anticalin") based on the human apolipoprotein D. Chembiochem 5:191–199

Vuyisich M, Beal PA (2002) Controlling protein activity with ligand-regulated RNA aptamers. Chem Biol 9:907–913

Walter G, Bussow K, Lueking A, Glokler J (2002) High-throughput protein arrays: prospects for molecular diagnostics. Trends Mol Med 8:250–253

Wang C, Zhang M, Yang G, Zhang D, Ding H, Wang H, Fan M, Shen B, Shao N (2003) Single-stranded DNA aptamers that bind differentiated but not parental cells: subtractive systematic evolution of ligands by exponential enrichment. J Biotechnol 102:15–22

White R, Rusconi C, Scardino E, Wolberg A, Lawson J, Hoffman M, Sullenger B (2001) Generation of species cross-reactive aptamers using "toggle" SELEX. Mol Ther 4:567–573

White RR, Sullenger BA, Rusconi CP (2000) Developing aptamers into therapeutics. J Clin Invest 106:929–934

Wibrand F (2004) A microplate-based enzymatic assay for the simultaneous determination of phenylalanine and tyrosine in serum. Clin Chim Acta 347:89–96

Wilson C, Nix J, Szostak J (1998) Functional requirements for specific ligand recognition by a biotin-binding RNA pseudoknot. Biochemistry 37:14410–14419

Wilson DS, Szostak JW (1999) In vitro selection of functional nucleic acids. Annu Rev Biochem 68:611–647

Winkler WC, Cohen-Chalamish S, Breaker RR (2002) An mRNA structure that controls gene expression by binding FMN. Proc Natl Acad Sci USA 99:15908–15913

Woodbury CP Jr, Venton DL (1999) Methods of screening combinatorial libraries using immobilized or restrained receptors. J Chromatogr B Biomed Sci Appl 725:113–137

Zaydan R, Dion M, Boujtita M (2004) Development of a new method, based on a bioreactor coupled with an L-lactate biosensor, toward the determination of a nonspecific inhibition of L-lactic acid production during milk fermentation. J Agric Food Chem 52:8–14

Zhang Z, Blank M, Schluesener HJ (2004) Nucleic acid aptamers in human viral disease. Arch Immunol Ther Exp (Warsz) 52:307–315

Zhu H, Snyder M (2003) Protein chip technology. Curr Opin Chem Biol 7:55–63

RNA Aptamers as Potential Pharmaceuticals Against Infections with African Trypanosomes

H. U. Göringer[1] (✉) · M. Homann[1] · M. Zacharias[2] · A. Adler[1]

[1]Genetics, Darmstadt University of Technology, Schnittspahnstr. 10, 64287 Darmstadt, Germany
goringer@hrzpub.tu-darmstadt.de

[2]Computational Biology, School of Engineering and Science, International University Bremen, Campus Ring 1, 28759 Bremen, Germany

1	Introduction	376
2	African Trypanosomes as Extracellular Parasites	378
2.1	The Variant Surface Architecture of African Trypanosomes	378
2.2	Trypanosomes and the Blood–Brain Barrier	379
3	Trypanosome-Specific Aptamers	380
3.1	RNA Aptamers that Recognize the Cell Surface of Live Trypanosomes	380
3.2	RNA Aptamers that Recognize the Variant-Surface Glycoprotein	384
4	Converting Trypanosome-Specific Aptamers in Pharmaceutical Reagents . .	386
4.1	Aptamer Core Structures and Serum Stability	386
4.2	"Piggy-Back Approach"	388
4.3	Future Developments	389
	References	390

Abstract Protozoal pathogens cause symptomatic as well as asymptomatic infections. They have a worldwide impact, which in part is reflected in the long-standing search for antiprotozoal chemotherapy. Unfortunately, effective treatments for the different diseases are by and large not available. This is especially true for African trypanosomiasis, also known as sleeping sickness. The disease is an increasing problem in many parts of sub-Saharan Africa, which is due to the lack of new therapeutics and the increasing resistance against traditional drugs such as melarsoprol, berenil and isometamidium. Considerable progress has been made over the past 10 years in the development of nucleic acid-based drug molecules using a variety of different technologies. One approach is a combinatorial technology that involves an iterative Darwinian-type in vitro evolution process, which has been termed SELEX for "systematic evolution of ligands by exponential enrichment". The procedure is a highly efficient method of identifying rare ligands from combinatorial nucleic acid libraries of very high complexity. It allows the selection of nucleic acid molecules with desired functions, and it has been instrumental in the identification of a number of synthetic DNA and RNA molecules, so-called aptamers that recognize ligands of different chemical origin. Aptamers typically bind their target with high affinity and high specificity and have successfully been converted into pharmaceutically active compounds. Here we summarize the recent examples of the SELEX technique within the context of identifying high-affinity RNA ligands against the surface of the protozoan parasite *Trypanosoma brucei*, which is the causative agent of sleeping sickness.

Keywords African trypanosomes · SELEX · Nucleic acid aptamers · African sleeping sickness

1
Introduction

Parasitic diseases are among the most devastating illnesses in the world. They cause suffering in hundreds of millions of people as well as an unknown number of wild and domestic animals. Malaria is responsible for the death of approximately 2.7 million people per year (Andreopoulos 2003) and schistosomiasis and Chagas disease severely affect millions of people in Asia, Africa and South America (Capron et al. 2002; Dias et al. 2002). The problem is amplified by the fact that the number of drugs for treating parasite infections is very small. Most of the available therapeutics were discovered decades ago and are not very effective (Pecoul et al. 1999). They suffer from a variety of problems such as acute toxicity, short duration of action and the emergence of resistant parasites (Kaiser et al. 2002). The process of developing new and improved antiparasitic drugs has been very slow. This is due to many factors, among them a significant lack of interest from the pharmaceutical industry as well as insufficient funding for research in parasitology in general (Reich and Govindaraj 1998; Trouiller et al. 2001). Lastly, the process has also been slow because the random screening methods of pharmaceutical components have not been very successful at identifying antiparasitic compounds.

A completely different approach of searching for novel pharmacological agents has emerged from the field of combinatorial chemistry (Lam and Renil 2002). The technology includes a variety of techniques by which very large numbers of structurally distinct molecules are synthesized in a time and resource-effective way. The resulting pool of molecules reflects a library of diverse three-dimensional structures that is subsequently scrutinized for its pharmacological or diagnostic potential. A specific subset of methods uses combinatorial nucleic acid libraries in conjunction with a selection scheme based on in vitro evolution principles. The process is termed SELEX, which stands for "systematic evolution of ligands by exponential enrichment" (Ellington and Szostak 1990; Tuerk and Gold 1990). The SELEX technique is based on the idea of using nucleic acids as therapeutic reagents rather than the small organic compounds used in the past (Gold 1995). Since DNA and RNA molecules adopt stable and intricately folded three-dimensional shapes (Conn and Draper 1998) they are capable of providing a scaffold for the interaction with functional side groups of a bound ligand. This is supported by the fact that many gene regulatory processes and a significant number of other biochemical reactions rely on the formation of complexes between nucleic acids and chemically complementary compounds. The molecular interactions are essentially based on surface recognition principles and include all forms of physical interaction such as hydrogen bonding, hydrophobic interaction and ionic contacts.

Randomized sequence pools of oligonucleotides can be synthesized by solid phase chemistry resulting in sequence complexities in the range of 10^{15} different molecules. This represents a structural diversity not matched by any other combinatorial technique and is one of the prime advantages of the SELEX method. A premise of the process is that the number of oligonucleotides in the synthesized pool is large enough to contain at least a few tertiary structures, which can provide a binding site or catalytic activity for a chosen target. Nucleic acid molecules resulting from SELEX experiments have been termed aptamers (Ellington and Szostak 1990), derived from the Latin word *aptus*, meaning "fitting". The affinities of aptamers for their binding targets are of comparable strength to those of antibodies for their antigens (Jayasena 1999). Typically, the equilibrium dissociation constants (K_d) are in the low nanomolar to high picomolar range. An additional advantage of the SELEX process is that even molecules of very low copy number in the starting library can be enriched and selected. Nucleic acids can be enzymatically amplified, and thus significant amounts of material can be created in every individual round of the SELEX experiment. Lastly, the iterative cycles of in vitro selection and enzymatic amplification mimic a Darwinian-type process driving the selection towards an evolutionarily optimized structural solution for the chosen interaction.

Aptamers have been selected for a number of different ligands (Wilson and Szostak 1999) including nucleic acids, polypeptides, sugar molecules, small organic compounds as well as entire cells (see aptamer database at http://aptamer.icmb.utexas.edu/). Aptamers have even been selected for molecules that do not normally interact with nucleic acids within their natural context. This includes antibodies, phospholipases, hormones and growth factors. In addition, technologies for the chemical synthesis of very large aptamer quantities have been developed (Pieken 1997), as have chemical modification protocols to improve the stabilities (Eaton and Pieken 1995) and circulating half-lives of the DNA and RNA molecules in blood plasma and other body fluids (Brody and Gold 2000). Aptamers can be readily adopted for diagnostic applications (Hesselberth et al. 2000), and in vivo experiments demonstrated that they generally exhibit low toxicity and immunogenicity characteristics. Nucleic acid-based therapeutics are now being tested in clinical trials (for a review see Nimjee et al. 2005), and the first aptamer drug has very recently been introduced into the pharmaceutical market (Gragoudas et al. 2004; Fine et al. 2005).

In the following, we will summarize the recent attempts to use the SELEX technology to identify high-affinity RNA ligands to live parasites and parasite-specific proteins. The described experiments are focused on African trypanosomes, the causative agent of sleeping sickness or African trypanosomiasis. We will discuss recent experimental data on how the trypanosome-specific aptamers were generated and how they can be converted into antiparasitic drugs, emphasizing the problems that RNA pharmaceuticals must overcome.

2
African Trypanosomes as Extracellular Parasites

2.1
The Variant Surface Architecture of African Trypanosomes

African trypanosomes are unicellular, uniflagellated protozoan parasites. They cause African trypanosomiasis or sleeping sickness, a chronic disease in humans as well as in wild and domestic animals. An estimated 50 million people in 36 African countries are at risk of an infection, and a total of about 500,000 newly infected cases per year has been estimated (Smith et al. 1998; Kioy et al. 2004). The disease is caused by two geographically distinct trypanosome subspecies: *Trypanosoma brucei rhodesiense* in East Africa and *T. brucei gambiense* in West Africa. Although the two parasites cause different clinical manifestations, both infections are ultimately fatal if untreated (Khaw and Panosian 1995; Kioy et al. 2004).

African trypanosomes are transmitted by tsetse flies and, as extracellular parasites, they multiply within the peripheral blood and the tissue fluids of the infected hosts (Fig. 1). During this bloodstream lifecycle stage, trypanosomes are covered with a layer of approximately 10 million molecules of a glycoprotein species known as variant surface glycoprotein (VSG). VSG molecules have a molecular mass of approximately 60 kDa, they homodimerize and are glycosylphosphatidylinositol (GPI)-anchored within the plasma membrane (Donelson 2003). The VSG surface induces a strong T cell-independent IgM response as well as a T cell-dependent B cell response, which elicits VSG-specific IgG (Sternberg 1998). The parasites, however, evade the host immune system by temporarily expressing different VSG variants (Rudenko et al. 1998). This phenomenon has been termed antigenic variation and has its molecular basis in the surface presentation of structurally polymorphic N-terminal domains of the different VSG variants. The trypanosome genome encodes a repertoire of about 1,000 different *vsg* genes, but only one VSG is expressed at a given time. Thus, the VSG surface acts as an exclusion barrier for larger molecules, such as antibodies, while its variable characteristics cause the inability of the infected host to clear the infection.

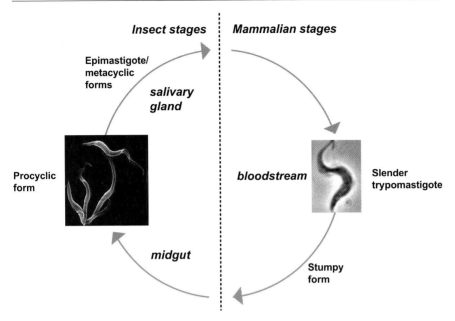

Fig. 1 Cartoon of the main stages of the *Trypanosoma brucei* lifecycle. The parasites replicate in the blood as slender trypomastigotes and differentiate into so-called stumpy forms at high cell densities. Transmission occurs during the blood meal of a tsetse fly, and the parasite is passaged from the midgut to the salivary glands of the insect vector. The lifecycle is characterized by changes in cell shape, cell cycle, metabolism and surface coat. It is completed after injection of metacyclic-stage parasites into the blood of another host organism

2.2
Trypanosomes and the Blood–Brain Barrier

Aside from the described immunological phenomenon, African sleeping sickness is characterized by a number of distinct neurological symptoms. They include disruption of sleep and extrapyramidal motor disturbances as well as neuropsychiatric changes (reviewed by Enanga et al. 2002). Although the histopathological reactions in the brain have been well described, the pathogenetic mechanisms behind the nervous system disease are still unclear. The blood–brain barrier (BBB), which separates circulating blood from the central nervous system, regulates the flow of materials to and from the brain. During the course of the disease, the integrity of the BBB becomes compromised and parasites cross the barrier. While the mechanism of this phenomenon is not understood, the following routes have been suggested: (1) entry of the parasite via the choroid plexus epithelium leading to the cerebrospinal fluid space; (2) entry via the cerebral capillary endothelium leading to the brain parenchyma and (3) penetration through disrupted tight junctions (Enanga et al. 2002; Londsdale-Eccles and Grab 2002). Morphologically, the BBB is constructed of apposed cerebral-endothelial capillary cells held together by so-called tight junctions. Tight junctions hold the brain microvascular endothelial cells closely together, thereby eliminating the gaps that usually occur between endothelial cells at other locations in the body. Central components of the tight junctions are transmembrane proteins such as occludin and claudins, and various membrane-associated guanylate kinases.

The picture is further complicated by the fact that trypanosomes react differently with different host species. When African trypanosomes gain access to the human brain, little or no damage to the barrier is observed. By contrast, chronic trypanosomiasis in rats is accompanied by extensive BBB damage. However, recent evidence demonstrated the direct translocation of *T. brucei* across the BBB in rats during the early stages of infections (Mulenga et al. 2001). Although extravascular trypanosomes were observed near capillaries, a generalized loss of proteins in the tight junction could not be identified.

Several chemotherapeutics are currently used to treat African sleeping sickness. Among them are compounds such as suramin, pentamidine, melarsoprol, eflornithine and, on an experimental basis, diminazene and nifurtimox (Stephenson and Wiselka 2000; Fairlamb 2003; Kioy et al. 2004). However, none of the therapeutic measures is very effective and no vaccination against trypanosome infections is available today. Therefore, new concepts and experimental strategies for developing novel drugs are required (Van Gompel and Vervoort 1997). Based on the described clinical evidence and based on the fact that most drugs, including many trypanocides, do not cross the BBB efficiently, it is important that new diagnostic and therapeutic approaches must address this problem as well.

3
Trypanosome-Specific Aptamers

3.1
RNA Aptamers that Recognize the Cell Surface of Live Trypanosomes

T. brucei was the first protozoan parasite that was targeted using SELEX technology. Homann and Göringer (1999) used African trypanosomes as a model system for the selection of aptamers to the surface of live parasites. The SELEX experiments were designed to identify high-affinity RNA ligands to variant as well as invariant molecules on the parasite surface. Such aptamers might interfere with surface protein function and have the potential to re-direct the immune response to the surface of the parasite. Other aptamers could potentially be used for the affinity isolation of previously unknown surface proteins.

Two separate SELEX experiments were performed in parallel using the bloodstream lifecycle stages of *T. brucei* strains MITat 1.2 and MITat 1.4 as targets (Cross 1975). The two parasite cell lines show variant specific characteristics, among them the stable expression of different VSG variants: VSG 221 in the case of MITat 1.2 and VSG 117 for MITat 1.4. The starting RNA library contained 2×10^{15} unique 85-mer RNA sequences with a central segment of 40 randomized nucleotide positions. In each cycle, living bloodstream stage cells were incubated with the pools of RNA molecules to allow the enrichment of RNA sequence variants that recognize exposed surface structures. After 12 cycles of binding, reverse transcription and amplification, the "winner" sequences were identified by cloning and sequencing. Together, 106 clones were analysed and three aptamer families were identified based on conserved sequence motifs and secondary structure elements (Fig. 2a).

Individual aptamers from each family were chosen for the identification of their interaction partners by zero-distance photo cross-linking (Fig. 2b). None of the aptamers was able to discriminate between MITat 1.2 and MITat 1.4 parasites, suggesting invariant surface elements as aptamer targets. One of the aptamers of family I (aptamer 2-16) was selected as a representative of this group and was characterized further. According to secondary structure prediction and enzymatic structure probing experiments, aptamer 2-16 folds into a pseudoknotted hairpin and assumes a compact, almost globular fold (Fig. 3). The ability to form this pseudoknotted hairpin seems to be an important determinant for the high-affinity interaction, since shortened, "non-pseudoknotted" variants of the RNA showed significantly reduced surface binding (Homann and Göringer 1999). Aptamer 2-16 bound to MITat 1.2 and MITat 1.4 cells with similar high affinities (K_d) of 60 nM. The aptamer target was identified by zero distance photo cross-linking as a 40–42 kDa pro-

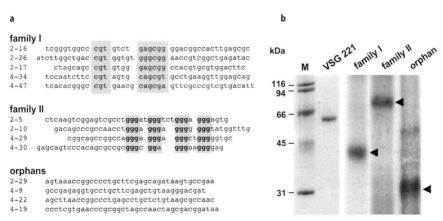

Fig. 2 a Aligned DNA sequences of RNA aptamer families specific for the *T. brucei* surface (Homann and Göringer 1999; Homann and Göringer 2001). Consensus sequence elements are shown in *shaded boxes*. **b** UV-crosslinking of radioactively labelled RNA aptamers from all three families to live parasites. The aptamers interact with different surface proteins as indicated by the *arrows*

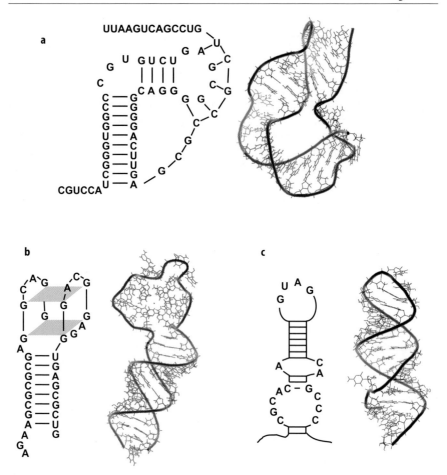

Fig. 3a–c Secondary structures of three *T. brucei*-specific RNA aptamers. **a** Aptamer 2-16 (Homann and Göringer 1999). **b** Aptamer N2 (unpublished). **c** Aptamer cl57 (Lorger et al. 2003). The 2D models were derived from secondary structure prediction and enzymatic probing experiments. The 3D models were created by molecular modelling followed by energy minimization

tein that is present on bloodstream-stage parasites but not on the surface of insect-stage trypanosomes. Using fluorescently labelled aptamer preparations in in situ localization experiments, the 42-kDa protein was identified as a component of the flagellar pocket (FP) of the parasite. The FP represents the main endo/exocytosis site of the trypanosome cell.

These results demonstrated that living parasite cells are suitable targets for SELEX experiments that are aimed at the identification of high-affinity ligands to surface components. The experiment further verified that specific RNA ligands can be selected without any knowledge of the cell's surface architecture

(Fig. 3) and it demonstrated the potential of SELEX technology to function as a mapping tool for cellular targets of unknown composition as suggested before (Morris et al. 1998; Ulrich et al. 2001).

Aside from their potential as identification tools, aptamers themselves may be used as therapeutic reagents against infectious protozoan parasites. In order to test the therapeutic potential of the selected RNA ligands (Homann and Göringer 1999), the above-described aptamer 2-16 was chosen for a further analysis with respect to its stability and fate after binding to the parasite cells.

Fluorescently labelled aptamer preparations were used to visualize RNA binding to the 42-kDa flagellar pocket polypeptide. Interaction with its target was followed by rapid endocytotic uptake and intracellular transport to the lysosome (Homann and Göringer 2001; Fig. 4). Co-localization experiments with transferrin suggested a receptor-mediated uptake pathway followed by vesicular routing to the lysosome. The aptamer was partially degraded during the uptake process; however, a core structure of about 50 nt proved significantly more stable towards RNase activities within the flagellar pocket and endosomal vesicles. Binding and uptake was sequence specific and was not observed with RNA molecules of random sequence. Thus, the specificity of binding and uptake suggested that aptamer 2-16 could be used for a trypanosome-specific delivery of RNA-coupled compounds to the lysosomal compartments of the parasite (see Sect. 4.2, "Piggy-Back Approach" below).

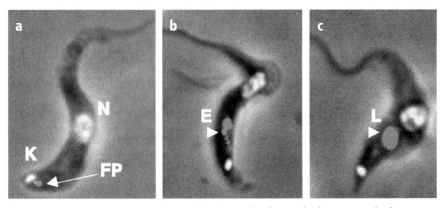

Fig. 4a–c Targeting aptamer-conjugated biotin molecules to the lysosome of African trypanosomes. Aptamer 2-16 (Homann and Göringer 1999; Homann and Göringer 2001) was covalently modified with a biotin moiety and the biotin group was detected with a fluorophore-conjugated anti-biotin antibody (*arrowheads*). **a** Initial binding occurs in the flagellar pocket (*FP*) of the parasite. **b** At later time points the signal is detected within endosomal vesicles (*E*) and finally accumulates (**c**) within the lysosome (*L*). *K* indicates the position of the kinetoplast; *N* represents the nucleus

3.2
RNA Aptamers that Recognize the Variant-Surface Glycoprotein

The aim of the SELEX experiment published by Lorger and co-workers (2003) was to select RNA aptamers that specifically recognize the VSG protein on the surface of bloodstream-stage African trypanosomes. The bloodstream lifecycle stage represents the infectious phase of the parasite, and VSG is the most abundant surface protein during that period. Since the parasite undergoes antigenic variation, which is characterized by the expression of different VSG variants, a selection scheme was designed to specifically target the structurally invariant domains of the surface proteins. Antigen-coupled versions of such "variant-independent" but VSG-specific aptamers should be able to re-direct the immune response of the infected host back to the surface of the parasite independently of the expressed VSG variant.

The starting material for the experiment was a combinatorial 2'-F-modified RNA library of 10^{14} different molecules. The variable domain was 40 nt long in order to restrict the molecular mass of the selected RNAs to a value around 30 kDa. Molecules of such size have been shown to be able to penetrate into the cavities between the surface-anchored VSG homodimers, thereby reaching structurally conserved parts of the protein (Blum et al. 1993). Modified RNA was used to ensure that the selected RNAs would have a reasonable serum stability (Pieken et al. 1991), as *T. brucei* is an extracellular blood parasite. Over the course of eight SELEX rounds, three RNA aptamer families and three orphan RNAs were selected. The desired specificity to exclusively recognize VSG molecules but different variants of the polypeptide was accomplished by performing a multi-target selection scheme. The first three selection rounds were carried out with a purified protein preparation of VSG variant 117 (Cross 1975). This step enriched for VSG-interacting aptamers and the resulting RNA pool was then challenged with live parasites expressing VSG117 on their surface (MITat 1.4 trypanosomes). This represented a de-selection step for RNAs that were not able to recognize the target protein within its natural, membrane-bound context. To drive the selection towards aptamers that were able to bind to more than one VSG variant, the remaining selection rounds were performed with a different variant (VSG221; Cross 1975). As before, a purified preparation of VSG221 was used first followed by selection rounds using live VSG221 parasites (strain MITat 1.2). All identified aptamers bound to VSG with affinities in the nanomolar or even subnanomolar concentration range. Importantly and as expected from the selection scheme, none of the RNAs were able to distinguish between the two VSG variants used during the experiment. Moreover, all aptamers bound with high affinity to VSG variants that were not used during the SELEX cycles (ILTat 1.1, AnTat 1.1) (Rice-Ficht et al. 1982; Michiels et al. 1983).

Thus, the experiment resulted in the selection of high affinity VSG-specific RNAs with no selectivity for a specific VSG variant. This indicated that the same or an almost identical aptamer binding site was present in all tested VSGs and further suggested that the aptamer/protein interaction takes place within a structurally conserved domain of the different VSGs. Lorger and co-workers confirmed, that the aptamer binding site resides within the lower half of the N-terminal VSG domain (Johnson and Cross 1979; Freymann et al. 1990), which indicated that the RNA molecules penetrated significantly into the molecular clefts of the VSG layer (Lorger et al. 2003).

Importantly, and as implemented in the selection strategy, the VSG-specific aptamers were not only able to bind to purified VSG preparations but also to the polypeptide on the surface of live trypanosomes. This was experimentally established by immunofluorescence microscopy using biotin-tethered RNA preparations and fluorescently labelled streptavidin (Lorger et al. 2003). The observed staining pattern confirmed that the biotinylated RNAs were bound over the parasite's entire cell surface, the image being indistinguishable from the surface distribution of the VSG molecules. Trypanosome strains that were

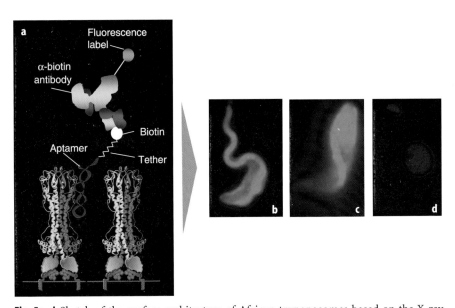

Fig. 5a–d Sketch of the surface architecture of African trypanosomes based on the X-ray structure of the N-terminal domains of a VSG221 homodimer (Freymann et al. 1990). The drawing depicts the arrangement of two glycosyl-phosphatidylinositol-anchored VSG homodimers and highlights binding of RNA aptamer cl57 (Lorger et al. 2003) to one VSG molecule at a binding site within its structurally conserved domain. The aptamer was covalently tethered to a biotin molecule in order to re-direct biotin-specific antibodies to the surface of the parasite. The aptamer was incubated with bloodstream stage *T. brucei* AnTat 1.1 parasites (**b**) or *T. congolense* BeNat1 cells (**c**). Detection of the biotin moiety was achieved using Cy3-conjugated anti-biotin antibodies as depicted in part **a**. In the absence of biotinylated aptamer cl57 no surface staining was observed (**d**); the nuclear and mitochondrial DNA were counterstained with H33342 (*blue*)

not used during the selection gave the same staining pattern as the two parasite strains that were used as targets, again demonstrating that the VSG specific aptamers are not variant-selective.

Since the aptamer-bound biotin group was accessible on the parasite cell surface, the authors performed an experiment to illustrate that antigen-coupled aptamers can be used to re-direct antibodies to the surface of the parasite. By using fluorophore-coupled anti-biotin antibodies they confirmed the binding of immunoglobulins to aptamer-decorated surfaces of *T. brucei* (Fig. 5). As before, the data showed that the entire surface of the parasite was covered with antibodies and no variation between different *T. brucei* variant strains was observed. Strikingly, the surface of *T. congolense* was also recognized by the chosen aptamer preparation. *T. congolense* represents a different trypanosome species, which expresses VSG molecules that resemble a specific subclass of the *T. brucei* VSGs (Urakawa et al. 1979). This implies that the selected aptamers might be used for other trypanosome species as well. Whether the described experiment can be performed in an animal model remains to be tested. As a first step in this direction, Lorger et al. (2003) determined that the RNAs do not bind to bovine serum albumin (BSA), the most abundant serum protein. In addition, the serum stability of the selected aptamers was analysed. The RNAs show serum half-lives of approximately 24 h, representing an encouraging value for an in vivo application.

4
Converting Trypanosome-Specific Aptamers in Pharmaceutical Reagents

4.1
Aptamer Core Structures and Serum Stability

The above-described SELEX experiments led to the identification of high affinity RNA ligands that target the trypanosomal surface and thus suggest several pharmacological approaches. In order to optimize the selected RNAs for their pharmaceutical use, the original molecules have to be converted from in vitro binding reagents to therapeutic compounds, which can be tested for their pharmacological effects in vivo. Within this context an important consideration is the cost of synthesis. Since aptamer-derived pharmaceuticals are chemically synthesized and therefore fairly expensive, the size of the aptamer is a critical feature and has to be reduced to its minimal active core. This can be experimentally addressed in so-called boundary experiments (Fitzwater and Polisky 1996). As already mentioned above, the minimal binding domain of aptamer 2-16 was determined as a 64-nt sequence (Homann et al. 2001), while the core structure of the VSG-specific aptamer was 44 nt long (Lorger et al. 2003).

Another critical characteristic of a pharmacologically active compound is its stability in biological fluids, especially in blood. The stability of nucleic acids is limited by their sensitivity towards nucleases, which are highly abundant in almost all in vivo environments. The in vitro half-life of a typical DNA oligonucleotide in human serum is 30–60 min, while that of a typical RNA molecule is in the range of only a few seconds. However, the stability of RNAs can be significantly increased by the incorporation of 2′-fluoro-, 2′-amino- or 2′-O-methyl-substituted ribonucleotides (Eaton and Pieken 1995). This is due to the fact that the naturally occurring cleavage of the phosphodiester backbone relies on the availability of a 2′-hydroxyl group. The 2′-OH enables the nucleophilic attack that leads to the formation of a 2′,3′-cyclic phosphodiester hydrolysis product, which further reacts to the corresponding 3′-nucleoside monophosphate.

A simple way of creating serum-stable aptamers is to introduce the modified nucleotides into the starting pool at the onset of the selection. This was done in the case of the VSG-specific aptamers by using 2′-F-modified pyrimidine nucleotides and led to RNA molecules with serum stabilities greater than 24 h (Lorger et al. 2003). However, aptamers can also be modified after the selection process, provided that the structure and function of the molecules remains unaffected by the modifications. This is not necessarily the case, given the critical role that 2′-hydroxyl groups can play in the structural organization of RNA. Aptamer 2-16 was originally selected as an unmodified RNA. The introduction of 2′ amino and/or 2′ fluoro modifications required a detailed analysis of the structural and functional consequences. C2′-modifications are known to affect the sugar puckering of nucleic acids. The preferred conformation of RNA helices is the A-form, which is characterized by a C3′-*endo* ribose conformation. Heteroatom modifications like 2′-amino-groups favour C2′-*endo* conformations, which is one of the major determinants of the "DNA-like" B-helix. In contrast, oligonucleotides substituted with 2′-O-methyl- or 2′-fluoro-groups shift the equilibrium towards the "RNA-like" C3′-*endo* form of the sugar.

Unmodified aptamer 2-16 RNA was characterized with a half-life of ≤ 5 s in human serum. The co-transcriptional incorporation of 2′-fluoro- and/or 2′-amino-substituted pyrimidine nucleotides increased the stability of the RNA up to a half-life of several days (unpublished data). A cell-binding analysis demonstrated, that 2′-amino-modifications led to a complete loss of the aptamer/parasite interaction, while 2′-fluoro-substitutions retained the binding capacity of the modified RNA. This observation is consistent with the above-described assumption that 2′-fluoro-substitutions are the preferred candidates for maintaining the structural integrity and thus function of modified aptamers. The binding affinity of 2′-fluoro-modified 2-16 RNA to parasite cells was marginally enhanced (K_d of 70 nM versus 60 nM for unmodified 2-16 RNA), suggesting a contribution of the 2′-F group in the interaction. In addition, 2′-F-modified 2-16 RNA bound to the flagellar pocket indistinguishable from unmodified aptamer preparations, and the molecules were internalized and transported to the lysosome via the same endosomal routing pathway. This was further confirmed in a structural analysis by chemical and enzyme probing experiments. The data confirmed that the overall structure of aptamer 2-16 was by and large unaltered by the 2′-substitutions, and as a consequence the RNA molecules had retained their functionality.

4.2
"Piggy-Back Approach"

Based on the characteristics of aptamer 2-16, which was shown to transport a covalently coupled compound to the lysosome of *T. brucei* (see Fig. 4), a strategy for interfering with the parasite involves the usage of molecules that become active within the acidic environment of the organelle and act as lysosomal toxins. Responsive hydrophobically associating polymers, pH-responsive pseudopeptides as well as membrane disturbing peptides are classes of molecules that possess the required characteristics (Eccleston et al. 2000; Tonge and Tighe 2001; Fig. 6).

Several strategies exist for the coupling of lysosomolytic compounds to RNA molecules and all of them require the introduction of functional groups into the different molecules. RNA molecules can be simply modified by enzymatic or chemical treatment. For example, a sulfhydryl group can be introduced at the 5′-end using T4 polynucleotide kinase (PNK) in the presence of ATP-γ-S. The coupling can then be achieved by a maleimide containing bifunctional crosslinker. Alternatively, the 3′-end of the RNA can be chemically oxidized using sodium periodate to create an aldehyde group, which reacts spontaneously with amino-groups at mild alkaline pH. The resulting labile Schiff base intermediate is converted into a stable secondary amide-bond at reducing conditions using sodium cyanoborhydride as the reducing agent. To achieve a high coupling rate, the stoichiometry of the reaction is a crucial criterion. For the latter coupling strategy, concentrations in the micromolar range for the 3′-oxidized RNA and a 100-fold molar excess of the amino group-containing reaction partner are necessary in order to obtain reasonable amounts of coupling product.

The above-described trypanosome-specific RNA aptamers typically exhibit a molecular mass of ≤ 30 kDa. Due to their small size they are usually rapidly cleared from an animal via the urinary system (Agrawal and Zhang 1997). Therefore, the in vivo half-life of an aptamer in the blood is not only determined by its resistance towards chemical or enzymatic cleavage, but is also constrained by the rate of systemic clearance. Increasing the molecular mass of nucleic acids has been shown to positively influence the retention time within living systems. In this way, the coupling of the RNA aptamers to a high molecular mass carrier compound such as polyethylene glycol (PEG) will have to be performed in order to increase the serum half-life of the RNAs (Bonora et al. 1997; Watson et al. 2000; Healy et al. 2004).

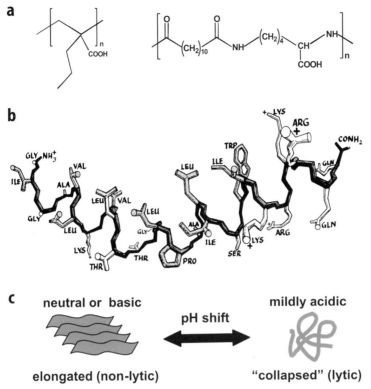

Fig. 6a–c Overview of pH-responsive compounds with membrane-disruptive activities. **a** The hydrophobically associating polymer poly (2-ethyl acrylic acid) (*left*) and the pseudopeptide poly (lysine dodecanamide) (*right*). **b** Melittin, a component of honeybee venom, is an example of an oligopeptide with membrane-disruptive activity. **c** At neutral or basic pH, these compounds form non-lytic elongated structures. At mildly acidic conditions, they assume a collapsed state with lytic activity

4.3
Future Developments

Finally, the described aptamers can be used as reagents for the identification of pharmaceutical lead compounds. As demonstrated by Green et al. (2001) high-throughput screening (HTS) assays can be performed to scrutinize small-molecule libraries for compounds with aptamer-displacing activity in competition binding experiments. Thus, instead of considering aptamers as drug leads themselves, they can be used for the identification of drug candidates from small-molecule libraries (Burgstaller et al. 2002). This approach has the potential to overcome several of the limitations of aptamer molecules (White et al. 2000) and can, for instance, be used for the identification of orally active compounds as well as molecules with low synthesis costs. Pilot competition screening experiments using the described VSG-specific aptamer and a selected set of small molecules have already been performed. The results indicate that multi-well formatted screening systems in combination with automated fluorescence or radioactive readouts can be performed.

With respect to the BBB problem, recent developments from several laboratories have established in vitro models for the BBB (Franke et al. 1999, 2000; Cecchelli et al. 2000; Gaillard et al. 2001). The systems are based on the co-culturing of brain capillary endothelial cells and astrocytes on permeable solid support membranes and have been characterized for a number of parameters including electrical resistance and expression of typical BBB markers (Gaillard et al. 2001). The systems likely provide a new tool to experimentally address the neurological aspects of African trypanosomiasis and might be helpful in assaying for interactions between trypanosomes and drugs within the host nervous system. Possibly, these in vitro systems can be tailored to select for a novel class of aptamers that can act as decoy molecules (Lebruska and Maher 1999; Mann and Dzau 2000) to inhibit binding and entry of the parasite into the brain capillary network.

Acknowledgements We would like to thank Markus Engstler for the graphical representation of the VSG dimer. The described research has been supported by the Howard Hughes Medical Institute (HHMI), the German Science Foundation (DFG) and the Dr. Illing Foundation. H.U.G. is an International Research Scholar of the HHMI.

References

Agrawal S, Zhang R (1997) Pharmacokinetics of oligonucleotides. Ciba Found Symp 209:60–75

Andreopoulos S (2003) Developing drugs for parasitic diseases. Science 300:430–431

Blum, ML, Down JA, Gurnett AM, Carrington M, Turner MJ, Wiley DC (1993) A structural motif in the variant surface glycoproteins of *Trypanosoma brucei*. Nature 362:603–609

Bonora GM, Ivanova E, Zarytova V, Burcovich B, Veronese FM (1997) Synthesis and characterization of high-molecular mass polyethylene glycol-conjugated oligonucleotides. Bioconjug Chem 8:793–797

Brody EN, Gold L (2000) Aptamers as therapeutic and diagnostic agents. J Biotechnol 74:5–13

Burgstaller P, Girod A, Blind M (2002) Aptamers as tools for target prioritization and lead identification. Drug Discov Today 7:1221–1228

Capron A, Capron M, Riveau G (2002) Vaccine development against schistosomiasis from concepts to clinical trials. Br Med Bull 62:139–148

Caruthers MH (1985) Gene synthesis machines: DNA chemistry and its uses. Science 230:281–285

Cecchelli R, Dehouck B, Descamp L, Fenart L, Buée-Scherrer V, Duhem C, Torpier G, Dehouck MP (2000) In vitro models of the blood-brain barrier and their use in drug development. In: The blood-brain barrier and drug delivery to the CNS. Eds. Begley DJ, Bradbury MW, Kreuter J. pp 65–75

Conn GL, Draper DE (1998) RNA structure. Curr Opin Struct Biol 8:278–285
Cross GAM (1975) Identification, purification and properties of clone-specific glycoprotein antigens constituting the surface coat of *Trypanosoma brucei*. Parasitology 71:393–417
Dias JC, Silveira AC, Schofield CJ (2002) The impact of Chagas disease control in Latin America: a review. Mem Inst Oswaldo Cruz 97:603–612
Donelson JE (2003) Antigenic variation and the African trypanosome genome. Acta Trop 85:391–404
Eaton BE, Pieken WA (1995) Ribonucleosides and RNA. Annu Rev Biochem 64:837–863
Eccleston ME, Kuiper M, Gilchrist FM, Slater NK (2000) pH-responsive pseudo-peptides for cell membrane disruption. J Control Release 69:297–307
Ellington AD, Szostak JW (1990) In vitro selection of RNA molecules that bind specific ligands. Nature 346:818–822
Enanga B, Burchmore RJS, Stewart ML, Barrett MP (2002) Sleeping sickness and the brain. Cell Mol Life Sci 59:845–858
Fairlamb AH (2003) Chemotherapy of human African trypanosomiasis: current and future prospects. Trends Parasitol 19:488–494
Fine SL, Martin DF, Kirkpatrick P (2005) Pegaptanib sodium. Nat Rev Drug Discov 4:187–8
Franke H, Galla HJ, Beuckmann CT (1999). An improved low-permeability in vitro model of the blood-brain barrier: transport studies on retinoids, sucrose, haloperidol, caffeine and mannitol. Brain Res 818:65–71
Franke H, Galla HJ, Beuckmann CT (2000). Primary cultures of brain microvessel endothelial cells: a valid and flexible model to study drug transport through the blood-brain barrier in vitro. Brain Res Protocols 5:248–256
Freymann D, Down J, Carrington M, Roditi I, Turner M, Wiley D. (1990) 2.9 Å resolution structure of the N-terminal domain of a variant surface glycoprotein from *Trypanosoma brucei*. J Mol Biol 216:141–460
Fitzwater T, Polisky B (1996). A SELEX primer. Methods Enzymol 267:275–301
Gaillard PJ, Voorwinden LH, Nielsen JL, Ivanov A, Atsumi R, Engman H, Ringbom C, de Boer AG, Breimer DD (2001). Establishment and functional characterization of an in vitro model of the blood-brain barrier, comprising a co-culture of brain capillary endothelial cells and astrocytes. Eur J Pharm Sci 12:215–222
Gold L (1995) Oligonucleotides as research, diagnostic, and therapeutic agents. J Biol Chem 270:13581–13584
Gragoudas ES, Adamis AP, Cunningham ET Jr, Feinsod M, Guyer DR (2004) Pegaptanib for neovascular age-related macular degeneration. N Engl J Med 351:2805–2816
Green LS, Bell C, Janjic N (2001) Aptamers as reagents for high-throughput screening. Biotechniques 30:1094–1110
Healy JM, Lewis SD, Kurz M, Boomer RM, Thompson KM, Wilson C, McCauley TG (2004) Pharmacokinetics and biodistribution of novel aptamer compositions. Pharm Res 21:2234–2246
Hesselberth J, Robertson MP, Jhaveri S, Ellington AD (2000) In vitro selection of nucleic acids for diagnostic applications. J Biotechnol 74:15–25
Homann M, Göringer HU (1999) Combinatorial selection of high affinity RNA ligands to live African trypanosomes. Nucl Acids Res 27:2006–2014
Homann M, Göringer HU (2001) Uptake and intracellular transport of RNA aptamers in African trypanosomes suggest therapeutic "piggy-back" approach. Bioorg Med Chem 9:2571–2580
Jayasena SD (1999) Aptamers: an emerging class of molecules that rival antibodies in diagnostics. Clin Chem 45:1628–1650

Johnson JG, Cross GA (1979) Selective cleavage of variant surface glycoproteins from *Trypanosoma brucei*. Biochem 178:689–697

Kaiser A, Gottwald A, Wiersch C, Maier W, Seitz HM (2002) The necessity to develop drugs against parasitic diseases. Pharmazie 57:723–728

Khaw M, Panosian CB (1995) Human antiprotozoal therapy: past, present, and future. Clin Microbiol Rev 8:427–439

Kioy D, Jannin J, Mattock N (2004) Human African trypanosomiasis. Nat Rev Microbiol 2:186–187

Lam KS, Renil M (2002) From combinatorial chemistry to chemical microarray. Curr Opin Chem Biol 6:353-358

Lebruska LL, Maher LJ 3rd (1999) Selection and characterization of an RNA decoy for transcription factor NF-kappa B. Biochemistry 38:3168–3174

Londsdale-Eccles JD, Grab DJ (2002) Trypanosome hydrolases and the blood-brain barrier. Trends Parasit 18:17–19

Lorger M, Engstler M, Homann M, Göringer HU (2003) Targeting the variable surface of African trypanosomes using VSG-specific, serum-stable RNA aptamers. Eukaryotic Cell 2:84–94

Mann MJ, Dzau VJ (2000) Therapeutic applications of transcription factor decoy oligonucleotides. J Clin Invest 106:1071–1075

Michiels F, Matthyssens G, Kronenberger P, Pays E, Dero B, Van Assel S, Darville M, Cravador A, Steinert M, Hamers R (1983) Gene activation and re-expression of a *Trypanosoma brucei* variant surface glycoprotein. EMBO J 2:1185–1192

Morris KN, Jensen KB, Julin CM, Weil M, Gold L (1998) High affinity ligands from in vitro selection: complex targets. Proc Natl Acad Sci USA 95:2902–2907

Mulenga C, Mhlanga JD, Kristensson K, Robertson B (2001) Trypanosoma brucei brucei crosses the blood-brain barrier while tight junction proteins are preserved in a rat chronic disease model. Neuropathol Appl Neurobiol. 27:77–85

Nimjee SM, Rusconi CP, Sullenger BA (2005) Aptamers: an emerging class of therapeutics. Annu Rev Med 56:187–188

Pecoul B, Chirac P, Trouiller P, Pinel J (1999) Access to essential drugs in poor countries: a lost battle? JAMA 281:361–367

Pieken WA, Olsen DB, Benseler F, Aurup H, Eckstein F (1991) Kinetic characterization of ribonuclease-resistant 2′-modified hammerhead ribozymes. Science 253:314–317

Pieken W (1997) Efficient process technologies for the preparation of oligonucleotides. Ciba Found Symp 209:218–222

Reich MR, Govindaraj R (1998) Dilemmas in drug development for tropical diseases. Experiences with praziquantel. Health Policy 44:1–18

Rice-Ficht AC, Chen KK, Donelson JE (1982) Point mutations during generation of expression-linked extra copy of trypanosome surface glycoprotein gene. Nature 298:676–679

Rudenko G, Cross M, Borst P (1998) Changing the end: antigenic variation orchestrated at the telomeres of African trypanosomes. Trends Microbiol 6:113–116

Smith DH, Pepin J, Stich AHR (1998) Human African trypanosomiasis: an emerging public health crisis. Brit Med Bull 54:341–355

Stephenson I, Wiselka M (2000) Drug treatment of tropical parasitic infections: recent achievements and developments. Drugs 60:985–995

Tonge SR, Tighe BJ (2001) Responsive hydrophobically associating polymers: a review of structure and properties. Adv Drug Deliv Rev 53:109–122

Trouiller P, Torreele E, Olliaro P, White N, Foster S, Wirth D, Pecoul B (2001) Drugs for neglected diseases: a failure of the market and a public health failure? Trop Med Int Health 6:945–951

Tuerk C, Gold L (1990) Systematic evolution of ligands by exponential enrichment: RNA ligands to bacteriophage T4 DNA polymerase. Science 249:505–510

Ulrich H, Alves MJ, Colli W (2001) RNA and DNA aptamers as potential tools to prevent cell adhesion in disease. Braz J Med Biol Res 34:295–300

Urakawa T, Eshita Y, Majiwa PA (1997) The primary structure of *Trypanosoma (Nannomonas) congolense* variant surface glycoproteins. Exp Parasit 85:215–224

Van Gompel A, Vervoort T (1997) Chemotherapy of leishmaniasis and trypanosomiasis. Curr Opin Infect Dis 10:469–474

Watson SR, Chang YF, O'Connell D, Weigand L, Ringquist S, Parma DH (2000) Anti-L-selectin aptamers: binding characteristics, pharmacokinetic parameters, and activity against an intravascular target in vivo. Antisense Nucleic Acid Drug Dev 10:63–75

White RR, Sullenger BA, Rusconi CP (2000) Developing aptamers into therapeutics. J Clin Invest 106:929–934

Wilson DS, Szostak JW (1999) In vitro selection of functional nucleic acids. Annu Rev Biochem 68:611–647

RNA Targeting Using Peptide Nucleic Acid

E. Nielsen

Department of Medical Biochemistry and Genetics, University of Copenhagen, The Panum Institute, Blegdamsvej 3c, 2200 Copenhagen N, Denmark
pen@imbg.ku.dk

1	Introduction	395
2	mRNA Targets	396
3	Telomerase	398
4	RNA Viruses	398
5	Cellular Delivery	399
6	In Vivo Bioavailability	400
7	Conclusion/Prospects	401
	References	401

Abstract The efforts towards peptide nucleic acid (PNA) drug discovery using cellular RNAs as molecular targets is briefly reviewed, with special emphasis on recent developments. Special attention is given to cellular delivery in vivo bioavailability and the possibilities of using PNA oligomers to (re)direct alternative splicing of pre-messenger (m)RNA.

Keywords Peptide nucleic acid (PNA) · Antisense · Drug discovery · Cellular delivery · Bioavailability

1
Introduction

Peptide nucleic acids (PNAs) were introduced in 1991 as a DNA mimic based on a pseudopeptide backbone (Nielsen et al. 1991; Fig. 1). It was immediately realized that this type of molecule could have multiple applications within molecular biology and drug discovery due to its unique chemical and structural properties, not least its ability to bind sequence complementary single-stranded RNA (and DNA) with high affinity and sequence specificity. Since the discovery of PNA, a large number of derivatives and analogues have been synthesized and characterized (Ganesh and Nielsen 2000), but it is still almost exclusively the original aminoethylglycine (aeg) PNA that is used for biological experiments.

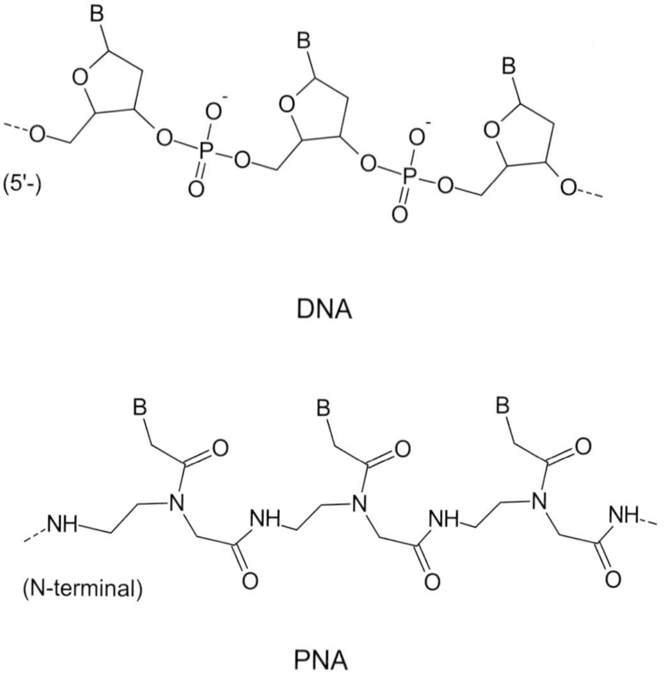

Fig. 1 Chemical structure of PNA compared to DNA

For more than the past 10 years, PNA oligomers have been studied intensely as sequence-specific inhibitors of RNA and DNA function with one of the major aims being the discovery of novel medical drugs. The present chapter will focus on RNA as a target exemplified by messenger (m)RNA, viral RNA and telomerase RNA, and the crucial issues of cellular delivery and in vivo bioavailability will also be discussed.

2
mRNA Targets

PNA can target RNA with high sequence specificity and affinity (Jensen et al. 1997). However, analogously to the situation with other antisense agents, only certain regions of any specific mRNA is accessible for hybridization in vivo. Furthermore, PNA–RNA hybrids are not substrates for any known enzymes (such as RNase H), and therefore any inhibition of the mRNA function presumably relies on direct steric blocking of mRNA function—such as translation initiation or elongation—or mRNA processing, most notably splicing. Consequently, some type of "gene-walk" is typically needed to identify sensitive mRNA sequence targets and definitely required to optimize the targets (e.g. Doyle et al. 2001; Dryselius et al. 2003; Siwkowski et al. 2004).

Many examples from a variety of systems in terms of genes and cell types have demonstrated that PNA blocking of both translation initiation when targeting the AUG start site or regions upstream from this as well as arrest of elongating ribosomes when targeting intra gene regions is feasible and effective using PNA oligomers ranging from 10–18 nucleobases (e.g. Koppelhus et al. 2002). In particular from in vitro cell-free experiments, the evidence suggests, however, that either very high affinity PNAs (e.g. triplex forming PNAs binding to homopurine targets) or special mRNA regions are required for effective translation arrest (Knudsen and Nielsen 1996; Sénamaud-Beaufort et al. 2003). On the other hand, indirect evidence (i.e. no detection of truncated protein product) for in vitro cell culture studies does indicate antisense effects of PNA targeting intragene regions (e.g. Shiraishi and Nielsen 2004) thereby suggesting a translation arrest mechanism.

Although some studies have concluded that PNA antisense targeting might rival short interfering (si)RNA techniques in potency (Liu et al. 2004), PNA may well find its greatest potential as a sequence specific modulator of splicing. Almost all eukaryotic mRNAs mature through a splicing process, and recent evidence suggests that alternative splicing is exploited extensively by Nature for regulation of gene expression. Indeed, a large variety of diseases are rooted in incorrect or defective mRNA splicing and may therefore eventually be treated using splicing correction agents (Cartegni and Krainer 2003; Kole et al. 2004). Correct splicing involves proper recognition of splice sites by the spliceosome, and incorrect splicing may arise when mutations create novel cryptic splice sites in the mRNA that are inadvertently utilized by the spliceosome. Steric

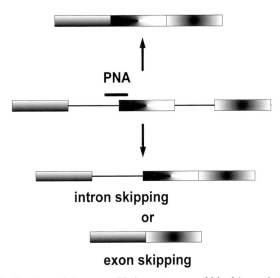

Fig. 2 Schematic drawing of the most likely outcomes of blocking a intron–exon splice junction of an mRNA

blocking of intro-exon junctions (splice sites) by non-RNase H active antisense agents may favour specific or novel splice variants of the mRNA (Fig. 2) or may block mutant cryptic splice sites, restoring natural splicing. Several recent reports have demonstrated the effectiveness of PNA oligomers as splice modulators both in vitro in cell culture as well as in vivo (Siwkowski et al. 2004; Sazani et al. 2002). Thus, by rational choice of splice target, it should be possible to exploit this technology to correct aberrant splicing (Sazani et al. 2002), augment beneficial splice variants of a gene and create novel, not naturally occurring splice variants of proteins (e.g. by "cutting out" certain catalytic or regulatory domains), as well as simply inhibiting expression of a functional protein products by inhibiting correct splicing altogether (Siwkowski et al. 2004).

3
Telomerase

Telomerase seems an obvious candidate for "antisense" targeting because this enzyme carries within it a template RNA for telomere synthesis. Indeed, it has been found that telomerase is very sensitive to inhibition by PNA (and other antisense agents) targeting this RNA (Shammas et al. 2004; Hamilton et al. 1999; Folini et al. 2003), and that such anti-telomerase PNAs decrease the viability of cancer cells in culture. Furthermore, it was recently found that photochemical internalization of unmodified anti-telomeraser PNAs significantly increases their potency, thereby opening the possibilities of an eventual drug/photodynamic combination therapy (Folini et al. 2003). It is hoped that novel anticancer agents may be discovered amongst such anti-telomerase PNAs.

4
RNA Viruses

The genome of RNA viruses has also proved a sensitive target for PNA. Whereas the ribosome is a very robust cellular machinery that apparently has evolved to unravel folded mRNA structures and therefore is usually not severely affected by bound antisense agents, reverse transcriptase is very sensitive to blockage by e.g. PNA bound to the virus RNA template (Chaubey et al. 2005). There may be good opportunity to exploit this finding in the discovery of novel drugs in the combat of acquired immunodeficiency syndrome (AIDS). Hepatitis B and C viruses have also been successfully targeted at the RNA level using PNA antisense in vitro and in cell culture systems (Robaczewska et al. 2005; Nulf and Corey 2004).

5
Cellular Delivery

It was clear from the onset of PNA research that cellular delivery was not trivial, as the PNA molecule is a largely hydrophilic, "non-small" molecule (typically with a molecular weight of 2,000–3,000 Da for antisense use). Therefore it is no surprise that simple PNA oligomers do not passively cross the lipid bilayer membrane of living cells (Wittung et al. 1995). It is in some cases possible to attain biological cellular effects of unmodified PNAs, but extremely high concentrations are then required (20–100 µM; Kaushik et al. 2002). Consequently, a delivery vehicle or system is desired for effective cellular delivery of antisense PNA, and a large variety of such agents and systems is now available. Broadly, they may be divided into two categories that depend on whether or not they require cationic lipids (liposomes). Cationic liposomes constitute the effective delivery vehicle of choice for anionic oligonucleotides, such as phosphodiester antisense agents and siRNA, but they are not effective for PNA delivery because these oligomers are not anionic, and therefore do not "self-assemble" with the cationic lipids. However, by forming a quasi-stable duplex with a partly complementary DNA oligonucleotide, quite effectively deliver the PNA–DNA complex into the cell, where the PNA following dissociation from the DNA may hybridize to the RNA (or DNA) target (Hamilton et al. 1999). This type of PNA delivery has been used extensively, in particular by the Corey group (e.g. Doyle et al. 2001; Liu et al. 2004). PNA oligomers conjugated to fatty acids can also be delivered via cationic lipids, but the conjugates have poor aqueous solubility and tend to aggregate (Ljungstrøm et al. 1999). However, it was recently discovered that PNA acridine conjugates are quite efficiently delivered by cationic lipids (Shiraishi and Nielsen 2004). Due to the protonation of the acridine at neutral pH, these compounds do not in general suffer from the severe solubility problems exhibited by the fatty acid conjugates.

A larger number of especially cationic peptides have been conjugated to PNA oligomers and reported to enhance "spontaneous" uptake significantly or even dramatically (Koppelhus et al. 2002; Lundberg and Langel 2003; Kilk et al. 2004; Kaihatsu et al. 2004). Originally it was reported that peptides such as pTat (RQIKIWFQNRRMKWKK, part of the Tat protein of HIV), pAntp (GRKKRRQRRRPPQ, part of the Drosophila antennapedia transcription factor and oligo arginine (e.g. Arg_9)) are taken up by cells via an energy-independent and non-carrier-mediated pathway. However, accumulating recent evidence clearly and unanimously points to the conclusion that the major port of cellular entry by these peptides is the endosomal/lysosomal pathway (Koppelhus et al. 2002; Gait 2003). Therefore, such PNA-peptide conjugates are indeed taken up effectively by the cells, but the majority of the material remains inactive in the endosomes/lysosomes from which they are only slowly and poorly released into the cytoplasm. Eventually most of the material is discarded from the cells via the lysosomal excretion. Consequently, it should be possible to increase the potency of such conjugates significantly by agents that are known to disrupt endosomes and/or lysosomes, such as chloroquine or certain photosensitizers (Folini et al. 2003). Specific cellular targeting is also possible using e.g. cancer cell-specific peptides (Mier et al. 2003) or other ligands, such as carbohydrates, that bind to specific cell receptors (van Rossenberg et al. 2003; Hamzavi et al. 2003). However, although it has not been demonstrated, this type of delivery is also expected to occur via endocytosis. Most interestingly, it was also recently reported that organic, lipophilic cations such as triphenylphosphonium are also facilitating cellular uptake when conjugated to PNA (Filipovska et al. 2004). Finally, it is noteworthy that a quite simple synthetic cationic peptide [(KFF)$_3$K] rather effectively delivers short PNAs (10–12 bases) to bacterial cells (*Escherichia coli* and to a certain extent also to *Staphylococcus aureus*) (Good et al. 2001; Nekhotiaeva et al. 2004). Using such systems, it may be possible to develop novel antibacterial drugs.

Nonetheless—even with the less-than-perfect delivery systems discovered to date and developed for PNA as described above—it is indeed possible to perform efficacy studies in cells in culture in order to optimize the gene and sequence target (gene screening and gene walk), and typically sub- to low micromolar concentrations of PNA are required to observe (molecular) biological effects.

6
In Vivo Bioavailability

Only a few studies have convincingly and supported by molecular biology evidence demonstrated gene-specific effects of PNA in animals. By far the most informative study was published by Sazani et al., who used a transgenic mouse carrying a green fluorescent protein (GFP) splice variant that can be corrected by antisense (Sazani et al. 2002). Thus tissues and cells, which the antisense agent is reaching, will activate GFP. This study unequivocally showed that systemically delivered PNA conjugated to four lysines did indeed exhibit antisense effects in a variety of tissues (e.g. kidney, liver, lung, muscle and heart), whereas unmodified PNA did not. This is in accordance with other preliminary pharmacokinetic studies showing that unmodified PNA is very quickly excreted through the kidneys (Kristensen 2002; Hamzavi et al. 2003). Thus, the limited information available so far clearly indicates that in order to improve in vivo bioavailability some kind of chemical modification to or formulation of the PNA seems to be necessary. This apparently may be accomplished by some cationic peptides [K$_4$ or (KFF)$_3$K] (Sazani et al. 2002; Kristensen 2002), but is also possible through cell- or tissue-specific targeting as illustrated by *N*-acetyl-galactosamine (GalNAc) modification for effective liver targeting (Hamzavi et al. 2003).

7
Conclusion/Prospects

Realistically, PNA medical drugs are still not immediately "in sight", although the number of encouraging biological results are steadily increasing, and promising new sensitive targets such as telomeric RNA, viral RNA, bacterial RNA and mRNA splice junctions are being discovered. Furthermore, it will be exciting to see efficacy and biological effects that the targets of microRNAs may bring. Clearly, however, the next major challenge is developing in vivo delivery methods and sufficient in vivo bioavailability for good in vivo efficacy of PNA oligomers. PNAs are born with extremely high biostability, but methods that provide wide tissue or, alternatively, very specific tissue distribution in combination with slow clearance are required. Hopefully the "chemical versatility" of PNA will underlie fast and effective solutions to these challenges.

References

Cartegni L, Krainer AR (2003) Correction of disease-associated exon skipping by synthetic exon-specific activators. Nat Struct Biol 10:120–125

Chaubey B, Tripathi S, Ganguly S, Harris D, Casale RA, Pandey VN (2005) A PNA-transportan conjugate targeted to the TAR region of the HIV-1 genome exhibits both antiviral and virucidal properties. Virology 331:418–428

Doyle DF, Braasch DA, Simmons CG, Janowski BA, Corey DR (2001) Inhibition of gene expression inside cells by peptide nucleic acids: effect of mRNA target sequence, mismatched bases, and PNA length. Biochemistry 40:53–64

Dryselius R, Aswasti SK, Rajarao GK, Nielsen PE, Good L (2003) The translation start codon region is sensitive to antisense PNA inhibition in Escherichia coli. Oligonucleotides 13:427–433

Filipovska A, Eccles MR, Smith RAJ, Murphy MP (2004) Delivery of antisense peptide nucleic acids (PNAs) to the cytosol by disulphide conjugation to a lipophilic cation. FEBS Lett 556:180–186

Folini M, Berg K, Millo E, Villa R, Prasmickaite L, Daidone MG, Benatti U, Zaffaroni N (2003) Photochemical internalization of a peptide nucleic acid targeting the catalytic subunit of human telomerase. Cancer Res 63:3490–3494

Gait MJ (2003) Peptide-mediated cellular delivery of antisense oligonucleotides and their analogues. Cell Mol Life Sci 60:844–853

Ganesh KN, Nielsen PE (2000) Peptide nucleic acids: analogs and derivatives. Curr Org Chem 4:931–943

Good L, Awasthi SK, Dryselius R, Larsson O, Nielsen PE (2001) Bactericidal antisense effects of peptide-PNA conjugates. Nat Biotechnol 19:360–364

Hamilton SE, Simmons CG, Kathiriya IS, Corey DR (1999) Cellular delivery of peptide nucleic acids and inhibition of human telomerase. Chem Biol 6:343–351

Hamzavi R, Dolle F, Tavitian B, Dahl O, Nielsen PE (2003) Modulation of the pharmacokinetic properties of PNA: preparation of galactosyl, mannosyl, fucosyl, N-acetylgalactosaminyl, and N-acetylglucosaminyl derivatives of aminoethylglycine peptide nucleic acid monomers and their incorporation into PNA Oligomers. Bioconjug Chem 14:941–954

Jensen KK, Ørum H, Nielsen PE, Nordén B (1997) Kinetics for hybridization of peptide nucleic acids (PNA) with DNA and RNA studied with the BIAcore technique. Biochemistry 36:5072–5077

Kaihatsu K, Huffman KE, Corey DR (2004) Intracellular uptake and inhibition of gene expression by PNAs and PNA-peptide conjugates. Biochemistry 43:14340–14347

Kaushik N, Basu A, Pandey VN (2002) Inhibition of HIV-1 replication by anti-transactivation responsive polyamide nucleotide analog. Antiviral Res 56:13–27

Kilk K, Elmquist A, Saar K, Pooga M, Land T, Bartfai T, Soomets U, Langel Ü (2004) Targeting of antisense PNA oligomers to human galanin receptor type 1 mRNA. Neuropeptides 38:316–324

Knudsen H, Nielsen PE (1996) Antisense properties of duplex- and triplex-forming PNAs. Nucleic Acids Res 24:494–500

Kole R, Vacek M, Williams T (2004) Modification of alternative splicing by antisense therapeutics. Oligonucleotides 14:65–74

Koppelhus U, Awasthi SK, Zachar V, Holst HU, Ebbesen P, Nielsen PE (2002) Cell-dependent differential cellular uptake of PNA, peptides, and PNA-peptide conjugates. Antisense Nucleic Acid Drug Dev 12:51–63

Kristensen E (2002) In vitro and in vivo studies on pharmacokinetics and metabolism of PNA constructs in rodents. In: Nielsen PE (ed) Peptide nucleic acids: methods and protocols. Humana Press, Totowa, pp 259–269

Liu Y, Braasch DA, Nulf CJ, Corey DR (2004) Efficient and isoform-selective inhibition of cellular gene expression by peptide nucleic acids. Biochemistry 43:1921–1927

Ljungstrøm T, Knudsen H, Nielsen PE (1999) Cellular uptake of adamantyl conjugated peptide nucleic acids. Bioconjug Chem 10:965–972

Lundberg P, Langel Ü (2003) A brief introduction to cell-penetrating peptides. J Mol Recognit 16:227–233

Mier W, Eritja R, Mohammed A, Haberkorn U, Eisenhut M (2003) Peptide-PNA conjugates: targeted transport of antisense therapeutics into tumors. Angew Chem Int Ed Engl 42:1968–1971

Nekhotiaeva N, Awasthi SK, Nielsen PE, Good L (2004) Inhibition of Staphylococcus aureus gene expression and growth using antisense peptide nucleic acids. Mol Ther 10:652–659

Nielsen PE, Egholm M, Berg RH, Buchardt O (1991) Sequence-selective recognition of DNA by strand displacement with a thymine-substituted polyamide. Science 254:1497–1500

Nulf CJ, Corey D (2004) Intracellular inhibition of hepatitis C virus (HCV) internal ribosomal entry site (IRES)-dependent translation by peptide nucleic acids (PNAs) and locked nucleic acids (LNAs). Nucleic Acids Res 32:3792–3798

Robaczewska M, Narayan R, Seigneres B, Schorr O, Thermet A, Podhajska AJ, Trepo C, Zoulim F, Nielsen PE, Cova L (2005) Sequence-specific inhibition of duck hepatitis B virus reverse transcription by peptide nucleic acids (PNA). J Hepatol 42:180–187

Sazani P, Gemignani F, Kang S-H, Maier MA, Manoharan M, Persmark M, Bortner D, Kole R (2002) Systemically delivered antisense oligomers upregulate gene expression in mouse tissues. Nat Biotechnol 20:1228–1233

Sénamaud-Beaufort C, Leforestier E, Saison-Behmoaras TE (2003) Short pyrimidine stretches containing mixed base PNAs are versatile tools to induce translation elongation arrest and truncated protein synthesis. Oligonucleotides 13:465–478

Shammas MA, Liu XH, Gavory G, Raney KD, Balasubramanian S, Reis RJS (2004) Targeting the single-strand G-rich overhang of telomeres with PNA inhibits cell growth and induces apoptosis of human immortal cells. Exp Cell Res 295:204–214

Shiraishi T, Nielsen PE (2004) Down-regulation of MDM 2 and activation of p53 in human cancer cells by antisense 9-aminoacridine-PNA (peptide nucleic acid) conjugates. Nucleic Acids Res 32:4893–4902

Siwkowski AM, Malik L, Esau CC, Maier MA, Wancewicz EV, Albertshofer K, Monia BP, Bennett CF, Eldrup AB (2004) Identification and functional validation of PNAs that inhibit murine CD40 expression by redirection of splicing. Nucleic Acids Res 32:2695–2706

van Rossenberg SMW, Sliedregt-Bol KM, Prince P, Van Berkel TJC, Van Boom JH, Van der Marel GA, Biessen EAL (2003) A targeted peptide Nucleic acid to down-regulate mouse microsomal triglyceride transfer protein expression in hepatocytes. Bioconjug Chem 14:1077–1082

Wittung P, Kajanus J, Edwards K, Nielsen P, Nordén B, Malmström BG (1995) Phospholipid membrane permeability of peptide nucleic acid. FEBS Lett 365:27–29

Locked Nucleic Acid: High-Affinity Targeting of Complementary RNA for RNomics

S. Kauppinen[1] · B. Vester[2] · J. Wengel[3] (✉)

[1] Department of Functional Genomics, Exiqon, Bygstubben 9, 2950 Vedbaek, Denmark

[2] Nucleic Acid Center, Department of Biochemistry and Molecular Biology, University of Southern Denmark, 5230 Odense M, Denmark

[3] Nucleic Acid Center, Department of Chemistry, University of Southern Denmark, 5230 Odense M, Denmark
jwe@chem.sdu.dk

1	Introduction	406
1.1	LNA: Hybridization Properties and Structural Features	407
1.2	Susceptibility of LNA to Nucleases	408
2	RNA Targeting by LNAzymes	410
3	LNA Antisense	411
4	LNA-Modified siRNA	413
5	Transcriptome Analysis Facilitated by LNA	414
6	MicroRNA Detection Using LNA Probes	415
7	Future Prospects	416
	References	417

Abstract Locked nucleic acid (LNA) is a nucleic acid analog containing one or more LNA nucleotide monomers with a bicyclic furanose unit locked in an RNA-mimicking sugar conformation. This conformational restriction is translated into unprecedented hybridization affinity towards complementary single-stranded RNA molecules. That makes fully modified LNAs, LNA/DNA mixmers, or LNA/RNA mixmers uniquely suited for mimicking RNA structures and for RNA targeting in vitro or in vivo. The focus of this chapter is on LNA antisense, LNA-modified DNAzymes (LNAzymes), LNA-modified small interfering (si)RNA (siLNA), LNA-enhanced expression profiling by real-time RT-PCR and detection and analysis of microRNAs by LNA-modified probes.

Keywords LNA · Locked nucleic acid · RNA targeting · LNA antisense · MicroRNA targeting

1
Introduction

The expanding inventory of sequence databases and the concomitant sequencing of more than 200 genomes representing all three domains of life—bacteria, archaea, and eukaryotes—have been the primary drivers in the process of deconstructing living organisms into comprehensive molecular catalogs of genes, transcripts, and proteins. The importance of the genetic variation within a single species has become apparent, extending beyond the completion of genetic blueprints of several important genomes, and a worldwide effort culminated in the publication of the human genome sequence in 2001 (Lander et al. 2001; Venter et al. 2001; Sachidanandam et al. 2001). Also, the increasing number of detailed, large-scale molecular analyses of transcription originating from the human and mouse genomes along with the recent identification of several types of non-protein-coding (nc)RNAs, such as small nucleolar RNAs, small interfering (si)RNAs, microRNAs (miRNAs), and antisense RNAs, indicates that the transcriptomes of higher eukaryotes are much more complex than originally anticipated (Wong et al. 2001; Kampa et al. 2004).

As a result of the central dogma—DNA makes RNA and RNA makes protein—RNAs have been considered as simple molecules that just translate the genetic information into protein. Recently, it has been estimated that although most of the genome is transcribed, almost 97% of the genome does not encode proteins in higher eukaryotes, but putative, non-coding RNAs (Wong et al. 2001). The ncRNAs appear to be particularly well suited for regulatory roles that require highly specific nucleic acid recognition. Therefore, the view of RNA is rapidly changing from a merely informational molecule to one that comprises a wide variety of structural, informational, and catalytic molecules in the cell.

The challenges of establishing genome function and understanding the layers of information hidden in the complex transcriptomes of higher eukaryotes call for novel, improved technologies for detection, quantification, and functional analysis of RNA molecules in complex nucleic acid samples. Locked nucleic acid (LNA) constitutes a new class of bicyclic high-affinity RNA analogs in which the furanose ring of the ribose sugar is chemically locked in an RNA-mimicking conformation by the introduction of a $O2',C4'$-methylene bridge (Koshkin et al. 1998b; Obika et al. 1998). Several studies have demonstrated that LNA-modified oligonucleotides exhibit unprecedented thermal stability when hybridized with their DNA and RNA target molecules (Koshkin et al. 1998b; Obika et al. 1998; Braasch and Corey 2001; Jacobsen et al. 2004; Petersen and Wengel 2003). Consequently, an increase in melting temperature (T_m value) of +1 to +8 °C per introduced LNA monomer against complementary DNA and of +2 to +10 °C per LNA monomer against complementary RNA compared to unmodified duplexes has been reported. The first sections of this chapter describe some basic properties of LNA, whereas the latter sections are focused on recent contributions in LNA-mediated RNA targeting in vitro and in vivo. For more general descriptions on the overall properties and biotechnological applications of LNA, including DNA diagnostics and double-stranded (ds)DNA targeting by triple helix formation, the reader is referred to recent reviews on LNA (Petersen and Wengel 2003; Jepsen and Wengel 2004; Vester and Wengel 2004).

1.1
LNA: Hybridization Properties and Structural Features

We have defined LNA as an oligonucleotide containing at least one LNA monomer, i.e., one 2′-O,4′-C-methylene-β-D-ribofuranosyl nucleotide (see Fig. 1; Singh et al. 1998; Koshkin et al. 1998b). The LNA monomers adopt N-type sugar puckers, also termed C3′-endo conformations (Singh et al. 1998; Obika et al. 1997). The vast majority of reports on LNA-mediated RNA targeting have dealt with mixmer LNA oligonucleotides, i.e., LNAs containing a limited number of LNA monomers in combination with other types of monomers, such as DNA, RNA, and 2′-O-Me-RNA monomers. We predict that this approach will also be important for future uses of LNA due to the following characteristics of LNA/DNA, LNA/RNA, and LNA/2′-O-Me-RNA mixmers: (1) synthetic convenience—standard methods for DNA oligomerization can be used for commercially available LNA, DNA and 2′-O-Me-RNA phosphoramidite building blocks (Pfundheller et al. 2005); (2) easy access—LNAs (fully modified or mixmers) are commercially available; (3) high-affinity and sequence-selective targeting of RNA molecules in vitro or in vivo (Singh et al. 1998; Koshkin et al. 1998b; Obika et al. 1998; Wengel 1999; Kumar et al. 1998; Braasch and Corey 2001); and (4) compatibility with standard modifiers and modifications, e.g., phosphorothioate linkages (Kumar et al. 1998).

A key feature of oligonucleotides containing LNA nucleotides—i.e., fully modified LNAs, LNA/DNA mixmers, LNA/RNA mixmers, etc.—is the very high thermal stability of duplexes towards complementary RNA or DNA (Singh et al. 1998; Koshkin et al. 1998b; Obika et al. 1998; Wengel 1999; Kumar et al. 1998; Braasch and Corey 2001). Notably, this increase in affinity goes hand in hand with preserved, or even improved, Watson-Crick base

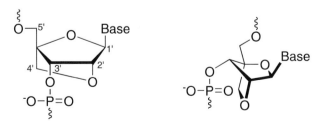

Fig. 1 The structure of locked nucleic acid (LNA) monomers

Duplexes	Melting temperature (°C)
5'-TTTTTT : *AAAAAA*	<10
5'-**TTTTT**T : *AAAAAA*	40
5'-GTGATATGC : 3'-*CACUAUACG*	28
5'-G**T**GA**T**A**T**GC : 3'-*CACUAUACG*	58
5'-**GTGATATGC** : 3'-*CACUAUACG*	74
5'-Biotin-T$_{20}$ (DNA oligo) : A_{20}	40
5'-Biotin-(**T**T)$_{10}$ (LNA_2.T): A_{20}	71
5'-Biotin-**T**$_{20}$ (LNA full) : A_{20}	>95

Fig. 2 Some examples of melting temperatures (T_m values) for hybridization of LNA and DNA oligonucleotides to complementary RNA sequences (Singh et al. 1998, Jacobsen et al. 2004). LNA monomers are shown in *capital boldface underlined letters*, DNA monomers in *capital letters*, and RNA monomers in *capital italics letters*

pairing selectivity. Examples of melting temperatures for LNAs complexed with RNA are shown in Fig. 2. LNA:LNA base pairing is also very strong (Koshkin et al. 1998a) and should be considered in relation to the risk of self-complementarity when designing LNAs for biophysical or biological experiments (see www.exiqon.com for LNA design tools). In general, the largest affinity increase per LNA monomer, and the optimum mismatch discrimination, is achieved for short oligonucleotides with a single or with several dispersed centrally positioned LNA monomers.

The structures of LNA:RNA duplexes have been studied by NMR spectroscopy showing similarities with the native nucleic acid duplexes, i.e., Watson-Crick base pairing, *anti* orientation of the nucleobases, base stacking, and a right-handed double helix conformation. A study including three 9-mer LNA/DNA:RNA hybrids, in which the LNA/DNA mixmer strand contained one, three, and nine LNA nucleotides, shows an increasing A-like character of the hybrids upon an increase in the LNA content of the LNA strand (Petersen et al. 2002; Nielsen et al. 2004). Thus, both the fully modified LNA:RNA duplex and the duplex with three LNA modifications were shown to adopt near canonical A-type duplex geometry, indicating that the LNA nucleotides tune the DNA nucleotides in the LNA strand conformationally to attain *N*-type sugar puckers (Petersen et al. 2002). An analogous, but localized, effect was observed in the LNA/DNA:RNA duplex with one LNA modification, where the LNA nucleotide was shown to perturb the sugar puckers of the neighboring DNA nucleotides, predominantly in the 3'-direction of the LNA nucleotide (Petersen et al. 2002).

1.2
Susceptibility of LNA to Nucleases

The basic biophysical properties of LNA make it attractive for diagnostic and therapeutic applications. This implies contact with various media containing nucleases such as serum or cellular media. In general, it is considered an advantage if modified oligonucleotides are resistant to degradation by nucleases. However, one exception is RNase H-mediated cleavage of the RNA target strand of a heteroduplex formed with an antisense oligonucleotide. As only very few fully modified oligonucleotides support RNase H-mediated cleavage, the so-called gapmer strategy using oligonucleotides composed of modified segments flanking a central DNA (or phosphorothioate DNA) segment that is compatible with RNase H activity is often used. In an early study, Wahlestedt et al. (2000) found that both an LNA/DNA/LNA gapmer, with a 6-nt DNA gap, and an LNA/DNA mixmer, with 6 DNA and 9 LNA nucleotides interspersed, elicited RNase H activity. However, no RNase H-mediated cleavage was observed with a fully modified 11-mer LNA or with an 11-mer LNA/DNA mixmer (Sørensen et al. 2002). Kurreck et al. (2002) investigated various LNA/DNA mixmers and gapmers and found that a gap of six DNA nucleotides is necessary for noteworthy RNase H activity, and that a gap of seven DNA nucleotides allows complete RNase H activity. Also Elmén et al. (2004) demonstrated that LNA/DNA mixmers activate RNase H when the mixmer has a DNA stretch of 6 nt. In accordance, Frieden et al. (2003a) concluded that a DNA gap size between 7 and 10 nt is optimal for LNA/DNA/LNA antisense gapmers. The overall conclusion is that antisense LNA oligonucleotides can be designed to elicit RNase H activity while still containing LNA monomers for improved binding and target accessibility (Jepsen and Wengel 2004).

With respect to other nucleases the situation is reversed, as resistance towards cleavage is advantageous. Complete stability against the $3'$-exonuclease snake venom phosphodiesterase (SVPD) was reported for a fully modified LNA (Frieden et al. 2003b) while a significant increase in $3'$-exonucleolytic stability was observed by blocking the $3'$-end with two LNA monomers. However, with only one penultimate LNA nucleotide or with a single LNA monomer in the middle of a sequence, no protection—or only a very minor protection—is induced (Morita et al. 2002; Lauritsen et al. 2003). S1 endonuclease susceptibility was also investigated, and fully modified LNA is very stable against S1 endonuclease, which is not the case for phosphodiester LNA/DNA/LNA gapmers. In a study of oligonucleotide stability in serum, LNA mixmers were found to be very stable, and LNA/DNA/LNA gapmers to be significantly more stable than DNA alone (Wahlestedt et al. 2000). Furthermore, Kurreck et al. (2002) reported that oligonucleotides with LNA nucleotides at the ends are more stable in human serum than the corresponding oligonucleotides composed of a phosphothioate DNA gaps with $2'$-O-methyl-RNA flanks. DNase I endonuclease degradation of end-modified 30-mer dsDNA showed that incorporation of one or two terminal LNA nucleotides ensures markedly increased stability (Crinelli et al. 2002). Similarly, Bal-31 exonucleolytic degradation of the same oligonucleotides showed that two terminal LNA nucleotides provide significant protection (Crinelli et al. 2002). The presence of a single $3'$-terminal LNA nucleotide significantly slowed degradation by the $3'$-$5'$ proofreading exonuclease of DNA polymerases Pfu and Vent (Di Giusto et al. 2004). From these studies it is clear that LNA nucleotides impose significant protection against nucleolytic degradation, especially if more than one LNA nucleotide is incorporated into an oligonucleotide.

2
RNA Targeting by LNAzymes

DNAzymes are catalytically active DNA molecules that are able to cleave RNA in a sequence-specific manner after binding to complementary sequences. The so-called 10–23 DNAzyme (Fig. 3) was found by in vitro selection (Santoro and Joyce 1997). It consists of a 15-nt catalytic core flanked by two binding arms that lead to the sequence selectivity of a given DNAzyme. LNAzymes are LNA-modified DNAzymes, and incorporation of two LNA nucleotides in each of the binding arms yielded an LNAzyme with a highly enhanced efficiency of RNA cleavage (Vester et al. 2002). Interestingly, cleavage of highly structured targets (a 58-nt RNA with known secondary structure and a 23S ribosomal RNA of 2904 nt) was shown to be significantly improved for LNAzymes compared to the corresponding unmodified DNAzymes, and multiple turnover cleavage reactions were observed both with a 17-nt minimal substrate and with a structured 58-nt substrate. These features, especially the improved RNA target accessibility, may be very significant for future uses. A similar approach was used by Schubert et al. (2003), who incorporated 3–4 LNA monomers at the ends of the binding arms and also observed a highly enhanced efficiency of RNA cleavage. Recently, it has been demonstrated that LNAzymes containing 3–4 LNA monomers at the ends of the binding arms cleave viral RNA structures that are resistant to hydrolysis by the corresponding unmodified DNAzyme, i.e., that efficient cleavage is correlated with improved binding affinity towards the target (Schubert et al. 2004).

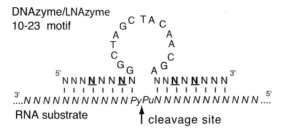

Fig. 3 Structure of 10–23 motif DNA- and LNAzymes. *Lines* indicate Watson-Crick base pairing with the RNA target sequence. LNA monomers are shown in *capital boldface underlined letters*, DNA monomers in *capital letters*, and RNA monomers in *capital italics letters*. The LNA monomers are introduced in the binding arms for enhanced RNA targeting and cleavage

One report on cellular activity of LNAzymes has been published. Fahmy and Khachigian (2004) used an LNAzyme design with two LNA monomers dispersed in each binding arm and they also included an 3′-3′ inverted thymidine monomer at the 3′-end to ensure nuclease stability. Based on a previous report (Khachigian et al. 2002) on inhibition of expression of EGR-1 by DNAzyme-mediated cleavage, Fahmy and Khachigian (2004) showed that serum-inducible smooth muscle cell proliferation was inhibited by greater than 50% at 20 nM LNAzyme, while no inhibition was evident by the corresponding DNAzyme at that concentration. Based on these data, it appears that the LNAzyme is superior at cleaving the EGR-1 transcript and inhibiting endogenous EGR-1 protein expression, SMC proliferation, and re-growth after injury.

LNAzymes containing LNA nucleotides in the binding arms display enhanced RNA affinity, allowing them to access RNA structures that cannot be targeted by the corresponding DNAzymes. As they furthermore show enhanced nuclease stability, and as efficient gene silencing in a cellular system has been reported, LNAzymes are promising novel agents for use in cells and for in vivo applications.

3
LNA Antisense

The term "antisense" is generally used for nucleic acid-based approaches that inhibit, in a sequence selective way, the processing of RNA from its transcription via messenger (m)RNA to protein, or the function of other forms of RNA. This includes, e.g., inhibition or alteration of splicing, translational arrest, or degradation of mRNA. By virtue of their intrinsic properties, oligonucleotides containing LNA nucleotides are obvious candidates for antisense-based gene silencing, and many previous LNA antisense studies are already reviewed (Petersen and Wengel 2003; Jepsen and Wengel 2004). Therefore, only selected examples are included herein supplemented with results from very recent studies.

The first in vivo antisense experiment with two different LNA sequences targeting DOR mRNA (coding for a receptor) in the central nervous system of living rats gave promising results (Wahlestedt et al. 2000). Upon direct injection of the antisense LNA oligonucleotides into the brain of living rats no tissue damage was observed, and a dose-dependent and highly efficacious knockdown of DOR was observed with both an LNA/DNA mixmer and an LNA/DNA/LNA gapmer. In another in vivo study, a fully modified LNA, targeting the RNA polymerase II gene product, inhibited tumor growth in mice and appeared non-toxic at doses less than 5 mg/kg per day (Fluiter et al. 2003). This study indicates LNA to be a much more potent class of antisense agents than the corresponding phosphothioate-DNA oligomers, which underlines the importance of differentiating between the variety of antisense chemistries available.

Several studies in vitro or in cells support the usefulness of LNA antisense for gene silencing (Obika et al. 2001; Hansen et al. 2003). One example of inhibition by LNA oligonucleotides involves telomerase (Elayadi et al. 2002), which is responsible for maintaining the telomere length from one generation to the next. Telomerase is a ribonucleoprotein that contains a protein domain and RNA with an 11-base sequence that binds telomeric DNA thereby guiding the addition of telomeric repeats. As telomerase is expressed in cancer cells but not in adjoining normal tissue, inhibition of telomerase will reduce tumor growth. LNA antisense constructs targeting the RNA moiety of telomerase are potent and selective inhibitors (Elayadi et al. 2002) with IC_{50} values of 10 nM for a 13-mer fully modified LNA, and a 13-mer LNA-DNA mixmer with 3 DNA monomers. By the introduction of two terminal phosphorothioate linkages, the potency was increased further by tenfold, but this was accompanied by a decreased match versus mismatch discrimination. Notably, even very short 8-mer fully modified LNAs are potent inhibitors (IC_{50} values of 2 nM and 25 nM with and without terminal phosphorothioate linkages, respectively). Also, in cells upon employing transfection 13-mer LNA oligomers induce inhibition of more than 80% of the telomerase activity (Elayadi et al. 2002). Short LNAs could be expected to cause problems due to binding to non-target nucleic acids, but no alteration of cell morphology was observed 7 days after transfection of 8-mer LNAs. Importantly, these results suggest that short LNAs may provide adequate affinity and sufficient selectivity for biologically relevant RNA targeting in cells.

Another important target for LNA antisense-based silencing is viral RNA. Effects of different LNAs on the interactions of the human immunodeficiency virus (HIV)-1 *trans*-activation responsive element (TAR) have recently been published (Arzumanov et al. 2003). Binding of oligonucleotides to TAR can inhibit Tat-dependent transcription thereby blocking full-length HIV transcription and hence viral replication, and chimeric sequences composed of LNA and 2′-O-methyl RNA nucleotides were shown to inhibit transcription in vitro (Arzumanov et al. 2003). Various LNA oligonucleotides were transfected into HeLa cells and derivatives with a minimum length of 12 residues showed 50% inhibition using nanomolar concentration of LNAs (Arzumanov et al. 2003), revealing the potential of LNA antisense oligonucleotides for in vivo targeting of RNA using non-RNase H dependent approaches. In another study, Elmén et al. (2004) demonstrated that LNA/DNA mixmers enhance the inhibition of HIV-1 genome dimerization and activate RNase H, show good uptake of the LNA/DNA mixmers in a T cell line, and inhibit replication of a clinical HIV-1 isolate.

A different antisense effect has been obtained by using LNA oligonucleotides to inhibit intron splicing. In an in vitro study, an 8-mer fully modified LNA and a 12-mer LNA/DNA chimera were shown to display a 50% inhibition at 150 nM and 30 nM, respectively, of a group I intron splicing in a transcription mixture from *Candida albicans* (Childs et al. 2002). The LNAs introduce misfolding of the RNA and consequently inhibition of the splicing process. In another study (Ittig et al. 2004), the antisense effects of tricyclo-DNA and LNA oligonucleotides on exon skipping were compared. Nuclear antisense effects of cyclophilin A pre-mRNA splicing by 9- to 15-mers fully modified oligonucleotides were investigated, and significant inhibition was observed for 11- to 15-mers tricyclo-DNAs and 13- to 15-mers LNAs with tricyclo-DNA being most potent.

It is cleat that LNA antisense—either LNA/DNA/LNA gapmers for RNase H activation or LNA mixmers for RNase H-independent activity—represents a favorable approach for gene silencing in vitro and in vivo, and reports indicate that LNA antisense rivals siRNA as the method of choice.

4
LNA-Modified siRNA

The discovery of the phenomenon of RNA interference (RNAi), in which dsRNA leads to the degradation of RNA that is homologous (Fire et al. 1998), has drawn much attention, as it mediates potent gene silencing in a number of different organisms and in mammalian cells. RNAi relies on a complex and ancient cellular mechanism that has probably evolved for protection against viral attack and mobile genetic elements.

A crucial step in the RNAi mechanism is the generation of siRNAs—dsRNAs that are about 22 nt each. The siRNAs lead to the degradation of homologous target RNA and might even lead to the production of more siRNAs against the same target RNA (Lipardi et al. 2001). In a recent review by Jepsen and Wengel (2004), LNA oligonucleotides in antisense experiments were compared with siRNA, leading to the conclusion that LNA antisense oligonucleotides combined with other modifications such as phosphothioate linkages might rival the currently very popular siRNA approach for gene silencing in vitro and in vivo. There is one study directly comparing the inhibitory effect of siRNA, LNA/DNA/LNA gapmers, phosphothioate-DNA, and $2'$-O-methyl-RNA on the expression of vanilloid receptor subtype 1 in cells (Grünweller et al. 2003). Both siRNA and LNA/DNA/LNA gapmers were found to be very efficient, and siRNA to be slightly more potent than the LNA/DNA/LNA gapmers. The siRNAs composed of dsRNA are themselves an obvious choice for incorporation of modified nucleotides for improved biostability and RNA targeting. Although the mechanisms of RNAi are not fully elucidated, a clearer picture of si/microRNA function and the usefulness of modified nucleotides in siRNA applications is emerging.

LNA monomers have been shown to be tolerated by the RNAi machinery and to provide thermal stability (Braasch et al. 2003). A new finding indicates that weak base pairing at the $5'$-end of the antisense strand is an important selection criterion for determining which siRNA strand will be used as template for mRNA degradation (Schwarz et al. 2003). Therefore the exact positioning and the overall number of LNA monomers will be very crucial for optimizing LNA-containing siRNA (siLNA). To address these issues, Elmén et al. (2005) systematically modified siRNA duplexes with LNA monomers and showed that siLNAs have substantially enhanced serum half-life compared to the corresponding siRNAs. Moreover, they provided evidence that the use of siLNAs reduces sequence-related off-target effects. Furthermore, they reported on improved efficacy of siLNAs on certain RNA motifs targeted against the severe acute respiratory syndrome coronavirus (SARS-CoV). These results emphasize siLNA's promise in converting siRNA from a functional genomics technology to a therapeutic platform.

5
Transcriptome Analysis Facilitated by LNA

Efficient selection of polyadenylated mRNA from eukaryotic cells and tissues is an essential step for a wide selection of functional genomics applications, including full-length complementary (c)DNA library construction and sequencing, Northern and dot blot analyses, gene expression profiling by microarrays, and quantitative RT-PCR. The key to successful selection of intact poly(A)$^+$RNA is a fast extraction of total RNA from cells and tissues using strong denaturing agents to disrupt the cells with the simultaneous denaturation of endogenous RNases followed by mRNA sample preparation from the extracted total RNA (Aviv and Leder 1972; Chirgwin et al. 1979; Chomczynski and Sacchi 1987). Since most eukaryotic mRNAs contain tracts of poly(A) tails at their 3' termini, polyadenylated mRNA can be selected by oligo(dT)-cellulose chromatography. Jacobsen et al. (2004) have recently reported on efficient isolation of intact poly(A)$^+$RNA using an LNA-substituted oligo(dT) affinity ligand, based on LNA-T's increased affinity to complementary poly(A) tracts. Poly(A)$^+$RNA could be isolated directly from 4 M guanidinium thiocyanate (GuSCN)-lysed *Caenorhabditis elegans* worm extracts as well as from lysed human K562 and K562/VCR leukemia cells using the LNA_2.T oligonucleotide (see Fig. 2) as an affinity probe, in which every second thymidine was substituted with LNA thymidine. In accordance with the significantly increased stability of the LNA_2.T-A duplexes in 4 M GuSCN, Jacobsen et al. (2004) obtained a 30- to 50-fold mRNA yield increase by using the LNA-substituted oligo(T) affinity probe compared to the DNA oligo(dT)-selected mRNA samples. The LNA_2.T affinity probe was furthermore shown to be highly efficient in isolation of poly(A)$^+$RNA in a low salt concentration range of 50 to 100 mM NaCl in the poly(A) binding buffer. The utility of the LNA oligo(T)-selected mRNA in quantitative real-time PCR was also demonstrated by analyzing the expression of the human *mdr1* gene in the two K562 human cell lines employing pre-validated Taqman assays (Jacobsen et al. 2004).

Large-scale expression analysis is increasingly important in many areas of biological research aiming at deciphering complex biological systems, thereby greatly facilitating the understanding of basic biological processes as well as human disease. Quantitative real-time RT-PCR has become the method of choice for accurate expression profiling of selected genes and validation of microarray data, and is especially suitable for analysis of low abundant mRNAs and small samples (Bustin 2000; Liss 2002). However, real-time RT-PCR is often hampered by the labor-intensive assay design and validation, which significantly lowers its throughput compared to genome-wide expression profiling by DNA microarrays. In response to this, Mouritzen et al. (2004) have developed a novel concept for quantitative real-time RT-PCR based on a library of 90 pre-validated dual-labeled LNA-enhanced detection probes, designated as ProbeLibrary. The probes are shortened to only 8–9 nt by substitution with LNA nucleotides, ensuring adequate duplex stability and compatibility in standard real-time RT-PCR assays. The use of short recognition sequences for the detection probes enables targeting of the frequently recurring short sequence tags in the human transcriptome, and this facilitated the re-use of the same LNA-enhanced probe in detection and quantification of many different transcripts. By careful selection of the most common 8- and 9-mer sequences in the human transcriptome, Mouritzen et al. (2004) constructed a library of 90 detection probes that detects 98% of all human transcripts. On average, each mRNA contains target sites for 16 ProbeLibrary probes, whereas the recognition sequence of each LNA-modified detection probe was found in more than 7,000 of the 38,556 human transcripts in the RefSeq database at the National Center for Biotechnical Information (NCBI). Similar LNA-based ProbeLibraries for expression profiling by real-time RT-PCR have recently been developed for mouse, rat, *Arabidopsis thaliana*, *Drosophila melanogaster*, and *C. elegans*. The use of LNA in microarrays has also been exploited. An optimal design for LNAs as capture probes for gene-expression profiling, together with a microarray study of *C. elegans* cytochrome expression, has been published (Tolstrup et al. 2003). The oligo design program, which might be useful in many applications, is freely available at http://lnatools.com/.

6
MicroRNA Detection Using LNA Probes

MicroRNAs (miRNAs) are 19- to 25-nt non-coding RNAs that are processed from longer endogenous hairpin transcripts by the enzyme Dicer (Ambros 2001; Ambros et al. 2003). To date, more than 1,500 microRNAs have been identified in invertebrates, vertebrates, and plants according to the miRNA registry database, and many miRNAs that correspond to putative genes have also been identified (Griffiths-Jones 2004). The high percentage of predicted miRNA targets acting as developmental regulators and the conservation of target sites suggests that miRNAs exhibit a wide variety of regulatory functions and exert significant effects on cell growth, development, and differentiation (Ke et al. 2003; Bartel 2004), including human development and disease.

The current view that miRNAs represent a hidden layer of gene regulation has resulted in high interest among researchers in the discovery of miRNAs, their targets, their expression, and their mechanism of action. Most miRNA researchers use Northern blot analysis combined with polyacrylamide gels to examine the expression of both the mature and precursor miRNAs, since it allows both quantitation of the expression levels and miRNA size determination (Lagos-Quintana et al. 2001; Reinhart et al. 2000; Lee and Ambros 2001). A major drawback of this method is its poor sensitivity, especially when monitoring expression of low-abundant miRNAs. In a recent paper, Valoczi et al. (2004) have described a new method for highly efficient detection of miRNAs by Northern blot analysis using LNA-modified oligonucleotide probes and demonstrated its significantly improved sensitivity by designing several LNA-modified oligonucleotide probes for detection of different miRNAs in mouse, *A. thaliana*, and *Nicotiana benthamiana*. They used the LNA/DNA mixmer probes in Northern blot analysis employing standard end-labeling techniques and hybridization conditions, and the sensitivity in detecting mature miRNAs by Northern blots is increased by at least tenfold compared to DNA probes, while it is simultaneously highly specific as demonstrated by the use of different single and double mismatched LNA probes (Valoczi et al. 2004). Besides being efficient as Northern probes, LNA-modified oligonucleotides have proved highly useful for in situ localization of miRNAs in cells and tissues. Accordingly, the temporal and spatial expression patterns of 115 conserved vertebrate miRNAs were determined in zebrafish embryos using LNA-modified oligonucleotide probes (Wienholds et al. 2005). Interestingly, most miRNAs were expressed in a highly tissue-specific manner during segmentation and later stages, but not early in development. These miRNAs may therefore play crucial roles in the maintenance of tissue identity or in differentiation (Wienholds et al. 2005).

7
Future Prospects

LNAs constitute an important addition to the tools available for biotechnology, as well as nucleic acid-based diagnostics and therapeutics. The remarkable hybridization properties of LNA, both with respect to its affinity and specificity, position it as an enabling molecule for molecular biology research and biotechnology innovation. The fact that LNA phosphoramidites and oligomers are commercially available, and that LNA nucleotides can be freely mixed with, e.g., DNA, RNA, and $2'$-*O*-Me-RNA monomers and standard probes, make LNA a highly flexible tool allowing fine-tuning of properties.

Short LNA oligonucleotides (fully modified) and short LNA/DNA and LNA/RNA mixmer oligonucleotides are uniquely suited for targeting of complementary RNA and DNA. LNA will likely have a significant impact especially with respect to in vitro and in vivo RNA analysis and interference with RNA-mediated processes. LNA antisense rivals siRNA for gene silencing, but a satisfactory answer to the demand for efficient cellular delivery in vivo remains to be developed for LNA antisense, as for any other therapeutic oligonucleotide that relies on hybridization to a nucleic acid target for biological activity. Simple molecular-scale autonomous programmable computers based on nucleic acid interactions have been demonstrated, and a distinct prospect is nucleic acid-based, autonomous biomolecular computers that logically analyze the levels of mRNAs or non-coding RNAs, such as miRNAs, and in response produce a molecule capable of affecting levels of gene expression (Benenson et al. 2004). The superb hybridization properties of LNA—allowing shortening of target sequences—and its nuclease stability could make LNA an important partner in such designs. LNA is a unique molecule, which is likely to make a significant impact on the future developments within many areas of biotechnology and medicine.

Acknowledgements The Nucleic Acid Center is funded by The Danish National Research Foundation for studies on nucleic acid chemical biology.

References

Ambros V (2001) MicroRNAs: tiny regulators with great potential. Cell 107:823–826

Ambros V, Lee RC, Lavanway A, Williams PT, Jewell D (2003) MicroRNAs and other tiny endogenous RNAs in C. elegans. Curr Biol 13:807–818

Arzumanov A, Stetsenko DA, Malakhov AD, Reichelt S, Sørensen MD, Babu BR, Wengel J, Gait MJ (2003) A structure-activity study of the inhibition of HIV-1 tat-dependent transactivation by mixmer 2′-O-methyl oligoribonucleotides containing locked nucleic acid (LNA), α-L-LNA, or 2′-thio-LNA residues. Oligonucleotides 13:435–453

Aviv H, Leder P (1972) Purification of biologically active globin messenger RNA by chromatography on oligothymidylic acid-cellulose. Proc Natl Acad Sci USA 69:1408–1412

Bartel DP (2004) MicroRNAs: genomics, biogenesis, mechanism and function. Cell 116:281–297

Benenson Y, Gil B, Ben-Dor U, Adar R, Shapiro E (2004) An autonomous molecular computer for logical control of gene expression. Nature 429:423–429

Braasch DA, Corey DR (2001) Locked nucleic acid (LNA): fine-tuning the recognition of DNA and RNA. Chem Biol 8:1–7

Braasch DA, Jensen S, Liu Y, Kaur K, Arar K, White MA, Corey DR (2003) RNA interference in mammalian cells by chemically-modified RNA. Biochemistry 8:7967–7975

Bustin SA (2000) Absolute quantification of mRNA using real-time reverse transcription polymerase chain reaction assays. J Mol Endocrinol 25:169–193

Childs JL, Disney MD, Turner DH (2002) Oligonucleotide directed misfolding of RNA inhibits Candida albicans group I intron splicing. Proc Natl Acad Sci USA 99:11091–11096

Chirgwin JM, Przybyla AE, MacDonald RJ, Rutter WJ (1979) Isolation of biologically active ribonucleic acid from sources enriched in ribonuclease. Biochemistry 18:5294–5299

Chomczynski P, Sacchi N (1987) Single-step method of RNA isolation by acid guanidinium thiocyanate-phenol-chloroform extraction. Anal Biochem 162:156–159

Crinelli R, Bianchi M, Gentilini L, Magnani M (2002) Design and characterization of decoy oligonucleotides containing locked nucleic acids. Nucleic Acids Res 30:2435–2443

De Giusto DA, King GC (2004) Strong positional preference in the interaction of LNA oligonucleotides with DNA polymerase and proofreading exonuclease activities: implications for genotyping assays. Nucleic Acids Res 32:e32

Elayadi AN, Braasch DA, Corey DR (2002) Implications of high-affinity hybridization by locked nucleic acid oligomers for inhibition of human telomerase. Biochemistry 41:9973–9981

Elmén J, Zhang HY, Zuber B, Ljungberg K, Wahren B, Wahlestedt C, Liang Z (2004) Locked nucleic acid containing antisense oligonucleotides enhance inhibition of HIV-1 genome dimerization and inhibit virus replication. FEBS Lett 578:285–290

Elmén J, Thonberg H, Ljungberg K, Frieden M, Westergaard M, Xu Y, Wahren B, Liang Z, Ørum H, Koch T, Wahlestedt C (2005) Locked nucleic acid (LNA) mediated improvements in siRNA stability and functionality. Nucleic Acids Res 33:439–447

Fahmy RG, Khachigian LM (2004) Locked nucleic acid modified DNA enzymes targeting early growth response-1 inhibit human vascular smooth muscle cell growth. Nucleic Acids Res 32:2281–2285

Fire A, Xu S, Montgomery MK, Kostas SA, Driver SE, Mello CC (1998) Potent and specific genetic interference by double-stranded RNA in Caenorhabditis elegans. Nature 391:806–811

Fluiter K, ten Asbroek LMA, de Wissel MB, Marit B, Jakobs ME, Wissenbach M, Olsson H, Olsen O, Oerum H, Baas F (2003) In vivo tumor growth inhibition and biodistribution studies of locked nucleic acid (LNA) antisense oligonucleotides. Nucleic Acids Res 31:953–962

Frieden M, Christensen SM, Mikkelsen ND, Rosenbohm C, Thrue CA, Westergaard M, Hansen HF, Ørum H, Koch T (2003a) Expanding the design horizon of antisense oligonucleotides with alpha-L-LNA. Nucleic Acids Res 31:6365–6372

Frieden M, Hansen HF, Koch T (2003b) Nuclease stability of LNA oligonucleotides and LNA-DNA chimeras. Nucleosides Nucleotides Nucleic Acids 22:1041–1043

Griffiths-Jones S (2004) The microRNA registry. Nucleic Acids Res 32:D109–D111

Grünweller A, Wyszko E, Bieber B, Jahnel R, Erdmann VA, Kurreck J (2003) Comparison of different antisense strategies in mammalian cells using locked nucleic acids, 2'-O-methyl RNA, phosphorothioates and small interfering RNA. Nucleic Acids Res 31:3185–3193

Hansen JB, Westergaard M, Thrue CA, Giwercman B, Oerum H (2003) Antisense knockdown of PKC-alpha using LNA-oligos. Nucleosides Nucleotides Nucleic Acids 22:1607–1609

Ittig D, Liu S, Renneberg D, Schuemperli D, Leumann CJ (2004) Nuclear antisense effects in cyclophilin A pre-mRNA splicing by oligonucleotides: a comparison of tricyclo-DNA with LNA. Nucleic Acids Res 32:346–353

Jacobsen Nana, Nielsen PS, Jeffares DC, Eriksen J, Ohlsson H, Arctander P, Kauppinen S (2004) Direct isolation of poly(A)+ RNA from 4 M guanidine thiocyanate-lysed cell extracts using locked nucleic acid-oligo(T) capture. Nucleic Acids Res 32:e64/1–e64/10

Jepsen JS, Wengel J (2004) LNA-antisense rivals siRNA for gene silencing. Curr Opin Drug Discov Devel 7:188–194

Kampa D, Cheng IJ, Kapranov P, Yamanaka M, Brubaker S, Cawley S, Drenkow J, Piccolboni A, Bekiranov S, Helt G, Tammana H, Gingeras TR (2004) Novel RNAs identified from an in-depth analysis of the transcriptome of human chromosomes 21 and 22. Genome Res 14:331–342

Ke X-S, Liu C-M, Liu D-P, Liang C-C (2003) MicroRNAs: key participants in gene regulatory networks. Curr Opin Chem Biol 7:516–523

Khachigian LM, Fahmy RG, Zhang G, Bobryshev YV, Kaniaros A (2002) c-Jun regulates vascular smooth muscle cell growth and neointima formation after arterial injury. Inhibition by a novel DNA enzyme targeting c-Jun. J Biol Chem 277:22985–22991

Koshkin AA, Nielsen P, Meldgaard M, Rajwanshi VK, Singh SK, Wengel J (1998a) LNA (locked nucleic acid): an RNA mimic forming exceedingly stable LNA:LNA duplexes. J Am Chem Soc 120:13252–13253

Koshkin AA, Singh SK, Nielsen P, Rajwanshi VK, Kumar R, Meldgaard M, Olsen CE, Wengel J (1998b) LNA (locked nucleic acids): synthesis of the adenine, cytosine, guanine, 5-methylcytosine, thymine and uracil bicyclonucleoside monomers, oligomerisation, and unprecedented nucleic acid recognition. Tetrahedron 54:3607–3630

Kumar R, Singh SK, Koshkin AA, Rajwanshi VK, Meldgaard M, Wengel J (1998) The first analogues of LNA (locked nucleic acids): Phosphorothioate-LNA and 2′-thio-LNA. Bioorg Med Chem Lett 8:2219–2222

Kurreck J, Wyszko E, Gillen C, Erdmann VA (2002) Design of antisense oligonucleotides stabilized by locked nucleic acids. Nucleic Acids Res 30:1911–1918

Lagos-Quintana M, Rauhut R, Lendeckel W, Tuschl T (2001) Identification of novel genes coding for small expressed RNAs. Science 294:853–858

Lander ES, Linton LM, Birren B, Nusbaum C, Zody MC, Baldwin J, Devon K, Dewar K, Doyle M, FitzHugh W, Funke R, Gage D, Harris K, Heaford A, Howland J, Kann L, Lehoczky J, LeVine R, McEwan P, McKernan K, Meldrim J, Mesirov J P, Miranda C, Morris W, Naylor J, Raymond C, Rosetti M, Santos R, Sheridan A, Sougnez C, Stange-Thomann N, Stojanovic N, Subramanian A, Wyman D, Rogers J, Sulston J, Ainscough R, Beck S, Bentley D, Burton J, Clee C, Carter N, Coulson A, Deadman R, Deloukas P, Dunham A, Dunham I, Durbin R, French L, Grafham D, Gregory S, Hubbard T, Humphray S, Hunt A, Jones M, Lloyd C, McMurray A, Matthews L, Mercer S, Milne S, Mullikin JC, Mungall A, Plumb R, Ross M, Shownkeen R, Sims S, Waterston RH, Wilson RK, Hillier LW, McPherson JD, Marra MA, Mardis ER, Fulton LA, Chinwalla AT, Pepin KH, Gish WR, Chissoe SL, Wendl MC, Delehaunty KD, Miner TL, Delehaunty A, Kramer JB, Cook LL, Fulton RS, Johnson DL, Minx PJ, Clifton SW, Hawkins T, Branscomb E, Predki P, Richardson P, Wenning S, Slezak T, Doggett N, Cheng JF, Olsen A, Lucas S, Elkin C, Uberbacher E, Frazier M, Gibbs RA, Muzny DM, Scherer SE, Bouck JB, Sodergren EJ, Worley KC, Rives CM, Gorrell JH, Metzker ML, Naylor SL, Kucherlapati RS, Nelson DL, Weinstock GM, Sakaki Y, Fujiyama A, Hattori M, Yada T, Toyoda A, Itoh T, Kawagoe C, Watanabe H, Totoki Y, Taylor T, Weissenbach J, Heilig R, Saurin W, Artiguenave F, Brottier P, Bruls T, Pelletier E, Robert C, Wincker P, Smith DR, Doucette-Stamm L, Rubenfield M, Weinstock K, Lee HM, Dubois J, Rosenthal A, Platzer M, Nyakatura G, Taudien S, Rump A, Yang H, Yu J, Wang J, Huang G, Gu J, Hood L, Rowen L, Madan A, Qin S, Davis RW, Federspiel NA, Abola AP, Proctor MJ, Myers RM, Schmutz J, Dickson M, Grimwood J, Cox DR, Olson MV, Kaul R, Raymond C, Shimizu N, Kawasaki K, Minoshima S, Evans GA, Athanasiou M, Schultz R, Roe BA, Chen F, Pan H, Ramser J, Lehrach H, Reinhardt R, McCombie WR, de la Bastide M, Dedhia N, Blocker H, Hornischer K, Nordsiek G, Agarwala R, Aravind L, Bailey JA, Bateman A, Batzoglou S, Birney E, Bork P, Brown DG, Burge CB, Cerutti L, Chen HC, Church D, Clamp M, Copley RR, Doerks T, Eddy SR, Eichler EE, Furey TS, Galagan J, Gilbert JG, Harmon C, Hayashizaki Y, Haussler D, Hermjakob H (2001) Initial sequencing and analysis of the human genome. Nature 409:860–921

Lauritsen A, Dahl BM, Dahl O, Vester B, Wengel J (2003) Methylphosphonate LNA: a locked nucleic acid with a methylphosphonate linkage. Bioorg Med Chem Lett 13:253–256

Lee RC, Ambros V (2001) An extensive class of small RNAs in Caenorhabditis elegans. Science 294:862–864

Lipardi C, Wei Q, Paterson BM (2001) RNAi as random degradative PCR: siRNA primers convert mRNA into dsRNAs that are degraded to generate new siRNAs. Cell 107:297–307

Liss B (2002) Improved quantitative real-time RT-PCR for expression profiling of individual cells. Nucleic Acids Res 30:17:e89

Morita K, Hasegawa C, Kaneko M, Tsutsumi S, Sone J, Ishikawa T, Imanishi T, Koizumi M (2002) 2'-O,4'-C-Ethylene-bridged nucleic acids (ENA): highly nuclease-resistant and thermodynamically stable oligonucleotides for antisense drug. Bioorg Med Chem Lett 12:73–76

Mouritzen P, Nielsen PS, Jacobsen N, Noerholm M, Lomholt C, Pfundheller HM, Ramsing NB, Kauppinen S, Tolstrup N (2004) The ProbeLibrary—expression profiling 99% of all human genes using only 90 dual-labeled real-time PCR probes. Biotechniques 37:492–495

Nielsen KE, Rasmussen J, Kumar R, Wengel J, Jacobsen JP, Petersen M (2004) NMR studies of fully modified Locked nucleic acid (LNA) hybrids: solution structure of an LNA:RNA hybrid and characterization of an LNA:DNA hybrid. Bioconjug Chem 15:449–457

Obika S, Nanbu D, Hari Y, Morio K-I, In Y, Ishida T, Imanishi T (1997) Synthesis of 2′-O,4′-C-methyleneuridine and -cytidine novel bicyclic nucleosides having a fixed C3′-endo sugar puckering. Tetrahedron Lett 38:8735–8738

Obika S, Nanbu D, Hari Y, Andoh J-I, Morio K-I, Doi T, Imanishi T (1998) Stability and structural features of the duplexes containing nucleoside analogues with a fixed N-type conformation, 2′-O,4′-C-methyleneribonucleosides. Tetrahedron Lett 39:5401–5404

Obika S, Hemamayi R, Masuda T, Sugimoto T, Nakagawa S, Mayumi T, Imanishi T (2001) Inhibition of ICAM-1 gene expression by antisense 2′,4′-BNA oligonucleotides. Nucleic Acids Res Suppl 1:145–146

Petersen M, Wengel J (2003) LNA: a versatile tool for therapeutics and genomics. Trends Biotechnol 21:74–81

Petersen M, Bondensgaard K, Wengel J, Petersen JP (2002) Locked nucleic acid (LNA) recognition of RNA: NMR solution structures of LNA:RNA hybrids. J Am Chem Soc 124:5974–5982

Pfundheller HM, Sorensen AM, Lomholt C, Johansen AM, Koch T, Wengel J (2005) Locked nucleic acid synthesis. Methods Mol Biol 288:127–146

Reinhart BJ, Slack FJ, Basson M, Pasquinelli AE, Bettinger JC, Rougvie AE, Horvitz HR, Ruvkun G (2000) The 21 nucleotide let-7 RNA regulates developmental timing in Caenorhabditis elegans. Nature 403:901–906

Sachidanandam R, Weissman D, Schmidt SC, Kakol JM, Stein LD, Marth G, Sherry S, Mullikin JC, Mortimore BJ, Willey DL, Hunt SE, Cole CG, Coggill PC, Rice CM, Ning Z, Rogers J, Bentley DR, Kwok PY, Mardis ER, Yeh RT, Schultz B, Cook L, Davenport R, Dante M, Fulton L, Hillier L, Waterston RH, McPherson JD, Gilman B, Schaffner S, Van Etten WJ, Reich D, Higgins J, Daly MJ, Blumenstiel B, Baldwin J, Stange-Thomann N, Zody MC, Linton L, Lander ES, Altshuler D et al. (2001) A map of human genome sequence variation containing 1.42 million single nucleotide polymorphisms. Nature 409:928–933

Santoro SW, Joyce GF (1997) A general purpose RNA-cleaving DNA enzyme. Proc Natl Acad Sci USA 94:4262–4266

Schubert S, Gul DC, Grunert HP, Zeichhardt H, Erdmann VA, Kurreck J (2003) RNA cleaving "10–23" DNAzymes with enhanced stability and activity. Nucleic Acids Res 31:5982–5992

Schubert S, Fürste JP, Werk D, Grunert H-P, Zeichhardt H, Erdmann VA, Kurreck J (2004) Gaining target access for deoxyribozymes. J Mol Biol 339:355–363

Schwarz DS, Hutvágner G, Du T, Xu Z, Aronin N, Zamore PD (2003) Asymmetry in the assembly of the RNAi enzyme complex. Cell 115:199–208

Singh SK, Nielsen P, Koshkin AA, Wengel J (1998) LNA (locked nucleic acids): synthesis and high-affinity nucleic acid recognition. Chem Commun 4:455–456

Sørensen MD, Kvaernø L, Bryld T, Håkansson AE, Verbeure B, Gaubert G, Herdewijn P, Wengel J (2002) α-L-ribo-configured locked nucleic acid (α -L-LNA): synthesis and properties. J Am Chem Soc 124:2164–2176

Tolstrup N, Nielsen PS, Kolberg JG, Frankel AM, Vissing H, Kauppinen S (2003) OligoDesign: optimal design of LNA (locked nucleic acid) oligonucleotide capture probes for gene expression profiling. Nucleic Acids Res 31:3758–3762

Valoczi A, Hornyik C, Varga N, Burgyan J, Kauppinen S, Havelda Z (2004) Sensitive and specific detection of microRNAs by northern blot analysis using LNA-modified oligonucleotide probes. Nucleic Acids Res 32:e175

Venter JC, Adams MD, Myers EW, Li PW, Mural RJ, Sutton GG, Smith HO, Yandell M, Evans CA, Holt RA, Gocayne JD, Amanatides P, Ballew RM, Huson DH, Wortman JR, Zhang Q, Kodira CD, Zheng XH, Chen L, Skupski M, Subramanian G, Thomas PD, Zhang J, Gabor Miklos GL, Nelson C, Broder S, Clark AG, Nadeau J, McKusick VA, Zinder N, Levine AJ, Roberts RJ, Simon M, Slayman C, Hunkapiller M, Bolanos R, Delcher A, Dew I, Fasulo D, Flanigan M, Florea L, Halpern A, Hannenhalli S, Kravitz S, Levy S, Mobarry C, Reinert K, Remington K, Abu-Threideh J, Beasley E, Biddick K, Bonazzi V, Brandon R, Cargill M, Chandramouliswaran I, Charlab R, Chaturvedi K, Deng Z, Di Francesco V, Dunn P, Eilbeck K, Evangelista C, Gabrielian AE, Gan W, Ge W, Gong F, Gu Z, Guan P, Heiman TJ, Higgins ME, Ji RR, Ke Z, Ketchum KA, Lai Z, Lei Y, Li Z, Li J, Liang Y, Lin X, Lu F, Merkulov GV, Milshina N, Moore HM, Naik AK, Narayan VA, Neelam B, Nusskern D, Rusch DB, Salzberg S, Shao W, Shue B, Sun J, Wang Z, Wang A, Wang X, Wang J, Wei M, Wides R, Xiao C, Yan C, Yao A, Ye J, Zhan M, Zhang W, Zhang H, Zhao Q, Zheng L, Zhong F, Zhong W, Zhu S, Zhao S, Gilbert D, Baumhueter S, Spier G, Carter C, Cravchik A, Woodage T, Ali F, An H, Awe A, Baldwin D, Baden H, Barnstead M, Barrow I, Beeson K, Busam D, Carver A, Center A, Cheng ML, Curry L, Danaher S, Davenport L, Desilets R, Dietz S, Dodson K, Doup L, Ferriera S, Garg N, Gluecksmann A, Hart B, Haynes J, Haynes C, Heiner C, Hladun S, Hostin D, Houck J, Howland T, Ibegwam C, Johnson J, Kalush F, Kline L, Koduru S, Love A, Mann F, May D, McCawley S, McIntosh T, McMullen I, Moy M, Moy L, Murphy B, Nelson K, Pfannkoch C, Pratts E, Puri V, Qureshi H, Reardon M, Rodriguez R, Rogers YH, Romblad D, Ruhfel B, Scott R, Sitter C, Smallwood M, Stewart E, Strong R, Suh E, Thomas R, Tint NN, Tse S, Vech C, Wang G, Wetter J, Williams S, Williams M, Windsor S, Winn-Deen E, Wolfe K, Zaveri J, Zaveri K, Abril JF, Guigo R, Campbell MJ, Sjolander KV, Karlak B, Kejariwal A, Mi H, Lazareva B, Hatton T, Narechania A, Diemer K, Muruganujan A, Guo N, Sato S, Bafna V, Istrail S, Lippert R, Schwartz R, Walenz B, Yooseph S, Allen D (2001) The sequence of the human genome. Science 291:1304–1351

Vester B, Wengel J (2004) LNA (locked nucleic acid): high affinity targeting of complementary RNA and DNA. Biochemistry 43:13233–13241

Vester B, Lundberg L, Sørensen MD, Babu BR, Douthwaite S, Wengel J (2002) LNAzymes: incorporation of LNA-type monomers into DNAzymes markedly increases RNA cleavage. J Am Chem Soc 124:13682–13683

Wahlestedt C, Salmi P, Good L, Kela J, Johnsson T, Hokfelt T, Broberger C, Porreca F, Lai J, Ren K, Ossipov M, Koshkin A, Jakobsen N, Skouv J, Oerum H, Jacobsen MH, Wengel J (2000) Potent and nontoxic antisense oligonucleotides containing locked nucleic acids. Proc Natl Acad Sci USA 97:5633–5638

Wengel J (1999) Synthesis of 3'-C- and 4'-C-branched oligonucleotides and the development of locked nucleic acid (LNA). Acc Chem Res 32:301–310

Wienholds E, Kloosterman WP, Miska E, Alvarez-Saavedra E, Berezikov E, de Bruijn E, Horvitz HR, Kauppinen S, Plasterk RHA (2005) MicroRNA expression in zebrafish embryonic development. Science 309:310–311

Wong GK, Passey DG, Yu J (2001) Most of the human genome is transcribed. Genome Res 11:1975–1977

Engineering RNA-Based Circuits

R. Narayanaswamy · A.D. Ellington (✉)

Institute for Cellular and Molecular Biology, University of Texas at Austin,
1 University Station A4800, Austin TX, 78712-0159, USA
andy.ellington@mail.utexas.edu

1	Introduction	423
2	RNA-Based Gene Regulation	424
2.1	Regulation via Base-Pairing	424
2.2	Shape-Based RNA Regulation	426
2.3	Catalytic Regulation	428
3	Optimizing the Functions of RNA Tools	431
4	Engineering Biological Circuits	432
4.1	Synthetic Biology	432
4.2	Designed RNA Circuits for Translation Control	434
4.3	Designed RNA Circuits for Transcription Control	435
5	The Path Forward: Engineered Circuits in Medicine	436
	References	438

Abstract Nucleic acids can modulate gene function by base-pairing, via the molecular recognition of proteins and metabolites, and by catalysis. This diversity of functions can be combined with the ability to engineer nucleic acids based on Watson-Crick base-pairing rules to create a modular set of molecular "tools" for biotechnological and medical interventions in cellular metabolism. However, these individual RNA-based tools are most powerful when combined into rational logical or regulatory circuits, and the circuits can in turn be evolved for optimal function. Examples of genetic circuits that control translation and transcription are herein detailed, and more complex circuits with medical applications are anticipated.

Keywords siRNA (small interfering RNA) · miRNA (microRNA) ·
RNAi (RNA interference) · UTR (untranslated region) · RBS (ribosome binding site)

1
Introduction

Our perspective of the "central dogma" in molecular biology has undergone significant changes over the years. While genetic information does flow from DNA to protein via RNA, the role of RNA as a messenger obscures its importance as a structural, catalytic, and regulatory molecule in cellular processes. Beyond the ability of nucleic acids to base-pair during replication and transcription, DNA and RNA can also fold into complex structures that enable them to bind ligands and catalyze diverse reactions (Breaker 2004). Naturally-occurring, regulatory RNAs rely on a variety of mechanisms, such as antisense binding (microRNAs), ligand-gated conformational changes (riboswitches), and catalysis (hammerhead cleavases, self-splicing introns). Indeed, given that biochemical and structural data indicate that the ribosome itself is a ribozyme (Moore and Steitz 2003; Steitz and Moore 2003), the appreciation of the role of RNA-based machines and pathways in cellular metabolism may only be beginning.

Concomitantly, the opportunities for using RNA to engineer or intervene in cellular processes are equally immense. This chapter will outline the natural roles of RNA in the regulation of gene expression, and will then emphasize accompanying advances in biotechnology and medicine that may make regulatory RNAs important tools for diagnosis and therapy. Looking to the future, we will consider the possibility that genetic circuits composed of regulatory RNAs can be used to engineer complex cellular behaviors and create "designer" drugs.

2
RNA-Based Gene Regulation

While base-pairing is an obvious and recurring theme in RNA-based regulation, the ability to recognize the shapes and chemistries of non-nucleic acid ligands and to catalyze reactions have also been observed. We will first examine examples of base-pairing, ligand recognition, and catalytic mechanisms in RNA biology in order to come up with an appropriate biotechnology tool set for the design and implementation of RNA-based genetic circuits.

2.1
Regulation via Base-Pairing

Base-pairing or antisense-based RNA regulation primarily involves small non-coding RNAs found in either bacteria (i.e., sRNAs) or eukaryotes (i.e., microRNAs; Storz and Wassarman 2004). In particular, there are a handful of well-studied sRNAs in *Escherichia coli*, such as dsrA, rhyB, and Spot42 (Masse et al. 2003; Gottesman 2004; Storz et al. 2004). Such bacterial small (s)RNAs typically function by binding to and modulating translation of a target mRNA. For example, RhyB is an sRNA that helps regulate iron metabolism in *E. coli* (Masse and Gottesman 2002). Upon iron starvation, ryhB sRNA base-pairs with and promotes the degradation of transcripts encoding iron containing enzymes. Under conditions of excess iron, the Fur repressor is active and stops transcription of, among other genes, ryhB. The ryhB and other sRNAs act in conjunction with the RNA chaperone Hfq (Zhang et al. 2003). The Hfq protein does not bind to a precise target sequence but tends to bind AU-rich sequences (Vytvytska et al. 2000; Brescia et al. 2003; Mikulecky et al. 2004). Since this sequence can at times overlap with the RNase E cleave sites (Mackie and Genereaux 1993; McDowall et al. 1994; Moll et al. 2003), it has been proposed that Hfq mediates binding of the sRNA to 5′ untranslated regions (UTRs) and then triggers their degradation by RNase E (Carpousis 2002). In contrast to this mechanism, gadY is an sRNA that pairs the 3′ UTR of the GadX mRNA and leads to an increase in its translation (Opdyke et al. 2004). The fact that small RNAs can interact with either 5′ or 3′ UTRs to yield either destabilization or enhancement of translation suggests that simple base-pairing interactions can potentially be used to generate a wide variety of regulatory circuits and phenotypes.

The sRNAs appear to be generally important for bacterial gene regulation. Computational predictions put the number of sRNAs in *E. coli* at 50–100 (Argaman et al. 2001; Carter et al. 2001; Rivas et al. 2001; Wassarman et al. 2001; Chen et al. 2002, 2004), a number that represents about 2% of the available protein-coding genes. More than a third of all these non-coding RNAs are likely bound by Hfq and act via base-pairing (Zhang et al. 2003). Similar *trans*-encoded antisense RNAs have recently been discovered in other bacteria as well. Examples include the non-coding RNAs PrrF1 and PrrF2 in *Pseudomonas aeruginosa* (Wilderman et al. 2004) that function in iron homeostasis and the Qrr1–4 small RNAs in *Vibrio harveyi* that are involved in regulating quorum sensing (Lenz et al. 2004). Several of the Qrr RNAs appear to target the same UTR and thus may contribute to complex regulatory patterns (or merely be functionally redundant).

Non-coding RNA-based gene regulation has also been shown to be prevalent in higher eukaryotes (Bartel and Chen 2004; Griffiths-Jones et al. 2005). The closest functional equivalent of small RNAs in eukaryotes is the microRNA or miRNA (Bartel 2004; Huppi et al. 2005; Sontheimer 2005). The phenomenon of miRNA-based gene regulation was first discovered when Fire and co-workers found that injecting double-stranded RNA into *Caenorhabditis elegans* silenced cognate genes (Seydoux et al. 1996; Fire et al. 1998; Ambros 2004). Since then, microRNA-based gene regulation has been unequivocally demonstrated in mammals, plants, fungi, and flies, and miRNAs have been implicated in metabolic processes ranging from the control of cell death and proliferation, to hematopoietic lineage differentiation in animals and leaf and flower development in plants.

Much like their bacterial counterparts, miRNAs are encoded and expressed separately from the genes they target, recognizing their target sequences by base-pairing. In plants, miRNAs perfectly base-pair with their target messenger (m)RNAs and direct ribonucleolytic cleavage (Tang et al. 2003; Tang and Zamore 2004). There are also some examples of animal miRNAs that repress their targets via this mechanism (Yekta et al. 2004), but most animal miRNAs appear to bind imperfectly to sites in 3' UTRs and inhibit translation sans cleavage. The current model for maturation and function of mammalian miRNA involves nuclear cleavage of a 60- to 70-nt primary miRNA hairpin, transport of the pre-miRNA into the cytoplasm, processing by an endoribonuclease known as Dicer to yield a staggered, approximately 22-nt miRNA duplex, and loading onto mRNA strand by an RNA-induced silencing complex (RISC) (Bartel 2004; Tomari and Zamore 2005). The miRNA processing pathway can be co-opted by exogenously introduced double-stranded (ds)RNAs or short hairpin (sh)RNAs to generate, via Dicer, small interfering (si)RNAs, as originally observed in *C. elegans*. The difference lies in the fact that designed siRNAs have perfect complementarity to the coding region of a target mRNA, while natural miRNAs are generally imperfectly base-paired with the 3' UTR of a target mRNA. This difference is thought to be responsible for promoting ribonucleolytic cleavage (siRNA) as opposed to translation repression (miRNA; Parker et al. 2005).

Silencing of target mRNAs is but one instance of how small dsRNAs can regulate metabolism in eukaryotes. RNA-directed DNA methylation (RdDM) and RNA interference (RNAi)-mediated heterochromatin formation are epigenetic processes that result in covalent modification of cytosines in DNA and histones, respectively. RdDM has been described mostly in plants, while RNAi-mediated heterochromatin formation has been reported in fission yeast (Matzke and Birchler 2005). This range of regulatory mechanisms underlines the importance of small RNAs in higher eukaryotes. Further elucidation of the identities of small RNAs, and their pathways and targets, should eventually contribute to a better understanding of the etiology of many disease states, such as cancer (Johnson et al. 2005).

As was alluded to above, the presence of natural, sequence-directed regulatory mechanisms has given rise to many possibilities for engineering regulation. Antisense-based therapies have been developed to selectively inhibit a variety of genes (Crooke 2004); one such molecule (Vitravene) has been commercialized as an antiviral agent. Similarly, siRNA-mediated silencing of target genes has obvious therapeutic potential, not least because siRNAs are able to turnover their mRNA substrates (Hutvagner and Zamore 2002). One estimate suggests that siRNA-based therapeutics could eventually account for 10% of the American drug market (Howard 2003). Beyond binding to individual genes, engineered antisense sequences can potentially be adapted to gene circuits, as will be described in more detail in Sects. 4.2 and 5.

2.2
Shape-Based RNA Regulation

RNA molecules have long been known to be able to form structures that can bind to proteins and thereby assist in gene regulation. An example of this type of regulation in bacteria is mediated by 6S RNA (Wassarman and

Storz 2000). The 6S RNA is transcribed as cells exit the logarithmic phase and enter the stationary phase. It functions by binding to and inhibiting the RNA polymerase sigma70 subunit, which mediates recognition and transcription of housekeeping genes during the stationary phase. Similarly, B2 mRNA in eukaryotic cells has been shown to repress transcription in response to heat shock by binding and inhibiting RNA polymerase II (Allen et al. 2004; Espinoza et al. 2004).

More recently, natural RNAs have also been discovered that interact with small metabolites. Many of these so-called riboswitches are found in the 5′ UTRs of bacterial mRNAs, and their regulatory mechanism is reminiscent of the attenuation of the Trp operon (Brantl 2004; Mandal and Breaker 2004b). For example, a riboswitch responsible for sensing flavin mononucleotide (FMN) levels is present in the 5′ leader of riboflavin mRNA (Winkler et al. 2002) in *Bacillus subtilis*. Low levels of the metabolite facilitate the formation of an antiterminator in the leader region and allow transcription to occur. High levels of FMN stabilize a RNA stem-loop structure that leads to transcription termination. Similar riboswitches in low G+C gram-positive bacteria, like *B. subtilis*, regulate glutamine, thiamine, S-adenosyl methionine (SAM), and other anabolic pathways. Although gene regulation largely occurs by repression of transcription, an adenine-specific riboswitch has been shown to activate transcription by disruption of a transcription terminator (Mandal and Breaker 2004a). Another instance of an "on" switch is the glycine riboswitch that has two RNA binding sites in tandem that cooperatively sense glycine. As a result, this riboswitch is able to respond rapidly to falling or rising levels of glycine by repressing (default state) or activating (metabolite-induced) the glycine cleavage operon (Mandal et al. 2004).

In addition to modulating transcription, riboswitches that bind to metabolites like thiamine and riboflavin function via metabolite-mediated stabilization of an RNA conformer that occludes the ribosome binding site (RBS). Although riboswitches have not yet been unequivocally identified in eukaryotes, introns and 3′ and 5′ UTRs have been implicated in metabolite sensing and regulation of RNA processing, transport, and stability (Sudarsan et al. 2003).

Prior to the discovery of riboswitches, the ability of nucleic acids to bind to small metabolites had been amply demonstrated by in vitro selection. A wide variety of selected binding sequences (aptamers) have now been generated (Jayasena 1999; Wilson and Szostak 1999; Cox et al. 2002). In general, the three-dimensional structures of aptamers have proved to be complex folds that surround and interact with ligands at multiple points. In this regard, while biotechnology anticipated Nature, natural RNAs form even more complex ligand-binding structures. The crystal structure of a hypoxanthine-bound riboswitch from *B. subtilis* (Batey et al. 2004) reveals a complex RNA fold. Several phylogenetically conserved residues participate in interactions with hypoxanthine (Fan et al. 1996; Zimmermann et al. 1997; Baugh et al. 2000) and this ligand is almost completely occluded from solvent. As with other riboswitches that regulate transcription, two RNA folds are possible (Mandal et al. 2003): a classic Rho-independent terminator at sufficiently high concentrations of hypoxanthine, or a stable antiterminator element that facilitates transcription at low intracellular hypoxanthine concentrations.

In vitro selected aptamers have been adapted to function as biosensors (Soukup and Breaker 2000; Hesselberth et al. 2003; Robertson et al. 2004) that respond to a variety of environmental cues and ligands. This ability to engineer conformers and conformational changes has also allowed shape-based RNA regulators to be introduced into cellular systems, as will be described in more detail in Sects. 3 and 4. Thus, both riboswitches and aptamers are potential tools for the further elaboration of synthetic genetic circuits.

2.3
Catalytic Regulation

The finding by Cech and colleagues that an intron from *Tetrahymena thermophila* (Cech et al. 1981) could catalyze the cleavage and joining of oligoribonucleotide substrates in a highly specific manner confirmed previous suspicions that RNA could act as a robust catalyst. Since then, ribozymes with a variety of reaction mechanisms and functions have been discovered or evolved (Doudna and Cech 2002). In particular, the hammerhead ribozyme, at roughly 40 nt, is one of the smallest naturally occurring ribozymes and participates in the rolling circle replication of some viroids by carrying out a self-cleavage reaction in which a $2'$ oxygen participates in the attack on the scissile phosphate. The crystal structure of the hammerhead (Pley et al. 1994) revealed an active site formed by the apposition of three short helices, and catalysis occurs at the conserved trihelical junction. Whether hammerhead catalysis requires divalent cations and is driven by global or local rearrangements of the ribozyme is still being explored (Blount and Uhlenbeck 2005), but the orientation of the functional groups in the ribozyme catalytic site seem to play an important role in site-specific strand scission. Other ribozymes found in nature [such as the Hepatitis delta virus (HDV) and the hairpin ribozymes] share similar reaction characteristics to the hammerhead, although their three-dimensional structures (Ferre-D'Amare et al. 1998; Rupert and Ferre-D'Amare 2001) are quite different. As we shall see, this diversity of sequences, structures, and functions that are already observed in natural ribozymes presages the fact that they can be readily engineered as components of genetic circuits.

The hammerhead ribozyme has already proved to be an excellent platform for engineering. Its structural simplicity allowed the cleavage site to be separated as a substrate from the catalytic core. The result was a *trans*-cleaving ribozyme that could, by modifying its substrate binding arms, be altered so as to cleave nearly any sequence that contained UH (any base but G) (Kore et al. 1998). The engineered hammerhead has all the catalytic features of a true enzyme—multiple turnover of the substrate and rapid catalytic enhancement ($\sim 10^9$-fold over spontaneous RNA hydrolysis). Similar nucleic acid cleavases have been selected from random pools, such as the X-motif ribozyme and the 10–23 deoxyribozyme (Santoro and Joyce 1997; Tang and Breaker 2000; Emilsson and Breaker 2002; Lazarev et al. 2003), and have also been engineered to carry out site-specific cleavage reactions. The ability of site-specific cleavases to function as therapies for viral or other diseases has been detailed in several reviews (Sullenger and Gilboa 2002; Khan and Lal 2003; Michienzi et al. 2003; Puerta-Fernandez et al. 2003; Citti and Rainaldi 2005). Indeed, ribozymes that may be important for the treatment of diseases such as acquired immunodeficiency syndrome (AIDS), hepatitis, cancer, and diabetes are in different phases of clinical trials. Their ultimate success will depend upon optimizing various factors such as relative ribozyme amounts in cells, co-localization with target RNAs, structural features of the target that influence its accessibility by the ribozyme, intracellular stability, and so forth. However, this success will also make them excellent tools for inclusion in real genetic circuits for molecular medicine.

While the group I and group II introns are not as advanced in their applications as the hammerhead, they can also be engineered for specific functions in vivo. The group I intron, that is normally spliced out in *cis*, has been engineered to splice in *trans* and repair defective mRNAs (Sullenger and Gilboa 2002; Byun et al. 2003). Upon base-pairing with a target mRNA, the group I intron is poised to transfer the corrected part of transcript it carries at its 3' end. Sullenger and colleagues have developed and used this method to displace the faulty region of the sickle cell globin transcript, and demonstrated the repair of up to 50% of sickle cell transcripts in vivo. The spliceosome, another complex natural ribozyme that functions in splicing pre-mRNA, can also be used to reprogram target mRNAs to *trans*-splice in a method is known as SMaRT (spliceosome-mediated RNA *trans*-splicing) (Mansfield et al. 2004). Briefly, a pre-*trans* splicing molecule (PTM) can be designed comprising: (1) a binding domain that is complementary to the intron in the target pre-mRNA that serves to position the PTM at the appropriate location in the transcript, (2) a splicing domain that contains splice sites equivalent to those in the *cis*-spliced target precursor, and (3) a new coding domain that will replace the "defective" exon or exons in the linear mRNA. When such an engineered PTM is introduced exogenously into a cell it binds to and non-covalently branches from a pre-mRNA and can then be spliced by the spliceosome to upstream exons and thereby repair faulty transcripts.

The group II intron from *Lactococcus lactis* has also been studied in great detail. The lariat form of the intron binds to an intron-encoded endonuclease/reverse transcriptase protein and forms a catalytically active ribonucleoprotein particle that has the potential to reverse splice into target genes. Targeting is determined in part by base-pairing between the intron and a gene; and by altering the sequence of the intron, the intron-ribonucleoprotein can be programmed to insert into almost any gene. This technology has been used to create insertional mutations on demand in the genomic DNAs of several gram-positive and gram-negative bacteria (Lambowitz and Zimmerly 2004; Perutka et al. 2004). Karberg et al. (2001) have demonstrated the feasibility of targeting reverse splicing group II introns to mammalian cells as well, opening up the possibility that engineered introns could be used to knock out human disease genes.

By coupling ligand-binding with nucleic acid catalysis it has proved possible to engineer allosteric ribozymes or "aptazymes" (Roth and Breaker 2004). Essentially, generating an aptazyme involves replacing non-critical portions of a ribozyme's secondary structure with a ligand-binding RNA, such as a selected aptamer. These two functional regions are then joined by a short structural element that is sometimes referred to as a "communication module" (Koizumi et al. 1999; Robertson and Ellington 1999, 2000; Soukup and Breaker 2000; Jose et al. 2001). Ligand-binding causes conformational changes in the allosteric domain that are transmitted to and modulate the function of the catalytic domain, in much the same way that ligand-binding changes the conformation and function of the riboswitches described in the previous section. For example, an anti-ATP aptamer selected from a random sequence pool was joined to the hammerhead ribozyme, yielding an allosteric catalyst that was activated 180-fold in the presence of ATP (Tang and Breaker 1997). A similar experiment with a selected ribozyme ligase yielded an allosteric catalyst that was activated 800-fold in the presence of ATP (Robertson and Ellington 1999). Effector dependence can also be engineered into larger ribozymes, such as the group I intron. For example, Kertsburg and Soukup (2002) joined an anti-theophylline aptamer to the *Tetrahymena* group I ribozyme, randomized the joining region, and selected for a functional communication module. The resultant group I aptazyme was rendered roughly 26-fold more active by theophylline in vitro. Thompson et al. (2002) have also engineered theophylline-responsive group I introns by fusing the anti-theophylline aptamer with a T4 thymidylate synthase (td) group I self-splicing intron. In this latter instance, the aptazymes exhibit theophylline-dependent behavior in *E. coli*. Recently, a natural aptazyme was discovered in which binding to glucosamine-6-phosphate (GlcN6P) led to the cleavage of the 5′ UTR of the glmS mRNA that codes for a GlcN6P biosynthetic enzyme. This natural aptazyme thus participates in a negative feedback loop that decreases functional transcript levels, GlmS protein production, and presumably flux through the pathway (Winkler et al. 2004).

Other aptazymes have been engineered that are regulated by sequence rather than small molecules (Komatsu et al. 2000). Taira and co-workers have successfully designed a dimeric hammerhead ribozyme linked through the stem II region called a maxizyme (Kuwabara et al. 1998; Iyo et al. 2002; Iyo et al. 2004). Dimer formation and activation of catalysis is induced only upon base-pairing to a specific substrate mRNA, which is in turn cleaved. Maxizymes have now been used in vivo for the specific cleavage of defective mRNAs.

For example, a mouse model for chronic myeloid leukemia was treated with a maxizyme that cleaved the chimeric BCR-ABL (breakpoint cluster region-Abelson murine leukemia viral oncogene homolog) chimeric transcript. Active dimer formation only occurred when the two substrate binding regions of the ribozyme interacted with the two portions of the chimeric transcript. In contrast, wild-type mRNAs could only bind one of the two portions of the maxizyme and were thus resistant to cleavage and retained their normal cellular function.

3
Optimizing the Functions of RNA Tools

The basic functionality of a number of natural and unnatural RNA molecules can be adapted to regulate cellular metabolism via base-pairing, shape recognition, and catalysis. While the basic mechanisms can be configured to respond to and act upon sequence targets (such as site-specific siRNAs or ribozymes), it is more difficult to optimize the functions of components. To overcome this difficulty, directed evolution has previously been used to select nucleic acid binding species (aptamers) and to optimize the function of ribozymes and aptazymes.

Nucleic acid aptamers have already been selected against a wide variety of targets, from small molecules to proteins to supramolecular structures such as viruses and cells (Famulok and Mayer 1999; Hesselberth et al. 2000; Marshall and Ellington 2000). Aptamers typically bind to their targets with extremely high affinities and have specificities that rival antibodies. For example, an aptamer that binds theophylline discriminates approximately 10,000-fold against the highly similar molecule caffeine, which differs by only a single methyl group (Jenison et al. 1994). Aptamers not only bind their targets in vitro but can inhibit or modulate function in vivo, making them excellent potential therapeutics (as reviewed by Nimjee et al. 2005). Aptamers have some advantages over antibodies in a therapeutic setting, such as an inherent lack of immunogenicity. However, the in vivo instability of aptamers is an obvious challenge with respect to their exogenous delivery for therapeutic applications. To get around this problem, modified nucleotides have been used to increase aptamer stability and half-life in vivo. For endogenous delivery and gene therapy applications, specialized constructs have been designed for the in vivo expression of aptamers. For example, Joshi and Prasad (2002) have expressed anti-human immunodeficiency virus (HIV) reverse transcriptase aptamers from a vector and flanked them with *cis*-cleaving hammerhead ribozymes in order to release the functional aptamer from the transcript. The aptamers were able to reduce HIV replication up to 99.5% in vivo.

Hammerhead ribozymes are already excellent therapeutics, even without directed evolution. However, directed evolution can be used to optimize targeting. Pan et al. (2003) have reported the in vitro selection of ribozymes that efficiently target human papillomavirus type 16 E6/E7 mRNA. The in vitro-selected ribozymes were also shown to have activity in cell culture while their catalytically inactive counterparts did not.

The directed evolution of ribozymes has also proved useful in functional genomics applications. Taira and colleagues (Wadhwa et al. 2004) employed a randomized, active hammerhead ribozyme library to screen for defects in muscle differentiation in myoblast cultures. The selected ribozymes were cloned, sequenced, and their putative targets identified by computationally scanning for appropriate cleavage sites in a genome. The mRNA targets were then validated by rationally targeting them with a hammerhead ribozyme or siRNA to determine if they would once again produce the desired phenotype. Suyama et al. (2004a,b), have carried out a similar experiment in which they injected pre-metastatic cells containing the ribozyme library into mice and then isolated pulmonary tumors from cells that had metastasized. Identification of mRNA targets that were cleaved by the ribozymes provided valuable information about what genes were actually involved in tumor metastasis. Of course, such large-scale reverse genetic studies can also potentially be performed using siRNA or shRNA libraries (Berns et al. 2004; Paddison et al. 2004; Silva et al. 2004; Dykxhoorn and Lieberman 2005; Sonnichsen et al. 2005) rather than ribozymes.

4
Engineering Biological Circuits

4.1
Synthetic Biology

We have seen in the previous section that RNA is a flexible tool that can be used to modulate cellular behavior in a predictable manner. These tools now enable us to go to the next step and rationally connect them to give a logical output. In general, the use of biological components in engineered genetic circuits has been termed "synthetic biology." In addition to creating circuits for their own sake, though, the engineered tools and quantitative practices of synthetic biology are becoming increasingly important for understanding natural biological systems and pathways. This is especially true given the recent emphasis on a related buzzword, "systems biology," which has come to mean the examination and modeling of the multitude of molecular interactions in a cell or organism. High-throughput experiments like microarrays, mass spectrometry, and systematic genetic deletions offer up a panoramic, integrated view of biological systems (DeRisi et al. 1997; Giaever et al. 2002; Pollack and

Iyer 2002; Aebersold and Mann 2003; Ghaemmaghami et al. 2003; Huh et al. 2003; Ranish et al. 2003; Almaas et al. 2004; Barabasi and Oltvai 2004; Fraser and Marcotte 2004; Lee et al. 2004; Raser and O'Shea 2004). The pathways and networks that are conjured up by a systems biology approach can in many instances be explicitly tested by resorting to perturbations using the tools of synthetic biology, or by the parallel construction of artificial genetic circuits that mimic natural systems.

A brief overview of some of the high points in synthetic biology will set the stage for a more detailed examination of how RNA tools have been adapted to complex genetic circuits. Perhaps the defining experiment in engineering artificial genetic circuits and the christening of the field of synthetic biology was the construction of the "repressilator" by Elowitz and Leibler (2000). In this circuit, a series of repressors formed an oscillator in which repressor A repressed the production of repressor B, B repressed C, and C repressed A. In engineering terms, each connection mimicked an electronic inverter and the ring oscillator should inherently display oscillatory behavior. The operation of the repressilator was reported by observing the induction of the GFP (green fluorescent protein) gene, which was also under the control of one of the repressors. One reason that synthetic biology has a slightly different flavor than more conventional genetic or biotechnology engineering is that the circuit is composed of well-defined modules (promoters, repressors, genes) from a common "toolbox." This is the paradigm that we also wish to apply to RNA tools and circuits, as described in Sects. 4.2 and 5.

Upon introduction of a repressilator circuit into *E. coli*, the cells displayed a periodic green readout with the oscillations spanning multiple cell divisions. Interestingly, the oscillatory behavior, despite being transmitted to the progeny cells, was somewhat erratic and out of phase with only a sub population (∼40%) of cells exhibiting it. This variability from the predicted repressilator dynamics has been attributed to stochastic effects resulting from the inherent noise associated with cellular gene expression (Blake et al. 2003). The noisy behavior of the circuit points up a fundamental problem in engineering genetic circuits: While it is easy to define modular tools, these tools will not necessarily work similarly or predictably in different genetic backgrounds, and the degree to which their behavior can be quantitatively predicted even in a single genetic background is questionable. These problems are actually a boon to engineering with RNA—rather than protein—tools, since base-pairing is inherently digital and the thermodynamics of base-pairing will be largely independent of the phylogenetic origin or current cellular milieu of a given RNA.

As has been previously mentioned with respect to the development of molecular tools, rational engineering can be augmented with directed evolution. This is also true for circuit optimization. Yokobayashi et al. (2002) used engineering principles to design a two-piece logic gate circuit in which the cI repressor from lambda phage was under the control of LacI, and an enhanced yellow fluorescent protein (EYFP) reporter gene was under control of cI. The circuit acted as a logical inverter: Addition of isopropyl thiogalactoside (IPTG) would relieve cI repression by LacI, which would in turn repress expression of EYFP. However, this relatively simple circuit failed to function in a cellular context. Leaky expression of cI even in the absence of IPTG was sufficient to abrogate the expression of EYFP. The authors resorted to sieving a mutagenized cI library for the desired output, and consequently selected clones of cI that more weakly repressed EYFP expression, but also required more IPTG to fully repress the circuit as a whole. This experiment demonstrated the feasibility of judiciously combining a rational engineering approach with a directed evolution approach to fine-tune the function of an artificial circuit inside cells.

The circuits considered so far have been largely composed of DNA- and protein-based biological parts (Kaern et al. 2003). We will now explore the use of the previously described RNA "toolbox" for gene regulation and circuit engineering.

4.2
Designed RNA Circuits for Translation Control

The construction of RNA tools in synthetic circuits is largely predicated on the notion that predictable secondary structural features can be rationally modified to be effector dependent or to interact with one another. As an example, Isaacs et al. (2004) have engineered modular riboregulators that can be effectively used to repress or activate the translation of a target mRNA. A prokaryotic gene cassette was constructed in which a reporter gene (GFP) was adjoined to a short 5' UTR sequence that was complementary to the ribosome-binding site. As a result, reporter gene transcripts are auto-inhibited by the formation of an anti-RBS stem-loop structure. This *cis* repression can in turn be relieved by expression of a small non-coding RNA in *trans* from a second promoter. The *trans*-activating RNA targets the anti-RBS stem with high specificity by base-pairing and unfolds the stem-loop structure, ultimately allowing ribosome entry and translation to occur.

Bayer and Smolke (2005) have further exploited base-pairing interactions with target mRNAs by engineering effector-dependent antisense RNAs that can bind and inhibit (rather than activate) translation. The so-called "antiswitch" was designed by combining an antisense domain against a target reporter mRNA (GFP) with an anti-theophylline aptamer. In the absence of theophylline, the antisense domain is occluded by the formation of a hairpin stem, similar to the anti-RBS strategy described above. Theophylline binding to the aptamer domain induces a conformational change that exposes the antisense domain, allowing it to bind to its target, thus resulting in decreased translation of the target.

Finally, Mulligan and co-workers (Yen et al. 2004) have used catalytic rather than antisense and shape recognition strategies to engineer translation control of a mammalian target gene. Several *cis*-cleaving hammerhead ribozymes were cloned into multiple locations in the 5′ UTR of a reporter gene, *lacZ*, and tested for their ability to silence the reporter. The most efficient version of their construct decreased LacZ levels by roughly 1,400-fold compared to an inactive ribozyme control. The authors then engineered ligand dependence into their *cis*-acting catalytic repressors by two approaches: Various types of antisense oligonucleotides were assayed, and it was found that the absolute levels of LacZ gene expression after addition of morpholino-oligonucleotides approached 50% of the gene expression levels produced by the inactive ribozyme (that is, the theoretical maximum). The authors then screened a small molecule library for LacZ production and found that toyocamycin, a nucleoside analog, could specifically inhibit hammerhead cleavage and restore reporter gene expression in both tissue culture and mouse models.

The advantages of RNA-based translation control circuits relative to protein-mediated strategies, such as the use of repressors like TetR, are that multiple regulatory domains can be independently engineered and modulated in parallel with one another. The size of the constructs employed makes them quite portable and should allow their incorporation into virtually any target gene or vector construct. Moreover, antisense binding, shape recognition, and RNA catalysis are all biochemical properties that are independent of the phylogenetic characteristics of the organism in which they are found, unlike proteins, which frequently require changes to their translation start sites, codon usage, and the introduction of processing signals, such as nuclear localization sequences. Each of these features can be seen in the examples cited above, re-emphasizing that RNA tools are ideal candidates for engineering circuits and therapeutic applications.

4.3
Designed RNA Circuits for Transcription Control

RNA-mediated control of transcription is also possible. O'Malley and colleagues (Lanz et al. 1999) isolated and functionally characterized a novel transcription co-activator termed steroid receptor RNA co-activator (SRA). Distinct RNA motifs within SRA were shown to play structural roles in protein–RNA complexes towards and could facilitate *trans*-activation by steroid hormone receptors (Lanz et al. 2002).

In an engineering feat that mimicked the discovery of RNA mediators of protein function, Buskirk and Liu (2004b) developed a three-hybrid, in vivo evolution strategy to select for an RNA-based transcription activator. Their library and selection design consisted of a plasmid with a 5′ leader sequence, a random N40 or a N80 region, two MS2 protein binding sites, and a transcription terminator. This library was transformed into a yeast strain encoding a DNA binding protein, LexA, fused to the MS2 protein. The His3 gene, which is essential for histidine biosynthesis, was placed under the control of a LexA operator. This selection scheme ensured that transcription activation and histidine production would only occur if an RNA mediator from the random sequence library was able to bridge DNA, LexA, MS2, and the MS2 operator to a transcription activator. A surprisingly large fraction of the library (∼0.2%) proved to be functional, indicating that there are either many transcriptional activators that could be non-specifically attracted to the DNA-bound bait or many RNA structures could attract a given transcription activator (or both). By increasing the stringency of selection and using a biased pool that favored the dominant clone, the authors were able to isolate transcription activators that rivaled the efficiency of known transcription activator proteins.

As was the case for translation regulation, it also proved possible to engineer a ligand-dependent RNA transcription regulator (Buskirk et al. 2004a). The previously evolved RNA mediator was fused to a previously selected RNA aptamer against tetramethylrosamine (TMR) via a randomized linker region. Ligand-dependent mediators were selected in the presence of TMR. Individual clones showed dose-dependent activation dynamics and up to about tenfold transcription activation in the presence of TMR.

While only transcription activation has been targeted so far, the role of RNA sequences in splicing, termination, and processing should afford multiple avenues for engineering additional regulatory steps prior to translation.

5
The Path Forward: Engineered Circuits in Medicine

The point to building circuits is not merely to have set pieces like the repressilator or to have slightly more novel means of regulating individual genes, such as the simple RNA-based circuits described above. Rather, it should be possible to engineer complex networks that have a variety of inputs and outputs, and that rationally compute functional behavior based on pre-set interactions.

The concept of DNA-based computation has existed for over a decade, since Adleman reported using DNA to solve combinatorial optimization problems such as a Hamiltonian path problem (Adleman 1994). Briefly, such DNA-based molecular computations depend on embedding variables or instructions in the DNA sequence and allowing such instructions to act upon themselves to produce an output based on Watson-Crick base pairing rules (Cox et al. 1999). These DNA circuits based purely on "hybridization logic" are highly parallel and energy efficient, but are computationally limited by high error rates. However, the computations that are typically required in biology are not known for their digital exactitude, but rather yield appropriate analog responses (such as a gradient of gene expression). Thus, DNA or RNA computations merely must be applied to the right sorts of problems: the questions normally posed by cellular survival, development, and behavior.

Designed nucleic acid-based networks have in fact been shown to be able to decipher complex inputs and produce a programmed output. Stojanovic et al. (2003) describe a molecular automaton (a device that can convert information from one form to another through a defined procedure), made of only deoxyribozymes, that plays a version of tic-tac-toe against a human opponent with flawless logic. An RNA-cleaving deoxyribozyme originally selected by Breaker and Joyce (1994) served as a starting point for engineering the automaton. Modular, sequence-sensing allosteric domains were introduced into the deoxyribozymes, allowing the authors to activate or inhibit the enzyme in the presence of complementary oligonucleotide effectors. Different combinations of 23 different allosteric deoxyribozymes were placed in the 9 wells (representing the tic-tac-toe board) and eight different effectors were used to signal moves made by a human player at each step (that is, a move to square 4 in a tic-tac-toe board would be represented by adding effector 4 to all the wells). The combinations of deoxyribozymes "hardwired" in each well represented a series of Boolean logical functions appropriate to that well, such as the production of an "x" (fluorescent signal) in corner well 1 in response to a move by a human player to a non-corner position like well 4, but no production of signal if the human player moved to a corner position like well 3. The result was an astonishing exhibition in which the DNA-based automaton could respond to a human move by processing the sequence input and cleaving an RNA substrate to yield a fluorescent product in only one well, always yielding the correctly played output or move. The level of molecular integration required in this tour de force was not easily achieved: Every allosteric deoxyribozyme had to be empirically tuned to maximize the signal-to-noise ratio in response to its effector. However, this demonstration clearly revealed the complex information processing capabilities available to nucleic acid-based circuits and their ability to respond to environmental signals.

The natural extension of these first two demonstrations would be to make drugs that are capable of processing themselves based on cellular or physiological inputs. Benenson et al. (2004) have now designed biomolecular automata capable of performing logical analysis of mRNA disease indicators in vitro. An antisense single-stranded (ss)DNA is the surrogate drug, and it is appended to a "diagnostic" module with a hairpin structure. The hairpin contains four sets of cleavage points, each of which represents a different diagnostic transition. Diagnostic mRNA whose levels would be changed in a certain disease state can base-pair with a cast of *trans* oligonucleotide regulators. Hybridization in turn causes a unique restriction enzyme to form at the termini of the combined diagnostic-therapeutic hairpin structure, leading to its cleavage (the enzyme being present in the reaction mix). Different disease indicators (different mRNA levels) are sequentially assessed in order to cleave the hairpin stem down to the antisense oligonucleotide alone, which is then released as a drug. In the absence of the disease state, normal mRNA sequences and levels will not lead to the release of the drug. While this particular implementation would be grossly impractical in a cellular scenario (given the large number of complex components that would have to be delivered), it is nonetheless a working demonstration of how molecular diagnostics and molecular therapeutics can be brought together in the future to create "smart" drugs.

While DNA computation with restriction enzymes does not fit well into a cellular environment, RNA computations based on antisense binding, shape recognition, and RNA catalysis would be self-contained and self-processing. Moreover, RNA may itself prove to be an intrinsically better drug than DNA, given the natural mechanisms already in place for processing miRNAs and siRNAs. By targeting specific functions during the maturation of specific mRNAs, RNA-based tools and modules would allow us to exercise a wide range of control over engineered genetic circuits. Activation of the modules could be either sequence- or ligand-dependent, and therefore could be seamlessly overlaid on extant metabolism, ultimately resulting in increasingly smarter therapeutics.

Moreover, we need not restrict ourselves to antisense-like drugs directed against sequence targets. Into the future, aptamer drugs that target proteins both inside and outside of the cell may be released or modified as a result of engineered genetic circuits. Sullenger and co-workers have already shown how aptamer therapeutics can be rationally controlled (Rusconi et al. 2002) by engineering an antidote (typically a complementary oligonucleotide) that denatures the aptamer. Aptamer–antidote pairs could be obviously adapted to a medically relevant genetic circuit similar to the one described above. Similarly, RNA tools need not work on the transcription or translation machinery alone. The ability of aptamers to uniquely recognize protein shapes and modification states such as aptamers that can recognize the phosphorylation state of the extracellular signal-regulated (ERK)2 kinase (Vaish et al. 2002, 2004), may allow the construction of "genetic" circuits that operate solely at the level of the cell's signal transduction machinery. Such circuits could be considered molecular "augments" of the natural machinery that would fine tune its function and make it more resilient to genetic or environmental insults.

References

Adleman LM (1994) Molecular computation of solutions to combinatorial problems. Science 266:1021–1024

Aebersold R, Mann M (2003) Mass spectrometry-based proteomics. Nature 422:198–207

Allen TA, Von Kaenel S, Goodrich JA, Kugel JF (2004) The SINE-encoded mouse B2 RNA represses mRNA transcription in response to heat shock. Nat Struct Mol Biol 11:816–821

Almaas E, Kovacs B, Vicsek T, Oltvai ZN, Barabasi AL (2004) Global organization of metabolic fluxes in the bacterium Escherichia coli. Nature 427:839–843

Ambros V (2004) The functions of animal microRNAs. Nature 431:350–355

Argaman L, Hershberg R, Vogel J, Bejerano G, Wagner EG, Margalit H, Altuvia S (2001) Novel small RNA-encoding genes in the intergenic regions of Escherichia coli. Curr Biol 11:941–950

Barabasi AL, Oltvai ZN (2004) Network biology: understanding the cell's functional organization. Nat Rev Genet 5:101–113

Bartel DP (2004) MicroRNAs: genomics, biogenesis, mechanism, and function. Cell 116:281–297

Bartel DP, Chen CZ (2004) Micromanagers of gene expression: the potentially widespread influence of metazoan microRNAs. Nat Rev Genet 5:396–400

Batey RT, Gilbert SD, Montange RK (2004) Structure of a natural guanine-responsive riboswitch complexed with the metabolite hypoxanthine. Nature 432:411–415

Baugh C, Grate D, Wilson C (2000) 2.8 Å crystal structure of the malachite green aptamer. J Mol Biol 301:117–128

Bayer TS, Smolke CD (2005) Programmable ligand-controlled riboregulators of eukaryotic gene expression. Nat Biotechnol 23:337–343

Benenson Y, Gil B, Ben-Dor U, Adar R, Shapiro E (2004) An autonomous molecular computer for logical control of gene expression. Nature 429:423–429

Berns K, Hijmans EM, Mullenders J, Brummelkamp TR, Velds A, Heimerikx M, Kerkhoven RM, Madiredjo M, Nijkamp W, Weigelt B, Agami R, Ge W, Cavet G, Linsley PS, Beijersbergen RL, Bernards R (2004) A large-scale RNAi screen in human cells identifies new components of the p53 pathway. Nature 428:431–437

Blake WJ, M KA, Cantor CR, Collins JJ (2003) Noise in eukaryotic gene expression. Nature 422:633–637

Blount KF, Uhlenbeck OC (2005) The structure-function dilemma of the hammerhead ribozyme. Annu Rev Biophys Biomol Struct 34:415–440

Brantl S (2004) Bacterial gene regulation: from transcription attenuation to riboswitches and ribozymes. Trends Microbiol 12:473–475

Breaker RR (2004) Natural and engineered nucleic acids as tools to explore biology. Nature 432:838–845

Breaker RR, Joyce GF (1994) A DNA enzyme that cleaves RNA. Chem Biol 1:223–229

Brescia CC, Mikulecky PJ, Feig AL, Sledjeski DD (2003) Identification of the Hfq-binding site on DsrA RNA: Hfq binds without altering DsrA secondary structure. RNA 9:33–43

Buskirk AR, Landrigan A, Liu DR (2004a) Engineering a ligand-dependent RNA transcriptional activator. Chem Biol 11:1157–1163

Buskirk AR, Ong YC, Gartner ZJ, Liu DR (2004b) Directed evolution of ligand dependence: small-molecule-activated protein splicing. Proc Natl Acad Sci U S A 101:10505–10510

Byun J, Lan N, Long M, Sullenger BA (2003) Efficient and specific repair of sickle beta-globin RNA by trans-splicing ribozymes. RNA 9:1254–1263

Carpousis AJ (2002) The Escherichia coli RNA degradosome: structure, function and relationship in other ribonucleolytic multienzyme complexes. Biochem Soc Trans 30:150–155

Carter RJ, Dubchak I, Holbrook SR (2001) A computational approach to identify genes for functional RNAs in genomic sequences. Nucleic Acids Res 29:3928–3938

Cech TR, Zaug AJ, Grabowski PJ (1981) In vitro splicing of the ribosomal RNA precursor of Tetrahymena: involvement of a guanosine nucleotide in the excision of the intervening sequence. Cell 27:487–496

Chen S, Lesnik EA, Hall TA, Sampath R, Griffey RH, Ecker DJ, Blyn LB (2002) A bioinformatics based approach to discover small RNA genes in the Escherichia coli genome. Biosystems 65:157–177

Chen S, Zhang A, Blyn LB, Storz G (2004) MicC, a second small-RNA regulator of Omp protein expression in Escherichia coli. J Bacteriol 186:6689–6697

Citti L, Rainaldi G (2005) Synthetic hammerhead ribozymes as therapeutic tools to control disease genes. Curr Gene Ther 5:11–24

Cox JC, Cohen DS, Ellington AD (1999) The complexities of DNA computation. Trends Biotechnol 17:151–154

Cox JC, Hayhurst A, Hesselberth J, Bayer TS, Georgiou G, Ellington AD (2002) Automated selection of aptamers against protein targets translated in vitro: from gene to aptamer. Nucleic Acids Res 30:e108

Crooke ST (2004) Progress in antisense technology. Annu Rev Med 55:61–95

DeRisi JL, Iyer VR, Brown PO (1997) Exploring the metabolic and genetic control of gene expression on a genomic scale. Science 278:680–686

Doudna JA, Cech TR (2002) The chemical repertoire of natural ribozymes. Nature 418:222–228

Dykxhoorn DM, Lieberman J (2005) The silent revolution: RNA interference as basic biology, research tool, and therapeutic. Annu Rev Med 56:401–423

Elowitz MB, Leibler S (2000) A synthetic oscillatory network of transcriptional regulators. Nature 403:335–338

Emilsson GM, Breaker RR (2002) Deoxyribozymes: new activities and new applications. Cell Mol Life Sci 59:596–607

Espinoza CA, Allen TA, Hieb AR, Kugel JF, Goodrich JA (2004) B2 RNA binds directly to RNA polymerase II to repress transcript synthesis. Nat Struct Mol Biol 11:822–829

Famulok M, Mayer G (1999) Aptamers as tools in molecular biology and immunology. Curr Top Microbiol Immunol 243:123–136

Fan P, Suri AK, Fiala R, Live D, Patel DJ (1996) Molecular recognition in the FMN-RNA aptamer complex. J Mol Biol 258:480–500

Ferre-D'Amare AR, Zhou K, Doudna JA (1998) Crystal structure of a hepatitis delta virus ribozyme. Nature 395:567–574

Fire A, Xu S, Montgomery MK, Kostas SA, Driver SE, Mello CC (1998) Potent and specific genetic interference by double-stranded RNA in Caenorhabditis elegans. Nature 391:806–811

Fraser AG, Marcotte EM (2004) A probabilistic view of gene function. Nat Genet 36:559–564

Ghaemmaghami S, Huh WK, Bower K, Howson RW, Belle A, Dephoure N, O'Shea EK, Weissman JS (2003) Global analysis of protein expression in yeast. Nature 425:737–741

Giaever G, Chu AM, Ni L, Connelly C, Riles L, Veronneau S, Dow S, Lucau-Danila A, Anderson K, Andre B, Arkin AP, Astromoff A, El-Bakkoury M, Bangham R, Benito R, Brachat S, Campanaro S, Curtiss M, Davis K, Deutschbauer A, Entian KD, Flaherty P, Foury F, Garfinkel DJ, Gerstein M, Gotte D, Guldener U, Hegemann JH, Hempel S, Herman Z, Jaramillo DF, Kelly DE, Kelly SL, Kotter P, LaBonte D, Lamb DC, Lan N, Liang H, Liao H, Liu L, Luo C, Lussier M, Mao R, Menard P, Ooi SL, Revuelta JL, Roberts CJ, Rose M, Ross-Macdonald P, Scherens B, Schimmack G, Shafer B, Shoemaker DD, Sookhai-Mahadeo S, Storms RK, Strathern JN, Valle G, Voet M, Volckaert G, Wang CY, Ward TR, Wilhelmy J, Winzeler EA, Yang Y, Yen G, Youngman E, Yu K, Bussey H, Boeke JD, Snyder M, Philippsen P, Davis RW, Johnston M (2002) Functional profiling of the Saccharomyces cerevisiae genome. Nature 418:387–391

Gottesman S (2004) The small RNA regulators of Escherichia coli: roles and mechanisms. Annu Rev Microbiol 58:303–328

Griffiths-Jones S, Moxon S, Marshall M, Khanna A, Eddy SR, Bateman A (2005) Rfam: annotating non-coding RNAs in complete genomes. Nucleic Acids Res 33:D121–D124

Hesselberth J, Robertson MP, Jhaveri S, Ellington AD (2000) In vitro selection of nucleic acids for diagnostic applications. J Biotechnol 74:15–25

Hesselberth JR, Robertson MP, Knudsen SM, Ellington AD (2003) Simultaneous detection of diverse analytes with an aptazyme ligase array. Anal Biochem 312:106–112

Howard K (2003) Unlocking the money-making potential of RNAi. Nat Biotechnol 21:1441–1446

Huh WK, Falvo JV, Gerke LC, Carroll AS, Howson RW, Weissman JS, O'Shea EK (2003) Global analysis of protein localization in budding yeast. Nature 425:686–691

Huppi K, Martin SE, Caplen NJ (2005) Defining and assaying RNAi in mammalian cells. Mol Cell 17:1–10

Hutvagner G, Zamore PD (2002) A microRNA in a multiple-turnover RNAi enzyme complex. Science 297:2056–2060

Isaacs FJ, Dwyer DJ, Ding C, Pervouchine DD, Cantor CR, Collins JJ (2004) Engineered riboregulators enable post-transcriptional control of gene expression. Nat Biotechnol 22:841–847

Iyo M, Kawasaki H, Taira K (2002) Construction of an allosteric trans-maxizyme targeting for two distinct oncogenes. Nucleic Acids Res Suppl 115–116

Iyo M, Kawasaki H, Taira K (2004) Maxizyme technology. Methods Mol Biol 252:257–265

Jayasena SD (1999) Aptamers: an emerging class of molecules that rival antibodies in diagnostics. Clin Chem 45:1628–1650

Jenison RD, Gill SC, Pardi A, Polisky B (1994) High-resolution molecular discrimination by RNA. Science 263:1425–1429

Johnson SM, Grosshans H, Shingara J, Byrom M, Jarvis R, Cheng A, Labourier E, Reinert KL, Brown D, Slack FJ (2005) RAS is regulated by the let-7 microRNA family. Cell 120:635–647

Jose AM, Soukup GA, Breaker RR (2001) Cooperative binding of effectors by an allosteric ribozyme. Nucleic Acids Res 29:1631–1637

Joshi P, Prasad VR (2002) Potent inhibition of human immunodeficiency virus type 1 replication by template analog reverse transcriptase inhibitors derived by SELEX (systematic evolution of ligands by exponential enrichment). J Virol 76:6545–6557

Kaern M, Blake WJ, Collins JJ (2003) The engineering of gene regulatory networks. Annu Rev Biomed Eng 5:179–206

Karberg M, Guo H, Zhong J, Coon R, Perutka J, Lambowitz AM (2001) Group II introns as controllable gene targeting vectors for genetic manipulation of bacteria. Nat Biotechnol 19:1162–1167

Kertsburg A, Soukup GA (2002) A versatile communication module for controlling RNA folding and catalysis. Nucleic Acids Res 30:4599–4606

Khan AU, Lal SK (2003) Ribozymes: a modern tool in medicine. J Biomed Sci 10:457–467

Koizumi M, Soukup GA, Kerr JN, Breaker RR (1999) Allosteric selection of ribozymes that respond to the second messengers cGMP and cAMP. Nat Struct Biol 6:1062–1071

Komatsu Y, Yamashita S, Kazama N, Nobuoka K, Ohtsuka E (2000) Construction of new ribozymes requiring short regulator oligonucleotides as a cofactor. J Mol Biol 299:1231–1243

Kore AR, Vaish NK, Kutzke U, Eckstein F (1998) Sequence specificity of the hammerhead ribozyme revisited; the NHH rule. Nucleic Acids Res 26:4116–4120

Kuwabara T, Warashina M, Tanabe T, Tani K, Asano S, Taira K (1998) A novel allosterically trans-activated ribozyme, the maxizyme, with exceptional specificity in vitro and in vivo. Mol Cell 2:617–627

Lambowitz AM, Zimmerly S (2004) Mobile group II introns. Annu Rev Genet 38:1–35

Lanz RB, McKenna NJ, Onate SA, Albrecht U, Wong J, Tsai SY, Tsai MJ, O'Malley BW (1999) A steroid receptor coactivator, SRA, functions as an RNA and is present in an SRC-1 complex. Cell 97:17–27

Lanz RB, Razani B, Goldberg AD, O'Malley BW (2002) Distinct RNA motifs are important for coactivation of steroid hormone receptors by steroid receptor RNA activator (SRA). Proc Natl Acad Sci U S A 99:16081–16086

Lazarev D, Puskarz I, Breaker RR (2003) Substrate specificity and reaction kinetics of an X-motif ribozyme. RNA 9:688–697

Lee I, Date SV, Adai AT, Marcotte EM (2004) A probabilistic functional network of yeast genes. Science 306:1555–1558

Lenz DH, Mok KC, Lilley BN, Kulkarni RV, Wingreen NS, Bassler BL (2004) The small RNA chaperone Hfq and multiple small RNAs control quorum sensing in Vibrio harveyi and Vibrio cholerae. Cell 118:69–82

Mackie GA, Genereaux JL (1993) The role of RNA structure in determining RNase E-dependent cleavage sites in the mRNA for ribosomal protein S20 in vitro. J Mol Biol 234:998–1012

Mandal M, Breaker RR (2004a) Adenine riboswitches and gene activation by disruption of a transcription terminator. Nat Struct Mol Biol 11:29–35

Mandal M, Breaker RR (2004b) Gene regulation by riboswitches. Nat Rev Mol Cell Biol 5:451–463

Mandal M, Boese B, Barrick JE, Winkler WC, Breaker RR (2003) Riboswitches control fundamental biochemical pathways in Bacillus subtilis and other bacteria. Cell 113:577–586

Mandal M, Lee M, Barrick JE, Weinberg Z, Emilsson GM, Ruzzo WL, Breaker RR (2004) A glycine-dependent riboswitch that uses cooperative binding to control gene expression. Science 306:275–279

Mansfield SG, Chao H, Walsh CE (2004) RNA repair using spliceosome-mediated RNA trans-splicing. Trends Mol Med 10:263–268

Marshall KA, Ellington AD (2000) In vitro selection of RNA aptamers. Methods Enzymol 318:193–214

Masse E, Gottesman S (2002) A small RNA regulates the expression of genes involved in iron metabolism in Escherichia coli. Proc Natl Acad Sci U S A 99:4620–4625

Masse E, Majdalani N, Gottesman S (2003) Regulatory roles for small RNAs in bacteria. Curr Opin Microbiol 6:120–124

Matzke MA, Birchler JA (2005) RNAi-mediated pathways in the nucleus. Nat Rev Genet 6:24–35

McDowall KJ, Lin-Chao S, Cohen SN (1994) A+U content rather than a particular nucleotide order determines the specificity of RNase E cleavage. J Biol Chem 269:10790–10796

Michienzi A, Castanotto D, Lee N, Li S, Zaia JA, Rossi JJ (2003) RNA-mediated inhibition of HIV in a gene therapy setting. Ann N Y Acad Sci 1002:63–71

Mikulecky PJ, Kaw MK, Brescia CC, Takach JC, Sledjeski DD, Feig AL (2004) Escherichia coli Hfq has distinct interaction surfaces for DsrA, rpoS and poly(A) RNAs. Nat Struct Mol Biol 11:1206–1214

Moll I, Afonyushkin T, Vytvytska O, Kaberdin VR, Blasi U (2003) Coincident Hfq binding and RNase E cleavage sites on mRNA and small regulatory RNAs. RNA 9:1308–1314

Moore PB, Steitz TA (2003) The structural basis of large ribosomal subunit function. Annu Rev Biochem 72:813–850

Nimjee SM, Rusconi CP, Sullenger BA (2005) Aptamers: an emerging class of therapeutics. Annu Rev Med 56:555–583

Opdyke JA, Kang JG, Storz G (2004) GadY, a small-RNA regulator of acid response genes in Escherichia coli. J Bacteriol 186:6698–6705

Paddison PJ, Silva JM, Conklin DS, Schlabach M, Li M, Aruleba S, Balija V, O'Shaughnessy A, Gnoj L, Scobie K, Chang K, Westbrook T, Cleary M, Sachidanandam R, McCombie WR, Elledge SJ, Hannon GJ (2004) A resource for large-scale RNA-interference-based screens in mammals. Nature 428:427–431

Pan WH, Xin P, Bui V, Clawson GA (2003) Rapid identification of efficient target cleavage sites using a hammerhead ribozyme library in an iterative manner. Mol Ther 7:129–139

Parker JS, Roe SM, Barford D (2005) Structural insights into mRNA recognition from a PIWI domain-siRNA guide complex. Nature 434:663–666

Perutka J, Wang W, Goerlitz D, Lambowitz AM (2004) Use of computer-designed group II introns to disrupt Escherichia coli DExH/D-box protein and DNA helicase genes. J Mol Biol 336:421–439

Pley HW, Flaherty KM, McKay DB (1994) Three-dimensional structure of a hammerhead ribozyme. Nature 372:68–74

Pollack JR, Iyer VR (2002) Characterizing the physical genome. Nat Genet 32 Suppl 515–521

Puerta-Fernandez E, Romero-Lopez C, Barroso-delJesus A, Berzal-Herranz A (2003) Ribozymes: recent advances in the development of RNA tools. FEMS Microbiol Rev 27:75–97

Ranish JA, Yi EC, Leslie DM, Purvine SO, Goodlett DR, Eng J, Aebersold R (2003) The study of macromolecular complexes by quantitative proteomics. Nat Genet 33:349–355

Raser JM, O'Shea EK (2004) Control of stochasticity in eukaryotic gene expression. Science 304:1811–1814

Rivas E, Klein RJ, Jones TA, Eddy SR (2001) Computational identification of noncoding RNAs in E. coli by comparative genomics. Curr Biol 11:1369–1373

Robertson MP, Ellington AD (1999) In vitro selection of an allosteric ribozyme that transduces analytes to amplicons. Nat Biotechnol 17:62–66

Robertson MP, Ellington AD (2000) Design and optimization of effector-activated ribozyme ligases. Nucleic Acids Res 28:1751–1759

Robertson MP, Knudsen SM, Ellington AD (2004) In vitro selection of ribozymes dependent on peptides for activity. RNA 10:114–127

Roth A, Breaker RR (2004) Selection in vitro of allosteric ribozymes. Methods Mol Biol 252:145–164

Rupert PB, Ferre-D'Amare AR (2001) Crystal structure of a hairpin ribozyme-inhibitor complex with implications for catalysis. Nature 410:780–786

Rusconi CP, Scardino E, Layzer J, Pitoc GA, Ortel TL, Monroe D, Sullenger BA (2002) RNA aptamers as reversible antagonists of coagulation factor IXa. Nature 419:90–94

Santoro SW, Joyce GF (1997) A general purpose RNA-cleaving DNA enzyme. Proc Natl Acad Sci U S A 94:4262–4266

Seydoux G, Mello CC, Pettitt J, Wood WB, Priess JR, Fire A (1996) Repression of gene expression in the embryonic germ lineage of C. elegans. Nature 382:713–716

Silva JM, Mizuno H, Brady A, Lucito R, Hannon GJ (2004) RNA interference microarrays: high-throughput loss-of-function genetics in mammalian cells. Proc Natl Acad Sci U S A 101:6548–6552

Sonnichsen B, Koski LB, Walsh A, Marschall P, Neumann B, Brehm M, Alleaume AM, Artelt J, Bettencourt P, Cassin E, Hewitson M, Holz C, Khan M, Lazik S, Martin C, Nitzsche B, Ruer M, Stamford J, Winzi M, Heinkel R, Roder M, Finell J, Hantsch H, Jones SJ, Jones M, Piano F, Gunsalus KC, Oegema K, Gonczy P, Coulson A, Hyman AA, Echeverri CJ (2005) Full-genome RNAi profiling of early embryogenesis in Caenorhabditis elegans. Nature 434:462–469

Sontheimer EJ (2005) Assembly and function of RNA silencing complexes. Nat Rev Mol Cell Biol 6:127–138
Soukup GA, Breaker RR (2000) Allosteric nucleic acid catalysts. Curr Opin Struct Biol 10:318–325
Steitz TA, Moore PB (2003) RNA, the first macromolecular catalyst: the ribosome is a ribozyme. Trends Biochem Sci 28:411–418
Stojanovic MN, Stefanovic D (2003) A deoxyribozyme-based molecular automaton. Nat Biotechnol 21:1069–1074
Storz, Wassarman KM (2004) An abundance of RNA regulators. Annu Rev Biochem 74:199–217
Storz G, Opdyke JA, Zhang A (2004) Controlling mRNA stability and translation with small, noncoding RNAs. Curr Opin Microbiol 7:140–144
Sudarsan N, Barrick JE, Breaker RR (2003) Metabolite-binding RNA domains are present in the genes of eukaryotes. RNA 9:644–647
Sullenger BA, Gilboa E (2002) Emerging clinical applications of RNA. Nature 418:252–258
Suyama E, Kawasaki H, Wadhwa R, Taira K (2004a) Cell migration and metastasis as targets of small RNA-based molecular genetic analyses. J Muscle Res Cell Motil 25:303–308
Suyama E, Wadhwa R, Kaur K, Miyagishi M, Kaul SC, Kawasaki H, Taira K (2004b) Identification of metastasis-related genes in a mouse model using a library of randomized ribozymes. J Biol Chem 279:38083–38086
Tang G, Zamore PD (2004) Biochemical dissection of RNA silencing in plants. Methods Mol Biol 257:223–244
Tang G, Reinhart BJ, Bartel DP, Zamore PD (2003) A biochemical framework for RNA silencing in plants. Genes Dev 17:49–63
Tang J, Breaker RR (1997) Rational design of allosteric ribozymes. Chem Biol 4:453–459
Tang J, Breaker RR (2000) Structural diversity of self-cleaving ribozymes. Proc Natl Acad Sci U S A 97:5784–5789
Thompson KM, Syrett HA, Knudsen SM, Ellington AD (2002) Group I aptazymes as genetic regulatory switches. BMC Biotechnol 2:21
Tomari Y, Zamore PD (2005) Perspective: machines for RNAi. Genes Dev 19:517–529
Vaish NK, Dong F, Andrews L, Schweppe RE, Ahn NG, Blatt L, Seiwert SD (2002) Monitoring post-translational modification of proteins with allosteric ribozymes. Nat Biotechnol 20:810–815
Vaish NK, Kossen K, Andrews LE, Pasko C, Seiwert SD (2004) Monitoring protein modification with allosteric ribozymes. Methods 32:428–436
Vytvytska O, Moll I, Kaberdin VR, von Gabain A, Blasi U (2000) Hfq (HF1) stimulates ompA mRNA decay by interfering with ribosome binding. Genes Dev 14:1109–1118
Wadhwa R, Yaguchi T, Kaur K, Suyama E, Kawasaki H, Taira K, Kaul SC (2004) Use of a randomized hybrid ribozyme library for identification of genes involved in muscle differentiation. J Biol Chem 279:51622–51629
Wassarman KM, Storz G (2000) 6S RNA regulates E. coli RNA polymerase activity. Cell 101:613–623
Wassarman KM, Repoila F, Rosenow C, Storz G, Gottesman S (2001) Identification of novel small RNAs using comparative genomics and microarrays. Genes Dev 15:1637–1651
Wilderman PJ, Sowa NA, FitzGerald DJ, FitzGerald PC, Gottesman S, Ochsner UA, Vasil ML (2004) Identification of tandem duplicate regulatory small RNAs in Pseudomonas aeruginosa involved in iron homeostasis. Proc Natl Acad Sci U S A 101:9792–9797
Wilson DS, Szostak JW (1999) In vitro selection of functional nucleic acids. Annu Rev Biochem 68:611–647

Winkler WC, Cohen-Chalamish S, Breaker RR (2002) An mRNA structure that controls gene expression by binding FMN. Proc Natl Acad Sci U S A 99:15908–15913

Winkler WC, Nahvi A, Roth A, Collins JA, Breaker RR (2004) Control of gene expression by a natural metabolite-responsive ribozyme. Nature 428:281–286

Yekta S, Shih IH, Bartel DP (2004) MicroRNA-directed cleavage of HOXB8 mRNA. Science 304:594–596

Yen L, Svendsen J, Lee JS, Gray JT, Magnier M, Baba T, D'Amato RJ, Mulligan RC (2004) Exogenous control of mammalian gene expression through modulation of RNA self-cleavage. Nature 431:471–476

Yokobayashi Y, Weiss R, Arnold FH (2002) Directed evolution of a genetic circuit. Proc Natl Acad Sci U S A 99:16587–16591

Zhang A, Wassarman KM, Rosenow C, Tjaden BC, Storz G, Gottesman S (2003) Global analysis of small RNA and mRNA targets of Hfq. Mol Microbiol 50:1111–1124

Selection of RNase-Resistant RNAs

S. Kainz[1] · R. Czaja[1] · T. Greiner-Stöffele[2] · U. Hahn[1] (✉)

[1]Institut für Biochemie und Lebensmittelchemie,
Abteilung Biochemie und Molekularbiologie, Universität Hamburg,
Martin-Luther-King-Platz 6, 20146 Hamburg, Germany
uli.hahn@uni-hamburg.de

[2]NWG "Protein Engineering", Universität Leipzig, c/o c-LEcta GmbH, Deutscher Platz 5b, 04103 Leipzig, Germany

1	Introduction	447
2	Selection of 'Stabilimers'-RNase T1 Resistant RNAs	449
2.1	Structure Predictions	451
2.2	Which Conformation Does RNA Have to Adapt to Be Cleaved by RNase T1?	451
2.3	Molecular Dynamics Simulation of RNA Tetraloops	453
3	Summary and Outlook	454
	References	455

Abstract All RNA types are susceptible to ribonuclease (RNase) digestion, which might be a serious problem for several *in vitro* and *in vivo* applications. RNase resistance can be reached through chemical modifications or the selection of stable secondary structures via SELEX (systematic evolution of ligands by exponential enrichment). This chapter focuses on the selection of natural RNase-resistant RNAs, enriched by a selection process in the presence of RNase T1. Results of these investigations led to the identification of a particular structural motif, the tetraloop. Further applications could be the advised use of such motifs in order to reach higher stability of RNA molecules.

Keywords Chemical modifications · Secondary structures · RNase A · RNase T1 · Tetraloop

1
Introduction

RNA in cells is, in a natural way, subject to a relatively high turnover rate. The main RNA types in cells are: (1) messenger RNA (mRNA), as a template for the protein biosynthesis, (2) ribosomal RNA (rRNA), representing the major component of ribosomes and due to its structural and catalytic function also tightly connected to the protein synthesis, (3) transfer RNA (tRNA), carrying activated amino acids to the ribosome, and (4) small nucleolar RNA (snoRNA), involved in a modification process of rRNA. Apart from these, there are types of RNA involved in several essential cellular functions. The addressed RNA turnover has the advantage that, in the case of mRNA, the cell is able to vary the protein pool very fast. In this special case, mRNA degradation plays a decisive role precisely because it influences the transcription rates and the efficiency of translation of protein genes. But there also exist RNA species that have a comparatively long endurance, e.g. tRNA. Its remarkable stability partly results from modifications that are made at its bases after transcription. tRNA is also remarkable for its ability to fold into a stable three-dimensional structure.

For the appliance of RNA types as aptamers or ribozymes in diagnostic or therapeutic approaches, it might be desirable in some cases to have RNA species with an extended half-life. Endonucleolytic degradation of oligoribonucleotides proceeds through the activation of the 2'-hydroxyl group by a general base for inline attack on the internucleotidic phosphorus. The result of this attack is a transesterification that divides the RNA into a 5'-terminal oligoribonucleotide containing a 2',3'-cyclic phosphodiester at its 3'-end and a 3'-terminal fragment with a free 5'-hydroxyl group.

A simple way to increase the stability of RNA molecules is the modification of the ribonucleotide sugar at the 2'-position. Common groups are, for example, 2'-amino-, 2'-fluoro-, 2'-methoxy- or 2'-azido-modifications. Such modified RNAs can easily be synthesised by *in vitro* transcription using the corresponding 2'-modified 5'-triphosphates. The resulting RNA is therefore protected against endonucleolytic degradation (Kubik et al. 1997; Pagratis et al. 1997; Schürer et al. 2001). Although these chemical modifications at the 2'-position increase the stability of RNA molecules against RNases, these molecules are not completely resistant to all nucleases studied.

Due to the fact that all nucleases obtain a chiral structure resulting in a stereospecific substrate recognition, RNases only accept substrate molecules in the correct chiral configuration. The so-called Spiegelmers (Klussmann et al. 1996; Nolte et al. 1996) are aptamers built of L-ribonucleotides and represent therefore a mirror image of natural RNA. Due to their unnatural structure Spiegelmers are no substrates for any usually occurring RNases and exhibit thereby an exceptional stability for *in vivo* applications. But these modifications all share one important disadvantage. The synthesis of these modified RNAs requires unnatural building blocks, which means that their *in vivo* expression is not possible.

For such an *in vivo* application, RNA might only be stabilised due to the robustness of special structural elements. As long as we do not know the 3D structure of an RNA molecule, we can only deal with probable secondary structures predicted with the aid of algorithms like Mfold (Walter et al. 1994; Zuker 1989) to get an idea about structural elements of a given molecule. Nucleotides that are located in double-stranded regions are often protected against digestion because many of the known ribonucleases are single-strand specific enzymes. Nevertheless, single-stranded regions also exist in RNA molecules showing an above-average resistance against RNases. Hairpins with GNRA tetraloops (N=A/C/G/U; R=A/G) are common structural motifs with an unusually high

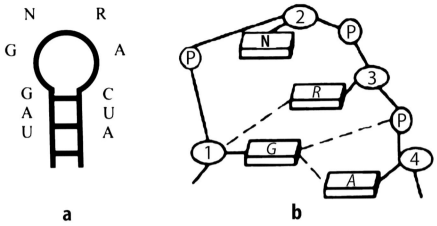

Fig. 1a Scheme of a GNRA loop (N=A/C/G/U; R=A/G). **b** Stabilising interactions in the GNRA motif (adapted from Jucker and Pardi 1995). These interactions are, in detail, (*1*) a GA base pair amongst the first and the fourth nucleotides of the loop, (*2*) a hydrogen bond between the amino proton of the guanine and the phosphate oxygen of the adenine, (*3*) an additional hydrogen bond between the 2′-hydroxyl group of the guanine and the N7 of the purine, and (*4*) stacking of the second, third and fourth loop nucleotides (*N, R, A*) (Jucker and Pardi 1995). This class of tetraloop is likely to be the most stable all known U-turns because of the additionally stabilisation by the GA base pair. This unusual feature can be used to stabilise RNA (and DNA) structures

stability (Woese et al. 1990). The GNRA tetraloop motif is stabilised by different interactions shown in detail in Fig. 1.

In order to get more insights into stabilisation of RNA by structural elements as well as to search for additional ribonuclease-resistant RNA molecules, a SELEX (systematic evolution of ligands by exponential enrichment) experiment in the presence of RNase T1 was performed.

2
Selection of 'Stabilimers'-RNase T1 Resistant RNAs

In the last few years, the number of publications using *in vitro* selection or *in vitro* evolution of nucleic acids has increased exponentially. Aptamers (Ellington and Szostak 1990) for numerous ligands have been selected by SELEX (Tuerk and Gold 1990). For special *in vivo* applications of RNA aptamers and intramers, it would be highly desirable to establish molecules with high RNase resistance. Starting from a combinatorial RNA library there were successfully selected unmodified 'normal' RNA molecules that were resistant against degradation by the two widely used RNases A and T1, which usually cleave single-stranded RNA after pyrimidines or guanines, respectively (Greiner-Stöffele et al. 2001).

The DNA starting pool was amplified by preparative PCR with the T7 RNA polymerase promoter sequence included within an overhanging primer. About 1.5 copies of the starting pool (2.8 nmol DNA) were subjected to T7 RNA polymerase transcription. With the resulting RNA—about 20 copies of the DNA starting pool—the first selection cycle was started by the addition of RNase T1. Table 1 shows the amounts of RNA before and after each selection cycle.

RNase T1 cleaves RNA specifically after guanosines in single-stranded regions. Assuming a statistical distribution of the four nucleotides and the occurrence of all guanosines in single-stranded areas, the enzyme would have the possibility to cleave 25% of all RNA bonds. The data showed that about 39% uncut phosphodiester bonds of the initial amount of 'RNA bonds' were found to be uncleaved, which meant that they were found in ethanol-precipitable oligonucleotides after the first selection cycle indicated by SYBR Green II. This dye principally indicates undigested RNA bonds due to its intercalation properties. The recovery of only 39% shows that the RNA was presumably well digested and only very small amounts of intact or partially digested RNA molecules were left. During the following selection cycles the amount of uncleaved and recoverable RNA increased to up to 64%.

Further analysis via agarose gel electrophoresis showed a shortening of the originally randomised region during the selection process. This could be proved by sequencing. Finally the randomised initially 60-nucleotide region was deleted down to six nucleotides between the two primer binding sites. Another remarkable result of this selection process was the fact that only one sequence had been selected. Different digestion tests followed and yielded not only a resistance of this RNA against RNase T1, which was the only selection criterion, but also against hydrolysis by RNase A. Incubation of this resulting RNase-resistant RNA—later named "stabilimer"—with human blood serum provided comparable results. More precise analyses of the stabilimer revealed its shortening by three or one bases after incubation with RNase A or T1,

Table 1 Applied and obtained RNA amounts in the SELEX cycles in the presence of RNase T1 (adapted from Greiner-Stöffele et al. 2001)

Cycle	Applied RNA (pmol)	Received RNA (pmol)[a]	Yield (%)
1	35,700	13,700	39
2	3,000	1,200	40
3	1,000	430	43
4	200	104	52
5	200	128	64

[a] Assuming the recovery of only full-length RNA molecules. Most of the RNA molecules are cut in two or more fragments, especially in the early cycles

respectively. The remaining portion stayed undigested even after 24 h and raising the incubation temperature significantly up to 50 °C.

2.1
Structure Predictions

To facilitate the interpretation of the results in the mentioned digestion experiments and due to the lack of exhaustive 3D structural information, a theoretical analysis had to be performed to get a possible explanation for the RNase resistance of the stabilimer.

In the first attempt, the stabilimer was analysed with the aid of Mfold, a software program predicting the secondary structure of RNA molecules. The result of this analysis is shown in Fig. 2.

The favoured structure contained, amongst other features, one loop— formed by nucleotides 25 to 42—that comprises a single-stranded region containing two guanines as well as one cytosine. Thus, in this loop the requirements for a possible digestion by RNases A or T1 were given. The secondary structure of this loop was predicted to be a GGCA tetraloop. This motif is convincingly the favourite element mediating the selected RNase resistance because of its increasing appearance in sequences during SELEX. Due to this fact, the question arose why RNases A and T1 are both unable to cleave within this GGCA tetraloop, although the necessary conditions seemed to be fulfilled.

2.2
Which Conformation Does RNA Have to Adapt to Be Cleaved by RNase T1?

A diversity of crystal structures of RNase T1 and its variants with inhibitors, reaction products, substrate analogues and the free enzyme has been solved over the years (Steyaert 1997). These data give evidence for a specific recognition site for the guanine base in RNase T1, built of the amino acids Tyr42,

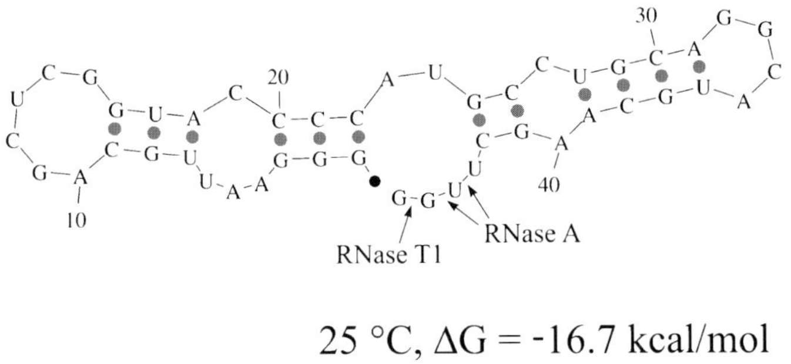

$$25\,°C, \Delta G = -16.7\,\text{kcal/mol}$$

Fig. 2 Secondary structure of the selected RNA molecule, predicted by the Mfold program (adapted from Greiner-Stöffele et al. 2001). *Arrows* mark possible cleavage sites for RNase T1 or RNase A at the 3' end

Asn43, Asn44, Tyr45, Glu46 and Asn98. Six hydrogen bonds exist between the protein and a bound guanine.

Only Glu46 is involved in hydrogen bond interactions with its side chain. The remaining four hydrogen bonds are realised between the base and the protein backbone (Fig. 3a). Other features of the bound guanosine must also be taken into account: The phosphate and the 2'-OH group, both indispensable for catalysis, are fixed by six H-bond interactions with the side chains of the amino acids Tyr38, His40, Glu58, Arg77 and His92 (Fig. 3b). Additionally the sandwich-like hydrophobic stacking interactions between the two tyrosines 42 and 45 are of significant influence on the enzyme substrate recognition (not shown). The contacts between the two His residues 40 and 92 and Glu58 are essential for catalysis.

More detailed information about the binding of the guanosine residue as a part of longer RNA molecules is currently unavailable. Therefore, only a molecular modelling approach is promising if one could try to explain why the stabilimer was not cut within putative single-stranded regions by the two tested RNases. The 3'-guanosine monophosphate (3'-GMP) was examined as part of a longer sequence and the conformation of this residue was investigated in this environment. The substrate analogue CH_3-$5'p_{(n-1)}$-$Pyr_{(n-1)}$-$5'p$-G-$3'p$-$Pyr_{(n+1)}$-$3'p_{(n+1)}$-CH_3, a trinucleoside-tetraphosphate analogue, was used in a complex with RNase T1 for different geometry optimisation tests. The chosen protein–substrate complex and applied distance constraints comprise several properties ensuring correct recognition of the central guanosine and necessary catalytic contacts. Analysis of different minimisations showed a variety of orientations of attached nucleotides, in contrast to the low divergence of the obtained backbone torsion angles of the guanosine residue (Fig. 4).

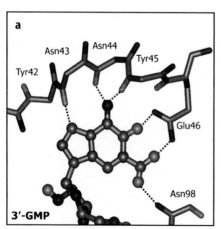 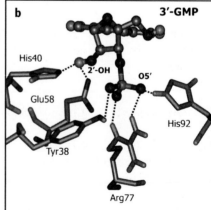

Fig. 3a,b Recognition and catalytic sites of RNase T1. **a** The binding site of RNase T1 for the guanine base (Tyr42–Tyr45 and Asn98 without side chains). **b** The amino acids involved in the binding of 3'-guanosine monophosphate (3'-GMP)

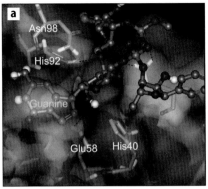 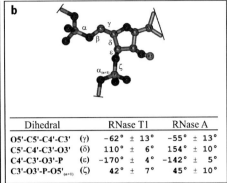

Fig. 4a,b Nucleic acid backbone conformation of CH_3-$5'p_{(n-1)}$-$Pyr_{(n-1)}$-$5'p$-G-$3'p$-$Pyr_{(n+1)}$-$3'p_{(n+1)}$-CH_3 bound to RNase T1

The sugar part of the guanosine exhibits a conformation between the C1'-endo and the C2'-exo puckering mode in all minimisations. In combination with the always observed *anti* orientation of the guanosine glycosyl bond, an unusual, but not impossible conformation for the combination of the puckering mode and χ was found (Saran et al. 1973).

To reflect at least partially the reaction steps following the binding of the substrate RNA, some additional constraints concerning the distances between the catalytically relevant groups were complemented in the calculations. As the result, the backbone conformation: $\gamma = -62°\pm13°$ ($-sc$), $\delta = 110°\pm6°$ ($+ac$), $\varepsilon = -170°\pm4°$ ($-ap$) and $\zeta = 42°\pm7°$ ($+sc$) was postulated for a productive binding of a guanosine within an RNA chain of at least three nucleotides to RNase T1.

As the selected RNA 46mer was also resistant against RNase A digestions, similar calculations were performed with an extended cytidine-3'-monophosphate (3'-CMP) sequence and RNase A (Zhang and Simon 2003). The results were comparable to the findings achieved for RNase T1. While δ and ε are shifted by about 45° and 30°, respectively, the dihedrals γ and ζ exhibited nearly the same values as for RNase T1 (Fig. 4b).

2.3
Molecular Dynamics Simulation of RNA Tetraloops

Molecular dynamics (MD) simulations have been performed in order to investigate whether such RNA chain conformations as described in the preceding section might be realised in the GGCA tetraloop developed from the secondary structure predictions. The program MCSYM (Major et al. 1993) was employed to transform the secondary structure of the nucleotides 25 to 42 into 3D structures. The investigators obtained 123 possible solutions building this structure. These motifs could be sorted into six different classes, concerning the nucleic acid backbone within the GGCA tetraloop. One representative of each class and a six-base single-stranded RNA chain as a cleavable reference were subjected to MD procedures focussing on the torsion angles γ and ζ. The nucleotides in the free RNA chain realised the necessary torsion angles, whereas in the GGCA loop limitations in the realisation of the angles could be observed. Plots of distinct torsion angles arising from MD simulations showed that both guanosines and the cytidine residue were not able to realise the conformation essential for productive enzyme substrate interaction. This demonstrates restrictions of the conformational degrees of freedom in the GGCA tetraloop that prevent the substrate nucleotides to position their 3'-phosphate group into the catalytic site of the RNases A and T1.

To prove the stability of some members of the GNRA tetraloop family, additional digestion experiments with 46mer mutants were performed. The GGCA tetraloop could be replaced via PCR mutagenesis by different representatives of the GNRA family (GAAA, GAGA and GCAA), and in all cases, NMR spectroscopy had determined the 3D structure. The resulting RNA molecules possessed the same stability as the originally selected sequence (Greiner-Stöffele et al. 2001).

3
Summary and Outlook

One RNA molecule resistant to RNase A and RNase T1 cleavage was selected from an RNA pool without additional modifications using SELEX. Secondary structure predictions suggested a GGCA tetraloop for a single-stranded part of the sequence. On the basis of conformational studies, certain values for different torsion angles were found, which have to be realised for proper enzyme-substrate interactions in the catalytic site in combination with an unrestricted positioning of the adjacent nucleotides. MD simulation studies showed that this essential combination of angle values can be achieved in a single-stranded RNA chain, but not if this chain is forced into a tetraloop. In simulations of diverse structures of the observed GGCA loop, there were always found restrictions of the conformational freedom in the combination of some angles. As a result, the possible nucleotide substrates—cytidine for RNase A and guanosine for RNase T1—cannot realise the conformation necessary for cleavage by these enzymes. Therefore, the resistance of the selected stabilimer at the site of the GGCA tetraloop could be explained.

These stabilimers probably could be used as stabilising modules in the generation of larger RNase-resistant RNA molecules, e.g. intramers. Molecules with these properties seem to be very helpful for medical applications. RNase-resistant RNA could also facilitate *in vitro* investigations, as the molecules might function in cell culture with the simultaneous use of fetal calf serum (FCS) or related substances.

Acknowledgements We are grateful to Michael Knauer for reading the manuscript.

References

Ellington AD, Szostak JW (1990) In vitro selection of RNA molecules that bind specific ligands. Nature 346:818–822

Greiner-Stöffele T, Forster HH, Hofmann HJ, Hahn U (2001) RNase-stable RNA: conformational parameters of the nucleic acid backbone for binding to RNase T1. Biol Chem 382:1007–1017

Jucker FM, Pardi A (1995) GNRA tetraloops make a U-turn. Rna 1:219–222

Klussmann S, Nolte A, Bald R, Erdmann VA, Furste JP (1996) Mirror-image RNA that binds D-adenosine. Nat Biotechnol 14:1112–1115

Kubik MF, Bell C, Fitzwater T, Watson SR, Tasset DM (1997) Isolation and characterization of 2′-fluoro-, 2′-amino-, and 2′-fluoro-/amino-modified RNA ligands to human IFN-gamma that inhibit receptor binding. J Immunol 159:259–267

Major F, Gautheret D, Cedergren R (1993) Reproducing the three-dimensional structure of a tRNA molecule from structural constraints. Proc Natl Acad Sci U S A 90:9408–9412

Nolte A, Klussmann S, Bald R, Erdmann VA, Furste JP (1996) Mirror-design of L-oligonucleotide ligands binding to L-arginine. Nat Biotechnol 14:1116–1119

Pagratis NC, Bell C, Chang YF, Jennings S, Fitzwater T, Jellinek D, Dang C (1997) Potent 2′-amino-, and 2′-fluoro-2′-deoxyribonucleotide RNA inhibitors of keratinocyte growth factor. Nat Biotechnol 15:68–73

Saran A, Pullman B, Perahia D (1973) Molecular orbital calculations on the conformation of nucleic acids and their constituents. VI. Conformation about the exocyclic C(4′)-C(5′) bond in alpha-nucleosides. Biochim Biophys Acta 299:497–499

Schürer H, Stembera K, Knoll D, Mayer G, Blind M, Forster HH, Famulok M, Welzel P, Hahn U (2001) Aptamers that bind to the antibiotic moenomycin A. Bioorg Med Chem 9:2557–2563

Steyaert J (1997) A decade of protein engineering on ribonuclease T1—atomic dissection of the enzyme-substrate interactions. Eur J Biochem 247:1–11

Tuerk C, Gold L (1990) Systematic evolution of ligands by exponential enrichment: RNA ligands to bacteriophage T4 DNA polymerase. Science 249:505–510

Walter AE, Turner DH, Kim J, Lyttle MH, Muller P, Mathews DH, Zuker M (1994) Coaxial stacking of helixes enhances binding of oligoribonucleotides and improves predictions of RNA folding. Proc Natl Acad Sci U S A 91:9218–9222

Woese CR, Winker S, Gutell RR (1990) Architecture of ribosomal RNA: constraints on the sequence of "tetra-loops". Proc Natl Acad Sci U S A 87:8467–8471

Zhang G, Simon AE (2003) A multifunctional turnip crinkle virus replication enhancer revealed by in vivo functional SELEX. J Mol Biol 326:35–48

Zuker M (1989) On finding all suboptimal foldings of an RNA molecule. Science 244:48–52

Subject Index

2′,5′-AS   155
2′-modification
– specific modification   226
2′-modifications   225
3′ untranslated region (UTR)   154
7SK RNA   55

acquired immunodeficiency syndrome (AIDS)   155, 158
active structural principle   328
African trypanosomes   378, 380, 384
– Trypanosoma brucei   378
AIDS   398
Air RNA   51, 52
algorithm   109
Alzheimer disease   60
aminoglycoside antibiotics   329, 330, 335
AML   187
Angelman syndrome   61
antennapedia   399
anti-angiopoietin-2 aptamer
– rat corneal micropocket angiogenesis assay   321
anti-IXa RNA aptamers   321
– antidote oligonucleotide   321
antibacterial   400
antibiotic   74, 75, 77, 79, 81, 83, 85, 87, 90
– amikacin   75
– aminoglycoside   74, 75, 77, 78, 81–90
– geneticin   81
– gentamicin   78
– hygromycin B   76, 83
– kanamycin A   75–78
– kanamycin B   75, 76
– lividomycin   76
– neamine   82
– neomycin B   75, 76, 78, 81, 84, 85, 89
– pactamycin   76, 83
– paromomycin   75, 76, 78, 81, 83, 84
– ribostamycin   76
– streptomycin   75–77, 89
– tetracyclin   76, 83
– tobramycin   75, 76, 78, 81
antibodies   359–361, 364, 365
antibody   328
antidote   362, 363
antigenic variation   378, 384
antiparasitic drug   376, 378
antisense   360
antisense agents   244
– target accessibility   244
antisense drugs   254
antisense oligonucleotide   262
antisense oligonucleotides   174, 246
– binding affinity   247
– cancer therapy   186
– chemical modification   254
– LNA   254
– modifications   178
– ribonuclease H   246
antisense transcript   57
antiviral therapy   132, 133
anto-angiopoietin-2 aptamer   321
APOBEC3G   160
aptamer   427
aptamer indentification   308
aptamers   306, 359–368
– advantages   359, 360, 363
– allosteric aptamers   361
– anti-thrombin aptamer   362
– applications in biotechnology and medicine   359, 366
– aptazymes   364
– chemical modifications   360, 361
– delivery   361

- disadvantages 361
- DNA aptamer 362
- in vitro selections 360
- nucleic acid 361
- nucleic acid aptamers 359
- properties 359, 361, 363
- RNA aptamer 362
- SELEX 360
- Spiegelmer 361
- starting libraries 360
- therapeutics 359, 361–363
- TNT aptamers 365
- VEGF aptamer 362
aptasensor 345, 354
aptazymes 307, 309
- hammerhead ribozyme 309
- RNA ligase 309
- self-alkylating ribozyme 309
- self-splicing intron 309
arginine 399
Argonaute 106
asymmetry 109

B cell chronic lymphocytic leukemia 63, 64
B2 RNA 56
Bcl-2 186
Beckwith-Wiedemann syndrome 52, 61
bioavailability 400
bioreactors 366
biosensor 342, 344, 359, 362, 365–368
- chip-based biosensor 351
- fibre optic biosensor 349
- fibre-optic field 365–367
- micromachined sensor 354
biostability 401
blood–brain barrier (BBB) 379
Burkitt lymphoma 64

c-myb 187
c-myc 55
Caenorhabditis elegans 152
cancer 52, 54, 60, 62, 64
- bladder 62
- breast 54, 60, 62
- colon 63
- colorectal 64
- lung 62, 63
- ovarian 54
- prostate 62–64
- uterus 54
carbohydrate 327, 328, 333, 335
cardiac myosin heavy chain 57
cationic lipid 111
cationic lipids 399
CCR5 159–161
$CD4^+$ 158–161
CD8 157
cell array 99
cellobiose 331
cellular delivery 399
centromeric 106
- regulation 106
chemotherapeutic 380
chimeric oligonucleotides 247
chromatin 106
CML 181
co-receptor 108
- CCR5 108
- CD4 108
combinatorial chemistry 376
commercial development 298
coxsackievirus B3 266, 281
CXCR4 159–161

delivery 183
deoxyribozyme 269
dextran 331
Dicer 106
DiGeorge syndrome 61, 62
disaccharides 331
DNA arrays 249
- accessible sites 250
DNA enzyme 269
DNA structure 1
- crystallographic analyses 5
- double helix 4
- evolution 5
- RNA—DNA hybrid 4
DNAzymes 175, 251, 269, 410
- 10–23 252
DOPE 155
DOSPA 155
double-stranded (ds)RNA 105
double-stranded RNAs (dsRNAs) 151–155, 160
Drosophila 109, 152, 156
- embryo 109
- extracts 109

Subject Index

ELISA 364
endosomal 399
engineering gene regulation 368
Epstein-Barr virus 58
Escherichia coli 400
eukaryotic translation-initiation factor 2C2 (eIF2C2) 154

FIV 110
fluorescence 349, 363, 365–367

gag 108
galactose 331
gapmer 264
gene discovery 205
gene therapy 292, 293, 295, 298
– adenoviral 292
– adenoviral vectors 295
– adenovirus 295
– naked DNA 292
– recombinant virus 293
– retroviral-incompetent vectors 294
– viral 292
gene walk 396, 400
genetic suppressor elements 249
GFP 160
glucose 331
glycoprotein 328
green fluorescent protein (GFP) 400

H-Ras 188
H1 promoter 156
H19 RNA 62
H9 cells 161
haematopoietic stem/progenitor cells 292, 293
hammerhead ribozyme 224–226, 231
– cleavage activity 225, 226
– design 231
– structure–function relationships 225
hammerhead ribozymes 198
HCV NS3 164
HCV NS5B 164
HCV replicon 164
heat shock 56
hepatitis B virus 267, 270, 278
hepatitis B virus (HBV) 163, 164
hepatitis C virus 265, 267, 270, 271, 279
hepatitis C virus (HCV) 163, 164
hepatitis virus 398

Hfq 11, 18, 21, 25
high-throughput analysis 364
high-throughput screening 389
histone 107
– deacetylation 107
HIV 289
HIV-1
– env 160
– gag 158–160, 163
– gp 120 160
– LTR 158
– nef 158, 160
– pol 163
– rev 158, 159
– tat 158, 159
– vif 158, 160
HIV-1 pNL4-3 158
HIV-2/SIV 110
Huh-7 cells 164
human cytomegalovirus 267
human immunodeficiency virus 108, 266, 267, 270, 271, 279
– (HIV)-1 108
human immunodeficiency virus type-1 (HIV-1) 157–161, 163
human papillomavirus 267
human papillomavirus type 16 (HPV-16) 163
human transcriptome 415
hybrid ribozymes 204

immunotherapy 189
indirect competitive assay 365
inflammatory diseases 185
influenza virus 276
influenza virus A 163, 165
– HA 165
– M1 165
– M2 165
– NA 165
– NP 165
– NS 165
– PA 165
– PB1 165
– PB2 165
– RNP 165
integrate 110
interferon (IFN) 155, 161, 162, 165
internal ribosomal-entry site (IRES) 164

intro-exon junctions   398
intron splicing   412
intron/exon   108
– splice junction   108

kanamycin B   329

lentivirus vector   157, 160
leukaemia   294
liposomes   399
LIT1 RNA   52
LNA   405
LNA antisense   411
LNA monomer   407
LNA/DNA/LNA gapmer   409
LNAzymes   251, 410
locked nucleic acid   405
locked nucleic acids   264
low molecular weight compounds   359, 362–364
lysosomal   399

Macugen   362, 363
MAGI   161
mannose   331
mantle cell lymphoma   63
maxizyme   203
messenger RNAs (mRNAs)   152–154, 156
metabolic disease   368
metabolism   18
– glycogen   22
– iron   19
methylation   106
micro RNA (miRNA)   58, 59
microarray technology   364
MicroRNAs   415
microRNAs   109
– miRNAs   109
microRNAs (miRNAs)   153, 154, 156
miRNA   415, 425
mixmer LNA oligonucleotides   407
molecular beacons   346, 347, 351
– aptamer beacons   346
mRNA   105, 181
– RNA mapping   182
– selecting target mRNA   181
multiple myeloma   64
multiplexing   108
– siRNA   108

N-acetyl-galactosamine   400
native gel electrophoresis   230
neomycin B   329
neuroblastoma   63
NF-κB   161, 162
nuclease-resistant RNA molecule (aptamers)   307, 310
– amino modified pyrimidines RNA   307
– amino modified RNA   316
– fluoro modified RNA   316
– fluro-modified pyrimidines RNA   307
– O-methyl modified RNA   316
nucleic acid library   375, 376
nucleic acids   359, 360, 365
nucleotide modification   264
nucleus   106

OCC-1   63
oligonucleotide   377, 387
oligonucleotide arrays   250
oligonucleotide libraries   246
– complexity   249
– hybridization   247
oligosaccharide   328, 334
Orthomyxoviridae   164

p24   159, 160
parasitic disease   376
PBMCs   160
PCR   108
peptide nucleic acid (PNA)   395
personalised medicine   368
pharmacokinetic   400
pharmacokinetics   362, 368
phosphorothioate   264
picornavirus   266
PKC-α   188
PKR   110, 155
poliovirus   163
post-SELEX modification   308, 315
– amino modified RNA   317
– aptamer-chip based biosensors   315
– dot blot assays   315
– flow cytometry   315
– fluorescence tagging   308, 315
– fluoro modified RNA   317
– L-RNA   318
– nuclease resistance   308, 317
– O-methyl modified RNA   317

Subject Index 461

- *O*-methylation   308
- Spiegelmers   308, 317
- target imaging   315
- truncation   308, 309
- Western-blot assays   315
post-transcriptional   106
- PTGS   106
Prader-Willi syndrome   61
pre-*let7* RNA   156
ProbeLibrary   415
processed pseudogene   57, 58
promoter   106
protein array   97
protein kinase C-α   235
protozoan parasite   376, 378, 380, 383
pseudotyped   110
- lentiviral vectors   110

quorum sensing   24, 28, 30
- two-component system   30

radioisotope-based systems   363
Raf protein   234
random fragmentation   253
- DNase I   253
- restriction enzymes   253
random oligonucleotide ligated libraries   248
Ras protein   234
receptor   110
- ligand   110
receptors   400
resistance   108
respiratory syncytial virus   277
retroviral vector   294
retroviruses   110
- Moloney murine leukemia virus   110
rev   108
reverse transcriptase   108
rhabdomyosarcoma   63
rhesus rotavirus (RRV)   163
riboswitch   427
riboswitches   11, 14, 361, 368
ribozyme   268, 428
- aptazyme   430
- group I intron   429
- group II intron   429
- hairpin   268
- hammerhead   268, 428
- maxizyme   430

ribozyme folding   229
ribozyme library   205, 207
ribozyme structure   226
ribozyme/substrate complex   228
- global structure   228
ribozymes   175, 250, 360, 361
- external guide sequences   251
- guide RNA   251
- hairpin   251
- hammerhead   250
- libraries   250
- M1   251
RISC   106
- RNA-induced silencing complex   106
RNA   74, 78–83, 85–90, 289, 290, 292, 296
- 4.5S RNA   86, 88
- 6S RNA   86, 88
- anti-sense   88
- antisense   289
- aptamers   289
- biocatalyst   79
- catalytic   87
- Dicer   292
- dsRNA   298
- Exportin 5   292
- group I intron   86, 89
- hairpin   89
- hammerhead   89
- HDV RNA   89
- HIV RNA   89
- microRNA   289, 291
- miRNA   79
- mRNA   80, 87
- ncRNA   74, 79
- PKR   291
- pre-mRNA   89
- processing   79
- protein interaction   89
- regulator   79
- ribozyme   89
- ribozymes   289, 290, 296, 298
- RNA aptamer   84, 85
- RNAi   289, 291, 296
- RNase P RNA   81, 86–88
- RRE   89
- rRNA   78, 81–83, 85
- scaffold   79
- siRNA   79, 292

- small hairpin RNA 289, 296
- splicing 89
- Spot 42 RNA 86, 88
- sRNA 79
- SRP RNA 88
- TAR 89
- tmRNA 86–88
- transfer-messenger 87
- translation 79
- transporter 79
- tRNA 80, 81, 85–87
RNA aptamer 384, 388
- prion CPvP^{sc}-binding aptamers 312
RNA aptamers 306, 313, 320
- age-related macular degeneration 321
- B52 protein in Drosophila 313
- clinical studies 320
- cocaine-displaceable RNA aptamers 311, 312
- complex targets 307
- disease-related proteins 307
- enzymes 307
- growth factors 306
- human immunodeficiency virus (HIV) 313
- in vivo applications 320
- intracellular targets 313
- intramers 313
- lymphocyte function-associated antigen (LFA)-1 313
- small organic molecules 306
- T4-DNA polymerase 306
RNA array 100
RNA interference 105, 118, 177, 272, 368, 413
- RNAi 105
RNA interference (RNAi) 151–155, 157–166, 224
RNA molecular switches 310
- allosteric-activated ribozyme 310
- control of gene expression 314
- ligand-activated aptamers 314
- ligand-inactivated intramers 314
- regulation of aptamer activity 314
RNA pharmaceutical 378
RNA polymerase III (pol III) 156, 157
RNA structure 1
- crystallographic analyses 5
- double helix 4

- evolution 5
- RNA—DNA hybrid 4
RNA-induced silencing complex (RISC) 224, 230
RNA-inducing silencing complex (RISC) 152, 153
RNAi 272, 413
RNAi suppressors 139
RNAs 11, 32
- (t)RNAs 26
- (tm)RNA 27
- mRNAs 11, 32, 33
- non-coding 21
- regulatory 11, 15, 31, 33
- small 11, 18, 21
RNase H 246, 396, 409
- accessible sites 246
- cleavage site 247
RNase III enzyme 152, 153
- Dicer 152–154, 156
RNase L 155
RNase resistance 449, 451
- RNases T1/A 449–451, 454
- stabilising modules 454
Rous sarcoma virus (RSV) 163

saturating 112
- RNAi pathway 112
SELEX 247, 306, 307, 328, 342, 375–377, 380, 382–384, 386, 449
- dimer formation 249
- directional library 249
- genomic libraries 248
- in vitro selection of nucleic acids 307
- primer-free libraries 249
- size selection 248
- stabilimer 450, 452
semliki forest virus (SFV) 163
sequence-walking 245
short hairpin RNA 273
short interfering RNA 272
short-interfering RNAs (siRNAs) 151–166
shRNA 201, 213, 273
Sialyl Lewis X 332
silencing 106
siLNA 414
siRNA 106, 197, 210, 230, 231, 252, 272, 413
- antisense 106

Subject Index

- chemical modifications   230
- design   232
- EPRIL   253
- genomic libraries   253
- green fluorescent protein   253
- off-target effects   252
- positive selection   254
- rational design   252
- REGS   253
- RNAi   106
- sequence requirements   252
- shRNA libraries   253
siRNA library   214
siRNA-expression vectors   200, 211
siRNAs   177
sleeping sickness   375, 376, 378–380
small hairpin RNA (shRNA)   156, 157, 160–165
small interfering (si)RNA   360
small interfering (si)RNAs   224
small interfering RNA   272
small molecules   359, 360, 362–364, 366, 368
- detection   359, 362–365, 368
small-interfering RNAs (siRNAs)   152
spectroscopy   363, 364
Spiegelmers
- anti-ghrelin L-aptamer   318
- anti-nociceptin/orphanin L aptamer   318
- Prader-Willi syndrome   318
splicing   398
splicing correction   397
sRNA   424
ssDNA   359
ssRNA   359
stability   109
- 3′-end   109
- 5′-end   109
- lower   109
Staphylococcus aureus   400
steroid receptor activator RNA   54
streptomycin   329
structure   448, 451
- molecular dynamics   453
- tetraloop   448, 451, 453
suppression   106, 110
- off-target   110
surface plasmon resonance   332, 335, 347, 363

synovial sarcoma   63
synthetic   108

T-cell   155, 160
TAR   108
- trans-activating response region   108
target mRNA   244
- accessible sites   244
- RNA structure   244
tat   108
Tat petide   399
telomerase   398, 412
therapeutic application   318
- anti-VEGF RNA aptamer   318, 319
- anticancer agent   319
- controlled-drug delivery   319
- DAG-NX213   319
- neovascularization   319
- NX-1838   319
- proliferative retinopathy   319
therapy targeting   238
thermodynamic   109
thermodynamic stability model   232
thermosensor   13, 14
thyroid hormone receptor   56
TLR7   162
TLR8   162
tobramycin   329
Toll-like receptor 3 (TLR3)   162
transcription   106
transcriptional   106
- TGS   106
transducer   344
transduction   292
transfection   98, 109
transgenes   110
transgenic mouse   400
translation arrest   397
translation initiation   397
trinitrotoluene (TNT)   364, 367
(t)RNA^{Val} 156
trypanosoma brucei   376
trypanosomias   390
trypanosomiasis   375, 378, 380
Tsix RNA   50
tumor suppressor gene   62
tumour–stroma interaction   237

U6 promoter   156, 158

vector   109, 110, 295

– lentiviral 110
virus replication 125–131, 133, 134
viruses 108
– hepatitis C 108
– influenza 108
– poliovirus dengue 108
– Semliki forest 108

Wnt signalling pathway 236

X-chromosome inactivation 49
Xist RNA 49, 50, 52